Cross-Sectional Atlas

Series Editor

Jin Seo Park, Department of Anatomy
Dongguk University School of Medicine
Gyeongju-Si, Kyongsang-bukto, Korea (Republic of)

This atlas series presents state of the art knowledge on sectional anatomy of human whole body. Since 2000, for this sectional anatomy, the sectioned images of nine human cadavers are being made in the Visible Korean project. In the sectioned images, horizontal, coronal, and sagittal planes of human body can be displayed in 48 bits color and pixel size of more than 0.04 mm x 0.04 mm.

In this atlas series, both macro-structures and micro-structures for anatomy can be identified by actual color and high resolution. Consequently, this atlas series makes it possible to systematically grasp morphological relationships among structures that cannot be realized in actual cadaver dissection. This atlas series will be of value not only for anatomists, neuroanatomists, and surgeons but also for medical students.

Jin Seo Park

Cross-Sectional Atlas of Human Upper Limbs

With 0.06-mm Pixel Size Color Images

Springer

Jin Seo Park
Department of Anatomy
Ajou University School of Medicine
Suwon, Korea (Republic of)

ISSN 3059-376X ISSN 3059-3778 (electronic)
Cross-Sectional Atlas
ISBN 978-981-95-1899-9 ISBN 978-981-95-1900-2 (eBook)
https://doi.org/10.1007/978-981-95-1900-2

This Springer imprint is published by the registered company Springer Nature Singapore Pte Ltd.
The registered company address is: 152 Beach Road, #21-01/04 Gateway East, Singapore 189721, Singapore

Contents

1 Introduction

1. Making the atlas of upper limbs from the sectioned images of male whole body

In the human upper limbs, there are various joints and numerous muscles, both large and small, that enable free movement. Around the shoulder, elbow, wrist, and phalangeal joints, the upper limbs move continuously and freely through the contraction and relaxation of muscles throughout the day, even during sleep. Due to these free movements, various diseases frequently occur around the joints. To diagnose diseases of the upper limbs, computed tomography (CT), magnetic resonance imaging (MRI), and ultrasonography are commonly used.

For basic education on CT, MRI, and ultrasonography of the upper limbs, not only medical students but also medical doctors study cross-sectional anatomy, learning the shape and location of normal structures using cross-sectional atlases [1, 2]. However, learners often find it challenging to understand sectional anatomy through these atlases because most of them consist of radiologic images in 8-bit or 16-bit grayscale. In grayscale images, distinguishing the appearance of structures is limited, even if the images have high pixel resolution. Therefore, atlases should be created using real-color sectioned images.

To address this necessity, the author of this atlas selected the true-color sectioned images of the Visible Korean project [3–5], which were in 48-bit true color with a pixel size of 0.06 mm × 0.06 mm. The high quality of these sectioned images has been proven through several scientific publications [3–5].

Notably, the author previously published *Cross-Sectional Atlas of the Human Head* in 2018 [6] and *Cross-Sectional Atlas of the Human Brainstem* in 2023 [7]. Building on this publishing experience, the *Cross-Sectional Atlas of the Human Upper Limb* was created using the sectioned images of a male whole body acquired in 2021 [5].

2. Making the atlas of upper limbs from the sectioned images of male whole body

In 2021, the axial sectioned images of a male whole body were presented. The intervals, pixel size, resolution, and color depth of these images were 0.5 mm (trunk), 0.06 mm × 0.06 mm, 8688 × 5792, and 48-bit true color, respectively (Fig. 1.1). Thanks to the high resolution and color depth, detailed anatomical information about the inner structures of the human body could be obtained from any region [5]. From the sectioned images of the whole body, the images containing only the right and left upper limbs, from the shoulder to the phalanges, were selected. Using Adobe Photoshop 2023, the trunk area in each selected image was removed (Fig. 1.2), and the remaining bilateral upper limbs were repositioned to the center of each image (Fig. 1.3). Through computer-assisted three-dimensional reconstruction, orthogonal sectioned images were generated from the axial sectioned images.

J. S. Park, *Cross-Sectional Atlas of Human Upper Limbs*, Cross-Sectional Atlas, https://doi.org/10.1007/978-981-95-1900-2_1

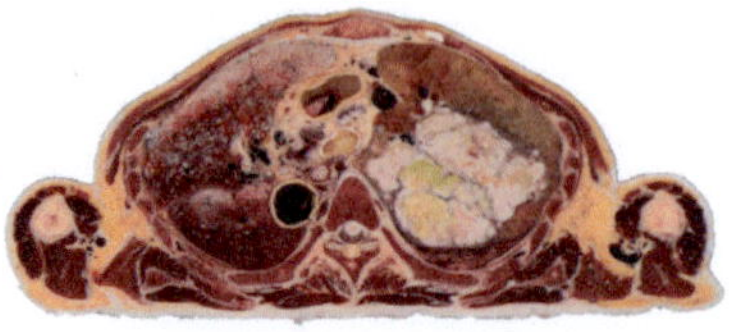

Fig. 1.1 Raw data:
Axial sectioned images of male whole body
Color depth 48 bits color
Resolution 8688 × 5792
Pixel size 0.06 mm × 0.06 mm

Fig. 1.2 Upper limb data with trunk area:
Removing trunk area in each sectioned image (black area)
Color depth 48 bits color
Resolution 8688 × 5792
Pixel size 0.06 mm × 0.06 mm

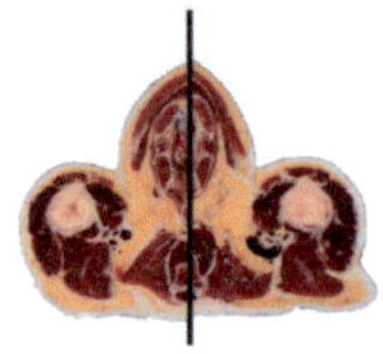

Fig. 1.3 Upper limb data without trunk:
Only upper limbs in the sectioned images Color depth 48 bits color
Resolution 4371 × 2240
Pixel size 0.06 mm × 0.06 mm

3. Coordinate system

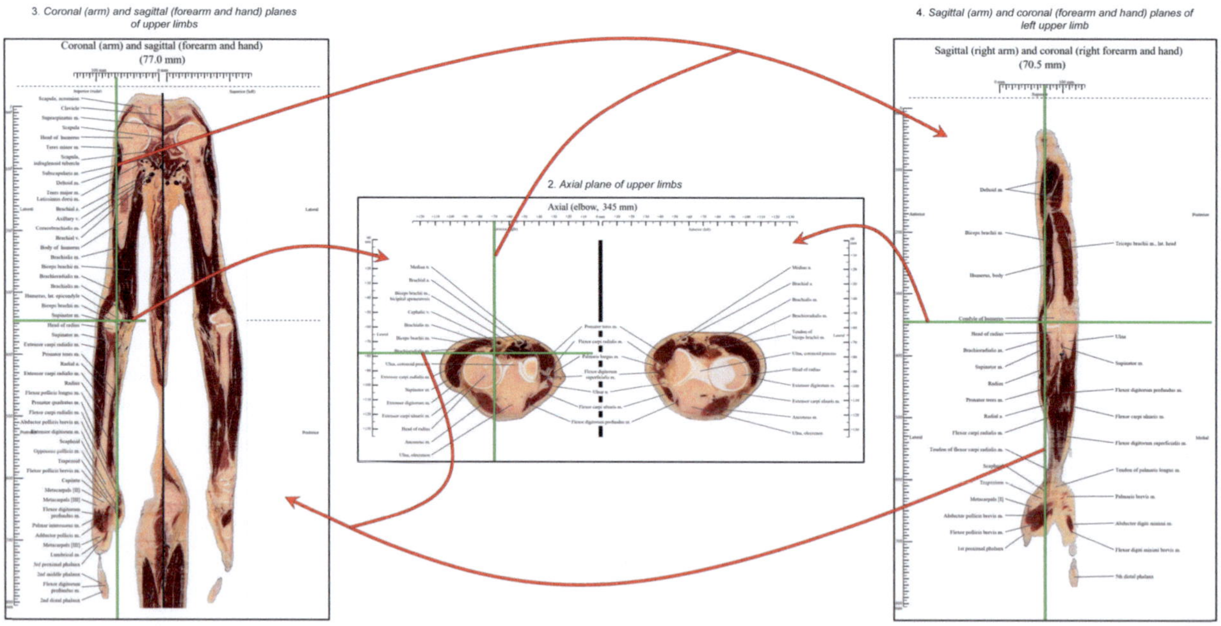

4. Terminology

All anatomical terms in this atlas were labeled according to the authorized anatomy terms from Terminologia Anatomica [8].

Acknowledgments This work was fully supported by Anatomage Inc, USA. We are deeply grateful for this assistance. This research was also supported by a Basic Science Research Program through the National Research Foundation of Korea (NRF) funded by the Ministry of Education (NRF-2021R1F1A1063044).

References

1. Naidich TP, et al. Duvernoy's Atlas of the Human Brain Stem and Cerebellum: High-Field MRI, Surface Anatomy, Internal Structure, Vascularization and 3D Sectional Anatomy. Wien (Vienna): Springer-Verlag; 2009.
2. DeArmond SJ, et al. Structure of the human brain: a photographic atlas. 3rd ed. New York: Oxford University Press; 1989.
3. Park JS, et al. Visible Korean human: improved serially sectioned images of the entire body. IEEE Trans Med Imaging. 2005;24:352–60.
4. Park JS, et al. A proposal of new reference system for the standard axial, sagittal, coronal planes of brain based on the serially-sectioned images. J Korean Med Sci. 2010;25:135–41.
5. You Y, Park JS, et al. Advanced-sectioned images obtained by microsectioning of the entire male body. Clin Anat. 2022;35:79–86.
6. Park JS. Cross-sectional atlas of the human head: with 0.1-mm pixel size color images. Germany: Springer; 2018.
7. You Y, Park JS. Cross-sectional atlas of human brainstem: with 0.06-mm pixel size color images. Germany: Springer; 2023.
8. Drake RL. Terminologia anatomica: international anatomical terminology by FIPAT. 2nd ed. Stuttgart: Thieme; 2011.

2 Axial Plane of Upper Limbs

Images of 0 mm ~ 75 mm, Shoulder, 5.0 mm intervals
Images of 80 mm ~ 320 mm, Arm, 5.0 mm intervals
Images of 325 mm ~ 345 mm, Elbow, 5.0 mm intervals
Images of 350 mm ~ 580 mm, Forearm, 5.0 mm intervals
Images of 585 mm ~ 625 mm, Wrist with expanded view, 2.5 mm intervals
Images of 630 mm ~ 805 mm, Hand, 5.0 mm intervals

All images, 0.06 mm × 0.06 mm sized-pixel
Abbreviation: ant., anterior; a., artery; inf., inferior; int., internal; lat., lateral; m., muscle; med., medial; n., nerve; nu., nucleus; post., posterior; sup., superior; tr., tract

J. S. Park, *Cross-Sectional Atlas of Human Upper Limbs*, Cross-Sectional Atlas, https://doi.org/10.1007/978-981-95-1900-2_2

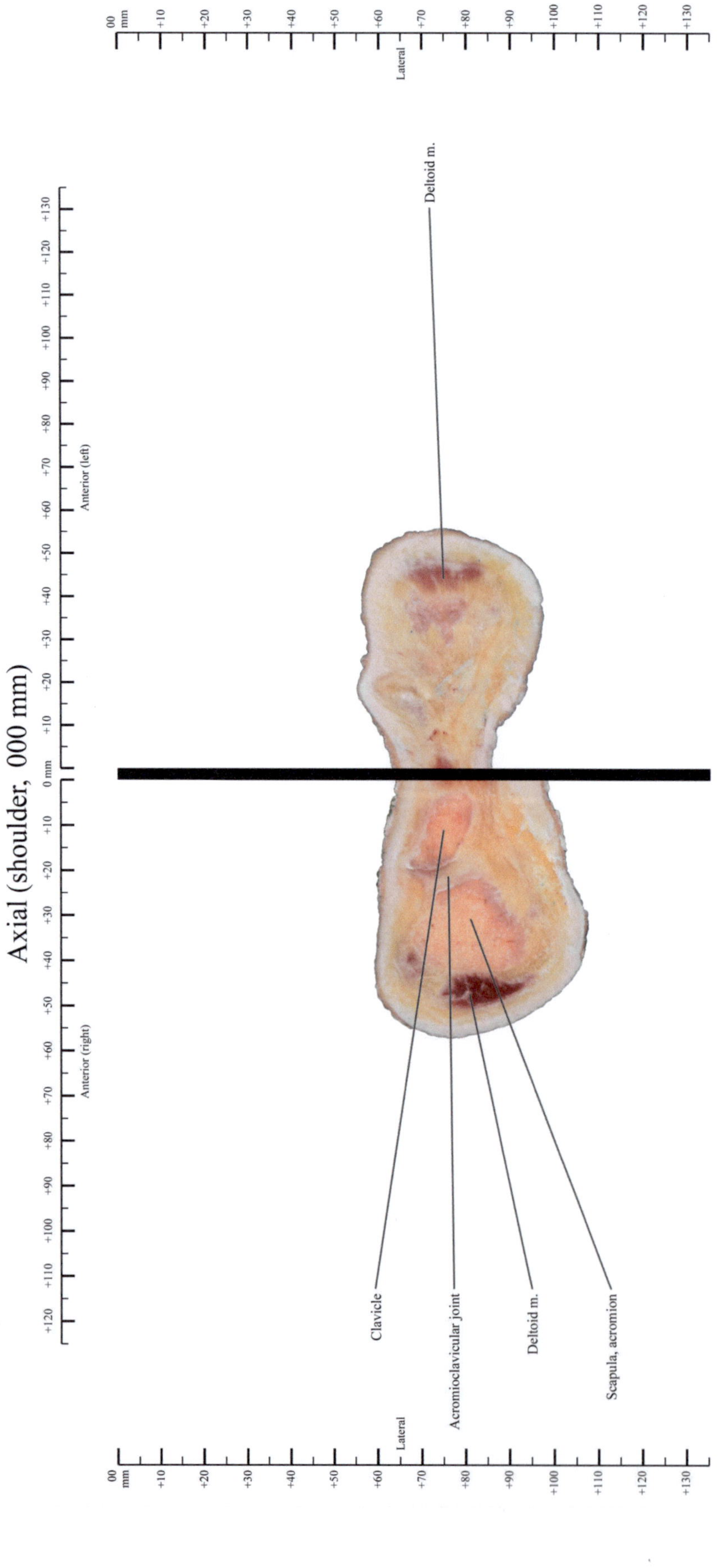
Axial (shoulder, 000 mm)
Anterior (left)
Anterior (right)
Lateral
Lateral
Deltoid m.
Clavicle
Acromioclavicular joint
Deltoid m.
Scapula, acromion

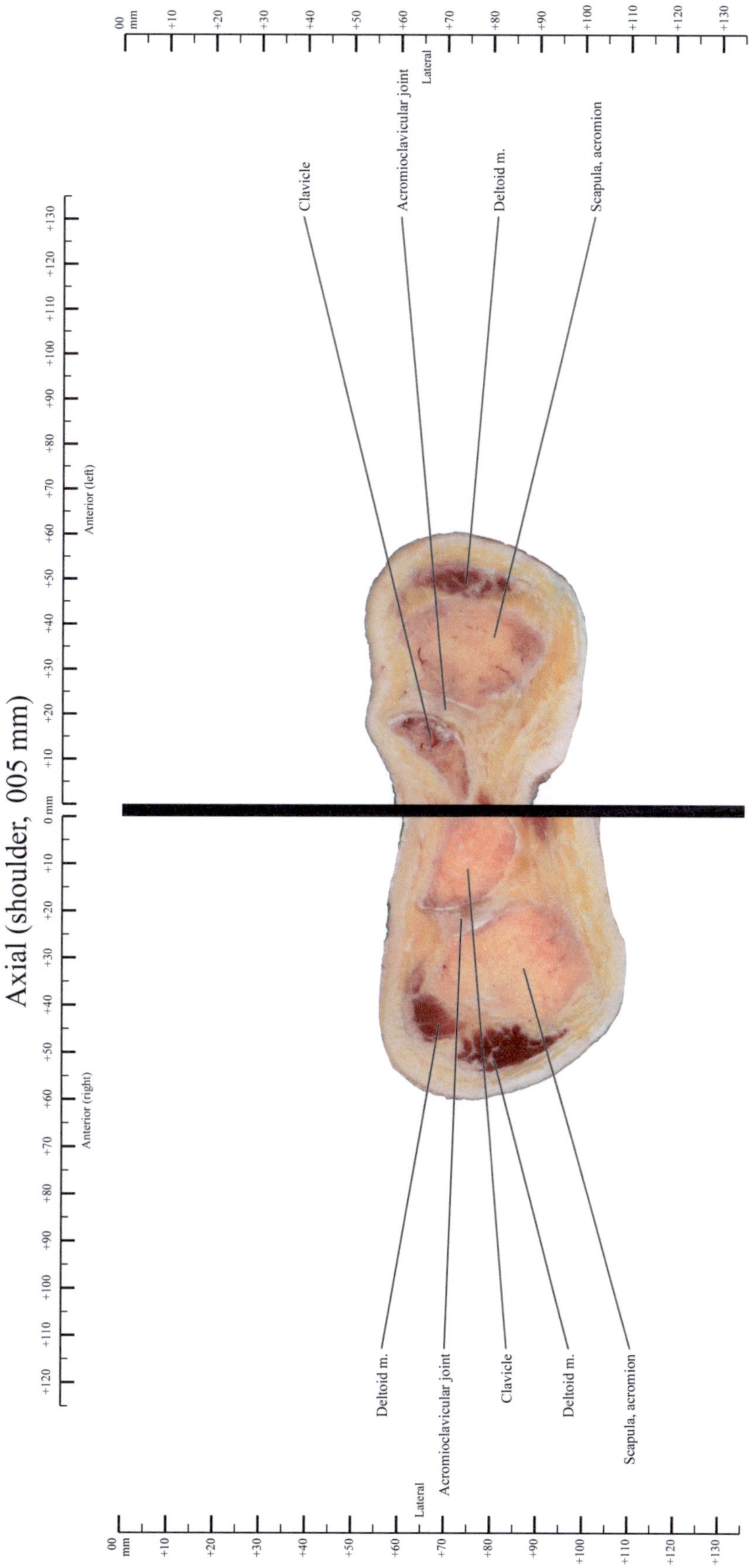
Axial (shoulder, 005 mm)
Anterior (left)
Anterior (right)
Lateral
Clavicle
Acromioclavicular joint
Deltoid m.
Scapula, acromion
Deltoid m.
Acromioclavicular joint
Clavicle
Deltoid m.
Scapula, acromion
Lateral

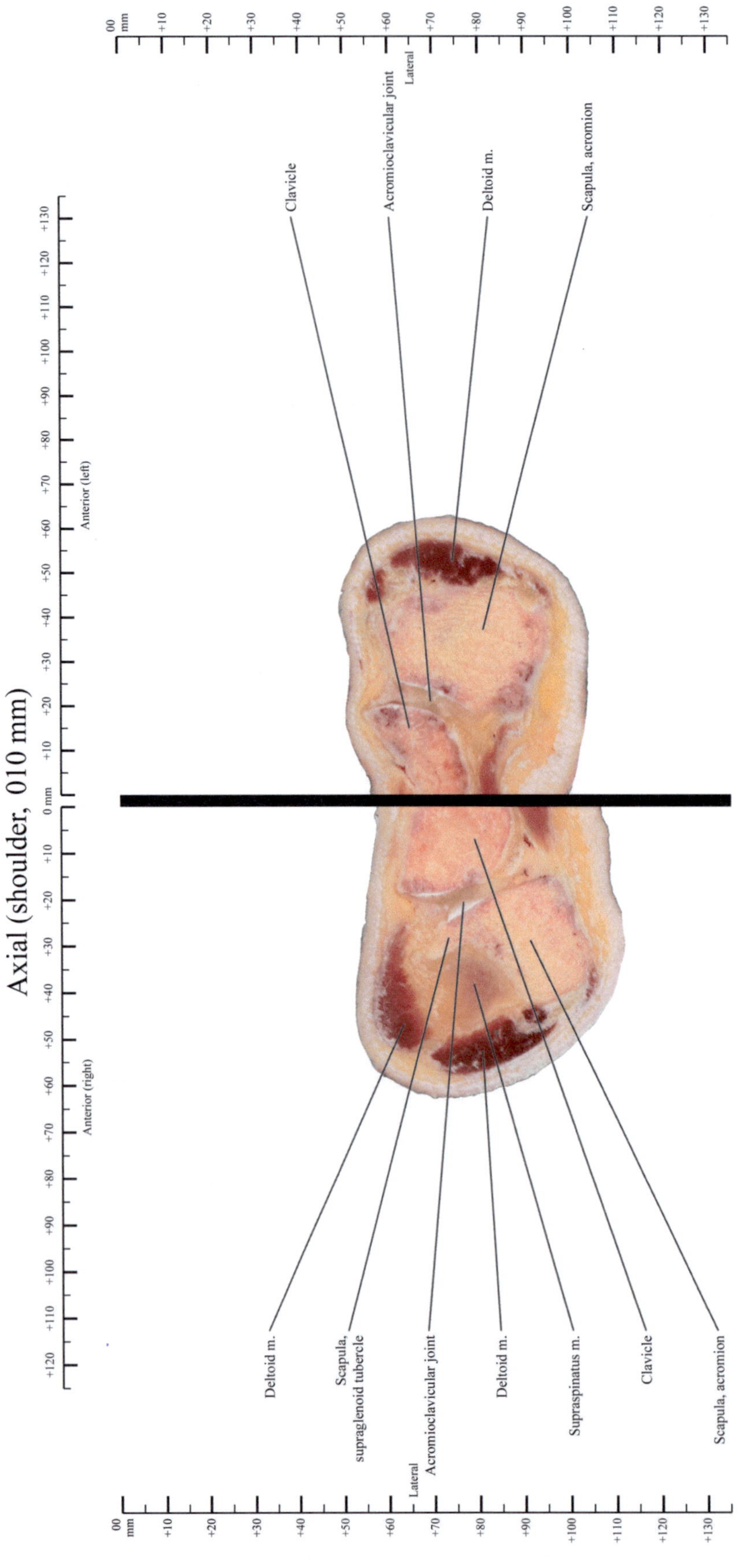
Axial (shoulder, 010 mm)
Anterior (left)
Anterior (right)
Lateral
Clavicle
Acromioclavicular joint
Deltoid m.
Scapula, acromion
Deltoid m.
Scapula, supraglenoid tubercle
Acromioclavicular joint
Deltoid m.
Supraspinatus m.
Clavicle
Scapula, acromion

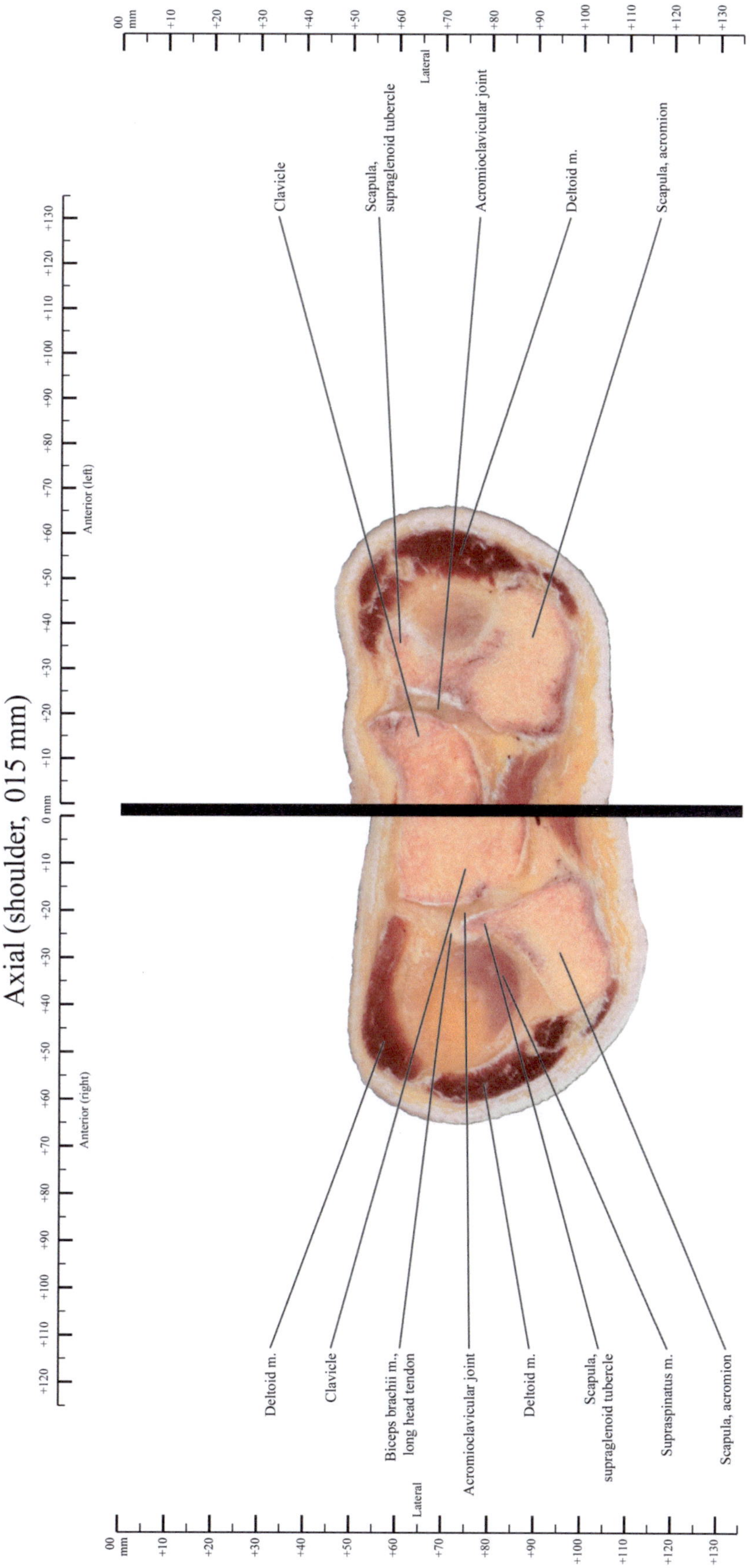
Axial (shoulder, 015 mm)
Anterior (left)
Anterior (right)
Lateral
Clavicle
Scapula, supraglenoid tubercle
Acromioclavicular joint
Deltoid m.
Scapula, acromion
Deltoid m.
Clavicle
Biceps brachii m., long head tendon
Acromioclavicular joint
Deltoid m.
Scapula, supraglenoid tubercle
Supraspinatus m.
Scapula, acromion
Lateral

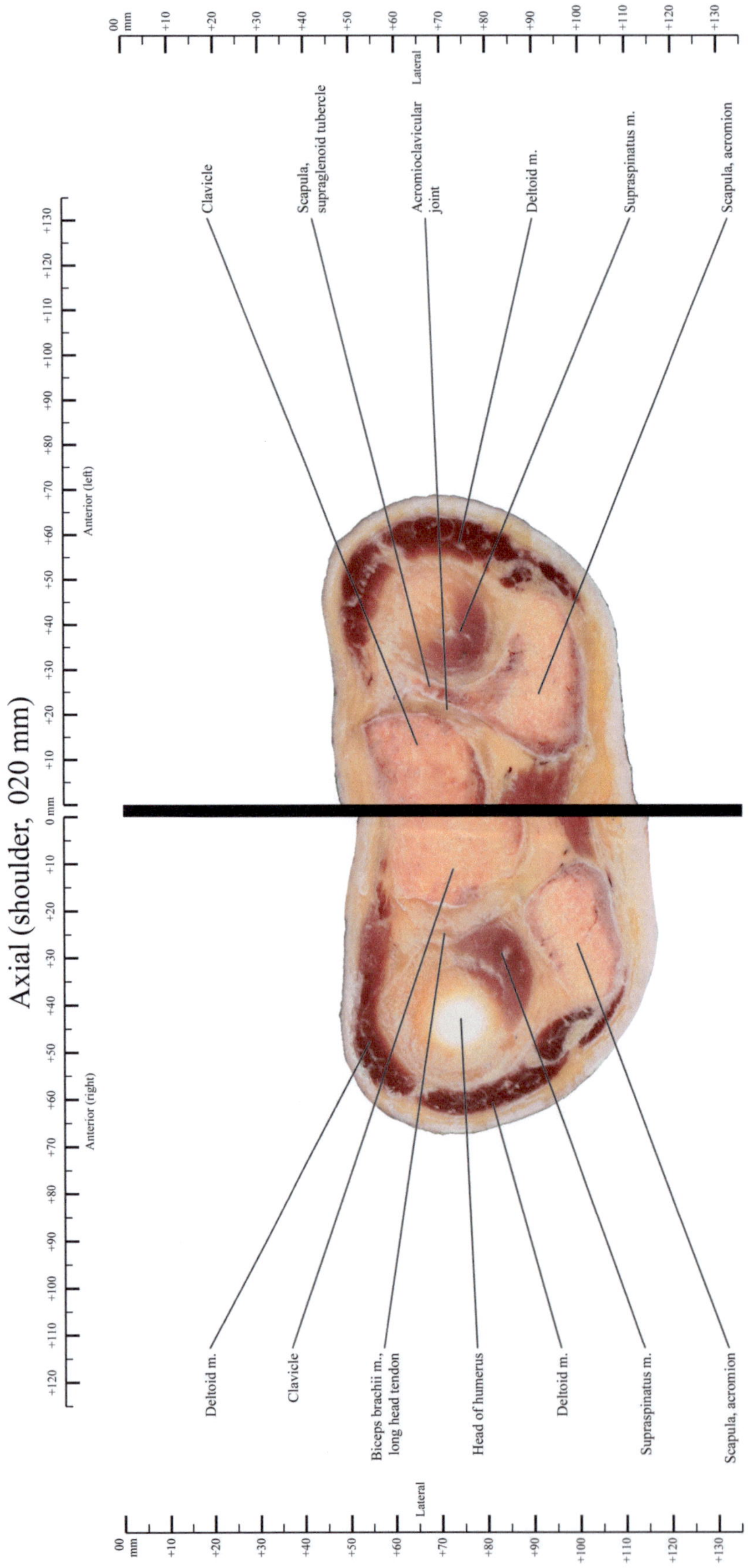
Axial (shoulder, 020 mm)
Anterior (left)
Anterior (right)
Lateral
Clavicle
Scapula, supraglenoid tubercle
Acromioclavicular joint
Deltoid m.
Supraspinatus m.
Scapula, acromion
Deltoid m.
Clavicle
Biceps brachii m., long head tendon
Head of humerus
Deltoid m.
Supraspinatus m.
Scapula, acromion

Axial (shoulder, 025 mm)

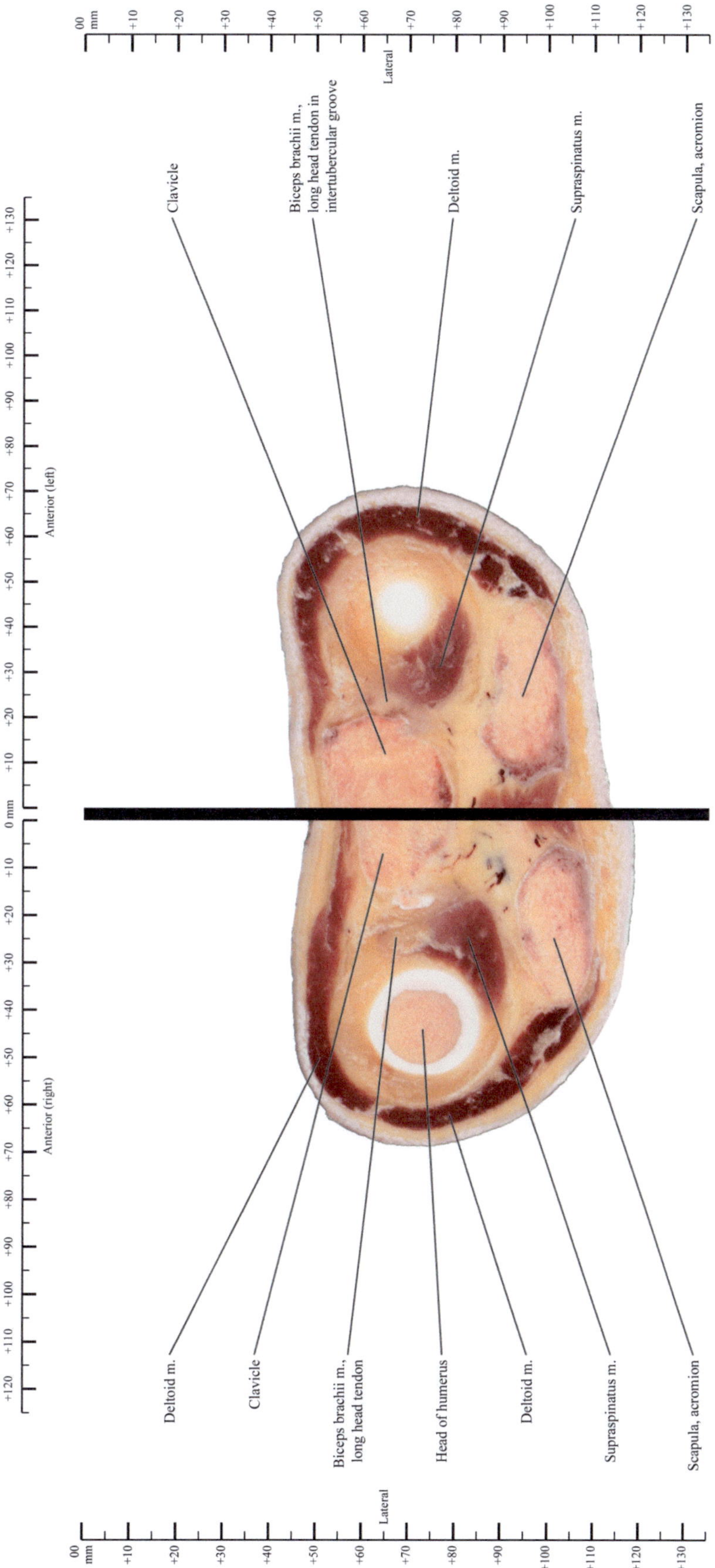

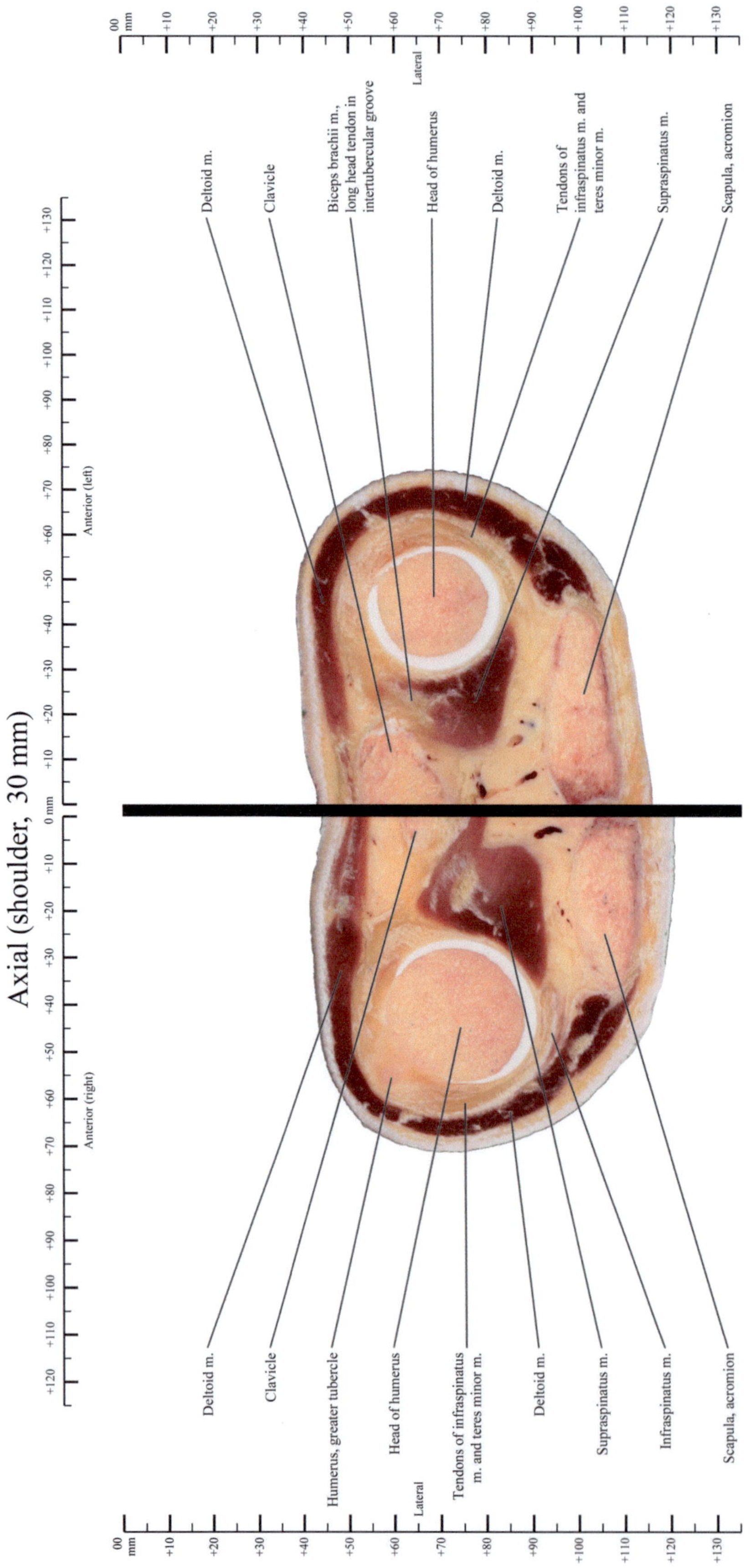
Axial (shoulder, 30 mm)
Anterior (left)
Anterior (right)
Lateral
Deltoid m.
Clavicle
Biceps brachii m., long head tendon in intertubercular groove
Head of humerus
Deltoid m.
Tendons of infraspinatus m. and teres minor m.
Supraspinatus m.
Scapula, acromion
Deltoid m.
Clavicle
Humerus, greater tubercle
Head of humerus
Tendons of infraspinatus m. and teres minor m.
Deltoid m.
Supraspinatus m.
Infraspinatus m.
Scapula, acromion

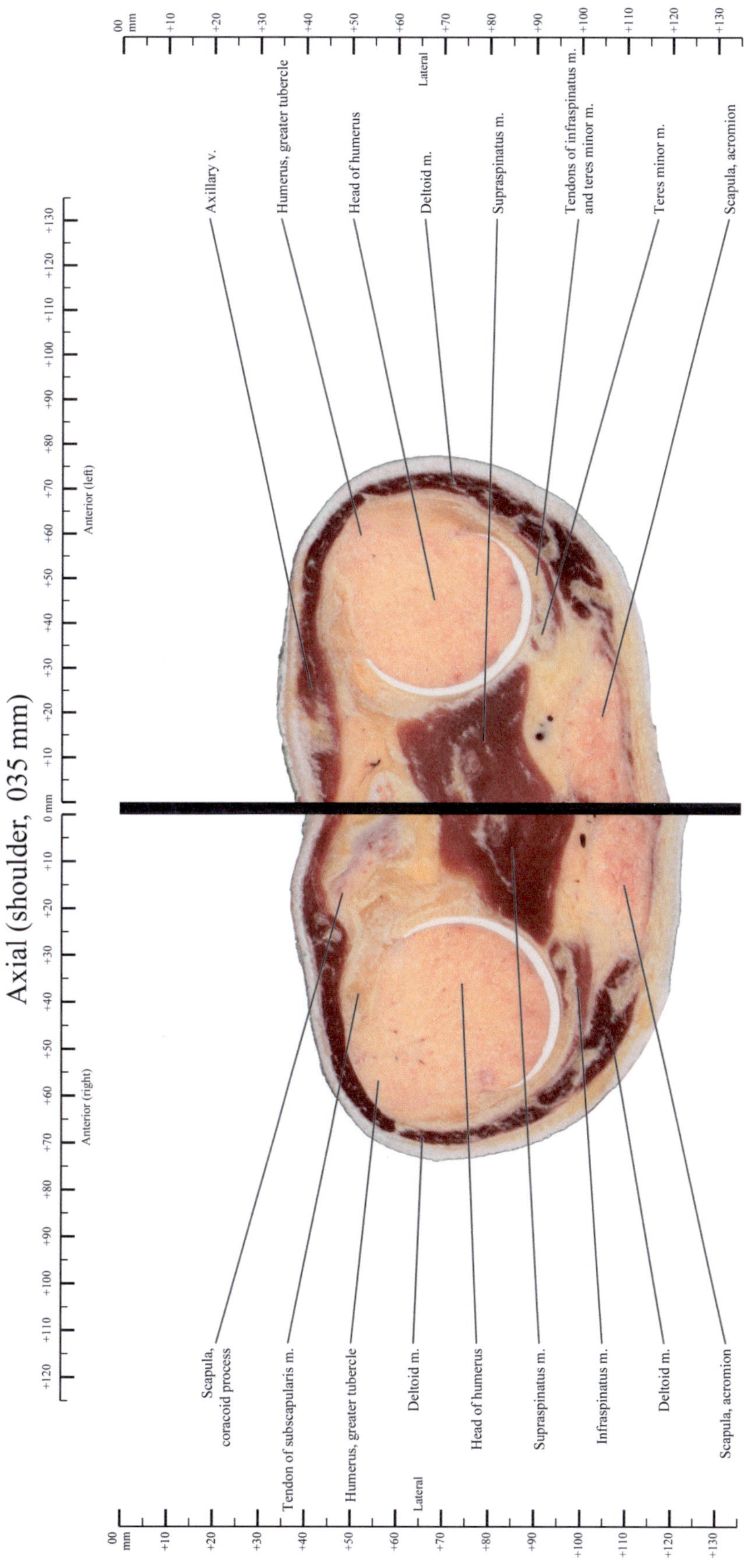
Axial (shoulder, 035 mm)
Anterior (left)
Anterior (right)
Lateral
Axillary v.
Humerus, greater tubercle
Head of humerus
Deltoid m.
Supraspinatus m.
Tendons of infraspinatus m. and teres minor m.
Teres minor m.
Scapula, acromion
Scapula, coracoid process
Tendon of subscapularis m.
Humerus, greater tubercle
Deltoid m.
Head of humerus
Supraspinatus m.
Infraspinatus m.
Deltoid m.
Scapula, acromion

Axial (shoulder, 040 mm)

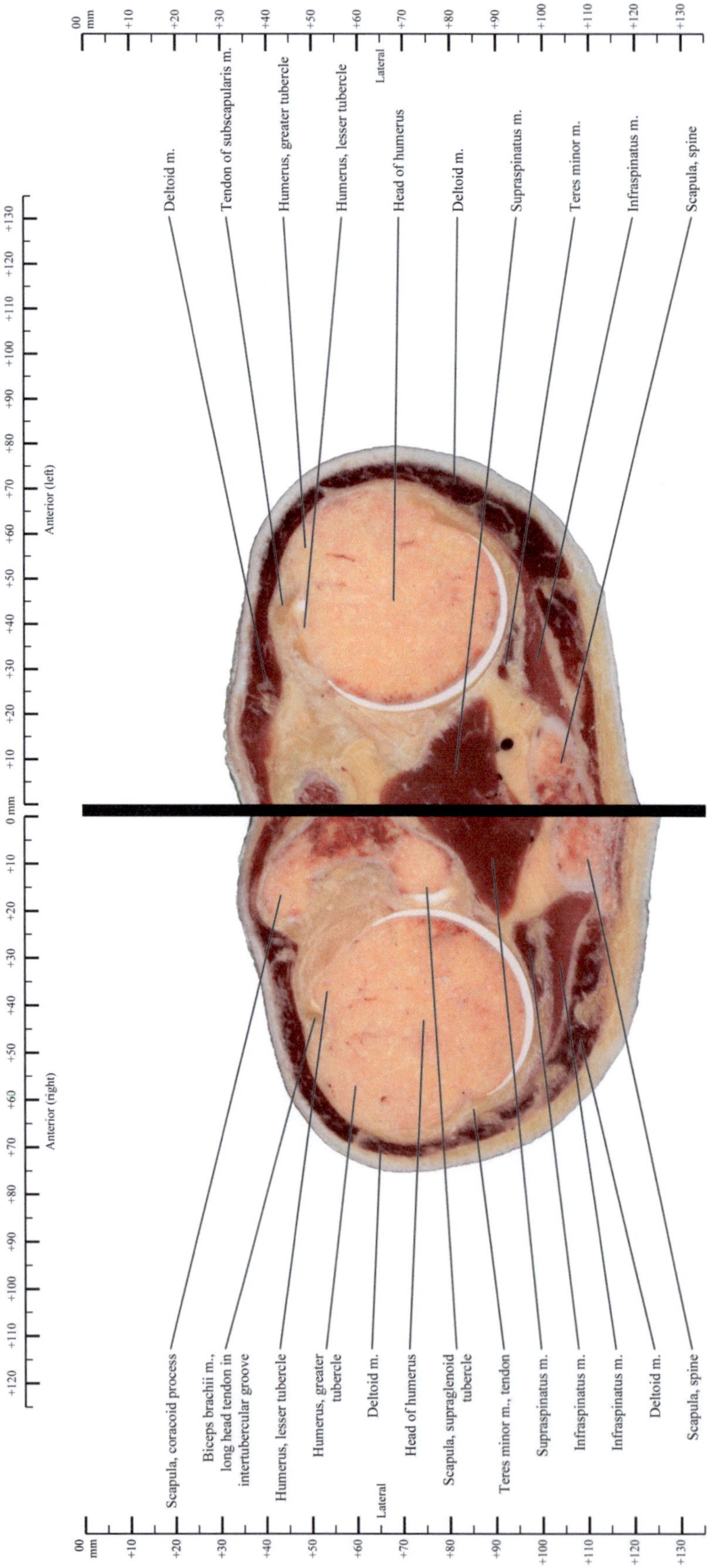

Axial (shoulder, 045 mm)

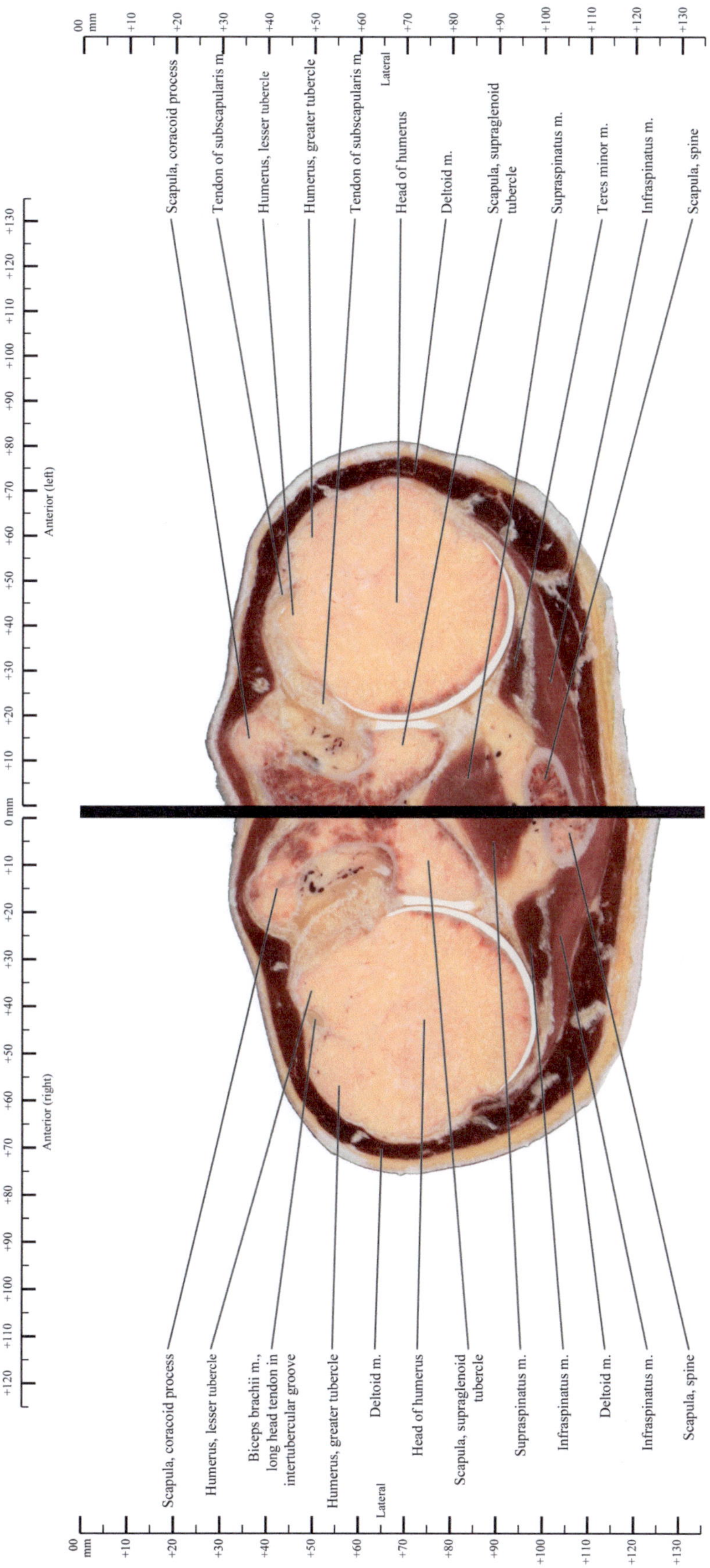
Scapula, coracoid process
Tendon of subscapularis m.
Humerus, lesser tubercle
Humerus, greater tubercle
Tendon of subscapularis m.
Lateral
Head of humerus
Deltoid m.
Scapula, supraglenoid tubercle
Supraspinatus m.
Teres minor m.
Infraspinatus m.
Scapula, spine
Anterior (left)
Anterior (right)
Scapula, coracoid process
Humerus, lesser tubercle
Biceps brachii m., long head tendon in intertubercular groove
Humerus, greater tubercle
Lateral
Deltoid m.
Head of humerus
Scapula, supraglenoid tubercle
Supraspinatus m.
Infraspinatus m.
Deltoid m.
Infraspinatus m.
Scapula, spine

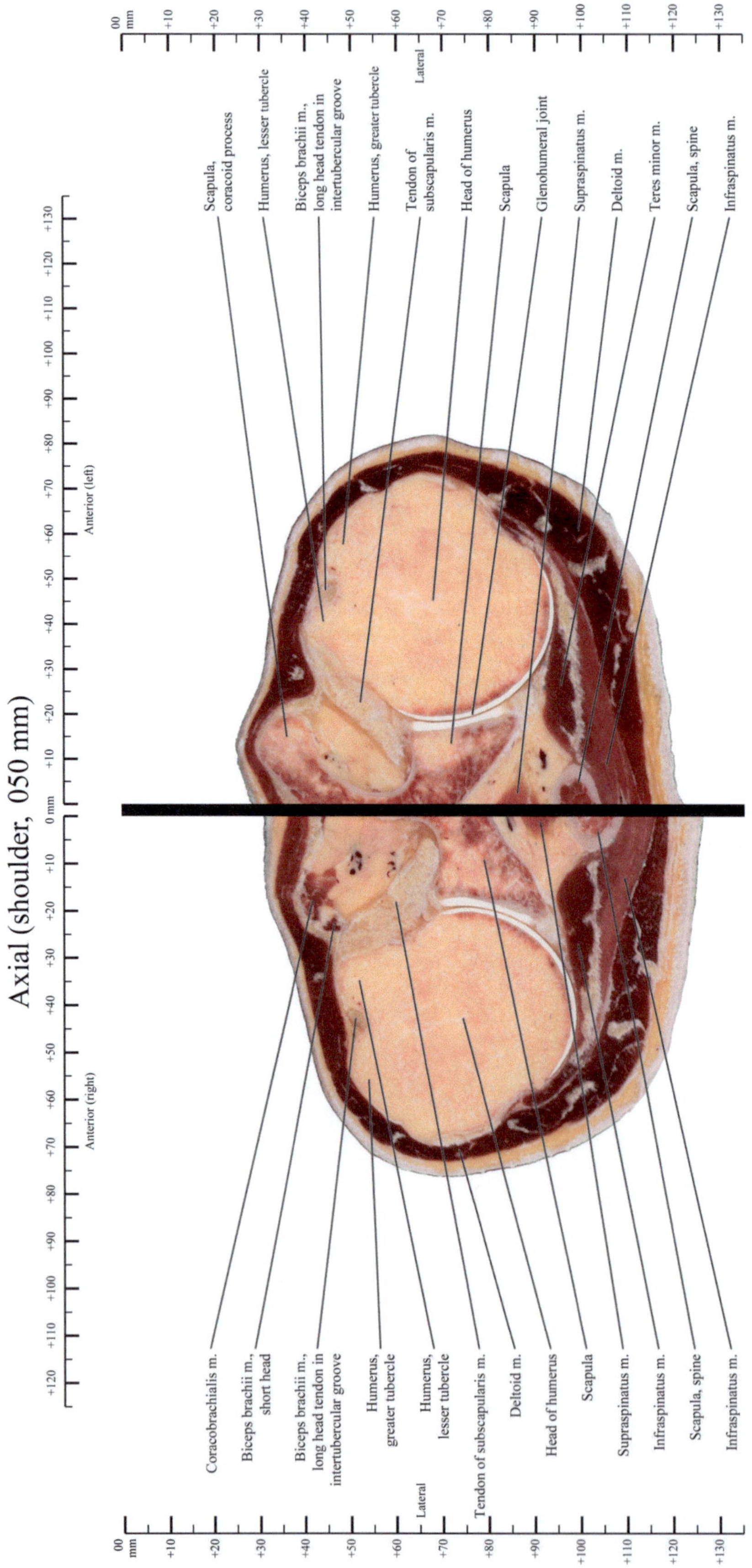
Axial (shoulder, 050 mm)
Anterior (left)
Anterior (right)
Lateral
Scapula, coracoid process
Humerus, lesser tubercle
Biceps brachii m., long head tendon in intertubercular groove
Humerus, greater tubercle
Tendon of subscapularis m.
Head of humerus
Scapula
Glenohumeral joint
Supraspinatus m.
Deltoid m.
Teres minor m.
Scapula, spine
Infraspinatus m.
Coracobrachialis m.
Biceps brachii m., short head
Biceps brachii m., long head tendon in intertubercular groove
Humerus, greater tubercle
Humerus, lesser tubercle
Tendon of subscapularis m.
Deltoid m.
Head of humerus
Scapula
Supraspinatus m.
Infraspinatus m.
Scapula, spine
Infraspinatus m.
Lateral

Axial (shoulder, 055 mm)

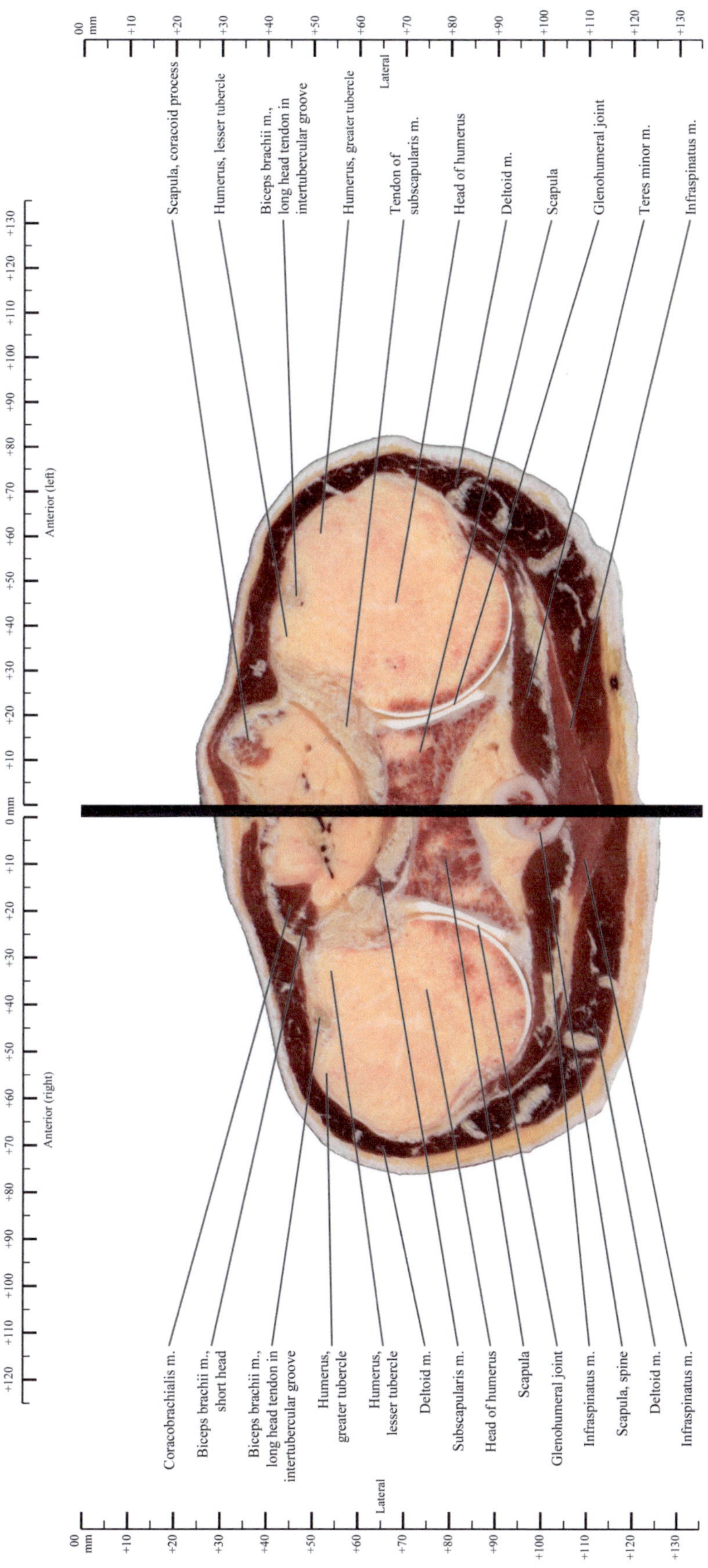

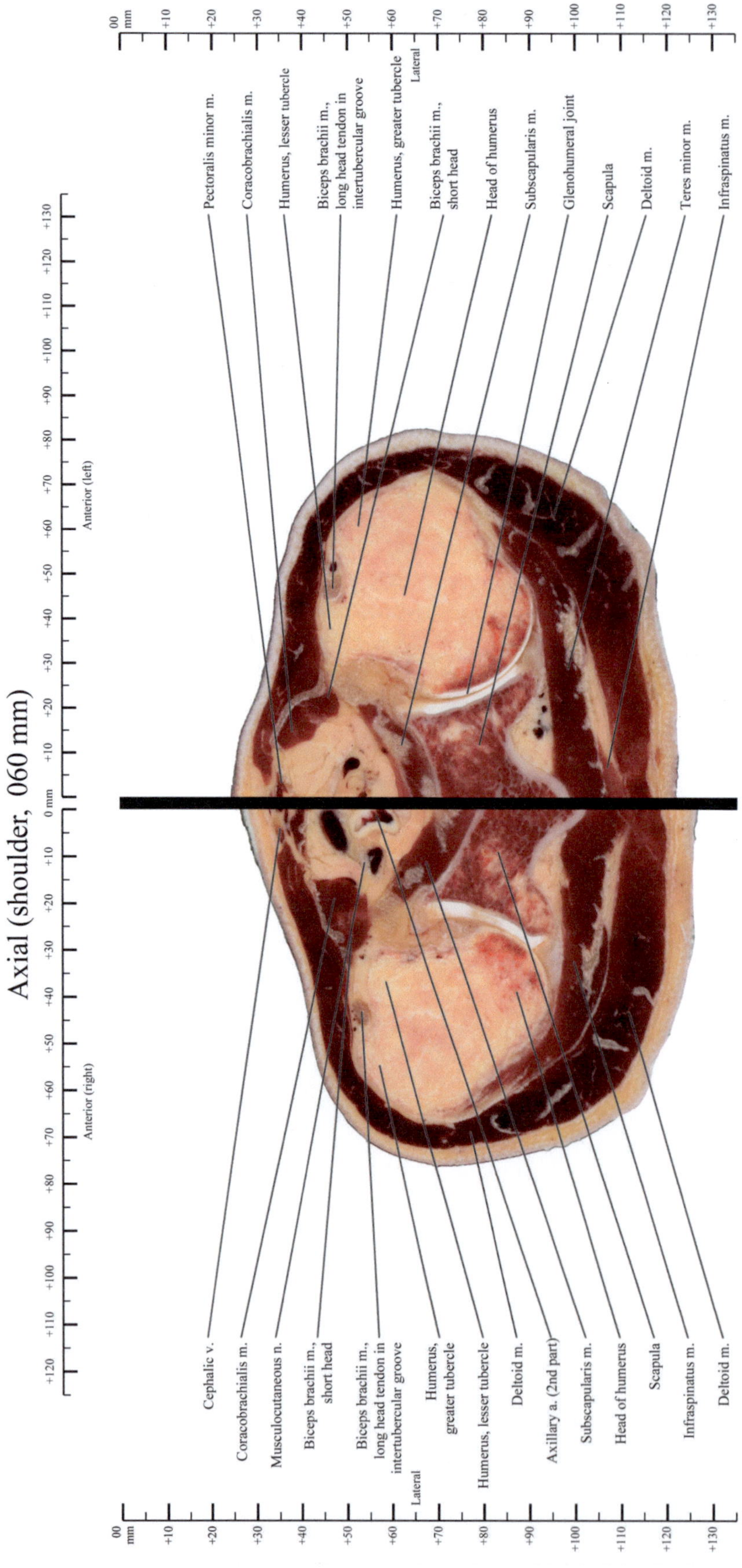
Axial (shoulder, 060 mm)
Anterior (left)
Anterior (right)
Lateral
Pectoralis minor m.
Coracobrachialis m.
Humerus, lesser tubercle
Biceps brachii m., long head tendon in intertubercular groove
Humerus, greater tubercle
Biceps brachii m., short head
Head of humerus
Subscapularis m.
Glenohumeral joint
Scapula
Deltoid m.
Teres minor m.
Infraspinatus m.
Cephalic v.
Coracobrachialis m.
Musculocutaneous n.
Biceps brachii m., short head
Biceps brachii m., long head tendon in intertubercular groove
Humerus, greater tubercle
Humerus, lesser tubercle
Deltoid m.
Axillary a. (2nd part)
Subscapularis m.
Head of humerus
Scapula
Infraspinatus m.
Deltoid m.

Axial (shoulder, 065 mm)

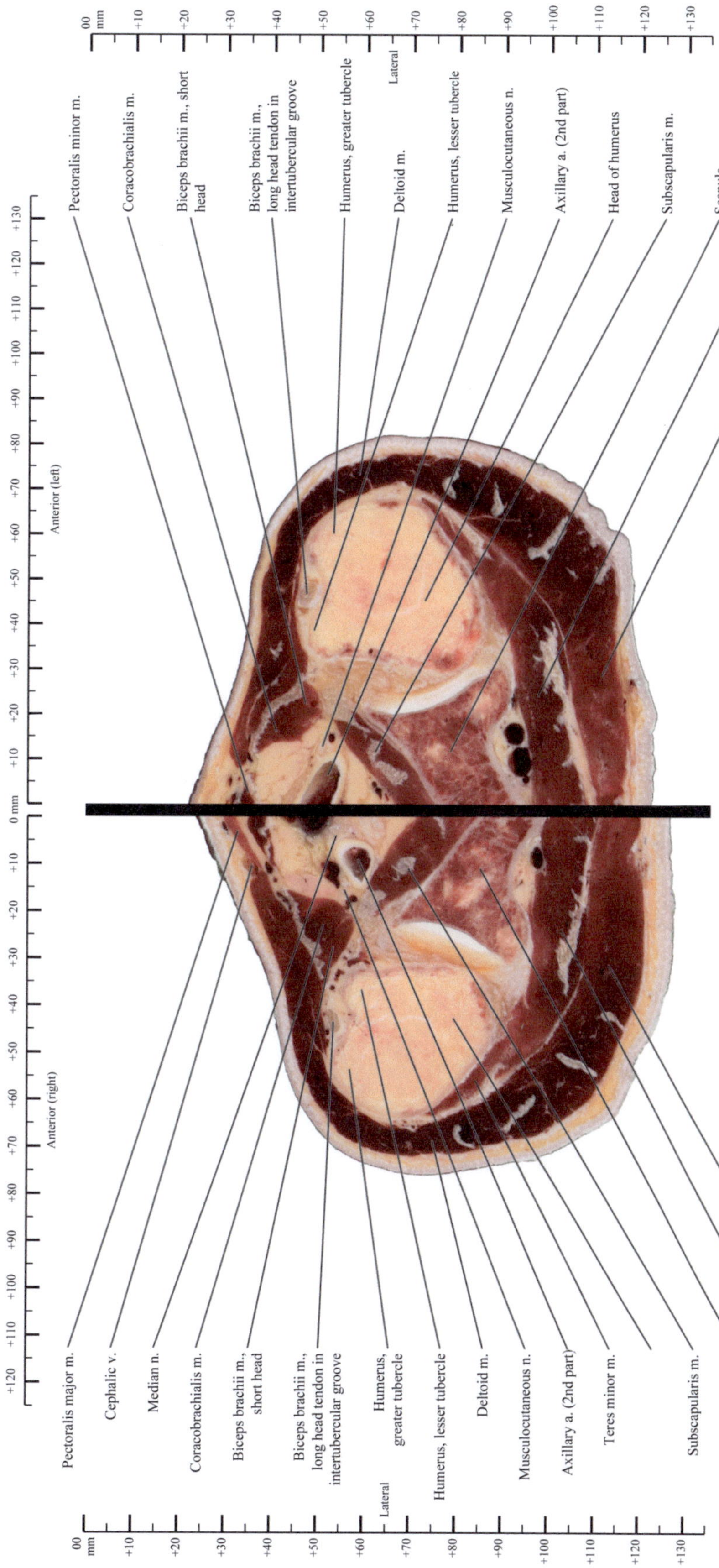

Axial (shoulder, 070 mm)

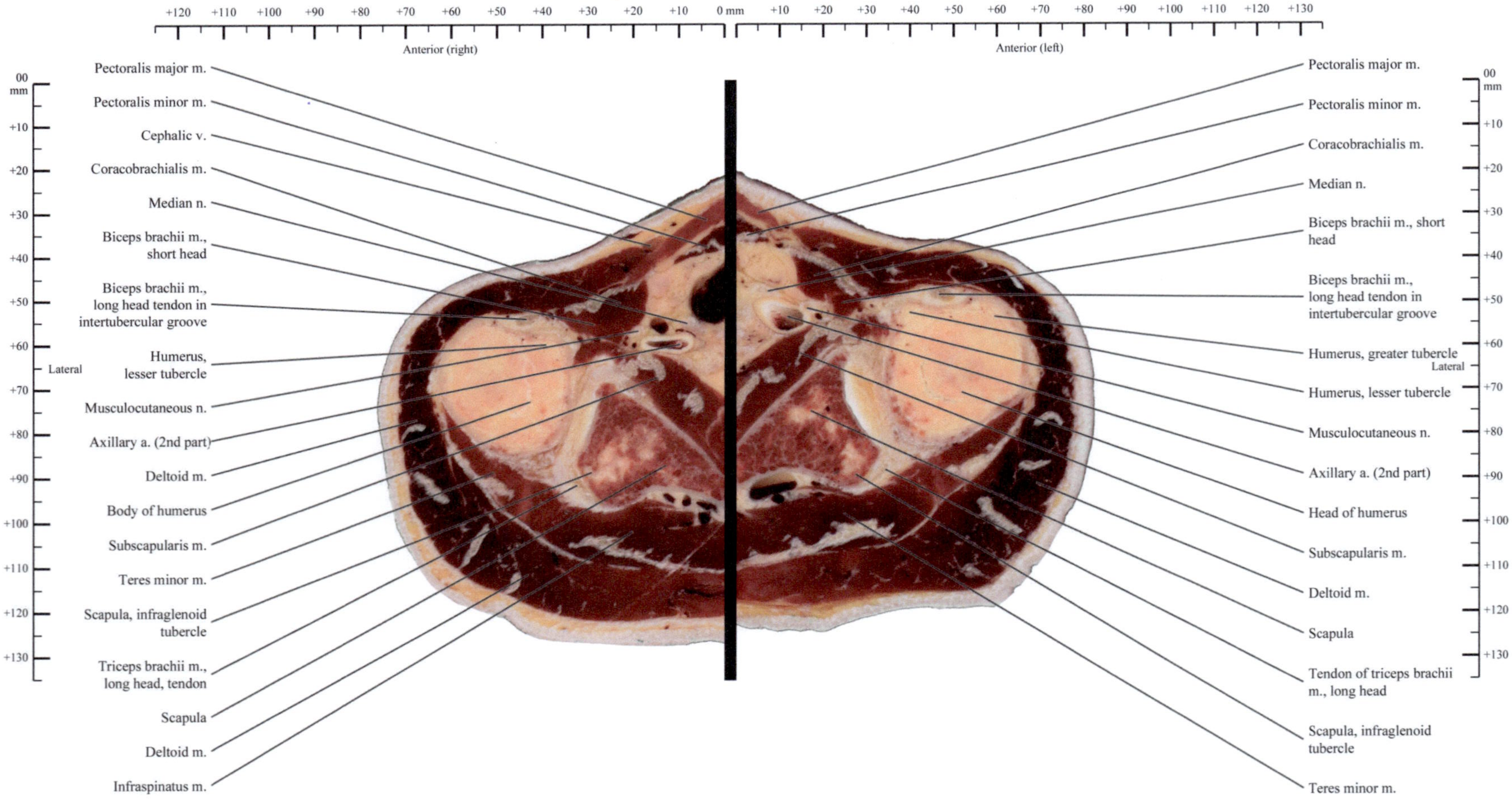

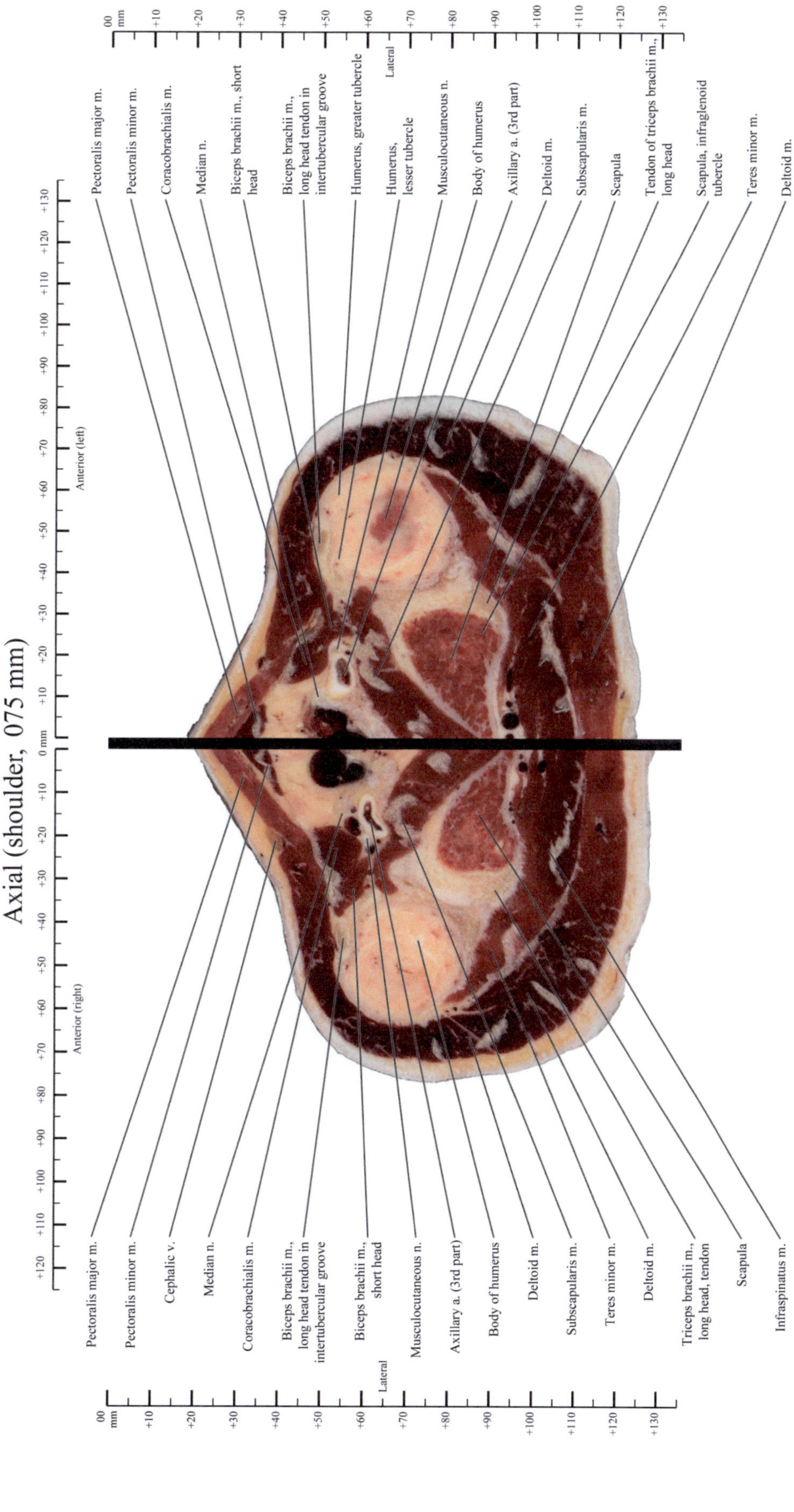
Axial (shoulder, 075 mm)
Anterior (left)
Anterior (right)
Lateral
Pectoralis major m.
Pectoralis minor m.
Coracobrachialis m.
Median n.
Biceps brachii m., short head
Biceps brachii m., long head tendon in intertubercular groove
Humerus, greater tubercle
Humerus, lesser tubercle
Musculocutaneous n.
Body of humerus
Axillary a. (3rd part)
Deltoid m.
Subscapularis m.
Scapula
Tendon of triceps brachii m., long head
Scapula, infraglenoid tubercle
Teres minor m.
Deltoid m.
Pectoralis major m.
Pectoralis minor m.
Cephalic v.
Median n.
Coracobrachialis m.
Biceps brachii m., long head tendon in intertubercular groove
Biceps brachii m., short head
Musculocutaneous n.
Axillary a. (3rd part)
Body of humerus
Deltoid m.
Subscapularis m.
Teres minor m.
Deltoid m.
Triceps brachii m., long head, tendon
Scapula
Infraspinatus m.

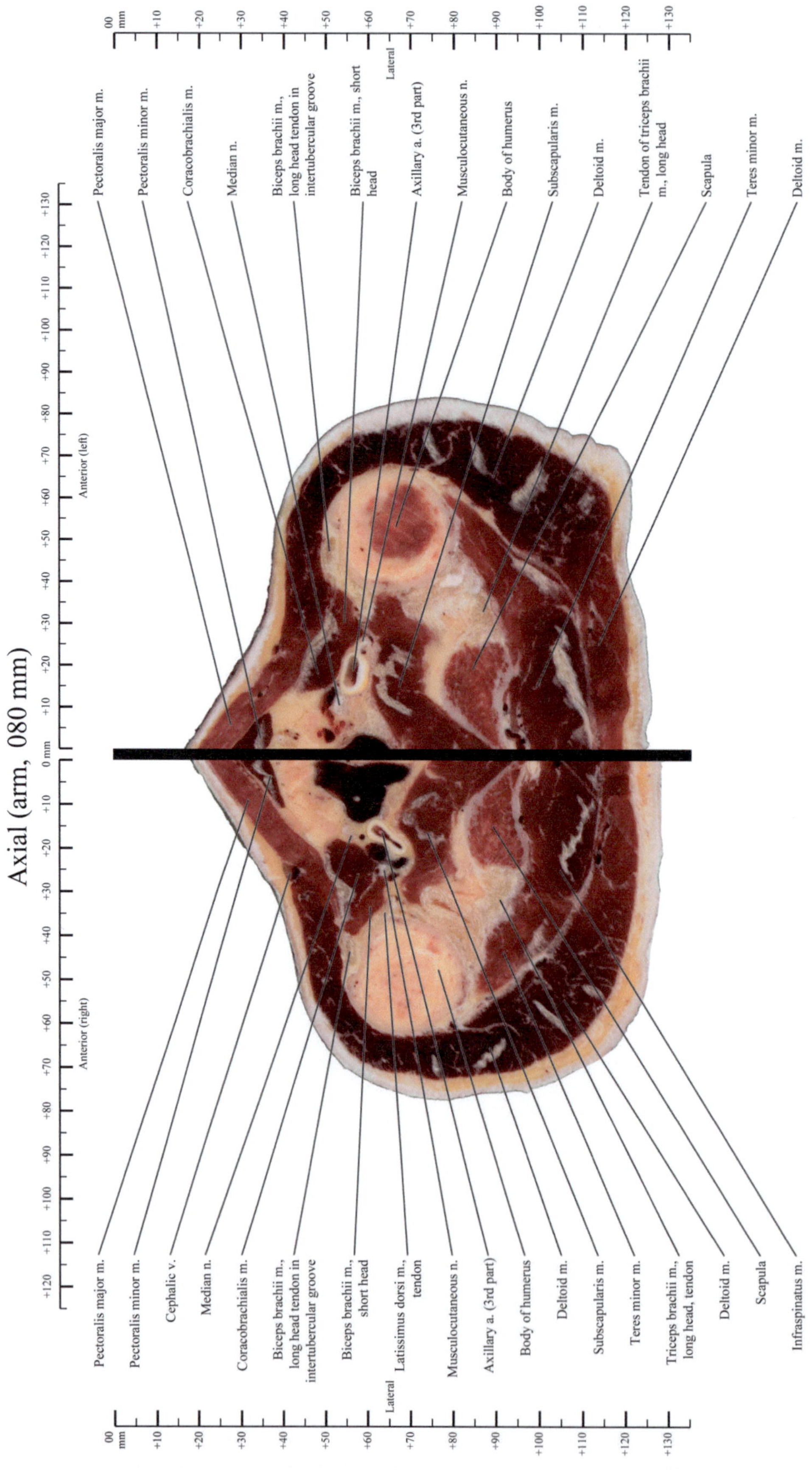
Axial (arm, 080 mm)
Anterior (left)
Anterior (right)
Lateral
Pectoralis major m.
Pectoralis minor m.
Coracobrachialis m.
Median n.
Biceps brachii m., long head tendon in intertubercular groove
Biceps brachii m., short head
Axillary a. (3rd part)
Musculocutaneous n.
Body of humerus
Subscapularis m.
Deltoid m.
Tendon of triceps brachii m., long head
Scapula
Teres minor m.
Deltoid m.
Pectoralis major m.
Pectoralis minor m.
Cephalic v.
Median n.
Coracobrachialis m.
Biceps brachii m., long head tendon in intertubercular groove
Biceps brachii m., short head
Latissimus dorsi m., tendon
Musculocutaneous n.
Axillary a. (3rd part)
Body of humerus
Deltoid m.
Subscapularis m.
Teres minor m.
Triceps brachii m., long head, tendon
Deltoid m.
Scapula
Infraspinatus m.

Axial (arm, 085 mm)

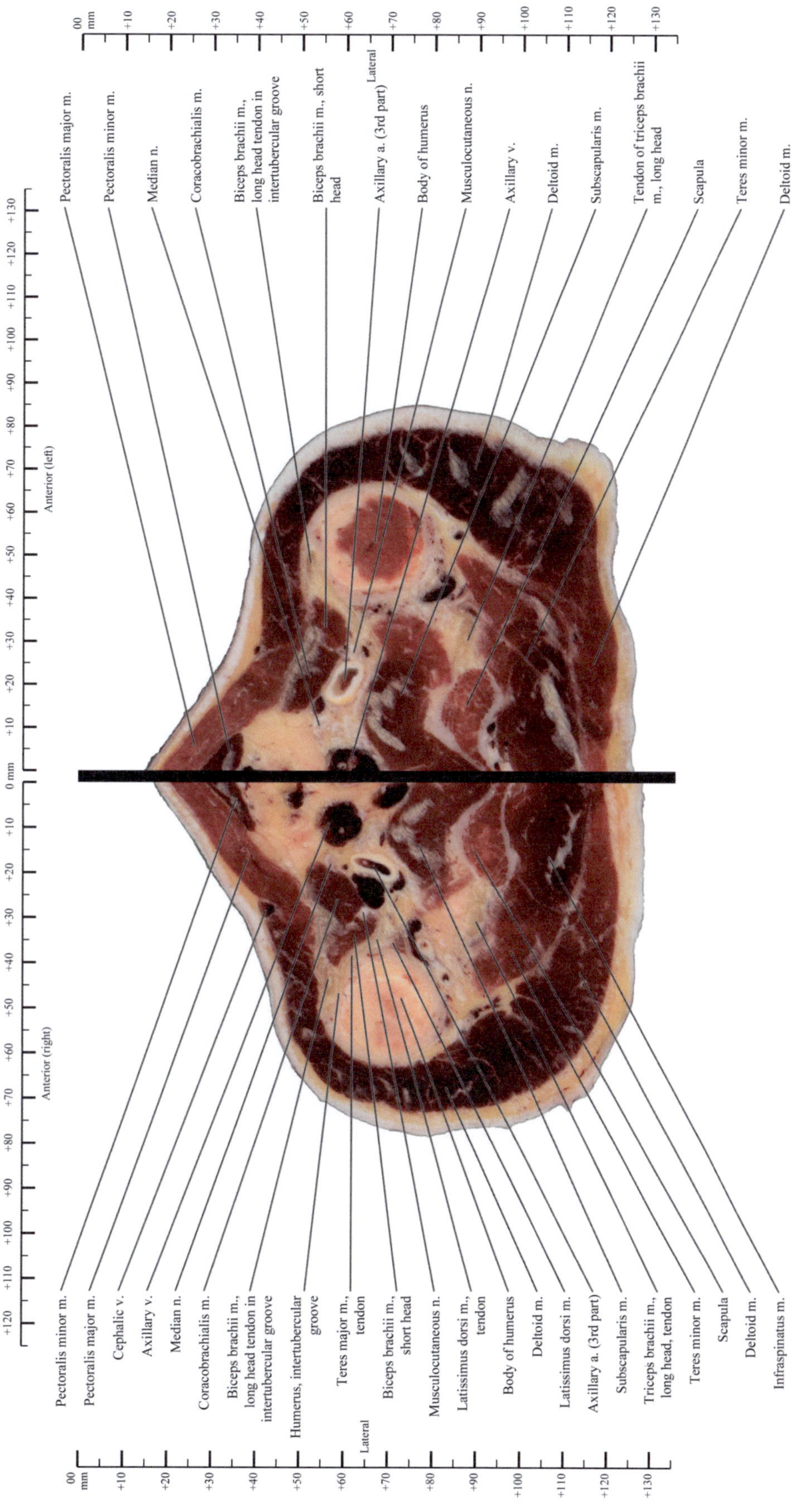

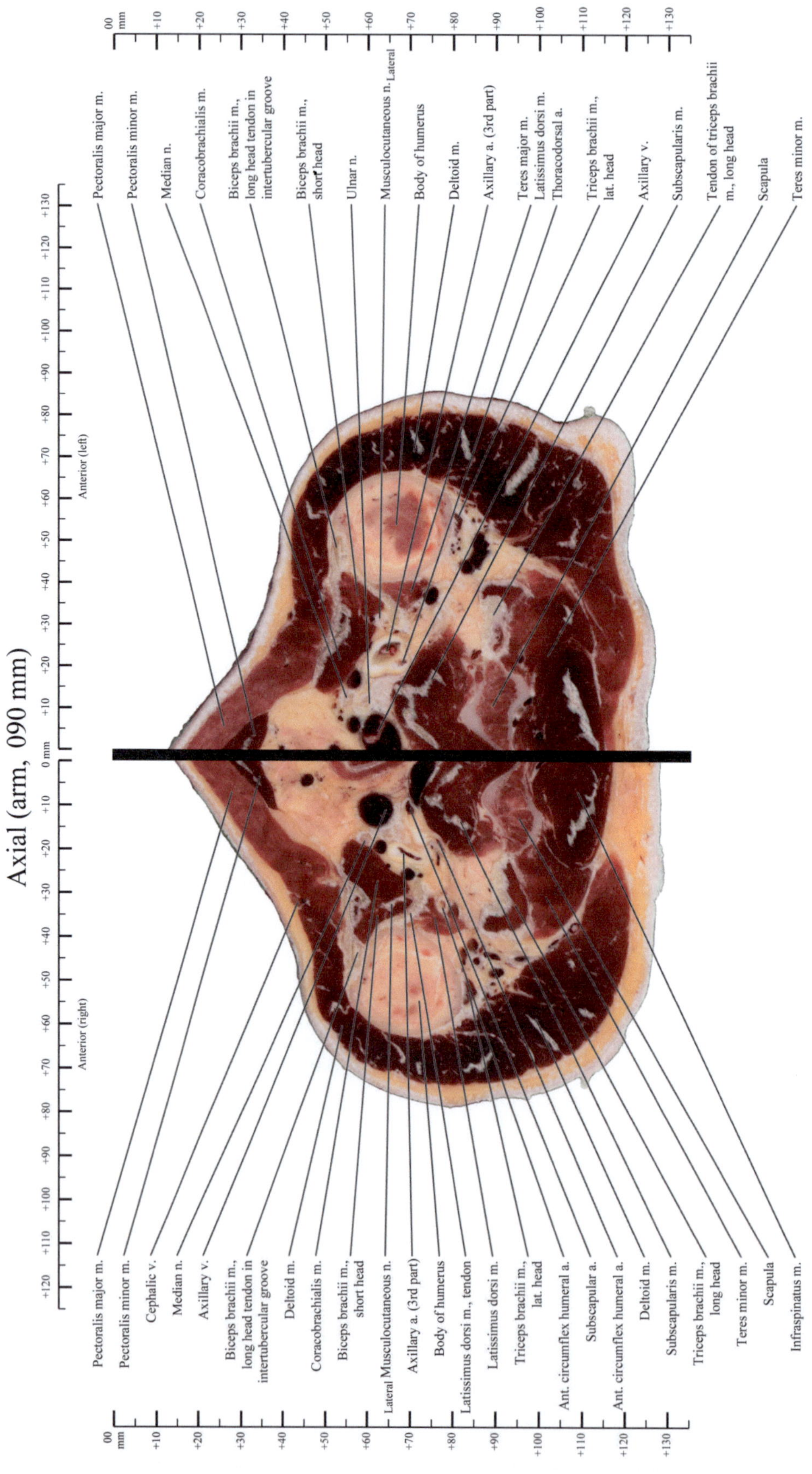

Axial (arm, 090 mm)
Anterior (right)
Anterior (left)
Pectoralis major m.
Pectoralis minor m.
Cephalic v.
Median n.
Axillary v.
Biceps brachii m., long head tendon in intertubercular groove
Deltoid m.
Coracobrachialis m.
Biceps brachii m., short head
Lateral Musculocutaneous n.
Axillary a. (3rd part)
Body of humerus
Latissimus dorsi m., tendon
Latissimus dorsi m.
Triceps brachii m., lat. head
Ant. circumflex humeral a.
Subscapular a.
Ant. circumflex humeral a.
Deltoid m.
Subscapularis m.
Triceps brachii m., long head
Teres minor m.
Scapula
Infraspinatus m.
Pectoralis major m.
Pectoralis minor m.
Median n.
Coracobrachialis m.
Biceps brachii m., long head tendon in intertubercular groove
Biceps brachii m., short head
Ulnar n.
Musculocutaneous n. Lateral
Body of humerus
Deltoid m.
Axillary a. (3rd part)
Teres major m.
Latissimus dorsi m.
Thoracodorsal a.
Triceps brachii m., lat. head
Axillary v.
Subscapularis m.
Tendon of triceps brachii m., long head
Scapula
Teres minor m.

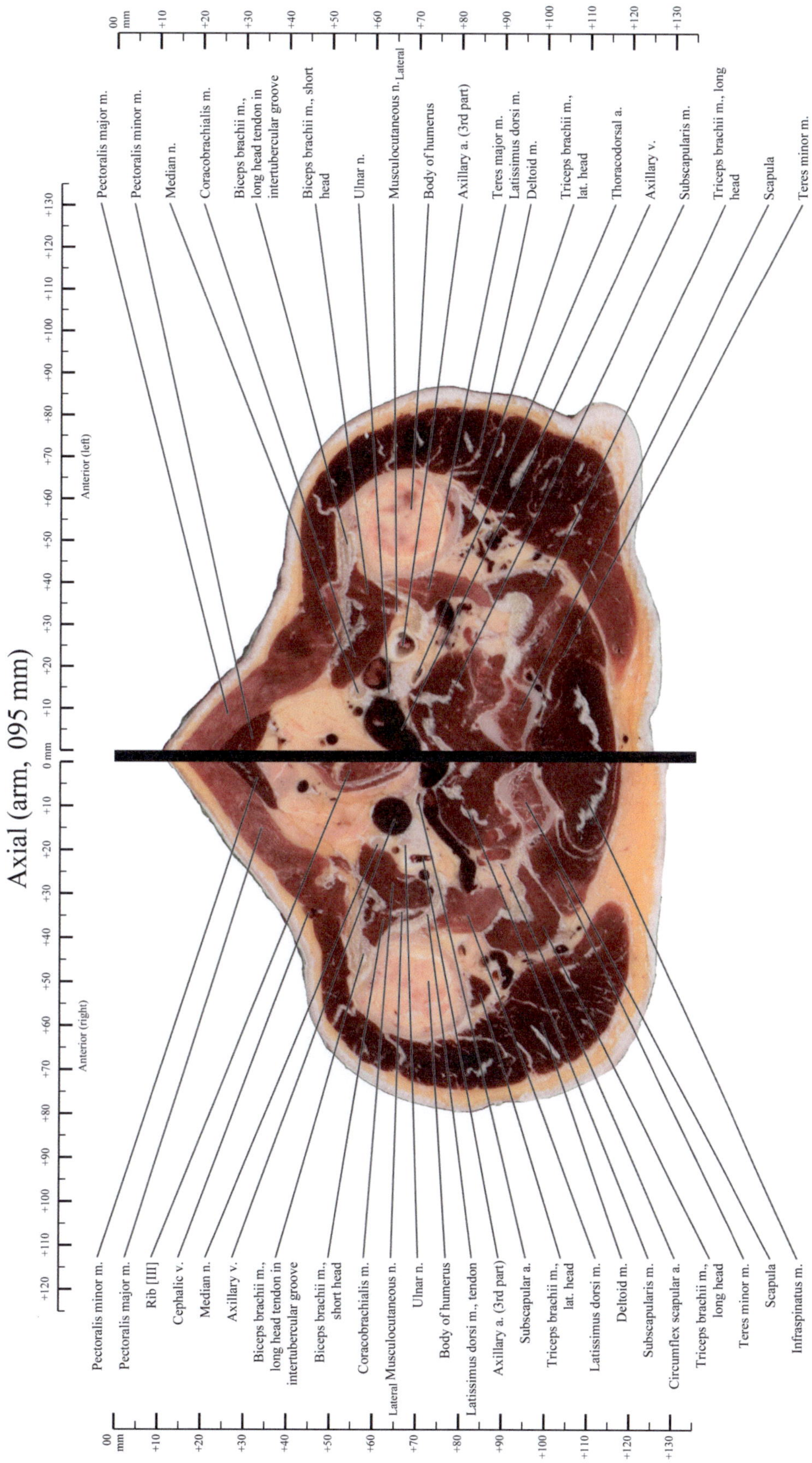

Axial (arm, 095 mm)

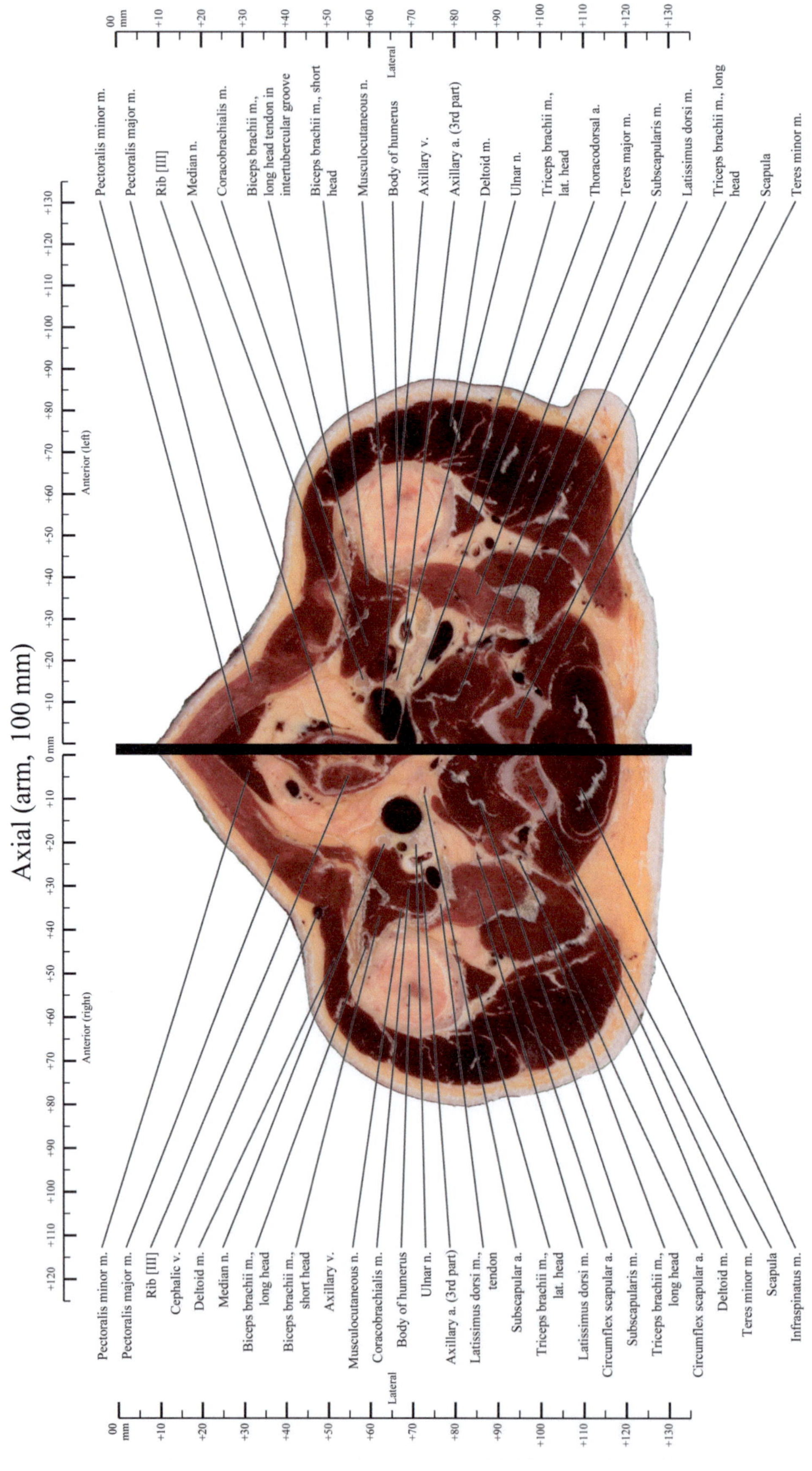
Axial (arm, 100 mm)
Anterior (left)
Anterior (right)
Lateral
Pectoralis minor m.
Pectoralis major m.
Rib [III]
Median n.
Coracobrachialis m.
Biceps brachii m., long head tendon in intertubercular groove
Biceps brachii m., short head
Musculocutaneous n.
Body of humerus
Axillary v.
Axillary a. (3rd part)
Deltoid m.
Ulnar n.
Triceps brachii m., lat. head
Thoracodorsal a.
Teres major m.
Subscapularis m.
Latissimus dorsi m.
Triceps brachii m., long head
Scapula
Teres minor m.
Pectoralis minor m.
Pectoralis major m.
Rib [III]
Cephalic v.
Deltoid m.
Median n.
Biceps brachii m., long head
Biceps brachii m., short head
Axillary v.
Musculocutaneous n.
Coracobrachialis m.
Body of humerus
Ulnar n.
Axillary a. (3rd part)
Latissimus dorsi m., tendon
Subscapular a.
Triceps brachii m., lat. head
Latissimus dorsi m.
Circumflex scapular a.
Subscapularis m.
Triceps brachii m., long head
Circumflex scapular a.
Deltoid m.
Teres minor m.
Scapula
Infraspinatus m.

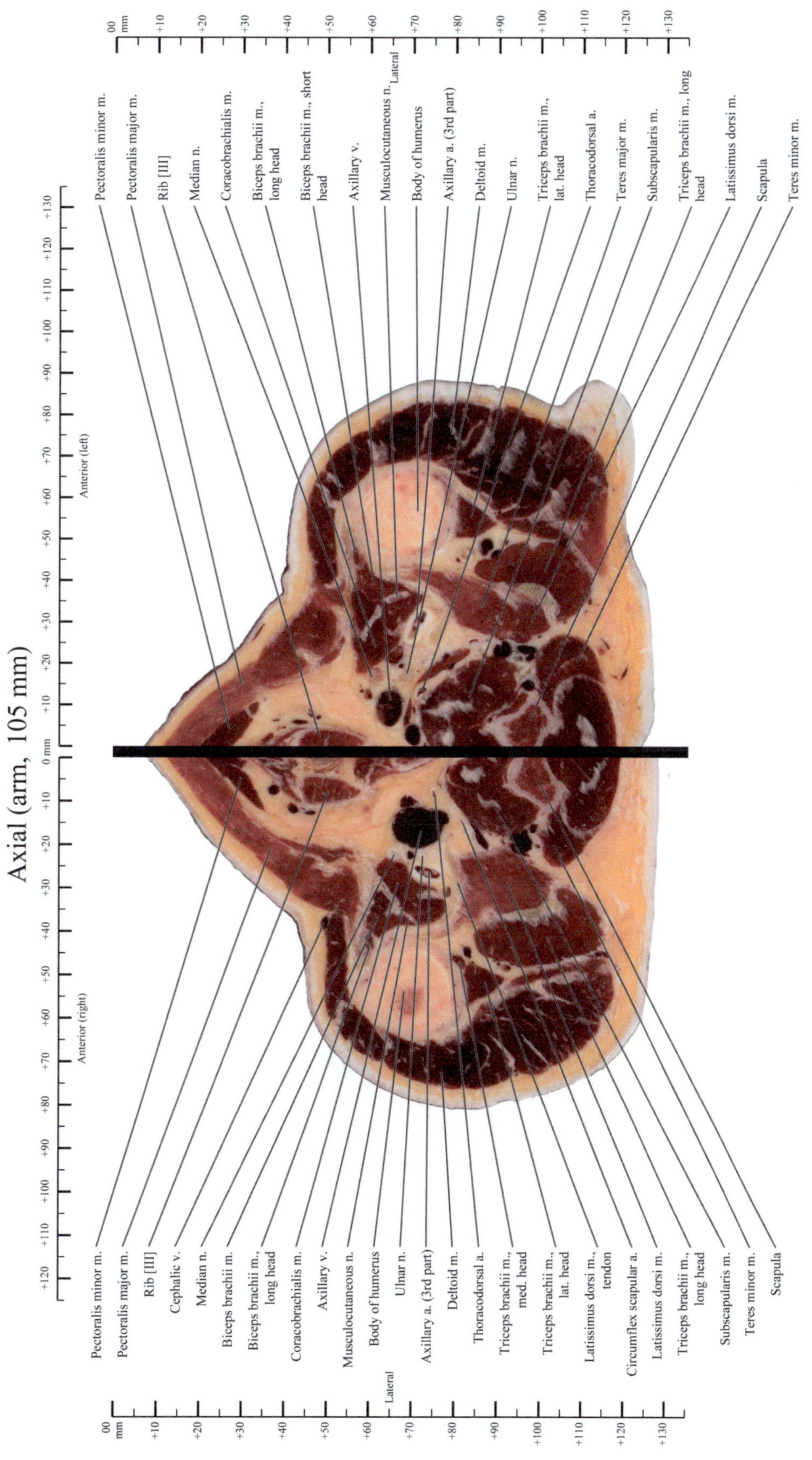
Axial (arm, 105 mm)
Anterior (left)
Anterior (right)
Pectoralis minor m.
Pectoralis major m.
Rib [III]
Median n.
Coracobrachialis m.
Biceps brachii m., long head
Biceps brachii m., short head
Axillary v.
Musculocutaneous n.
Lateral
Body of humerus
Axillary a. (3rd part)
Deltoid m.
Ulnar n.
Triceps brachii m., lat. head
Thoracodorsal a.
Teres major m.
Subscapularis m.
Triceps brachii m., long head
Latissimus dorsi m.
Scapula
Teres minor m.
Pectoralis minor m.
Pectoralis major m.
Rib [III]
Cephalic v.
Median n.
Biceps brachii m.
Biceps brachii m., long head
Coracobrachialis m.
Axillary v.
Musculocutaneous n.
Body of humerus
Ulnar n.
Axillary a. (3rd part)
Deltoid m.
Thoracodorsal a.
Triceps brachii m., med. head
Triceps brachii m., lat. head
Latissimus dorsi m., tendon
Circumflex scapular a.
Latissimus dorsi m.
Triceps brachii m., long head
Subscapularis m.
Teres minor m.
Scapula

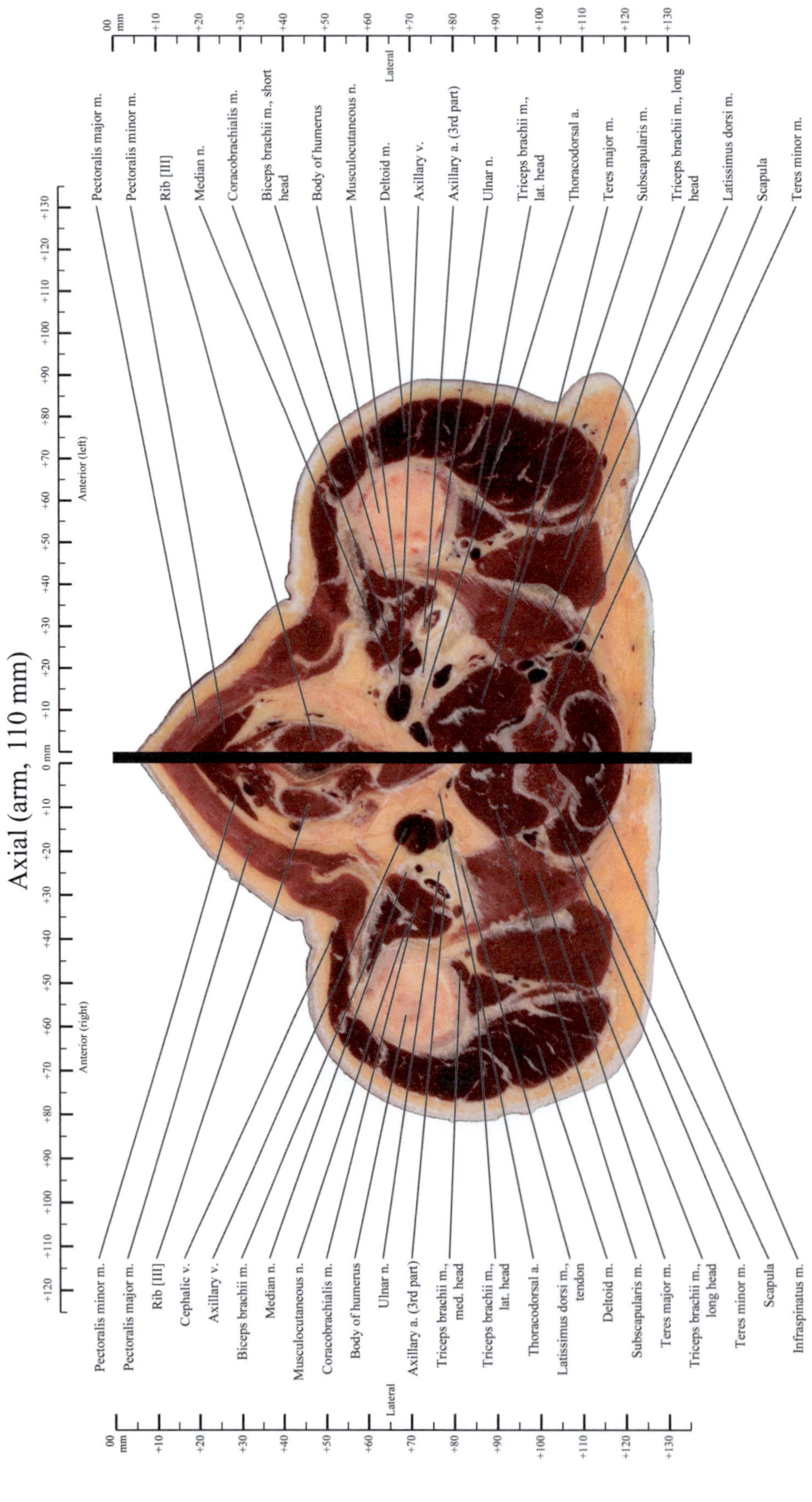
Axial (arm, 110 mm)
Anterior (left)
Anterior (right)
Lateral
Lateral
Pectoralis major m.
Pectoralis minor m.
Rib [III]
Median n.
Coracobrachialis m.
Biceps brachii m., short head
Body of humerus
Musculocutaneous n.
Deltoid m.
Axillary v.
Axillary a. (3rd part)
Ulnar n.
Triceps brachii m., lat. head
Thoracodorsal a.
Teres major m.
Subscapularis m.
Triceps brachii m., long head
Latissimus dorsi m.
Scapula
Teres minor m.
Pectoralis minor m.
Pectoralis major m.
Rib [III]
Cephalic v.
Axillary v.
Biceps brachii m.
Median n.
Musculocutaneous n.
Coracobrachialis m.
Body of humerus
Ulnar n.
Axillary a. (3rd part)
Triceps brachii m., med. head
Triceps brachii m., lat. head
Thoracodorsal a.
Latissimus dorsi m., tendon
Deltoid m.
Subscapularis m.
Teres major m.
Triceps brachii m., long head
Teres minor m.
Scapula
Infraspinatus m.
0 mm
00 mm

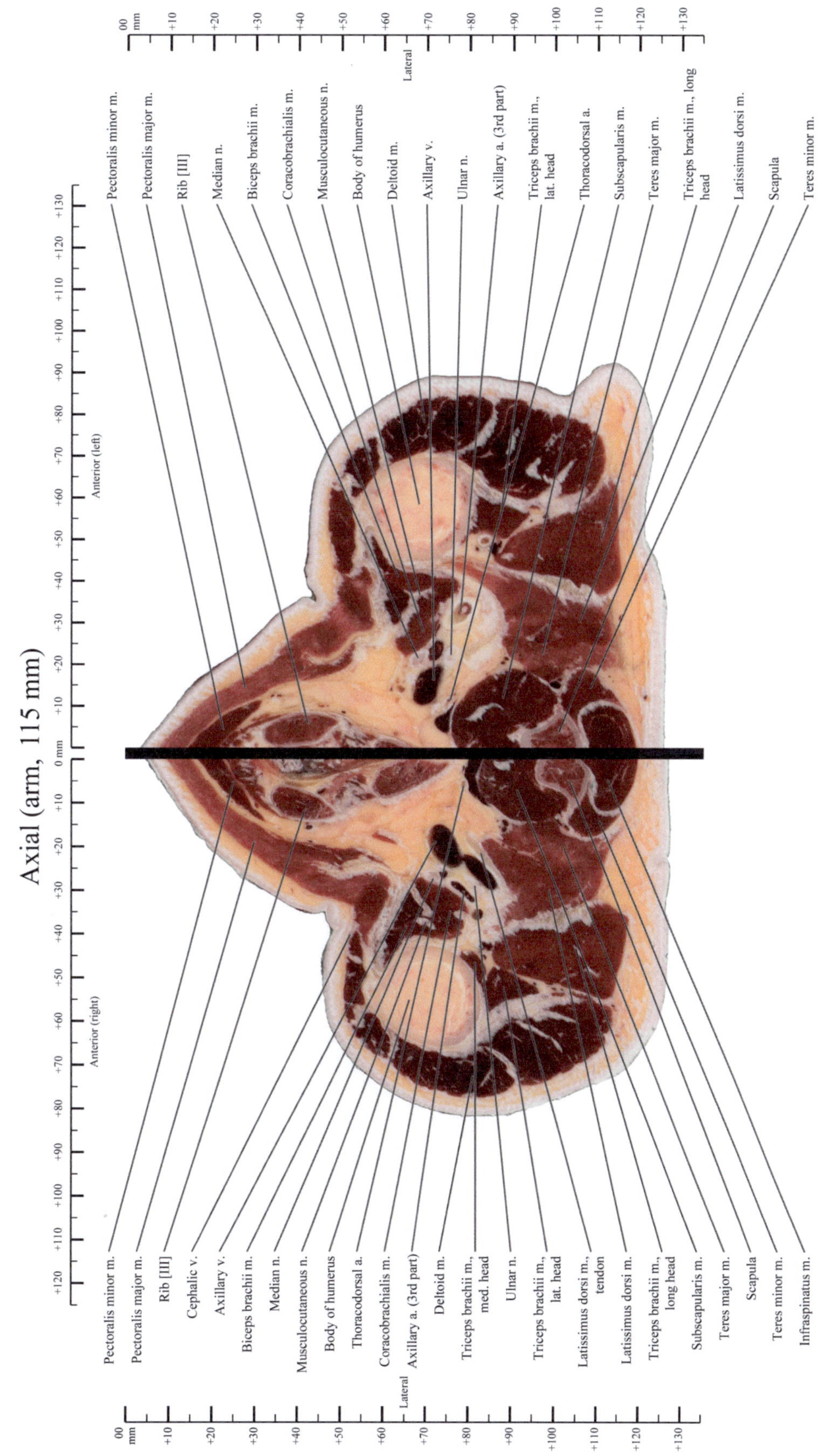
Axial (arm, 115 mm)
Anterior (left)
Anterior (right)
Lateral
Pectoralis minor m.
Pectoralis major m.
Rib [III]
Median n.
Biceps brachii m.
Coracobrachialis m.
Musculocutaneous n.
Body of humerus
Deltoid m.
Axillary v.
Ulnar n.
Axillary a. (3rd part)
Triceps brachii m., lat. head
Thoracodorsal a.
Subscapularis m.
Teres major m.
Triceps brachii m., long head
Latissimus dorsi m.
Scapula
Teres minor m.
Pectoralis minor m.
Pectoralis major m.
Rib [III]
Cephalic v.
Axillary v.
Biceps brachii m.
Median n.
Musculocutaneous n.
Body of humerus
Thoracodorsal a.
Coracobrachialis m.
Axillary a. (3rd part)
Deltoid m.
Triceps brachii m., med. head
Ulnar n.
Triceps brachii m., lat. head
Latissimus dorsi m., tendon
Latissimus dorsi m.
Triceps brachii m., long head
Subscapularis m.
Teres major m.
Scapula
Teres minor m.
Infraspinatus m.

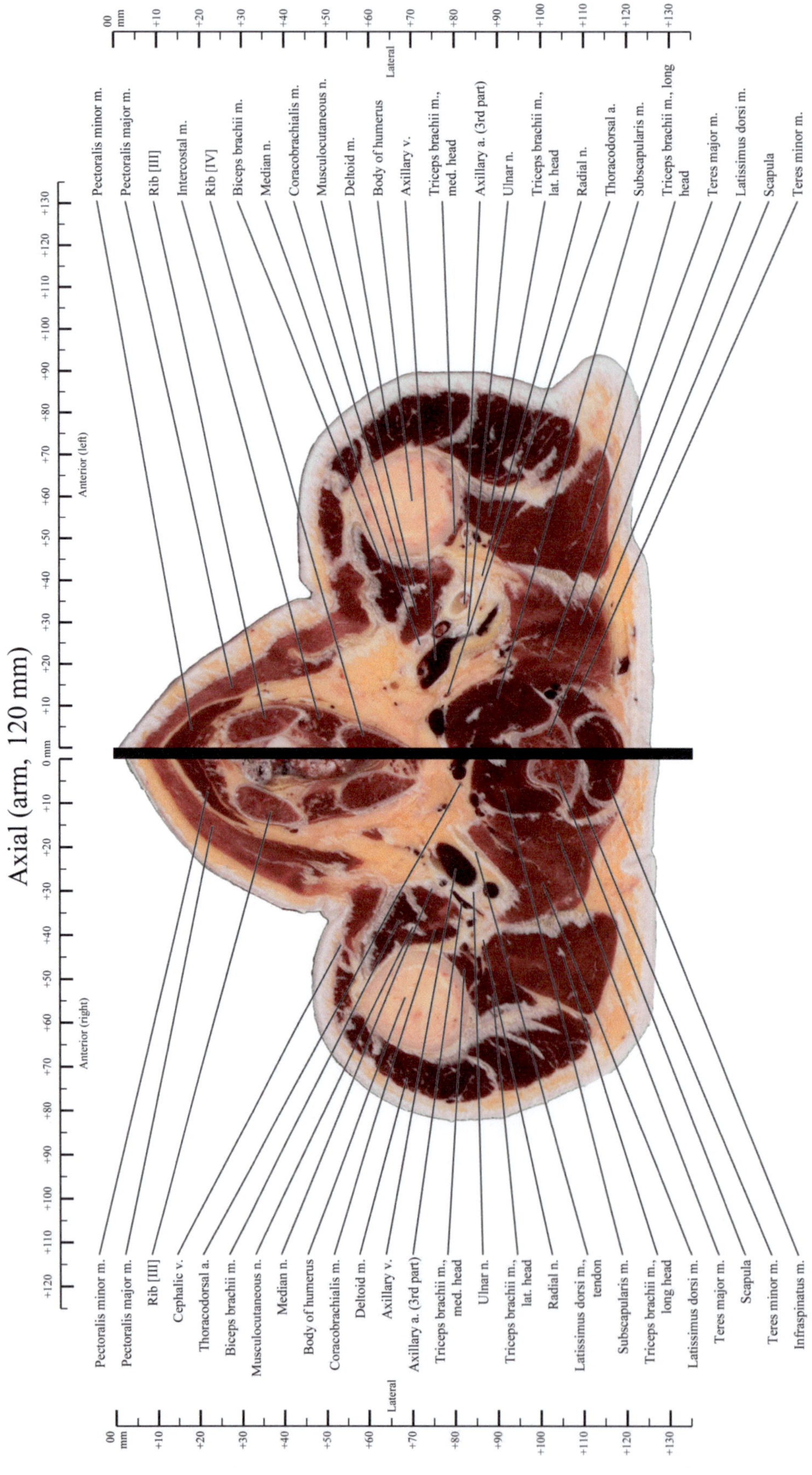
Axial (arm, 120 mm)
Anterior (left)
Anterior (right)
Lateral
Pectoralis minor m.
Pectoralis major m.
Rib [III]
Intercostal m.
Rib [IV]
Biceps brachii m.
Median n.
Coracobrachialis m.
Musculocutaneous n.
Deltoid m.
Body of humerus
Axillary v.
Triceps brachii m., med. head
Axillary a. (3rd part)
Ulnar n.
Triceps brachii m., lat. head
Radial n.
Thoracodorsal a.
Subscapularis m.
Triceps brachii m., long head
Teres major m.
Latissimus dorsi m.
Scapula
Teres minor m.
Pectoralis minor m.
Pectoralis major m.
Rib [III]
Cephalic v.
Thoracodorsal a.
Biceps brachii m.
Musculocutaneous n.
Median n.
Body of humerus
Coracobrachialis m.
Deltoid m.
Axillary v.
Axillary a. (3rd part)
Triceps brachii m., med. head
Ulnar n.
Triceps brachii m., lat. head
Radial n.
Latissimus dorsi m., tendon
Subscapularis m.
Triceps brachii m., long head
Latissimus dorsi m.
Teres major m.
Scapula
Teres minor m.
Infraspinatus m.

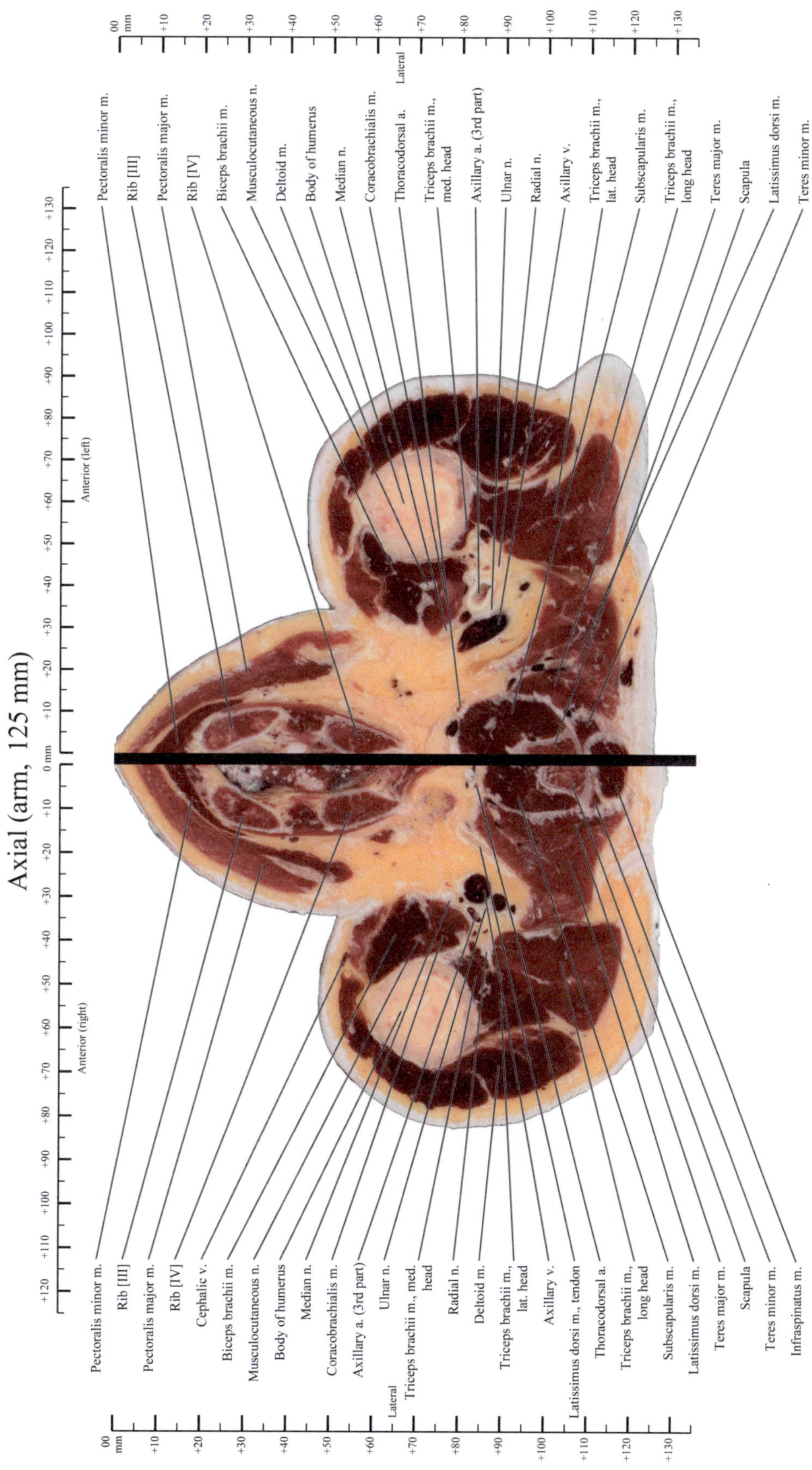
Axial (arm, 125 mm)
Anterior (left)
Anterior (right)
Lateral
Pectoralis minor m.
Rib [III]
Pectoralis major m.
Rib [IV]
Biceps brachii m.
Musculocutaneous n.
Deltoid m.
Body of humerus
Median n.
Coracobrachialis m.
Thoracodorsal a.
Triceps brachii m., med. head
Axillary a. (3rd part)
Ulnar n.
Radial n.
Axillary v.
Triceps brachii m., lat. head
Subscapularis m.
Triceps brachii m., long head
Teres major m.
Scapula
Latissimus dorsi m.
Teres minor m.
Pectoralis minor m.
Rib [III]
Pectoralis major m.
Rib [IV]
Cephalic v.
Biceps brachii m.
Musculocutaneous n.
Body of humerus
Median n.
Coracobrachialis m.
Axillary a. (3rd part)
Ulnar n.
Triceps brachii m., med. head
Radial n.
Deltoid m.
Triceps brachii m., lat. head
Axillary v.
Latissimus dorsi m., tendon
Thoracodorsal a.
Triceps brachii m., long head
Subscapularis m.
Latissimus dorsi m.
Teres major m.
Scapula
Teres minor m.
Infraspinatus m.

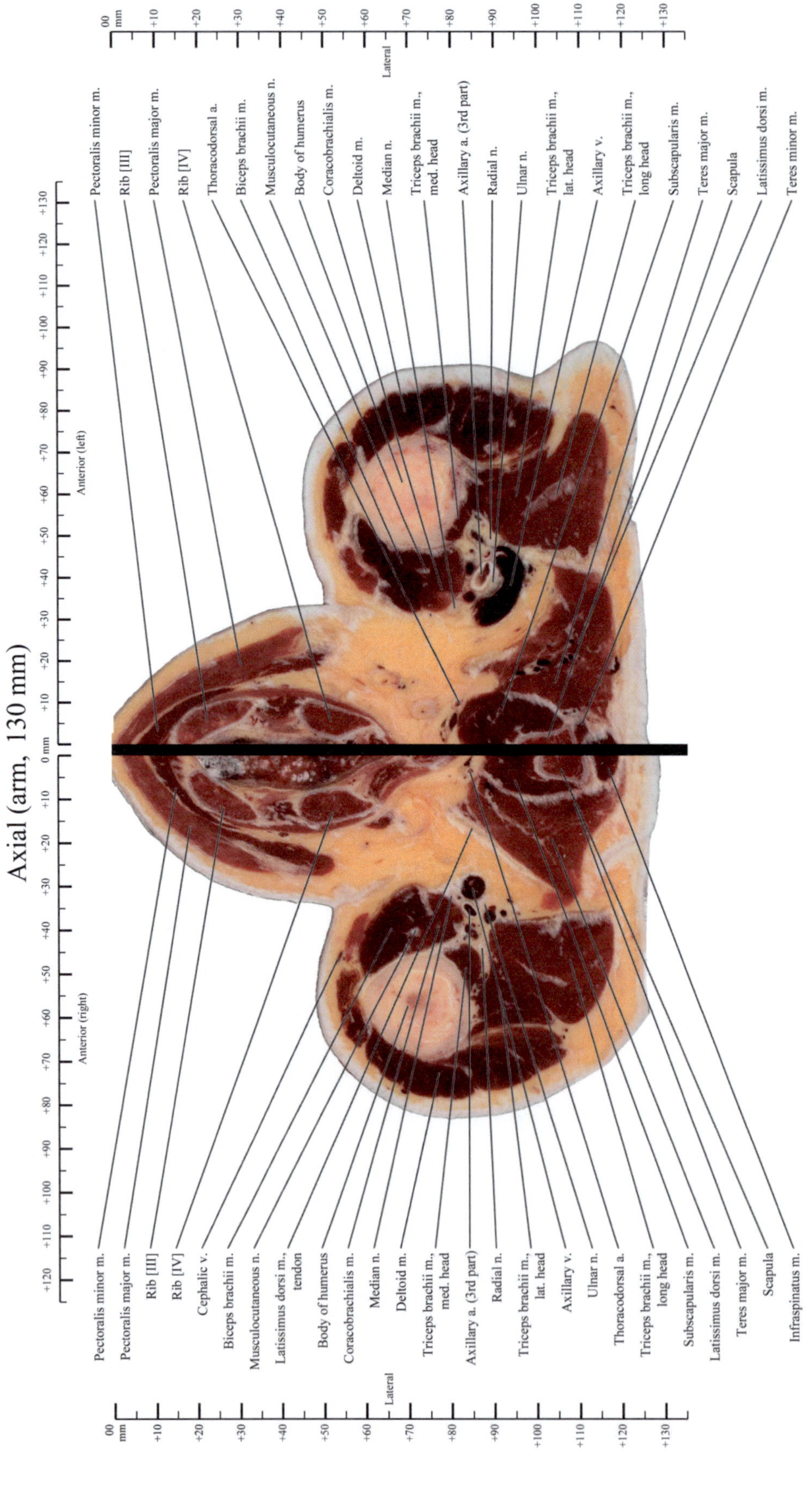
Axial (arm, 130 mm)
Anterior (left)
Anterior (right)
Lateral
Pectoralis minor m.
Rib [III]
Pectoralis major m.
Rib [IV]
Thoracodorsal a.
Biceps brachii m.
Musculocutaneous n.
Body of humerus
Coracobrachialis m.
Deltoid m.
Median n.
Triceps brachii m., med. head
Axillary a. (3rd part)
Radial n.
Ulnar n.
Triceps brachii m., lat. head
Axillary v.
Triceps brachii m., long head
Subscapularis m.
Teres major m.
Scapula
Latissimus dorsi m.
Teres minor m.
Pectoralis minor m.
Pectoralis major m.
Rib [III]
Rib [IV]
Cephalic v.
Biceps brachii m.
Musculocutaneous n.
Latissimus dorsi m., tendon
Body of humerus
Coracobrachialis m.
Median n.
Deltoid m.
Triceps brachii m., med. head
Axillary a. (3rd part)
Radial n.
Triceps brachii m., lat. head
Axillary v.
Ulnar n.
Thoracodorsal a.
Triceps brachii m., long head
Subscapularis m.
Latissimus dorsi m.
Teres major m.
Scapula
Infraspinatus m.

Axial (arm, 135 mm)

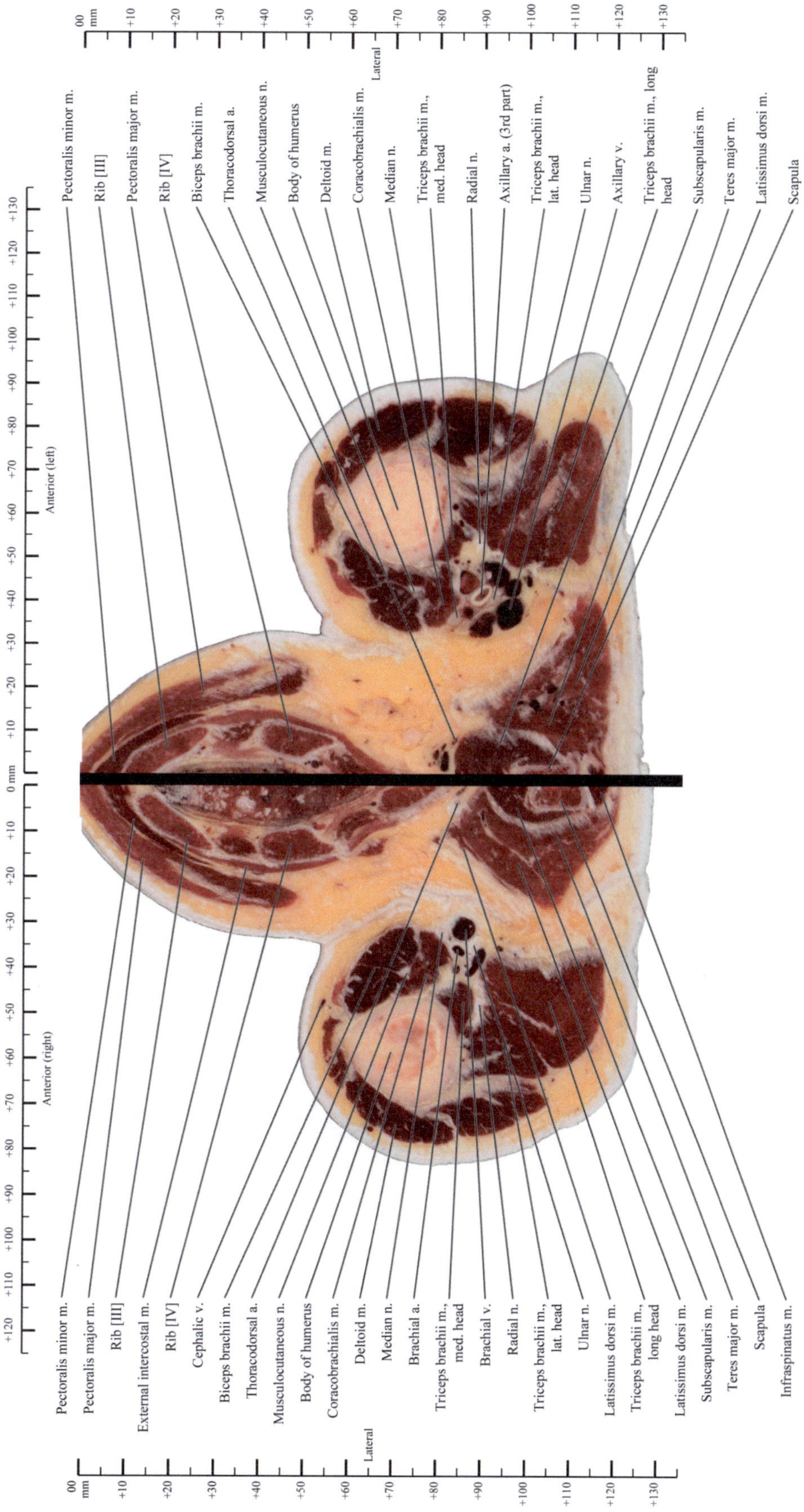

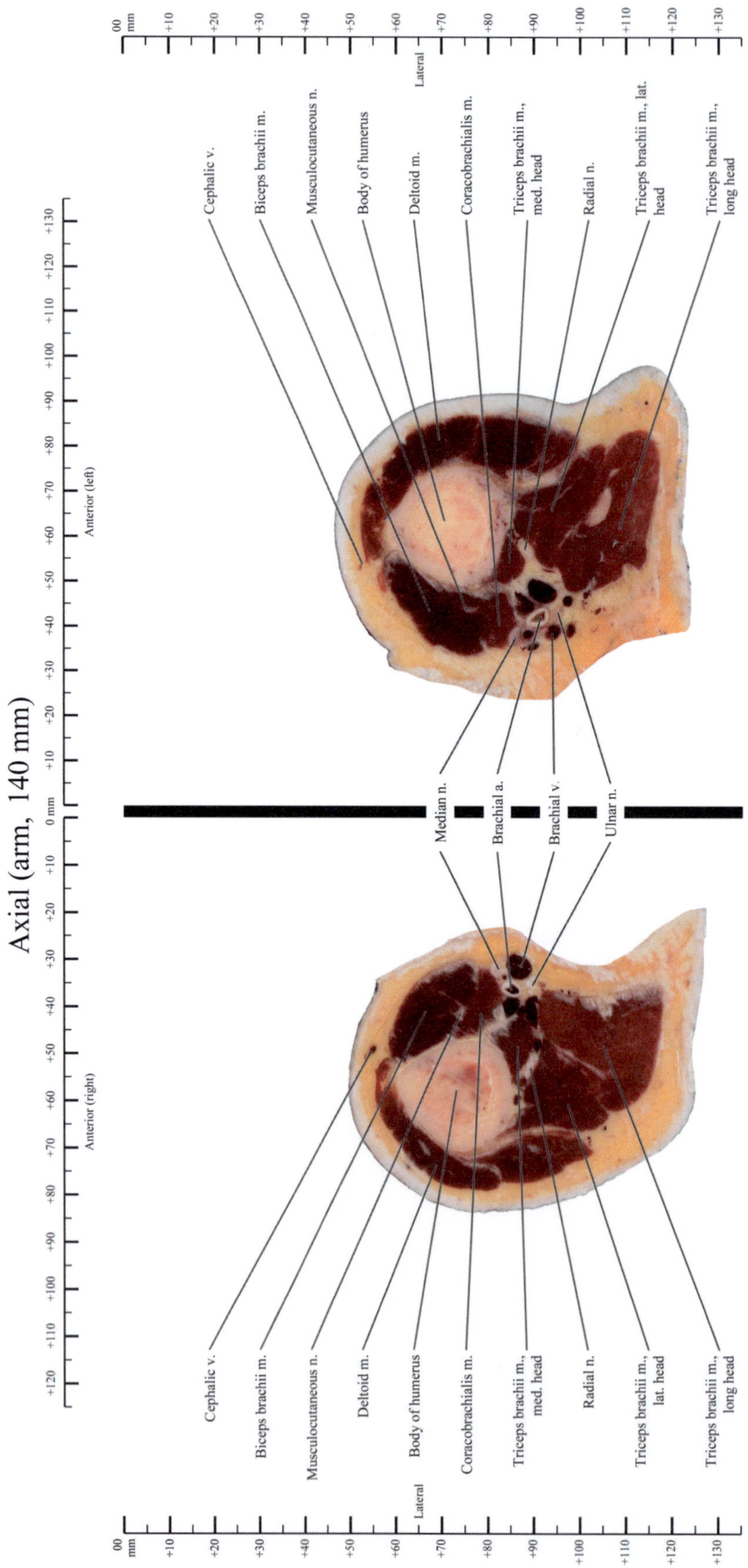

Axial (arm, 140 mm)
Anterior (left)
Anterior (right)
Lateral
Cephalic v.
Biceps brachii m.
Musculocutaneous n.
Body of humerus
Deltoid m.
Coracobrachialis m.
Triceps brachii m., med. head
Radial n.
Triceps brachii m., lat. head
Triceps brachii m., long head
Median n.
Brachial a.
Brachial v.
Ulnar n.
Lateral
Cephalic v.
Biceps brachii m.
Musculocutaneous n.
Deltoid m.
Body of humerus
Coracobrachialis m.
Triceps brachii m., med. head
Radial n.
Triceps brachii m., lat. head
Triceps brachii m., long head

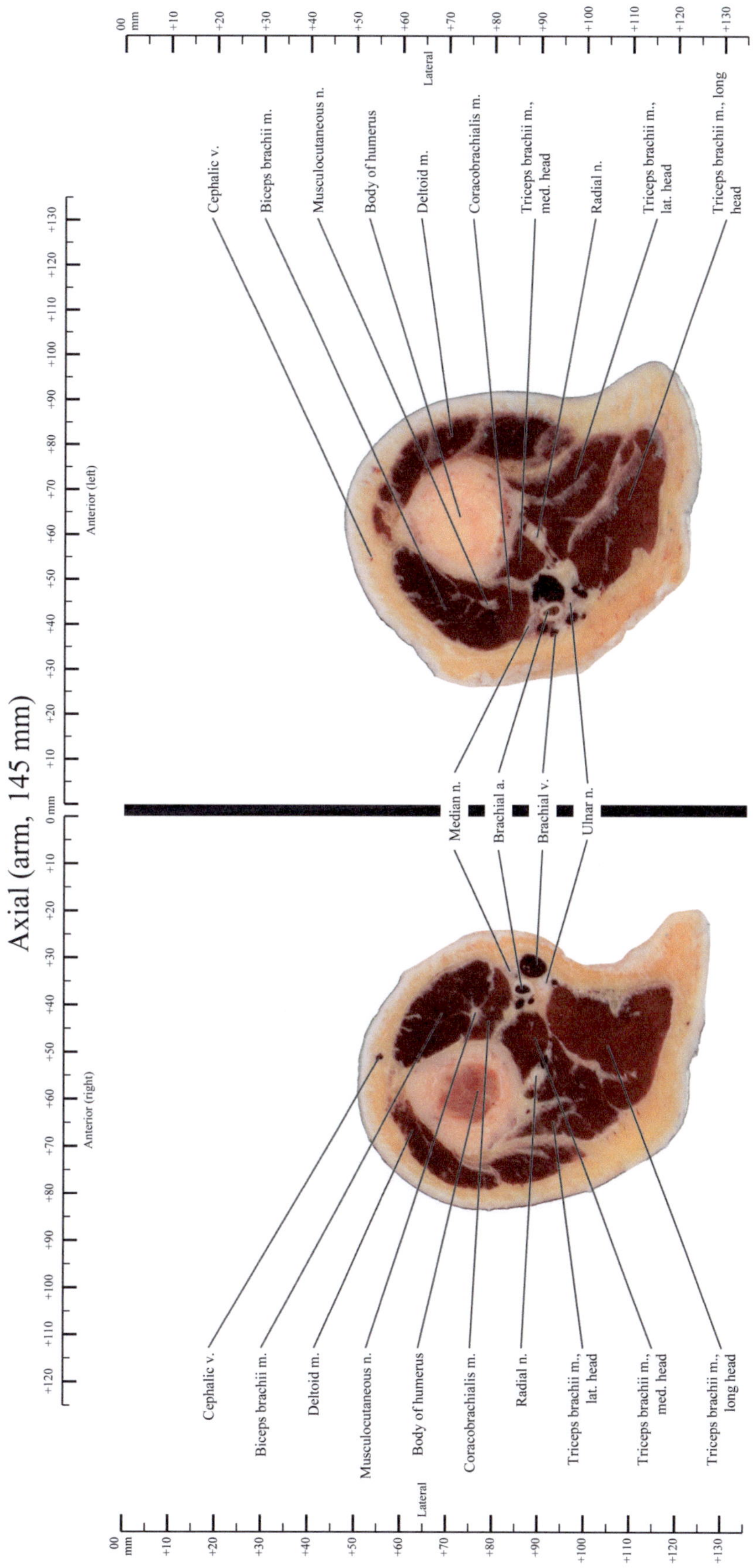
Axial (arm, 145 mm)
Anterior (left)
Anterior (right)
Lateral
Cephalic v.
Biceps brachii m.
Musculocutaneous n.
Body of humerus
Deltoid m.
Coracobrachialis m.
Triceps brachii m., med. head
Radial n.
Triceps brachii m., lat. head
Triceps brachii m., long head
Median n.
Brachial a.
Brachial v.
Ulnar n.
Cephalic v.
Biceps brachii m.
Deltoid m.
Musculocutaneous n.
Body of humerus
Coracobrachialis m.
Radial n.
Triceps brachii m., lat. head
Triceps brachii m., med. head
Triceps brachii m., long head

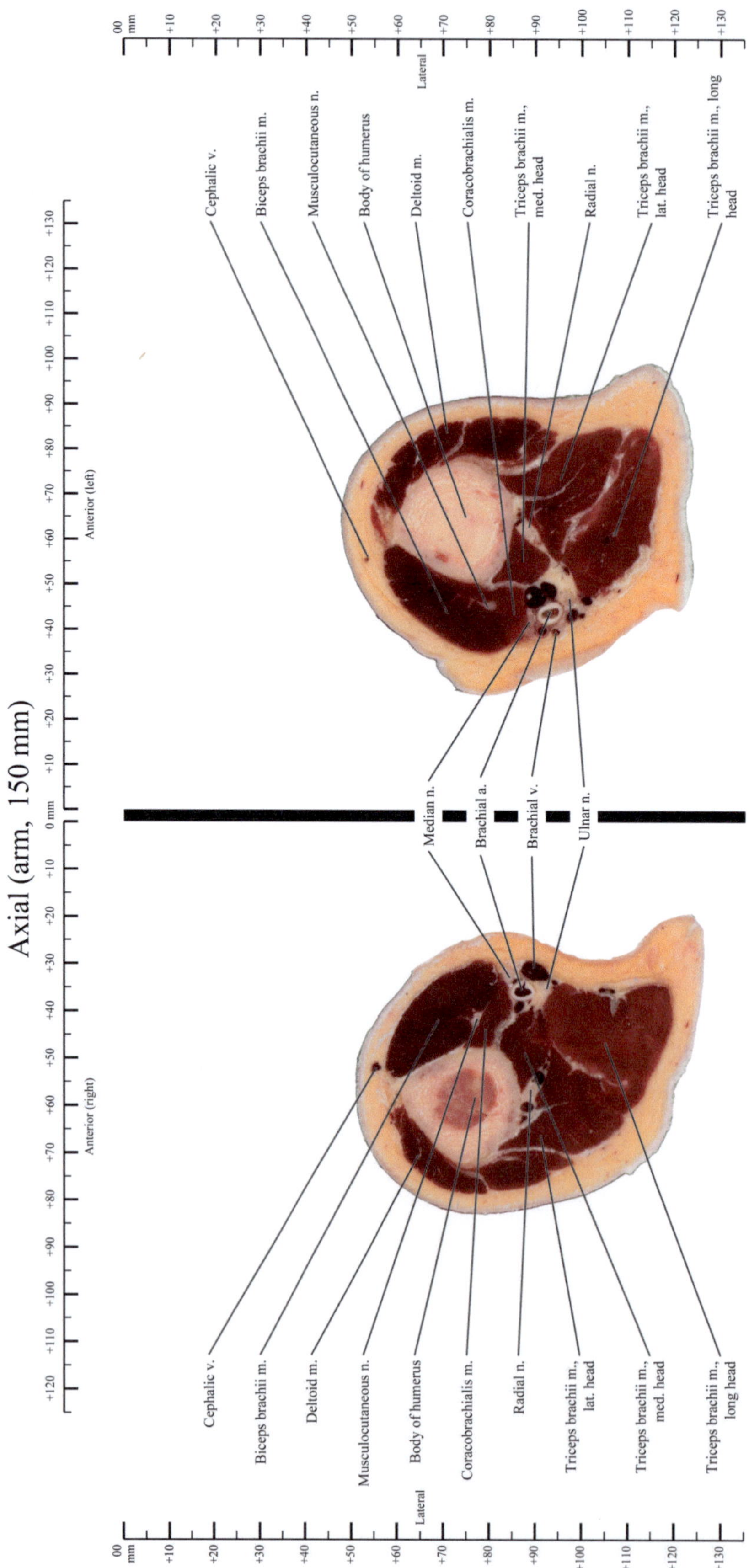
Axial (arm, 150 mm)
Anterior (left)
Anterior (right)
Lateral
Cephalic v.
Biceps brachii m.
Musculocutaneous n.
Body of humerus
Deltoid m.
Coracobrachialis m.
Triceps brachii m., med. head
Radial n.
Triceps brachii m., lat. head
Triceps brachii m., long head
Median n.
Brachial a.
Brachial v.
Ulnar n.
Cephalic v.
Biceps brachii m.
Deltoid m.
Musculocutaneous n.
Body of humerus
Coracobrachialis m.
Radial n.
Triceps brachii m., lat. head
Triceps brachii m., med. head
Triceps brachii m., long head

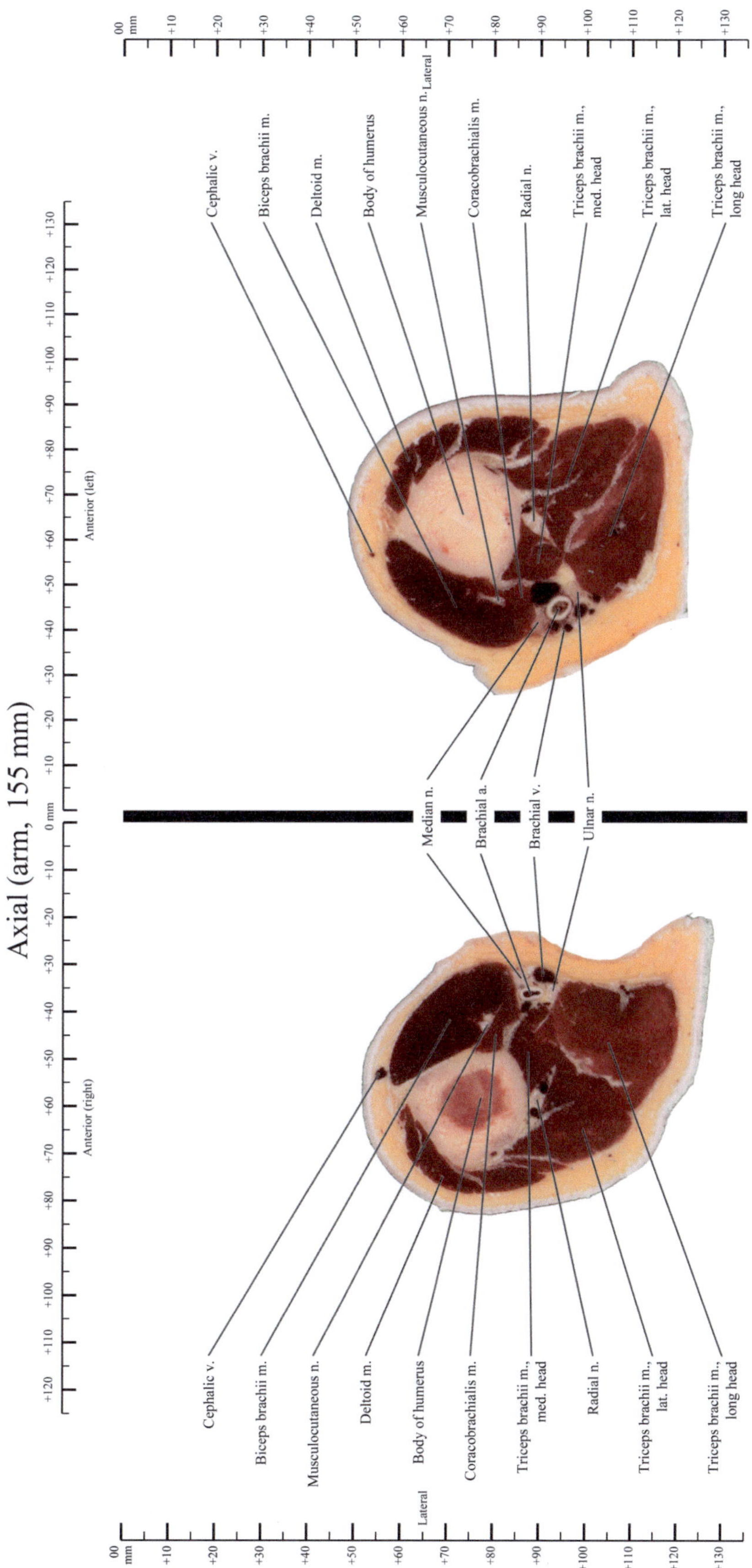
Axial (arm, 155 mm)
Anterior (left)
Anterior (right)
Lateral
Cephalic v.
Biceps brachii m.
Deltoid m.
Body of humerus
Musculocutaneous n.
Coracobrachialis m.
Radial n.
Triceps brachii m., med. head
Triceps brachii m., lat. head
Triceps brachii m., long head
Median n.
Brachial a.
Brachial v.
Ulnar n.

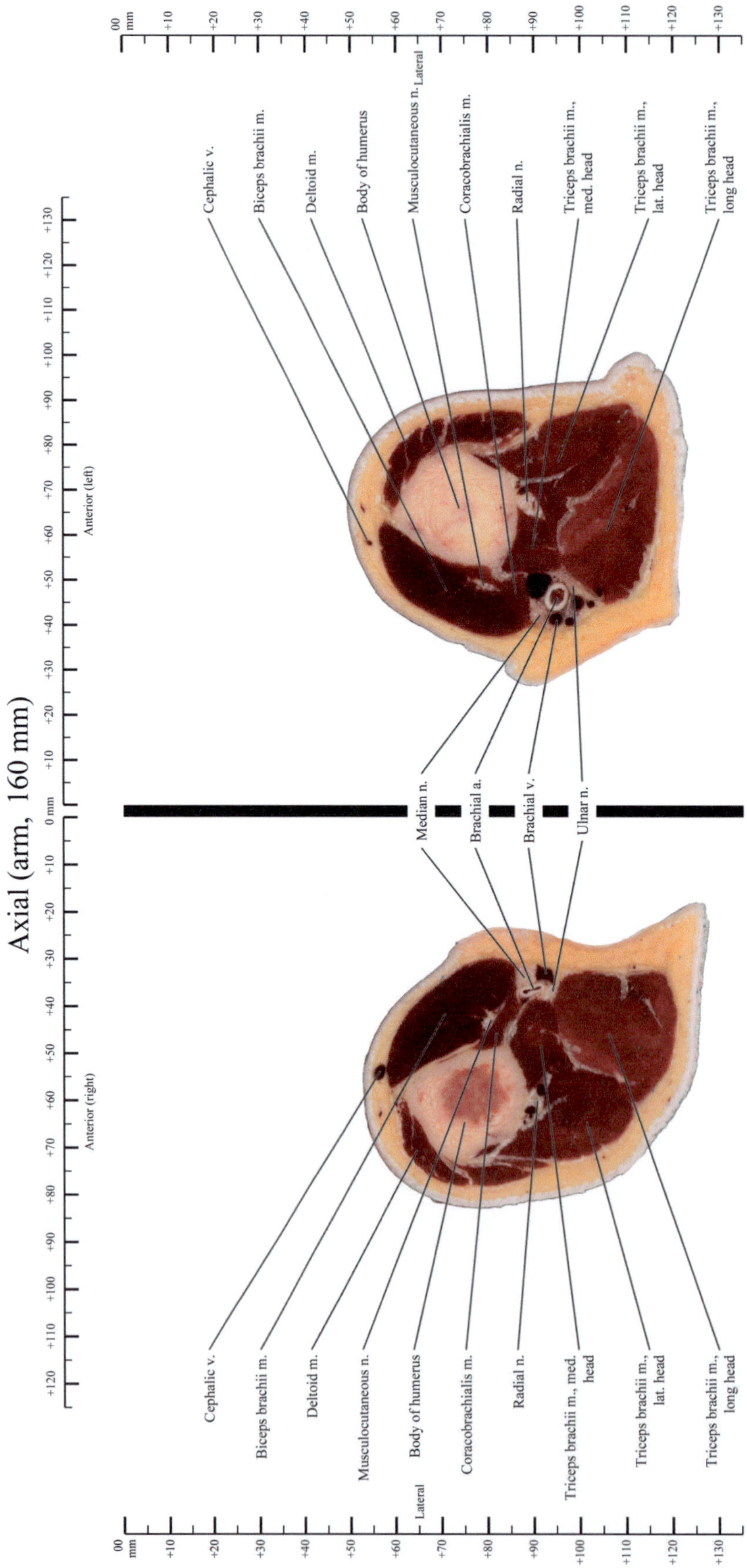
Axial (arm, 160 mm)
Anterior (left)
Anterior (right)
Lateral
Cephalic v.
Biceps brachii m.
Deltoid m.
Body of humerus
Musculocutaneous n.
Coracobrachialis m.
Radial n.
Triceps brachii m., med. head
Triceps brachii m., lat. head
Triceps brachii m., long head
Median n.
Brachial a.
Brachial v.
Ulnar n.

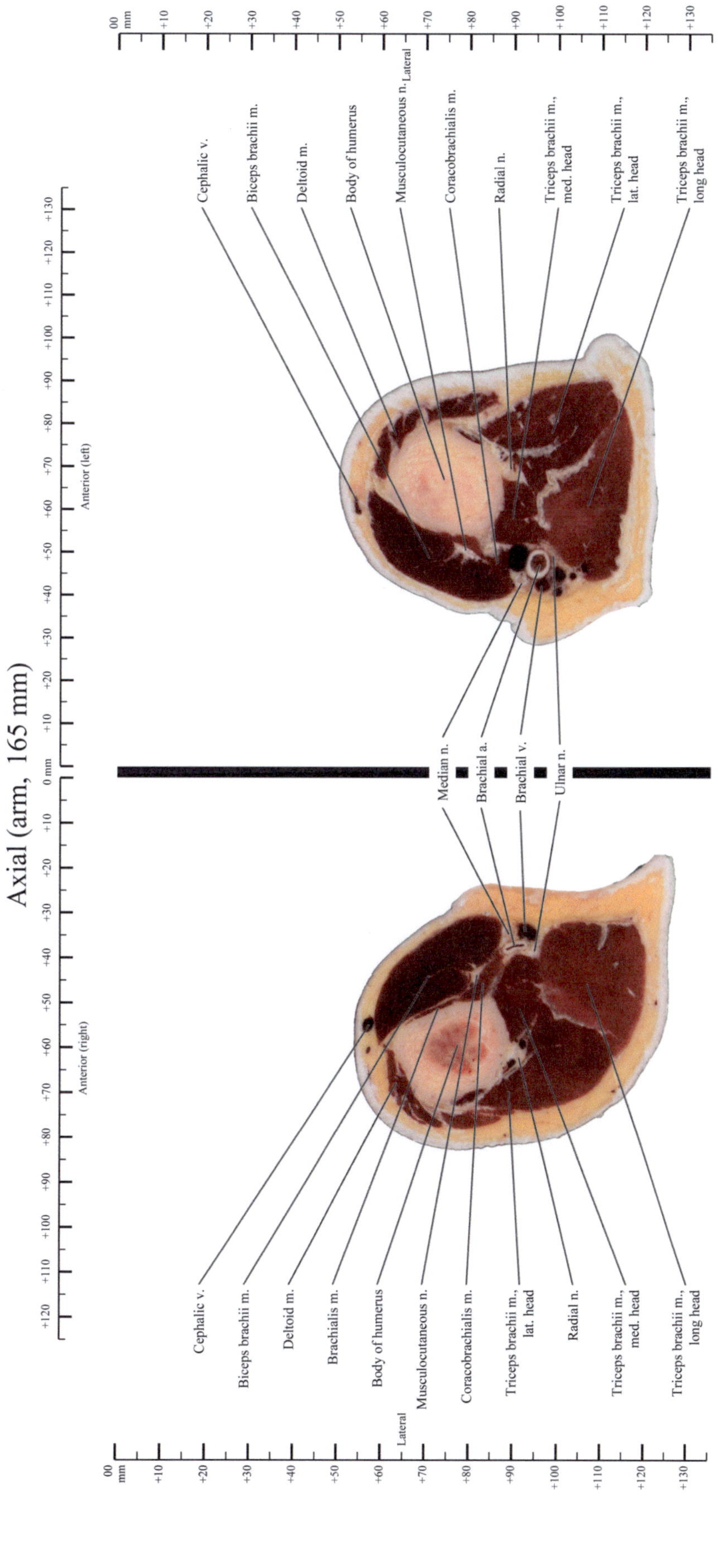

Axial (arm, 165 mm)
Anterior (left)
Anterior (right)
Lateral
Cephalic v.
Biceps brachii m.
Deltoid m.
Body of humerus
Musculocutaneous n.
Coracobrachialis m.
Radial n.
Triceps brachii m., med. head
Triceps brachii m., lat. head
Triceps brachii m., long head
Median n.
Brachial a.
Brachial v.
Ulnar n.
Brachialis m.

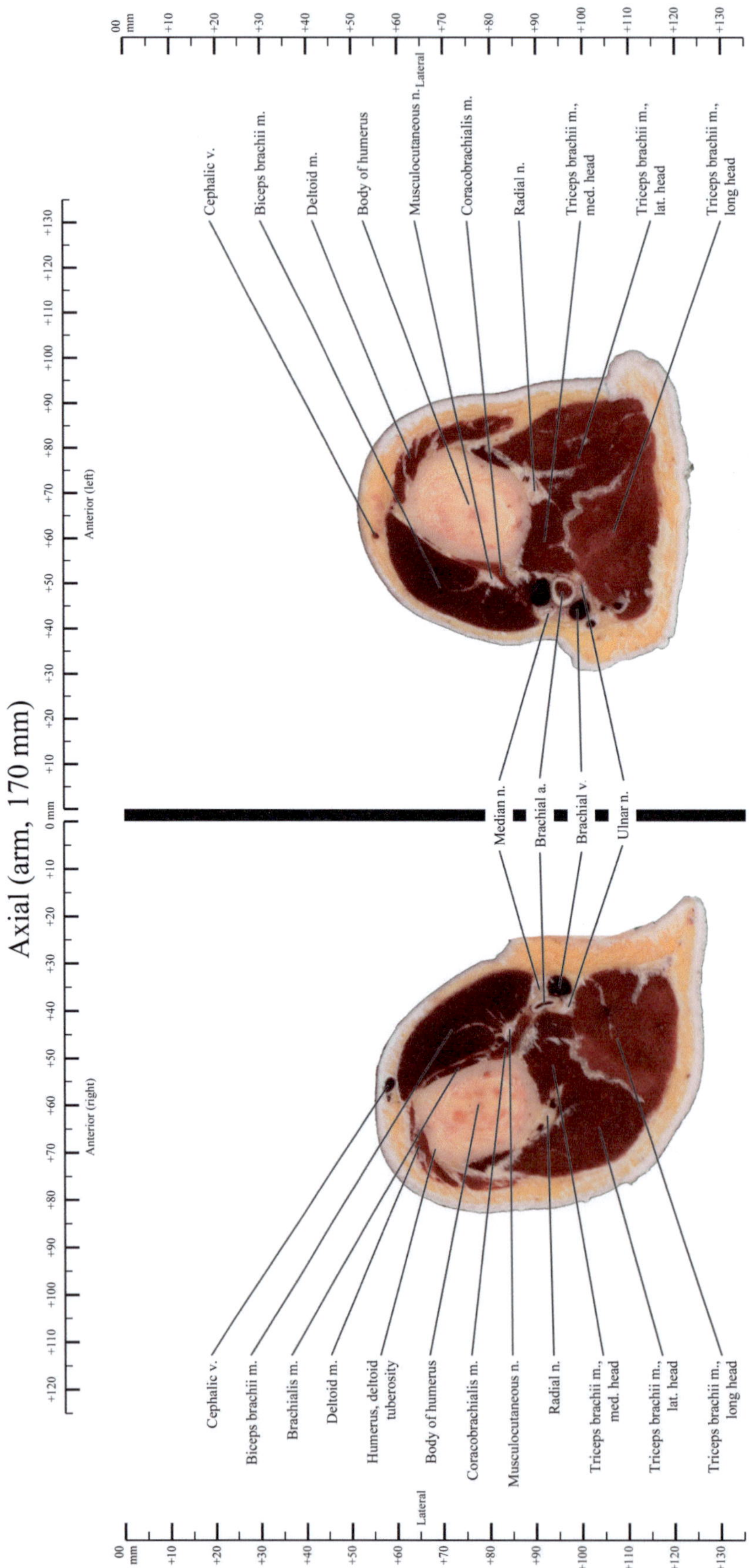
Axial (arm, 170 mm)
Anterior (left)
Anterior (right)
Lateral
Cephalic v.
Biceps brachii m.
Deltoid m.
Body of humerus
Musculocutaneous n.
Coracobrachialis m.
Radial n.
Triceps brachii m., med. head
Triceps brachii m., lat. head
Triceps brachii m., long head
Median n.
Brachial a.
Brachial v.
Ulnar n.
Brachialis m.
Humerus, deltoid tuberosity
0 mm
00 mm

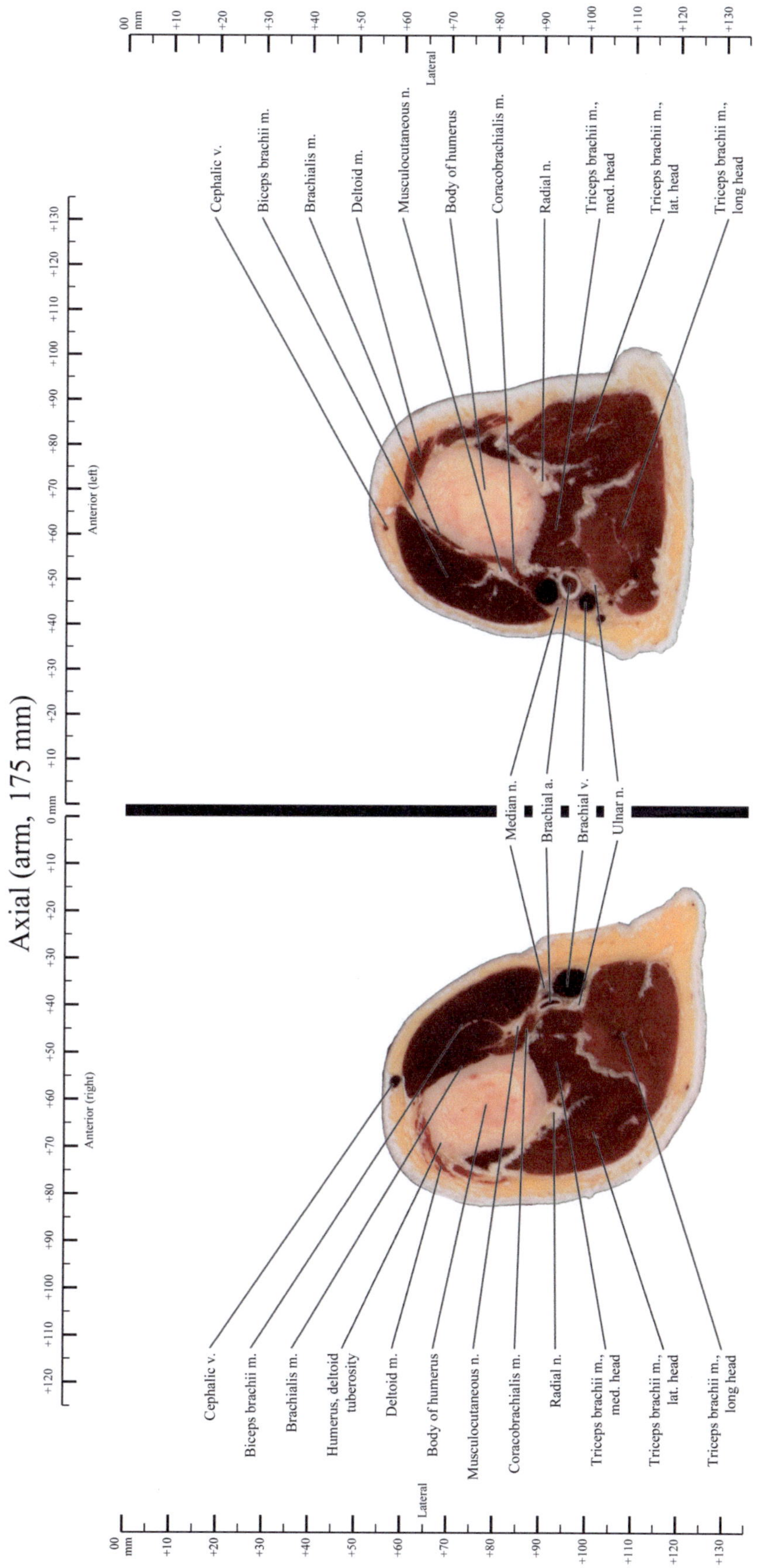
Axial (arm, 175 mm)
Anterior (left)
Anterior (right)
Lateral
Cephalic v.
Biceps brachii m.
Brachialis m.
Deltoid m.
Musculocutaneous n.
Body of humerus
Coracobrachialis m.
Radial n.
Triceps brachii m., med. head
Triceps brachii m., lat. head
Triceps brachii m., long head
Median n.
Brachial a.
Brachial v.
Ulnar n.
Humerus, deltoid tuberosity

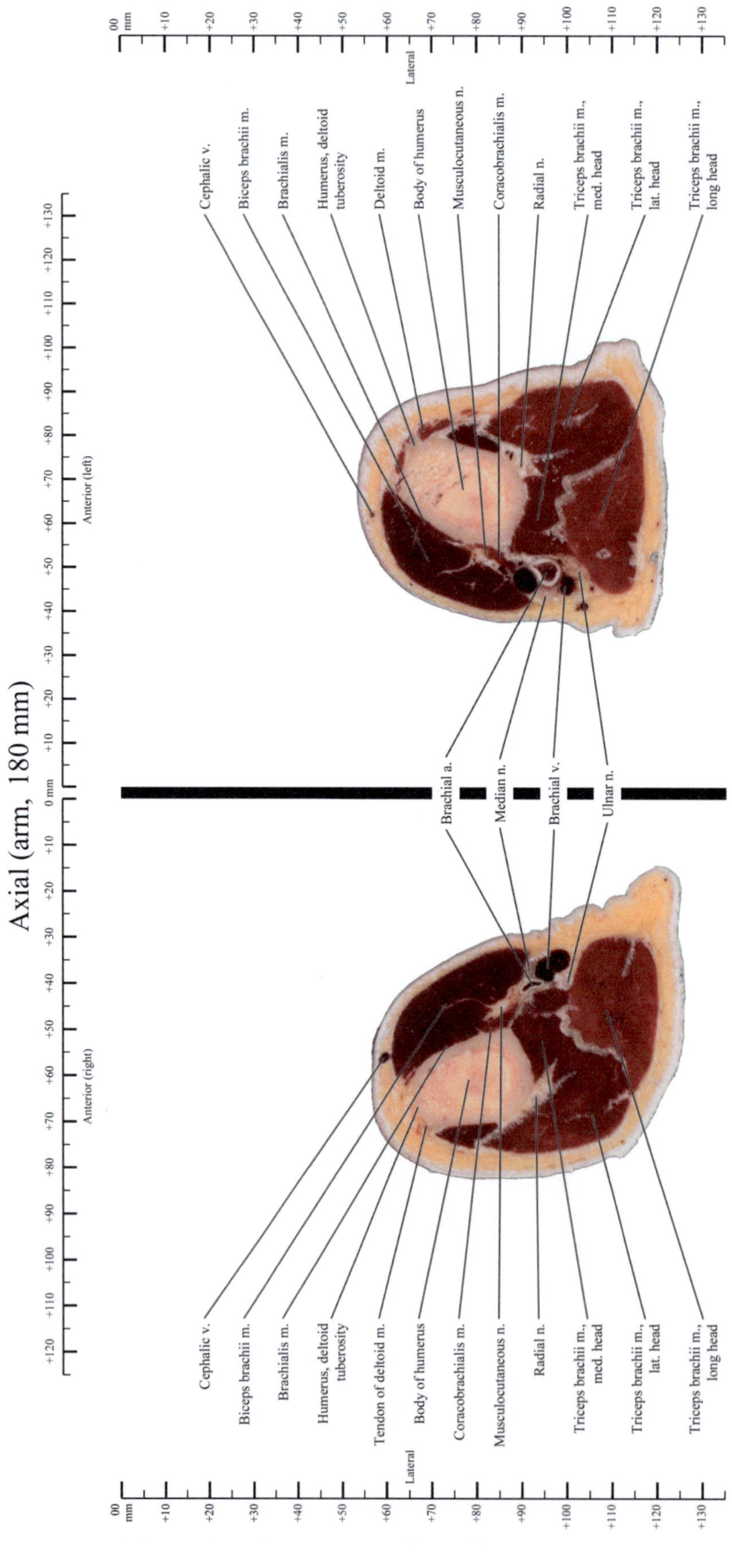
Axial (arm, 180 mm)
Anterior (left)
Anterior (right)
Lateral
Cephalic v.
Biceps brachii m.
Brachialis m.
Humerus, deltoid tuberosity
Deltoid m.
Tendon of deltoid m.
Body of humerus
Musculocutaneous n.
Coracobrachialis m.
Radial n.
Triceps brachii m., med. head
Triceps brachii m., lat. head
Triceps brachii m., long head
Brachial a.
Median n.
Brachial v.
Ulnar n.

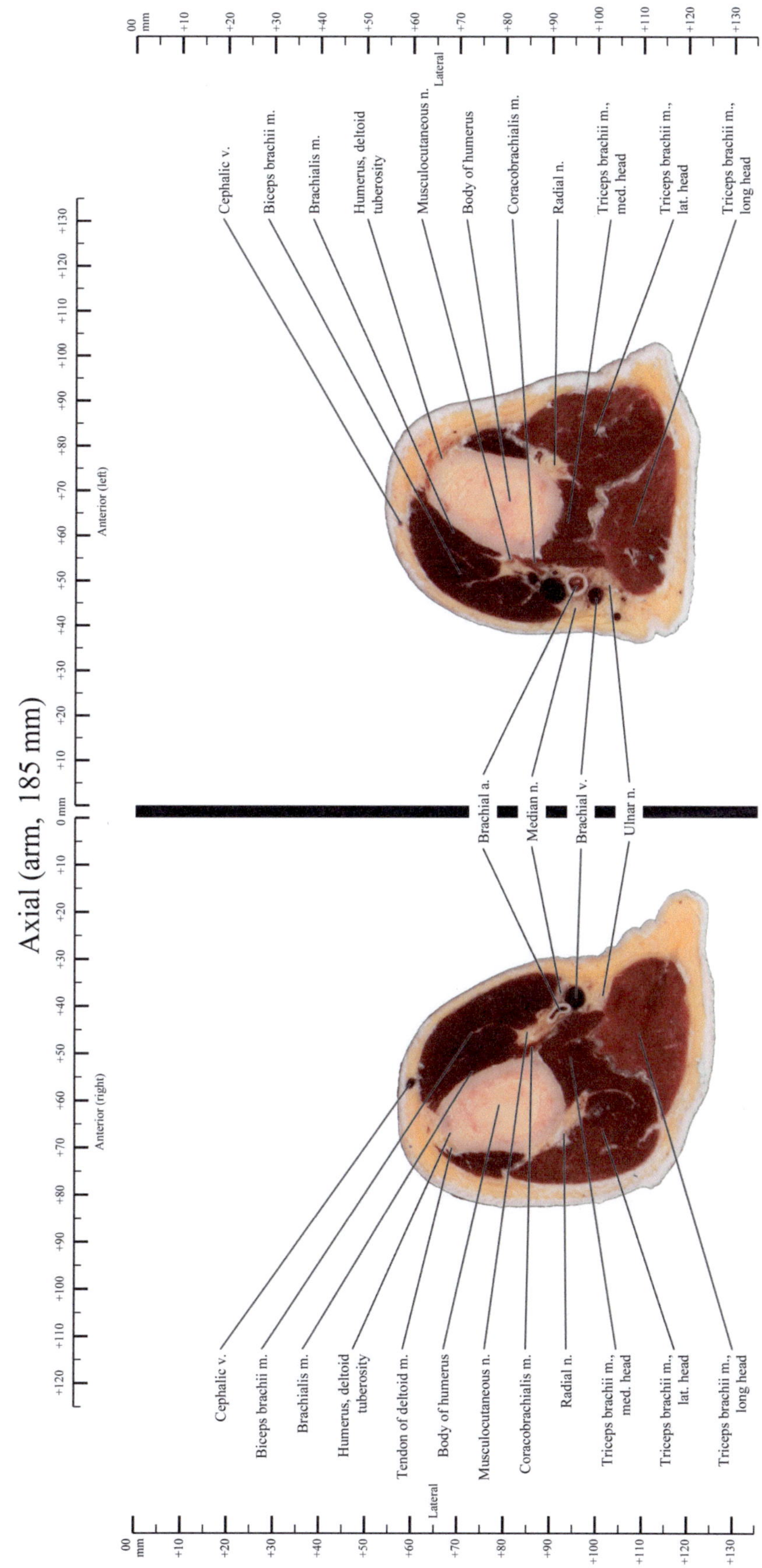
Axial (arm, 185 mm)
Anterior (left)
Anterior (right)
Lateral
Cephalic v.
Biceps brachii m.
Brachialis m.
Humerus, deltoid tuberosity
Musculocutaneous n.
Body of humerus
Coracobrachialis m.
Radial n.
Triceps brachii m., med. head
Triceps brachii m., lat. head
Triceps brachii m., long head
Brachial a.
Median n.
Brachial v.
Ulnar n.
Tendon of deltoid m.

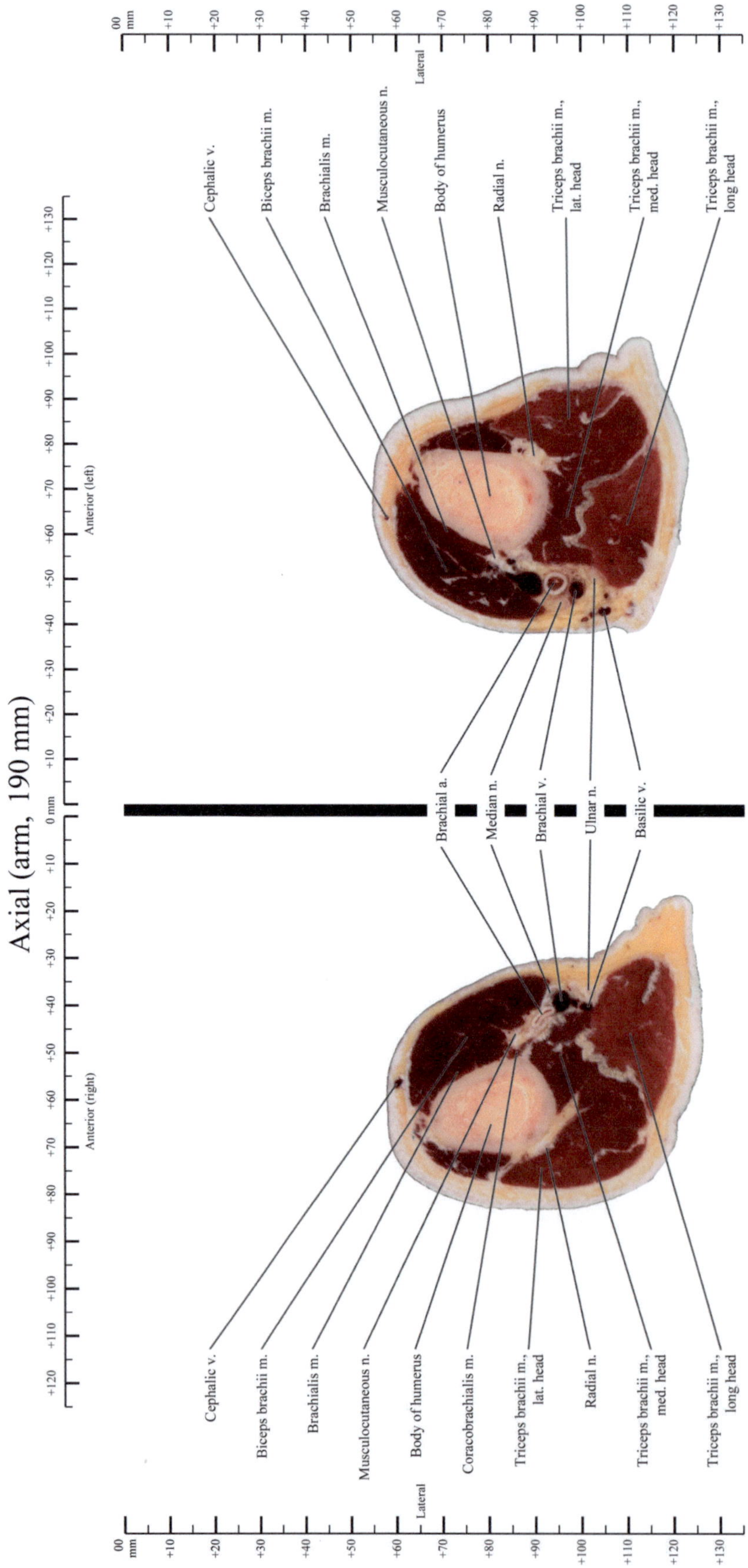
Axial (arm, 190 mm)
Anterior (left)
Anterior (right)
Lateral
Cephalic v.
Biceps brachii m.
Brachialis m.
Musculocutaneous n.
Body of humerus
Radial n.
Triceps brachii m., lat. head
Triceps brachii m., med. head
Triceps brachii m., long head
Brachial a.
Median n.
Brachial v.
Ulnar n.
Basilic v.
Coracobrachialis m.
00 mm
0 mm

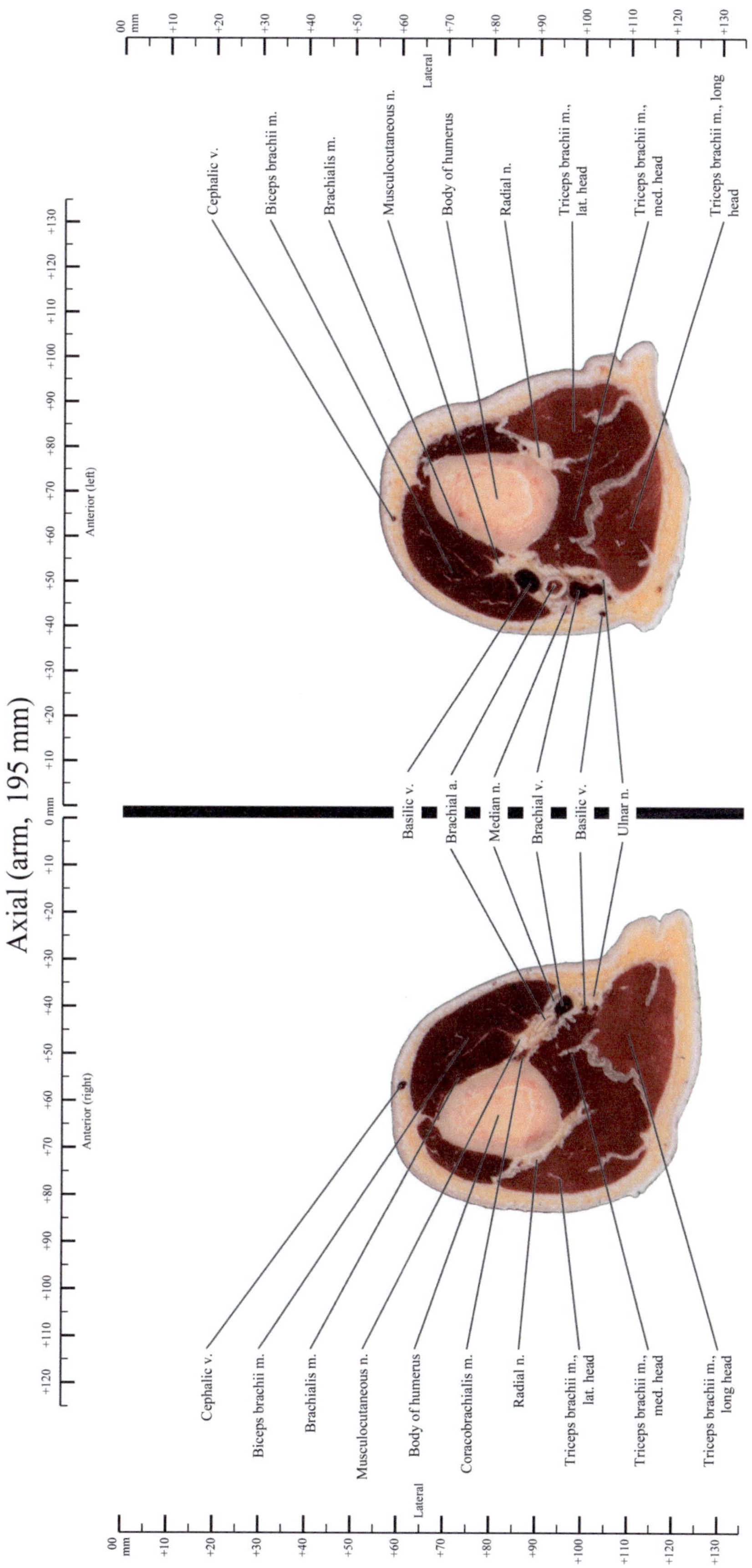
Axial (arm, 195 mm)
Anterior (left)
Anterior (right)
Lateral
Lateral
Cephalic v.
Biceps brachii m.
Brachialis m.
Musculocutaneous n.
Body of humerus
Radial n.
Triceps brachii m., lat. head
Triceps brachii m., med. head
Triceps brachii m., long head
Basilic v.
Brachial a.
Median n.
Brachial v.
Basilic v.
Ulnar n.
Cephalic v.
Biceps brachii m.
Brachialis m.
Musculocutaneous n.
Body of humerus
Coracobrachialis m.
Radial n.
Triceps brachii m., lat. head
Triceps brachii m., med. head
Triceps brachii m., long head

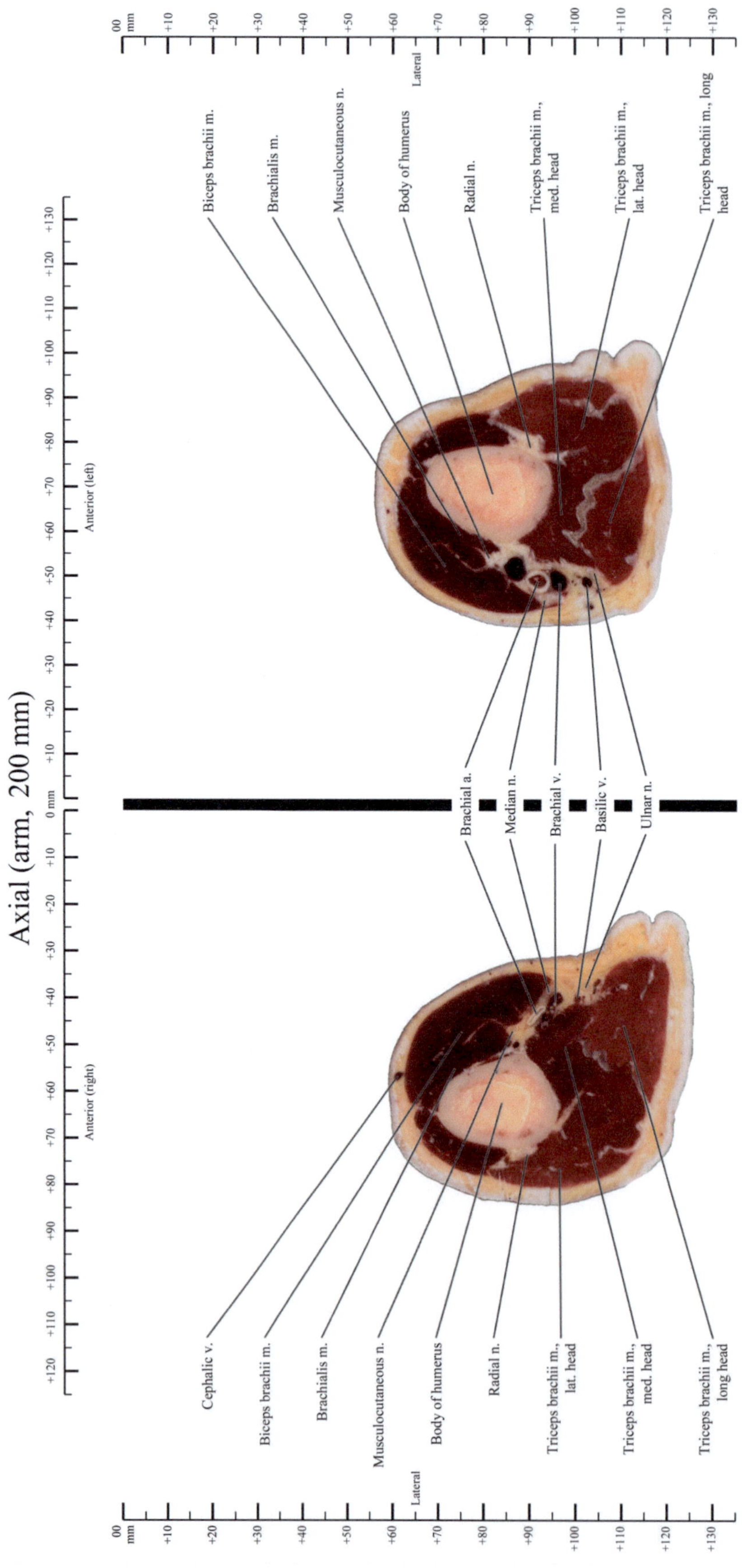
Axial (arm, 200 mm)
Anterior (left)
Anterior (right)
Lateral
Lateral
Biceps brachii m.
Brachialis m.
Musculocutaneous n.
Body of humerus
Radial n.
Triceps brachii m., med. head
Triceps brachii m., lat. head
Triceps brachii m., long head
Brachial a.
Median n.
Brachial v.
Basilic v.
Ulnar n.
Cephalic v.
Biceps brachii m.
Brachialis m.
Musculocutaneous n.
Body of humerus
Radial n.
Triceps brachii m., lat. head
Triceps brachii m., med. head
Triceps brachii m., long head
0 mm
00 mm

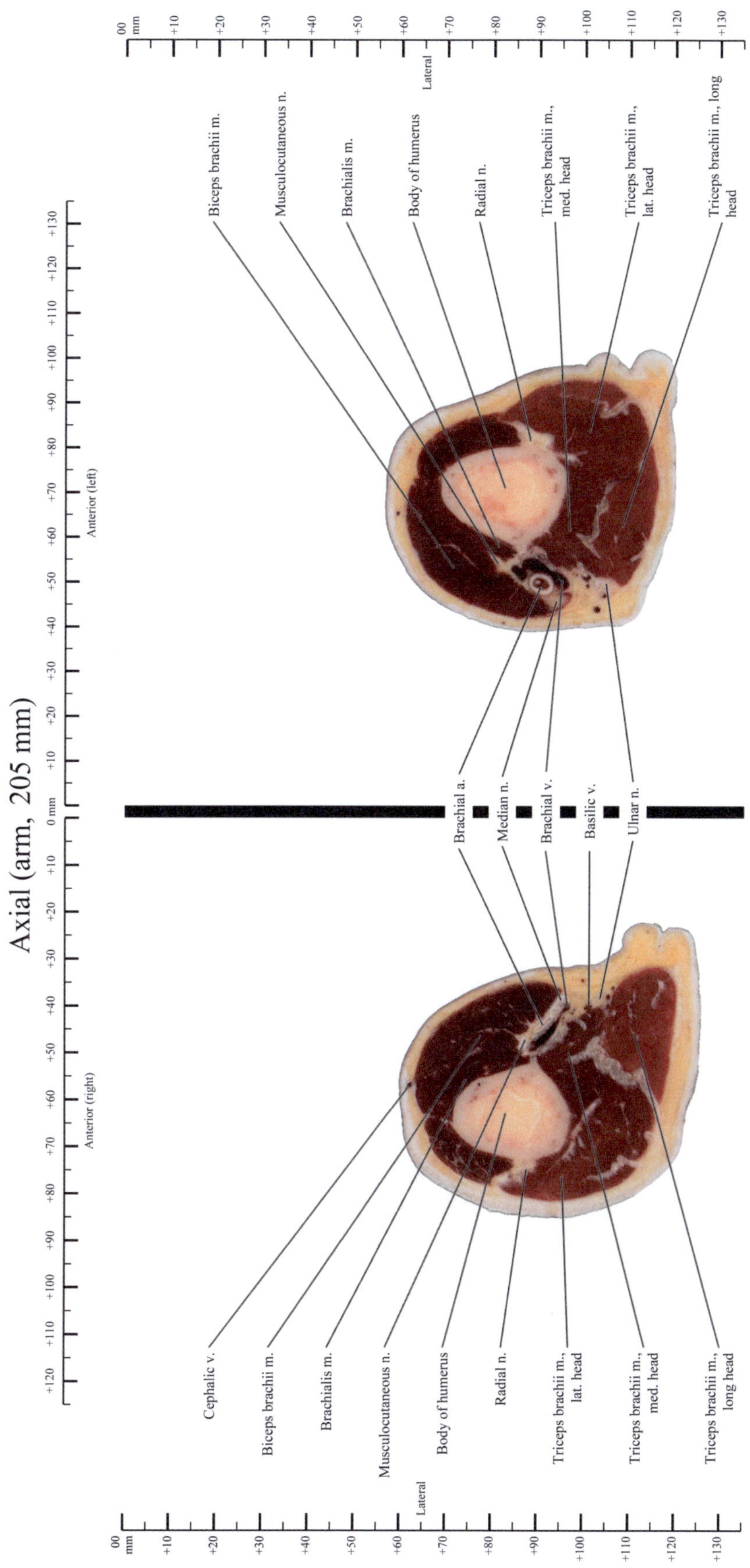
Axial (arm, 205 mm)
Anterior (left)
Anterior (right)
Lateral
Biceps brachii m.
Musculocutaneous n.
Brachialis m.
Body of humerus
Radial n.
Triceps brachii m., med. head
Triceps brachii m., lat. head
Triceps brachii m., long head
Brachial a.
Median n.
Brachial v.
Basilic v.
Ulnar n.
Cephalic v.

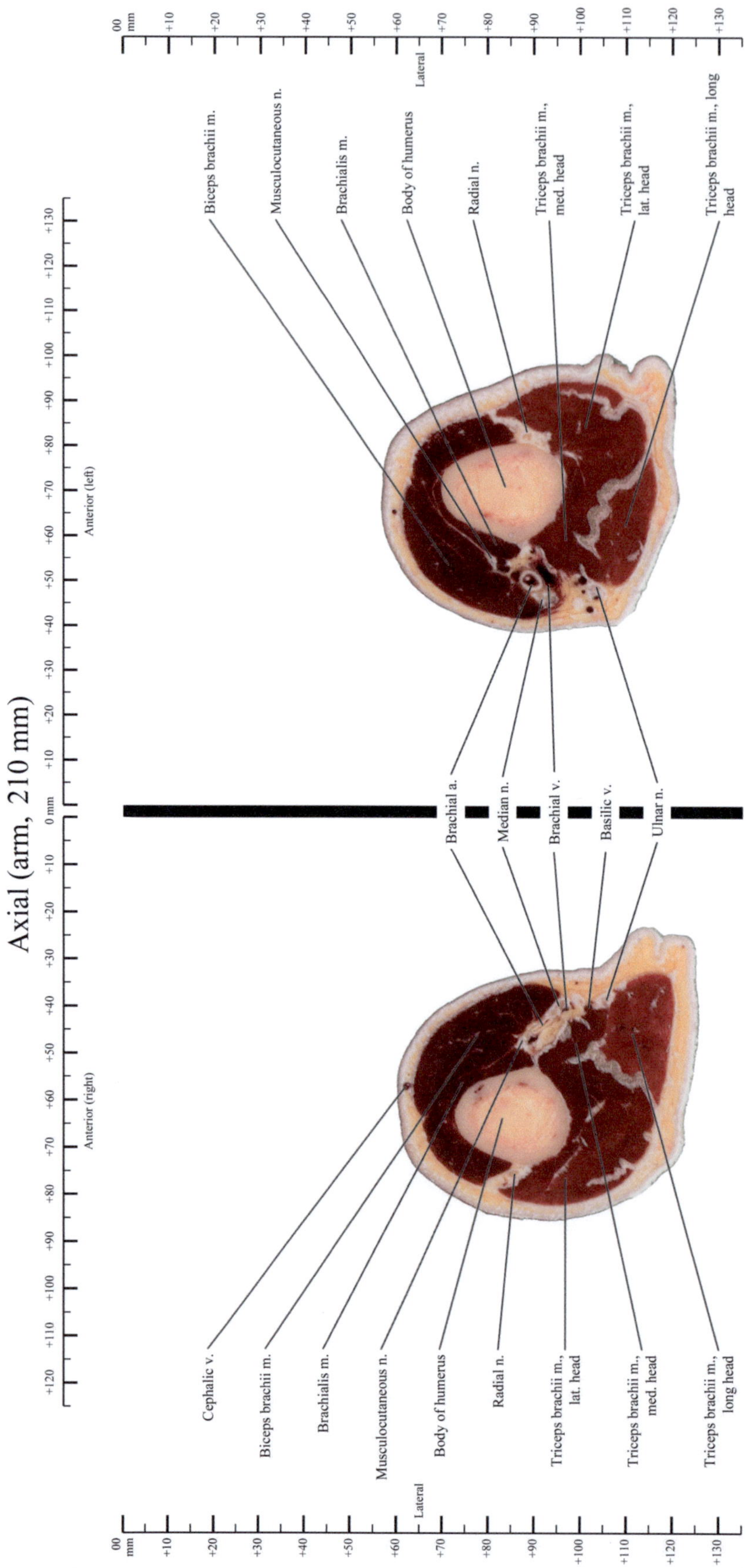

Axial (arm, 210 mm)
Anterior (left)
Anterior (right)
Lateral
Lateral
Biceps brachii m.
Musculocutaneous n.
Brachialis m.
Body of humerus
Radial n.
Triceps brachii m., med. head
Triceps brachii m., lat. head
Triceps brachii m., long head
Brachial a.
Median n.
Brachial v.
Basilic v.
Ulnar n.
Cephalic v.
Biceps brachii m.
Brachialis m.
Musculocutaneous n.
Body of humerus
Radial n.
Triceps brachii m., lat. head
Triceps brachii m., med. head
Triceps brachii m., long head

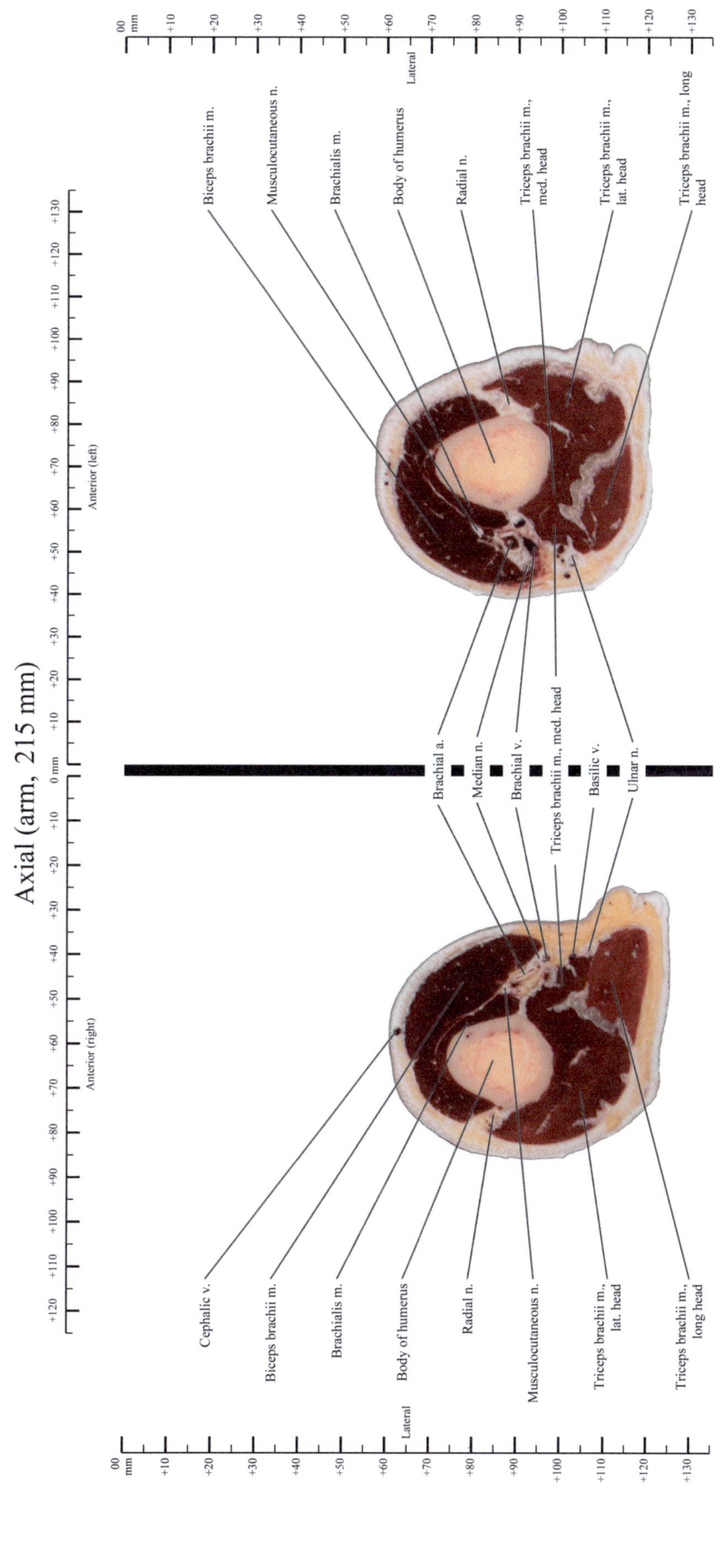
Axial (arm, 215 mm)
Anterior (left)
Anterior (right)
Lateral
Biceps brachii m.
Musculocutaneous n.
Brachialis m.
Body of humerus
Radial n.
Triceps brachii m., med. head
Triceps brachii m., lat. head
Triceps brachii m., long head
Brachial a.
Median n.
Brachial v.
Triceps brachii m., med. head
Basilic v.
Ulnar n.
Cephalic v.
Biceps brachii m.
Brachialis m.
Body of humerus
Radial n.
Musculocutaneous n.
Triceps brachii m., lat. head
Triceps brachii m., long head

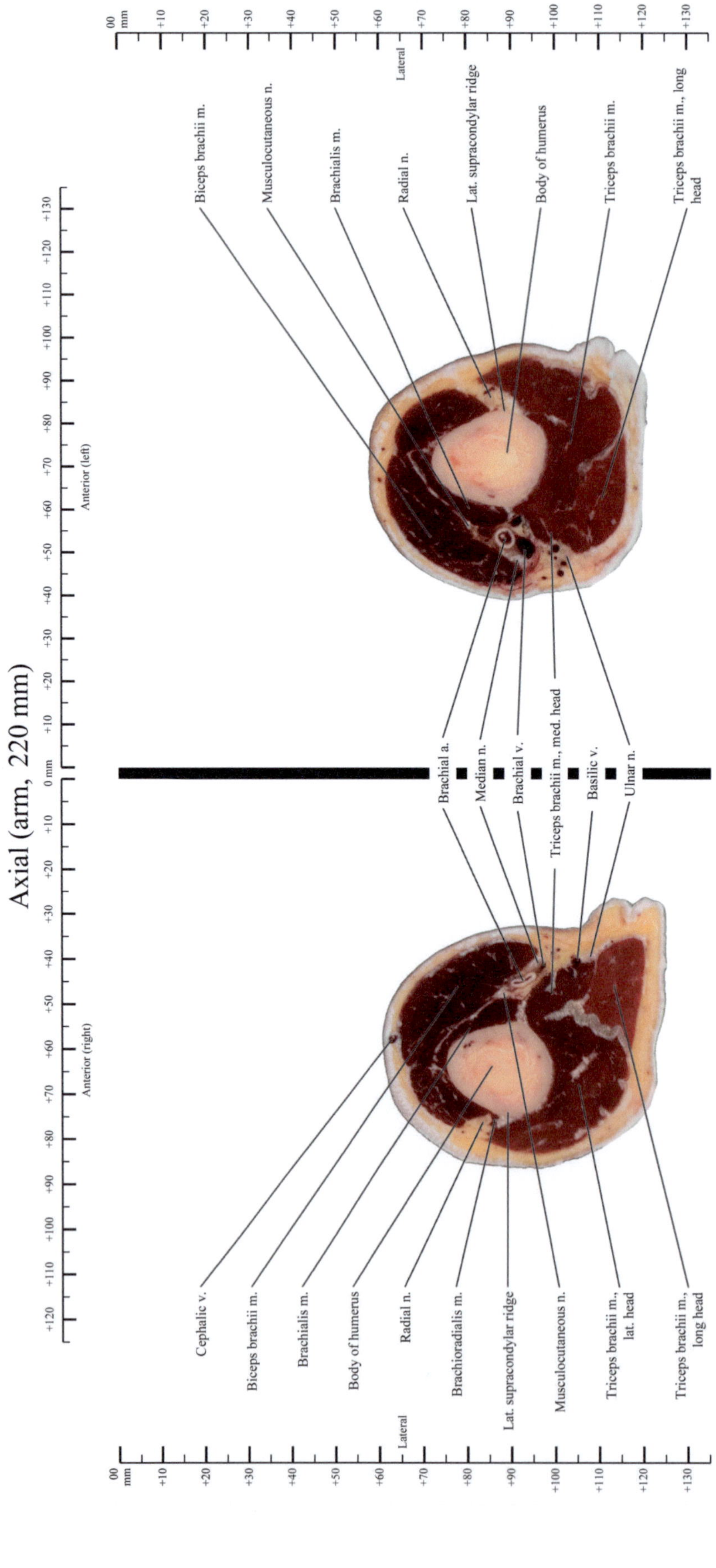
Axial (arm, 220 mm)
Anterior (left)
Anterior (right)
Lateral
Biceps brachii m.
Musculocutaneous n.
Brachialis m.
Radial n.
Lat. supracondylar ridge
Body of humerus
Triceps brachii m.
Triceps brachii m., long head
Brachial a.
Median n.
Brachial v.
Triceps brachii m., med. head
Basilic v.
Ulnar n.
Cephalic v.
Biceps brachii m.
Brachialis m.
Body of humerus
Radial n.
Brachioradialis m.
Lat. supracondylar ridge
Musculocutaneous n.
Triceps brachii m., lat. head
Triceps brachii m., long head
00 mm
0 mm

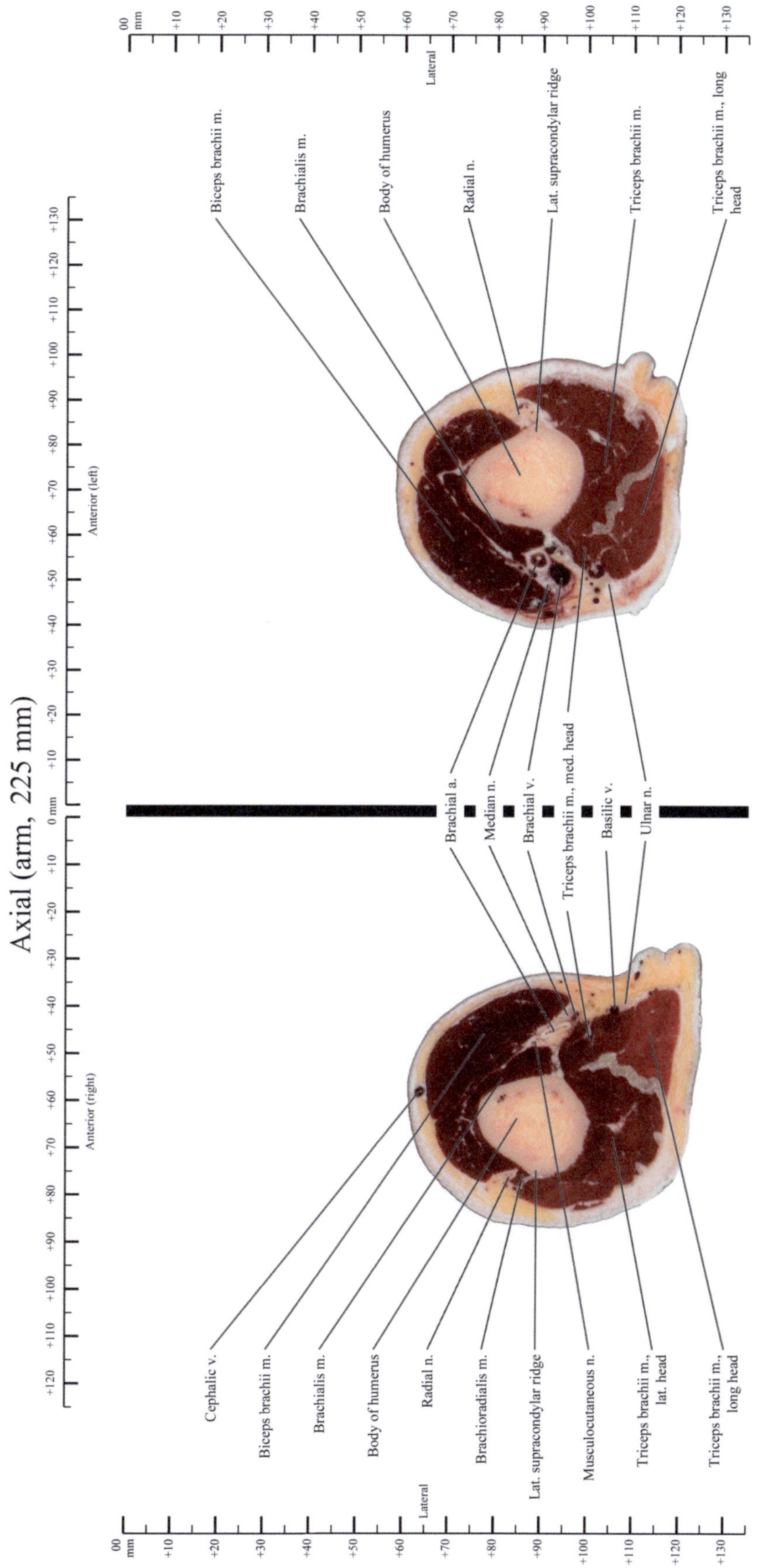
Axial (arm, 225 mm)
Anterior (left)
Anterior (right)
Lateral
Biceps brachii m.
Brachialis m.
Body of humerus
Radial n.
Lat. supracondylar ridge
Triceps brachii m.
Triceps brachii m., long head
Brachial a.
Median n.
Brachial v.
Triceps brachii m., med. head
Basilic v.
Ulnar n.
Cephalic v.
Biceps brachii m.
Brachialis m.
Body of humerus
Radial n.
Brachioradialis m.
Lat. supracondylar ridge
Musculocutaneous n.
Triceps brachii m., lat. head
Triceps brachii m., long head

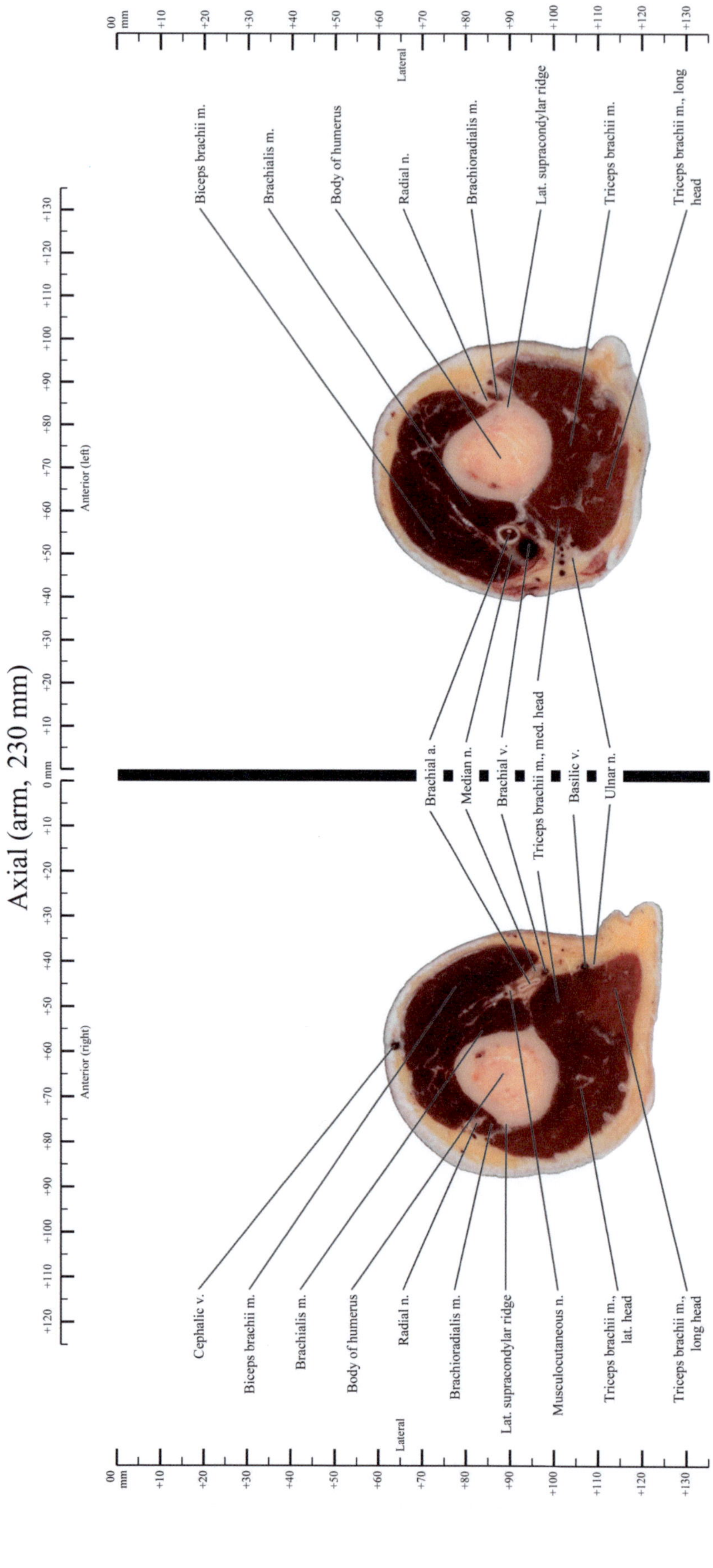
Axial (arm, 230 mm)
Anterior (left)
Anterior (right)
Lateral
Biceps brachii m.
Brachialis m.
Body of humerus
Radial n.
Brachioradialis m.
Lat. supracondylar ridge
Triceps brachii m.
Triceps brachii m., long head
Brachial a.
Median n.
Brachial v.
Triceps brachii m., med. head
Basilic v.
Ulnar n.
Cephalic v.
Musculocutaneous n.
Triceps brachii m., lat. head

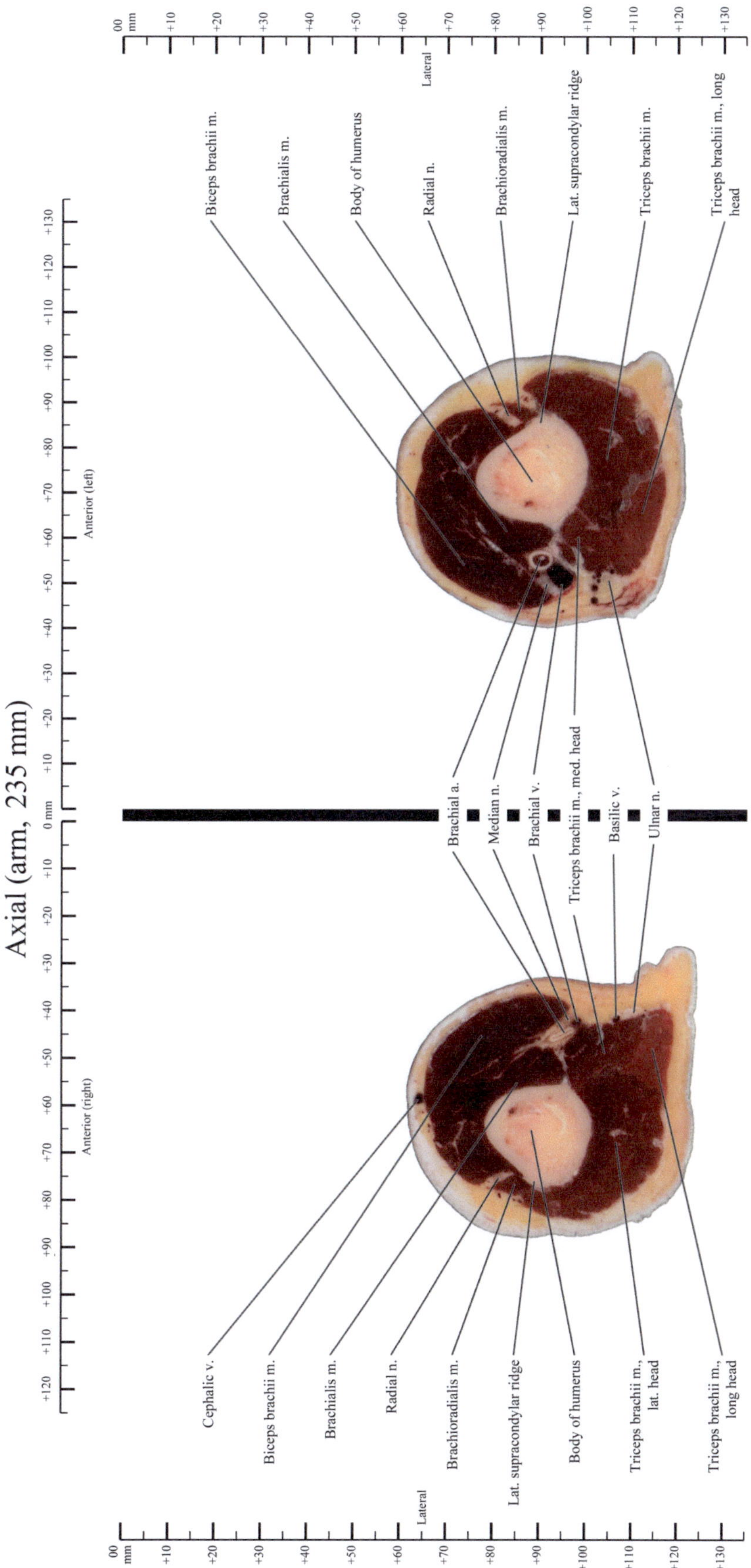
Axial (arm, 235 mm)
Anterior (left)
Anterior (right)
Lateral
Biceps brachii m.
Brachialis m.
Body of humerus
Radial n.
Brachioradialis m.
Lat. supracondylar ridge
Triceps brachii m.
Triceps brachii m., long head
Brachial a.
Median n.
Brachial v.
Triceps brachii m., med. head
Basilic v.
Ulnar n.
Cephalic v.
Triceps brachii m., lat. head

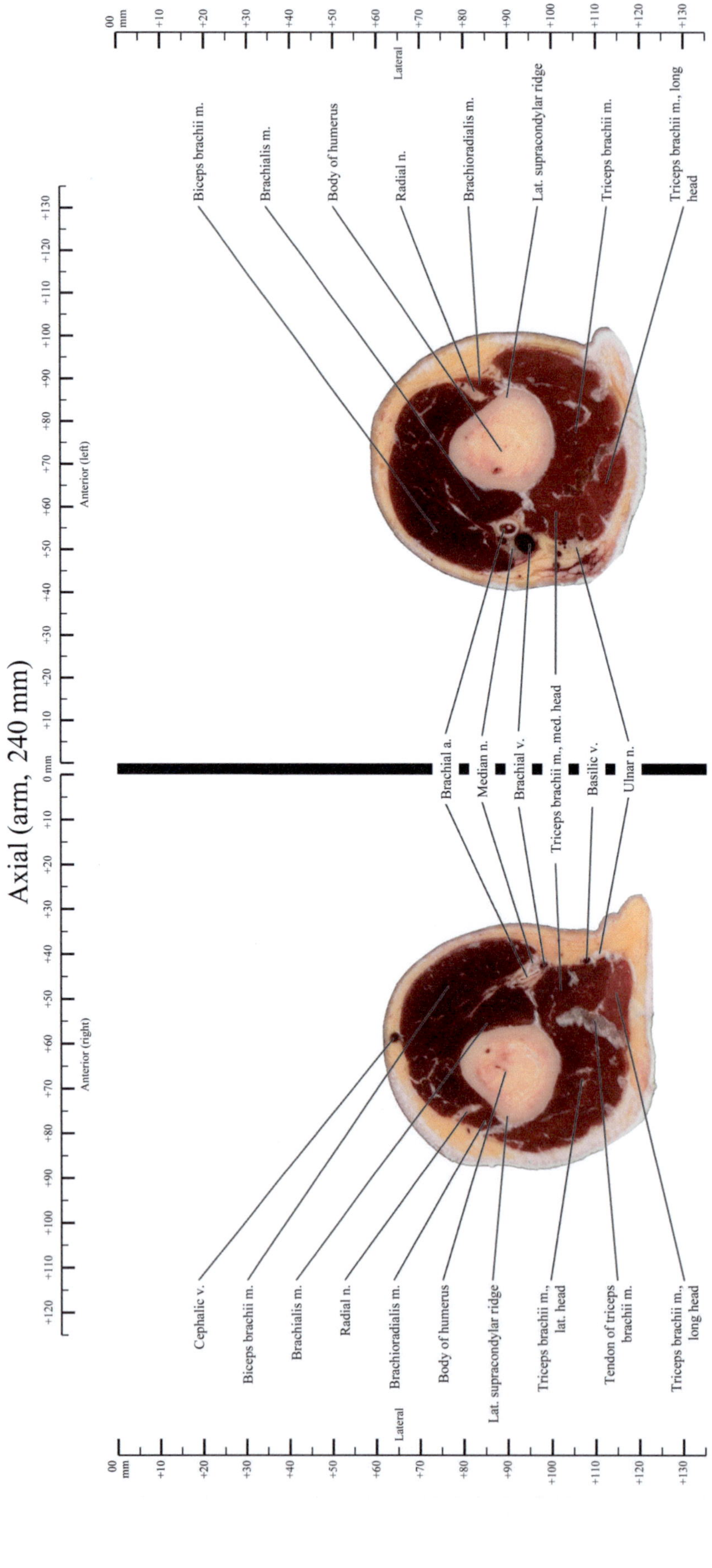
Axial (arm, 240 mm)
Anterior (left)
Anterior (right)
Lateral
Lateral
Biceps brachii m.
Brachialis m.
Body of humerus
Radial n.
Brachioradialis m.
Lat. supracondylar ridge
Triceps brachii m.
Triceps brachii m., long head
Brachial a.
Median n.
Brachial v.
Triceps brachii m., med. head
Basilic v.
Ulnar n.
Cephalic v.
Biceps brachii m.
Brachialis m.
Radial n.
Brachioradialis m.
Body of humerus
Lat. supracondylar ridge
Triceps brachii m., lat. head
Tendon of triceps brachii m.
Triceps brachii m., long head

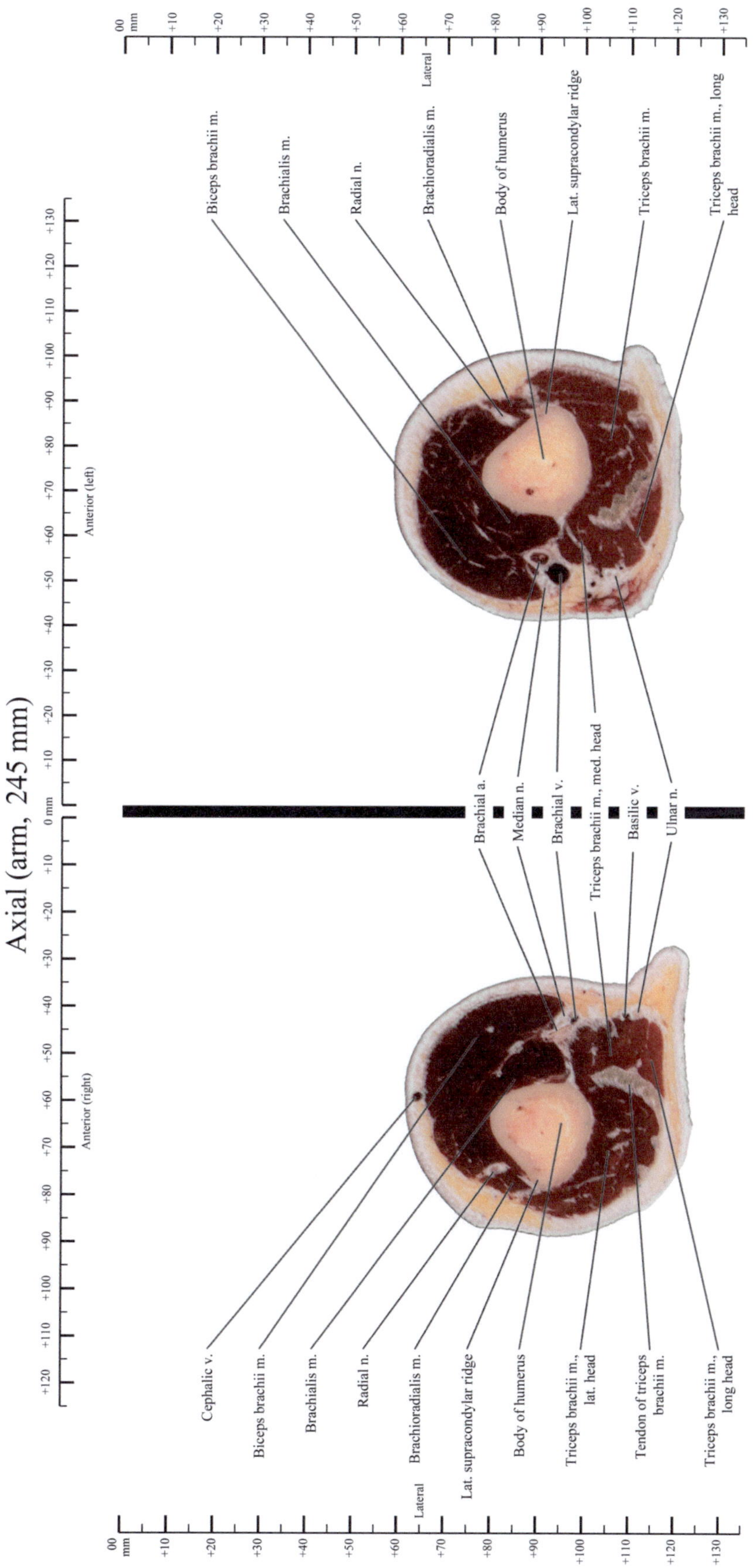
Axial (arm, 245 mm)
Anterior (left)
Anterior (right)
Lateral
Biceps brachii m.
Brachialis m.
Radial n.
Brachioradialis m.
Body of humerus
Lat. supracondylar ridge
Triceps brachii m.
Triceps brachii m., long head
Brachial a.
Median n.
Brachial v.
Triceps brachii m., med. head
Basilic v.
Ulnar n.
Cephalic v.
Triceps brachii m., lat. head
Tendon of triceps brachii m.

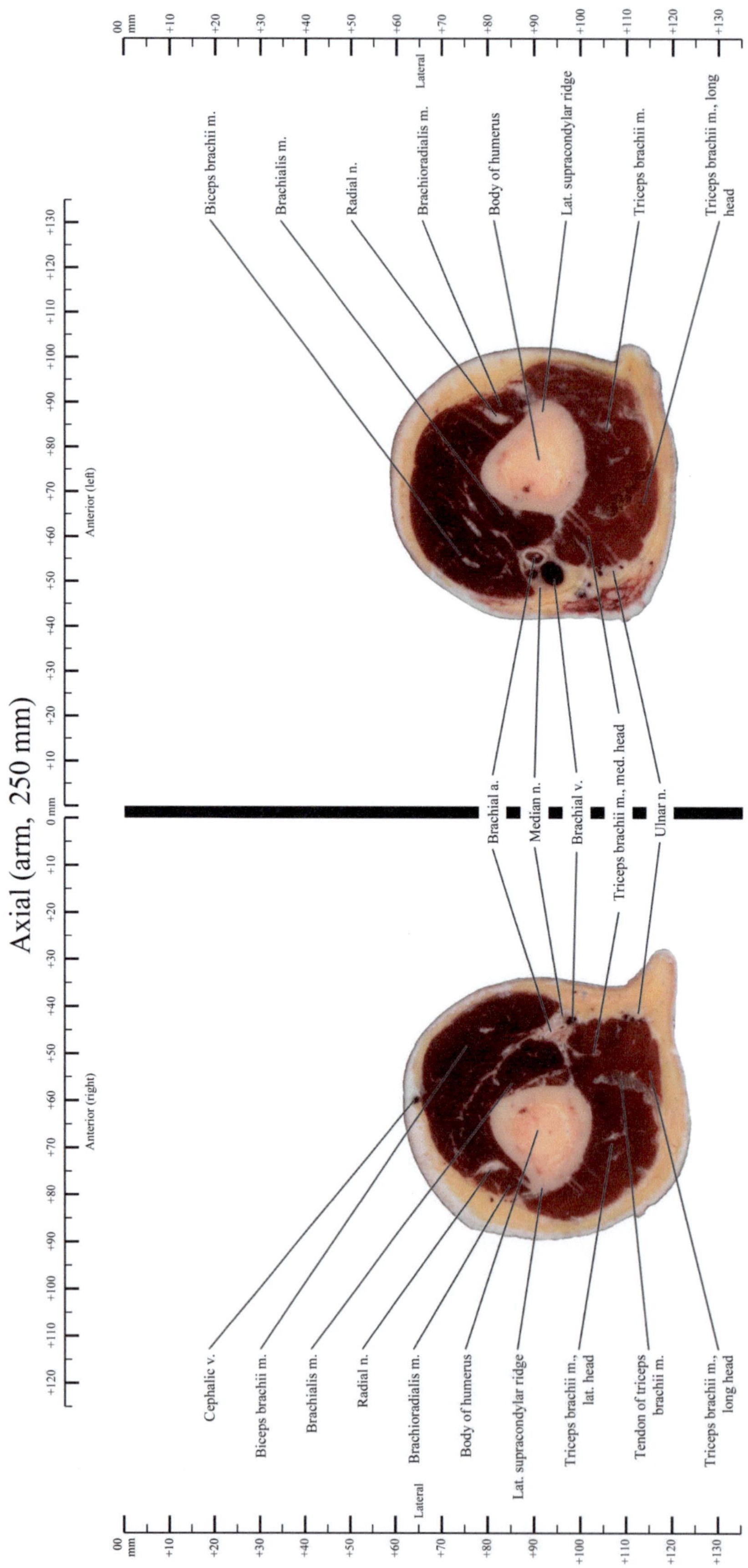
Axial (arm, 250 mm)
Anterior (left)
Anterior (right)
Lateral
Biceps brachii m.
Brachialis m.
Radial n.
Brachioradialis m.
Body of humerus
Lat. supracondylar ridge
Triceps brachii m.
Triceps brachii m., long head
Brachial a.
Median n.
Brachial v.
Triceps brachii m., med. head
Ulnar n.
Cephalic v.
Triceps brachii m., lat. head
Tendon of triceps brachii m.
Triceps brachii m., long head

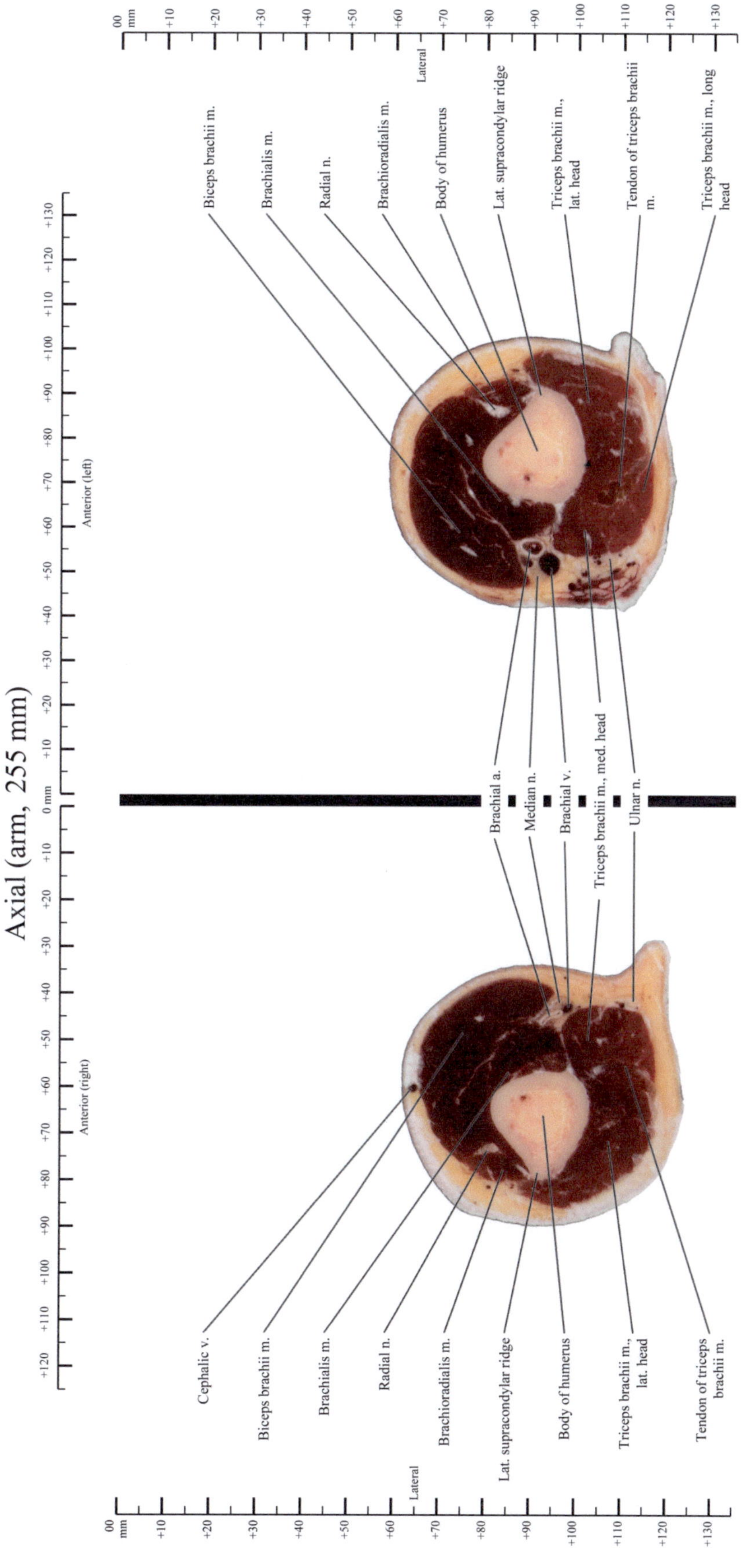
Axial (arm, 255 mm)
Anterior (left)
Anterior (right)
Lateral
Lateral
Biceps brachii m.
Brachialis m.
Radial n.
Brachioradialis m.
Body of humerus
Lat. supracondylar ridge
Triceps brachii m., lat. head
Tendon of triceps brachii m.
Triceps brachii m., long head
Brachial a.
Median n.
Brachial v.
Triceps brachii m., med. head
Ulnar n.
Cephalic v.
Biceps brachii m.
Brachialis m.
Radial n.
Brachioradialis m.
Lat. supracondylar ridge
Body of humerus
Triceps brachii m., lat. head
Tendon of triceps brachii m.

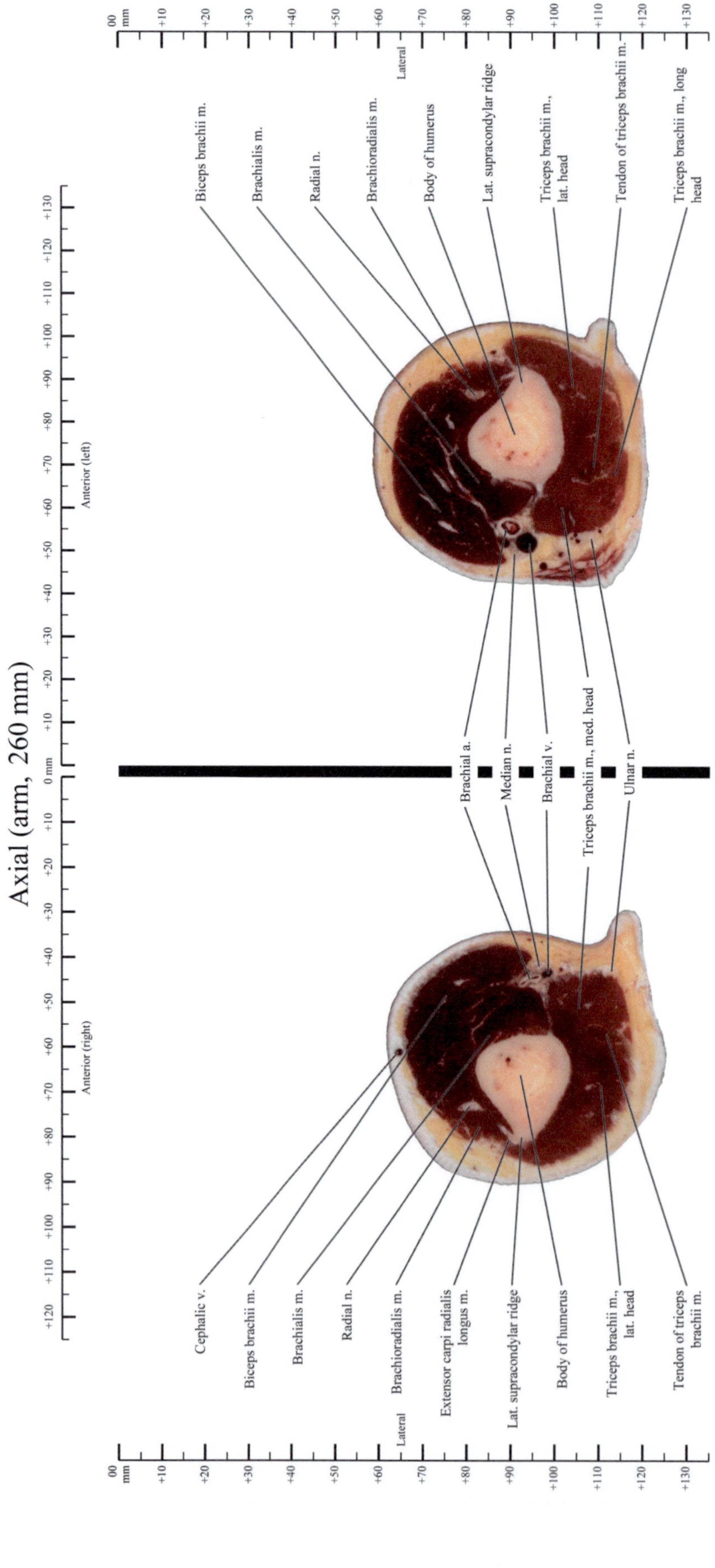
Axial (arm, 260 mm)
Anterior (left)
Anterior (right)
Lateral
Biceps brachii m.
Brachialis m.
Radial n.
Brachioradialis m.
Body of humerus
Lat. supracondylar ridge
Triceps brachii m., lat. head
Tendon of triceps brachii m.
Triceps brachii m., long head
Brachial a.
Median n.
Brachial v.
Triceps brachii m., med. head
Ulnar n.
Cephalic v.
Extensor carpi radialis longus m.

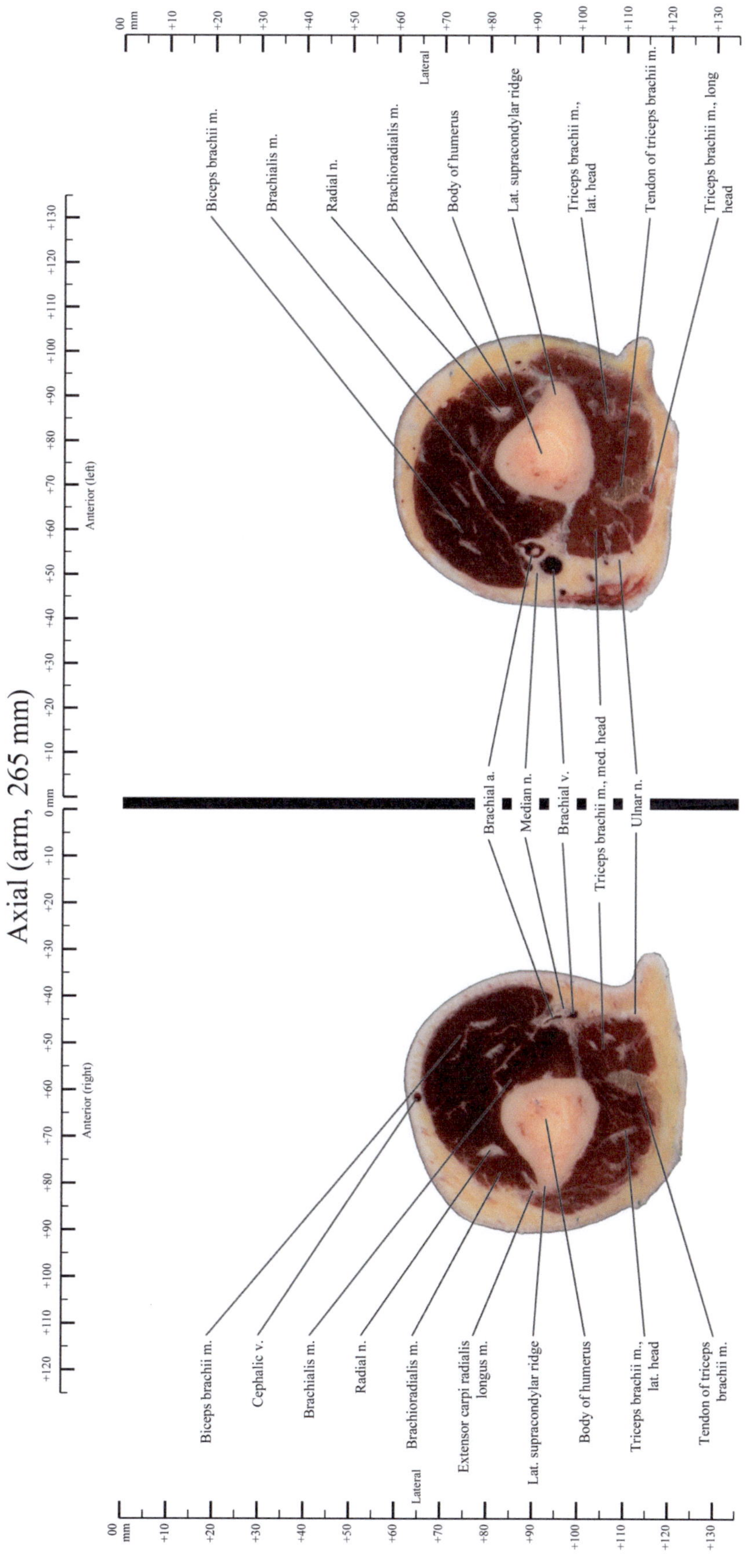
Axial (arm, 265 mm)
Anterior (left)
Anterior (right)
Lateral
Biceps brachii m.
Brachialis m.
Radial n.
Brachioradialis m.
Body of humerus
Lat. supracondylar ridge
Triceps brachii m., lat. head
Tendon of triceps brachii m.
Triceps brachii m., long head
Brachial a.
Median n.
Brachial v.
Triceps brachii m., med. head
Ulnar n.
Cephalic v.
Extensor carpi radialis longus m.
Tendon of triceps brachii m.

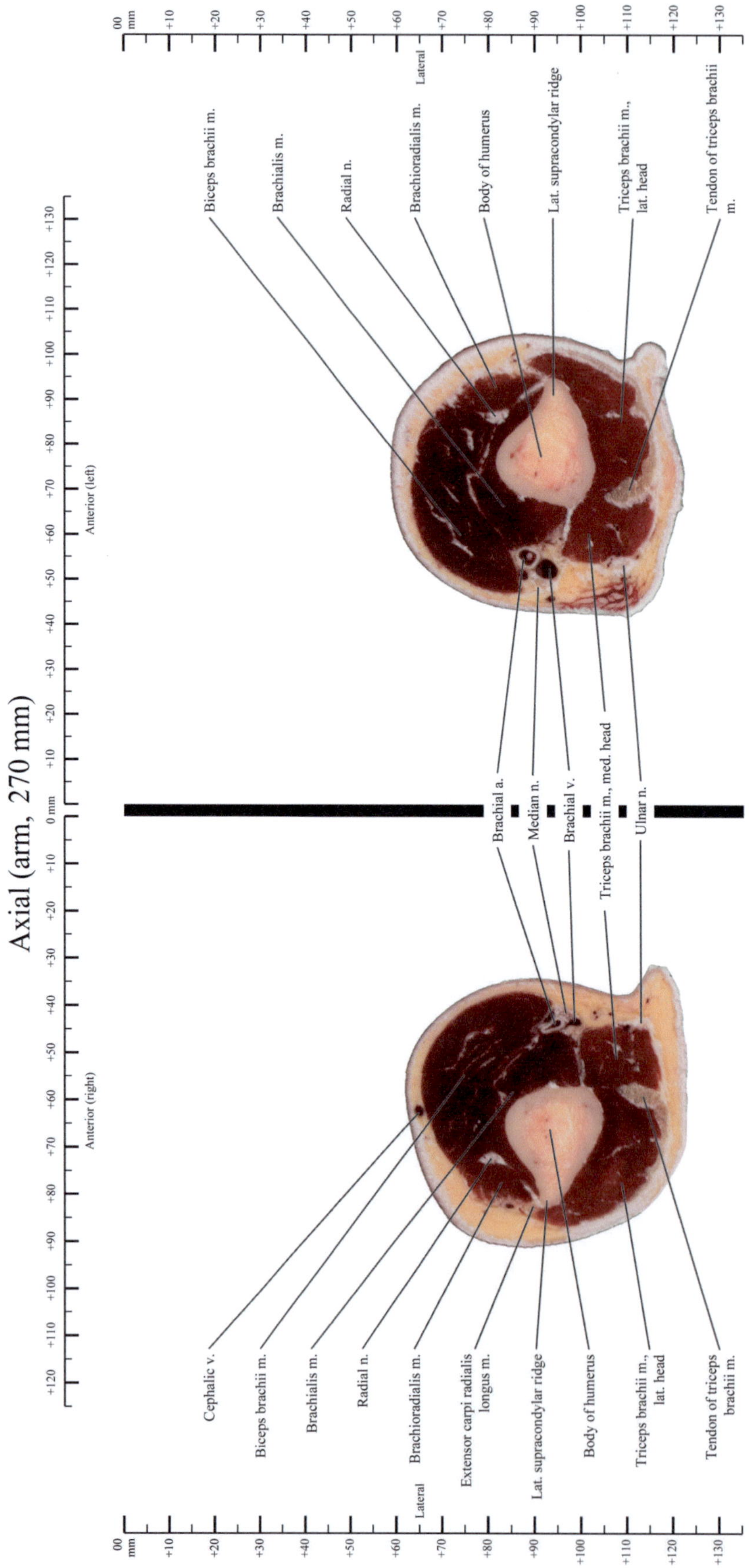
Axial (arm, 270 mm)
Anterior (left)
Anterior (right)
Lateral
Biceps brachii m.
Brachialis m.
Radial n.
Brachioradialis m.
Body of humerus
Lat. supracondylar ridge
Triceps brachii m., lat. head
Tendon of triceps brachii m.
Brachial a.
Median n.
Brachial v.
Triceps brachii m., med. head
Ulnar n.
Cephalic v.
Biceps brachii m.
Brachialis m.
Radial n.
Brachioradialis m.
Extensor carpi radialis longus m.
Lat. supracondylar ridge
Body of humerus
Triceps brachii m., lat. head
Tendon of triceps brachii m.
Lateral

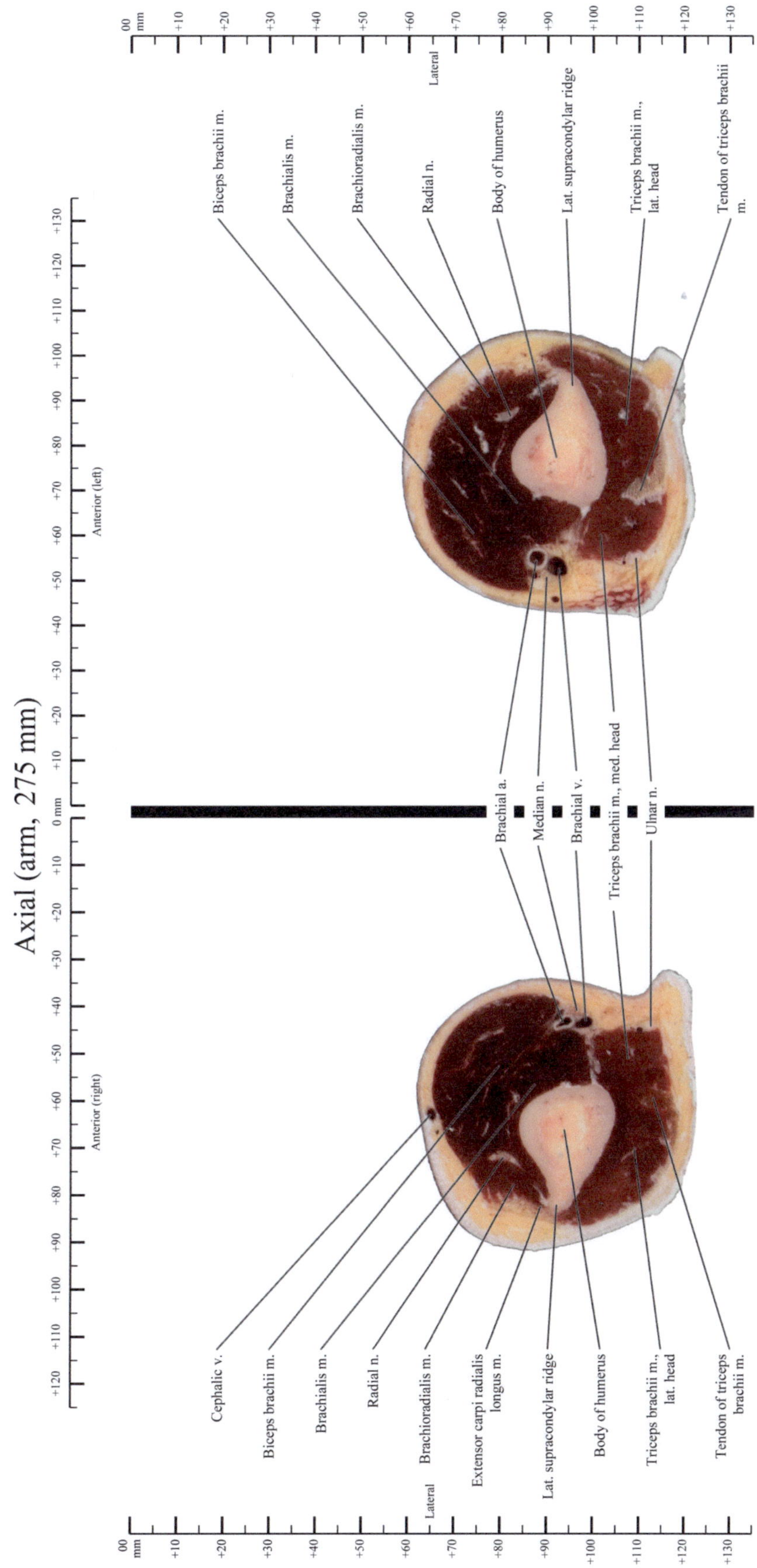
Axial (arm, 275 mm)
Anterior (left)
Anterior (right)
Lateral
Biceps brachii m.
Brachialis m.
Brachioradialis m.
Radial n.
Body of humerus
Lat. supracondylar ridge
Triceps brachii m., lat. head
Tendon of triceps brachii m.
Brachial a.
Median n.
Brachial v.
Triceps brachii m., med. head
Ulnar n.
Cephalic v.
Extensor carpi radialis longus m.

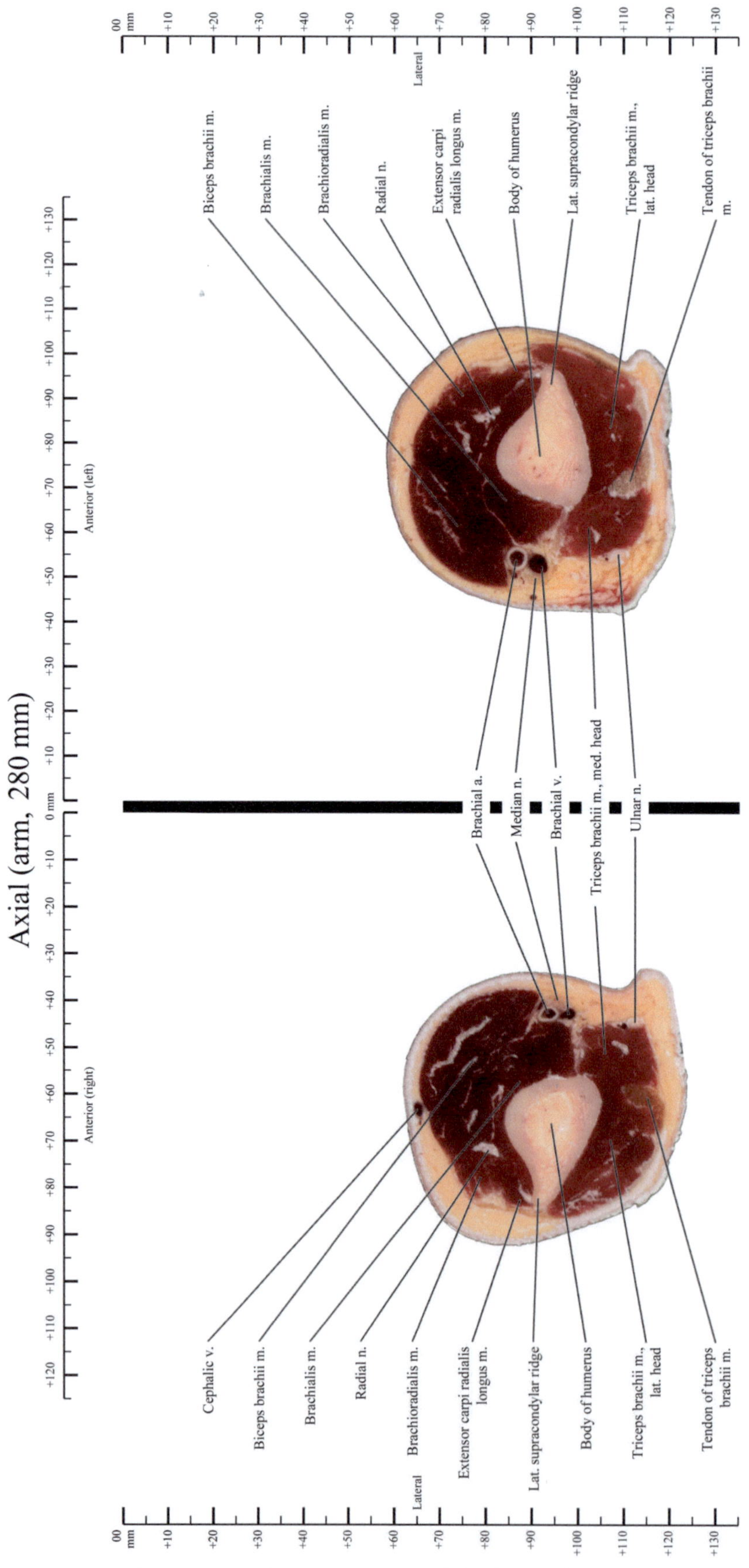
Axial (arm, 280 mm)
Anterior (left)
Anterior (right)
Lateral
Biceps brachii m.
Brachialis m.
Brachioradialis m.
Radial n.
Extensor carpi radialis longus m.
Body of humerus
Lat. supracondylar ridge
Triceps brachii m., lat. head
Tendon of triceps brachii m.
Brachial a.
Median n.
Brachial v.
Triceps brachii m., med. head
Ulnar n.
Cephalic v.

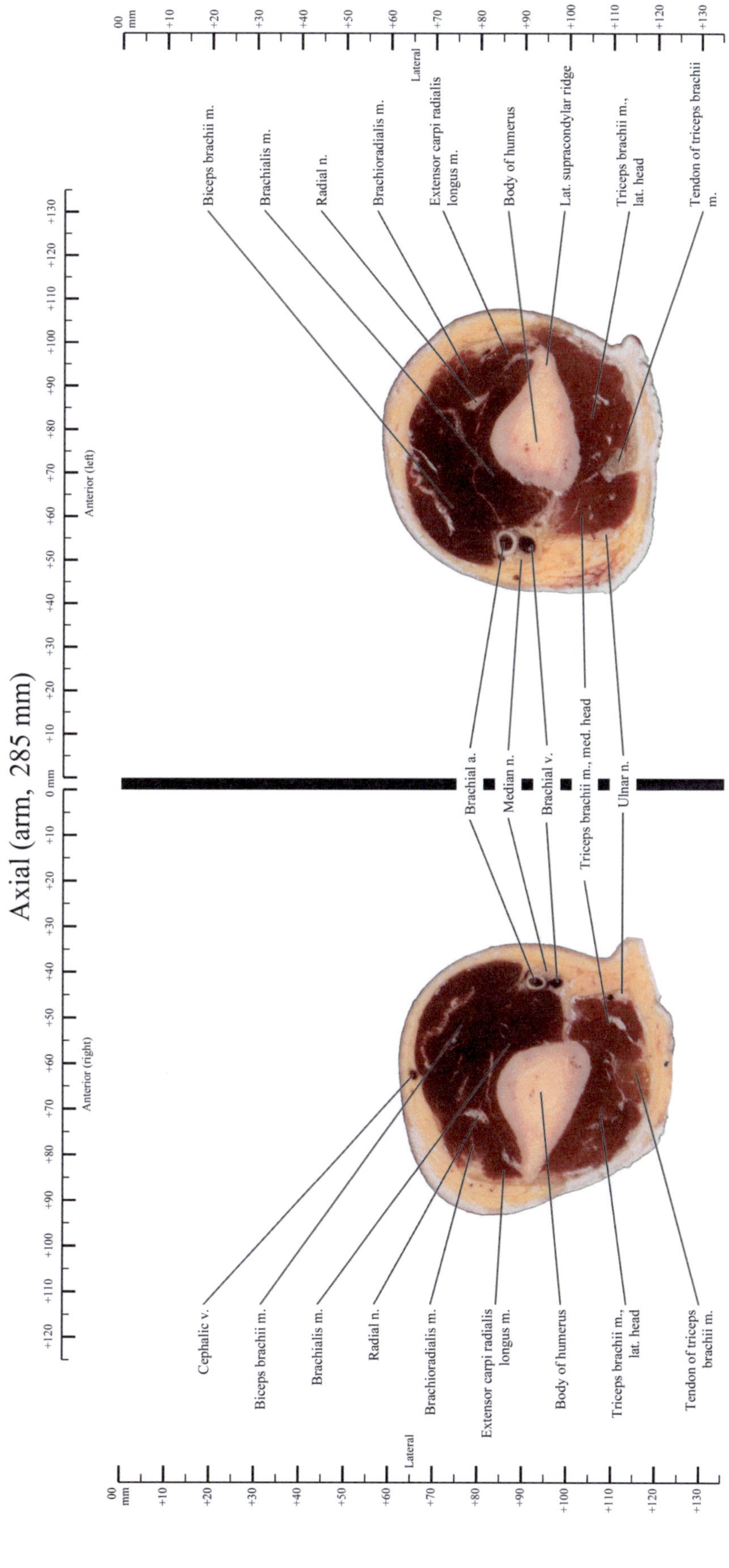

Axial (arm, 285 mm)
Anterior (left)
Anterior (right)
Lateral
Biceps brachii m.
Brachialis m.
Radial n.
Brachioradialis m.
Extensor carpi radialis longus m.
Body of humerus
Lat. supracondylar ridge
Triceps brachii m., lat. head
Tendon of triceps brachii m.
Brachial a.
Median n.
Brachial v.
Triceps brachii m., med. head
Ulnar n.
Cephalic v.

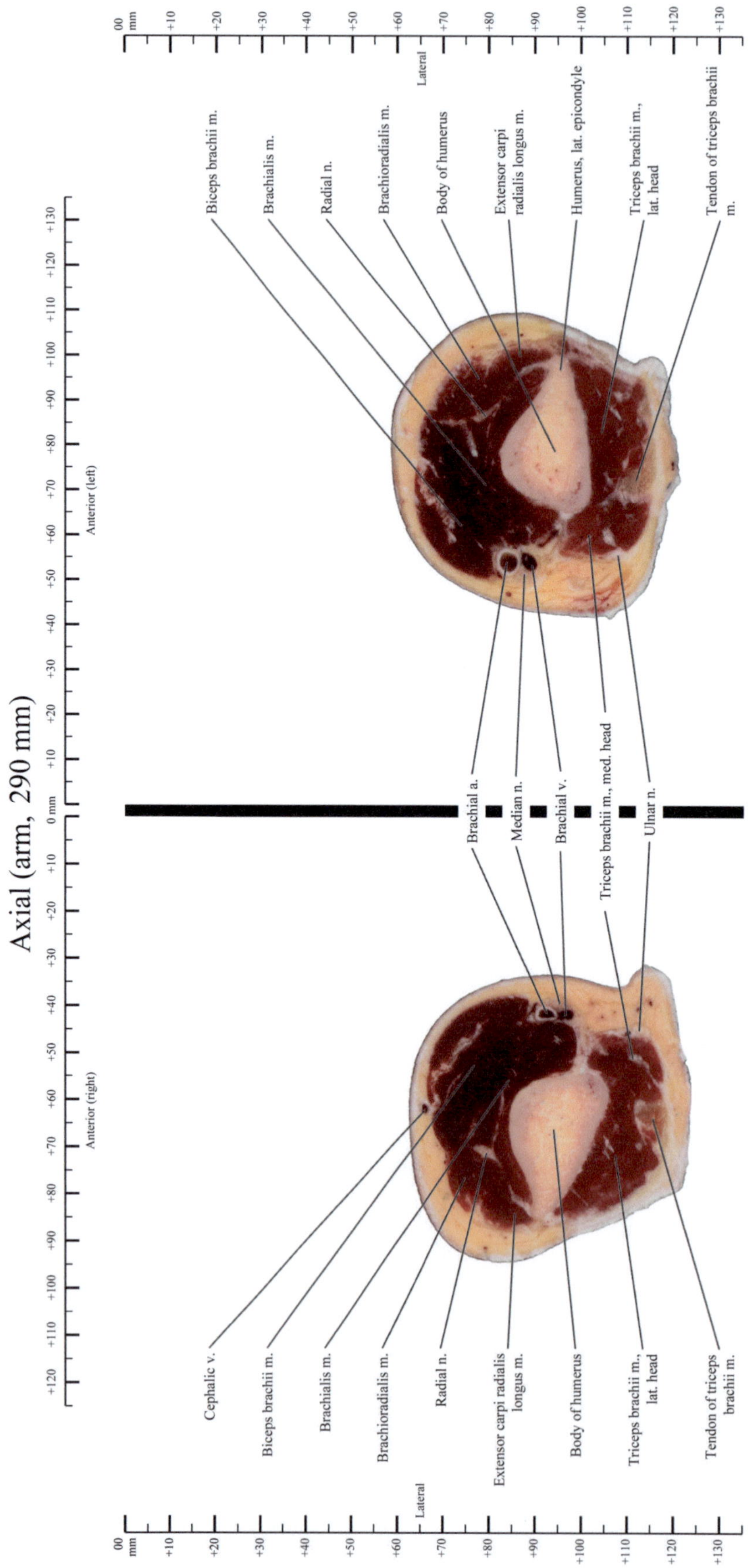
Axial (arm, 290 mm)
Anterior (left)
Anterior (right)
Lateral
Biceps brachii m.
Brachialis m.
Radial n.
Brachioradialis m.
Body of humerus
Extensor carpi radialis longus m.
Humerus, lat. epicondyle
Triceps brachii m., lat. head
Tendon of triceps brachii m.
Brachial a.
Median n.
Brachial v.
Triceps brachii m., med. head
Ulnar n.
Cephalic v.
Biceps brachii m.
Brachialis m.
Brachioradialis m.
Radial n.
Extensor carpi radialis longus m.
Body of humerus
Triceps brachii m., lat. head
Tendon of triceps brachii m.

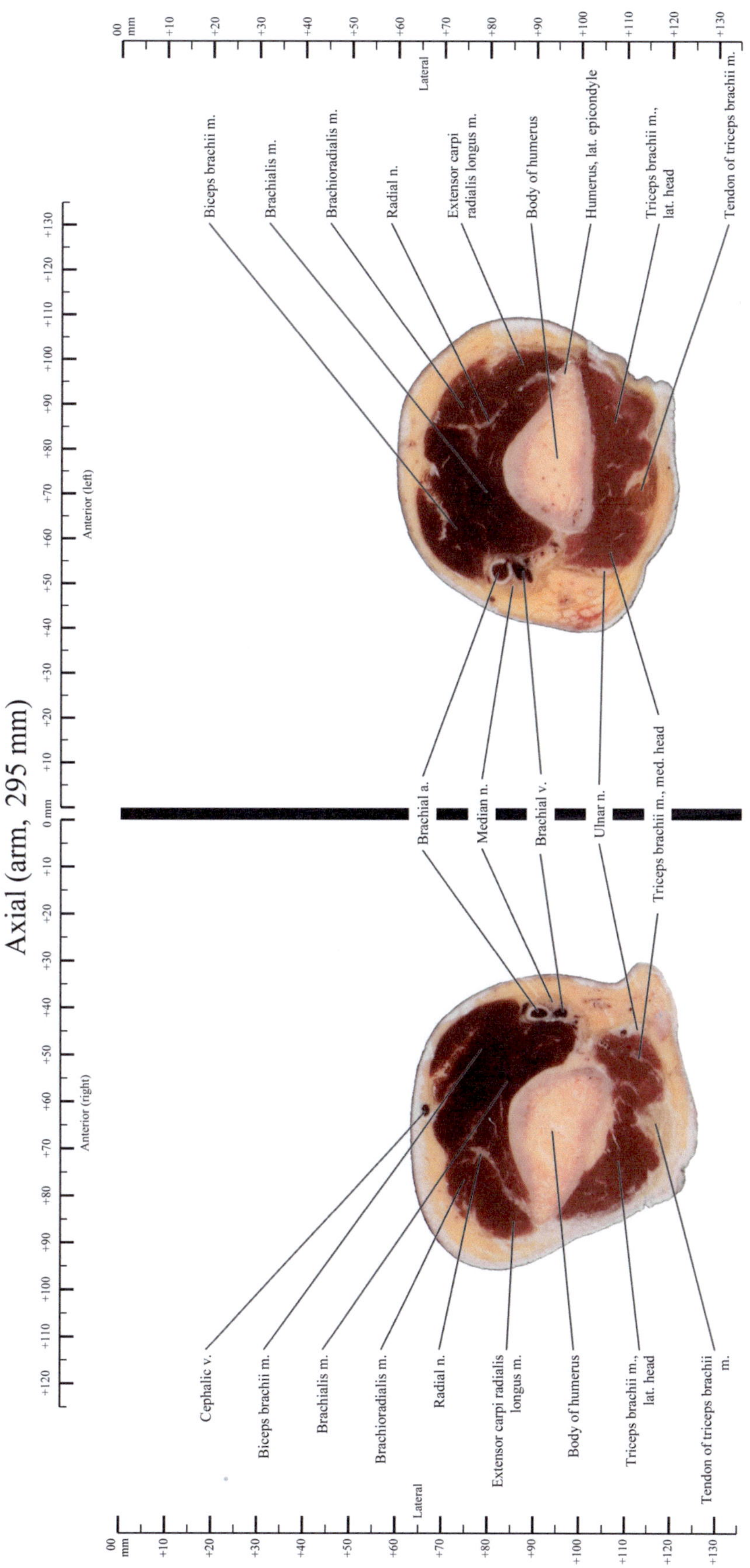
Axial (arm, 295 mm)
Anterior (left)
Lateral
Biceps brachii m.
Brachialis m.
Brachioradialis m.
Radial n.
Extensor carpi radialis longus m.
Body of humerus
Humerus, lat. epicondyle
Triceps brachii m., lat. head
Tendon of triceps brachii m.
Brachial a.
Median n.
Brachial v.
Ulnar n.
Triceps brachii m., med. head
Anterior (right)
Lateral
Cephalic v.
Biceps brachii m.
Brachialis m.
Brachioradialis m.
Radial n.
Extensor carpi radialis longus m.
Body of humerus
Triceps brachii m., lat. head
Tendon of triceps brachii m.

Axial (arm, 300 mm)

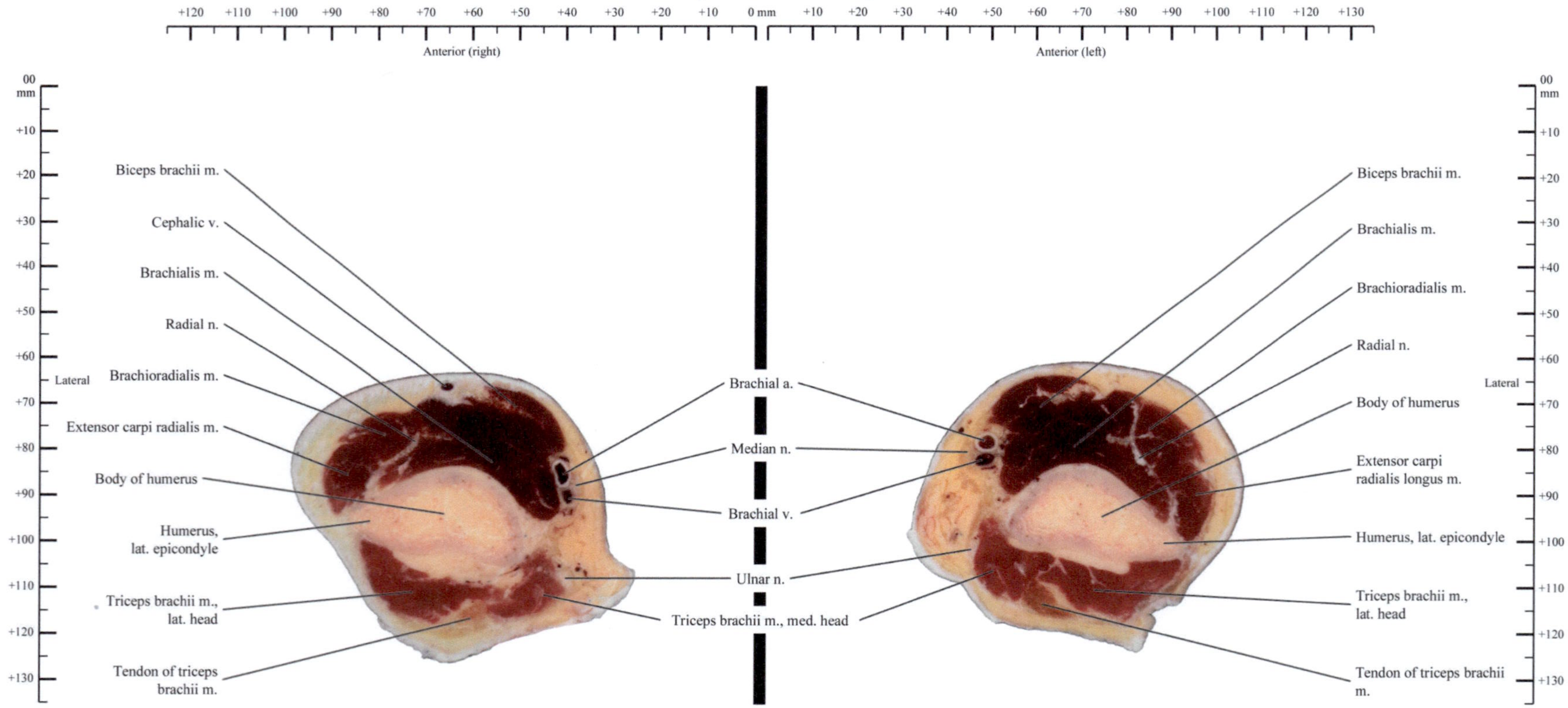

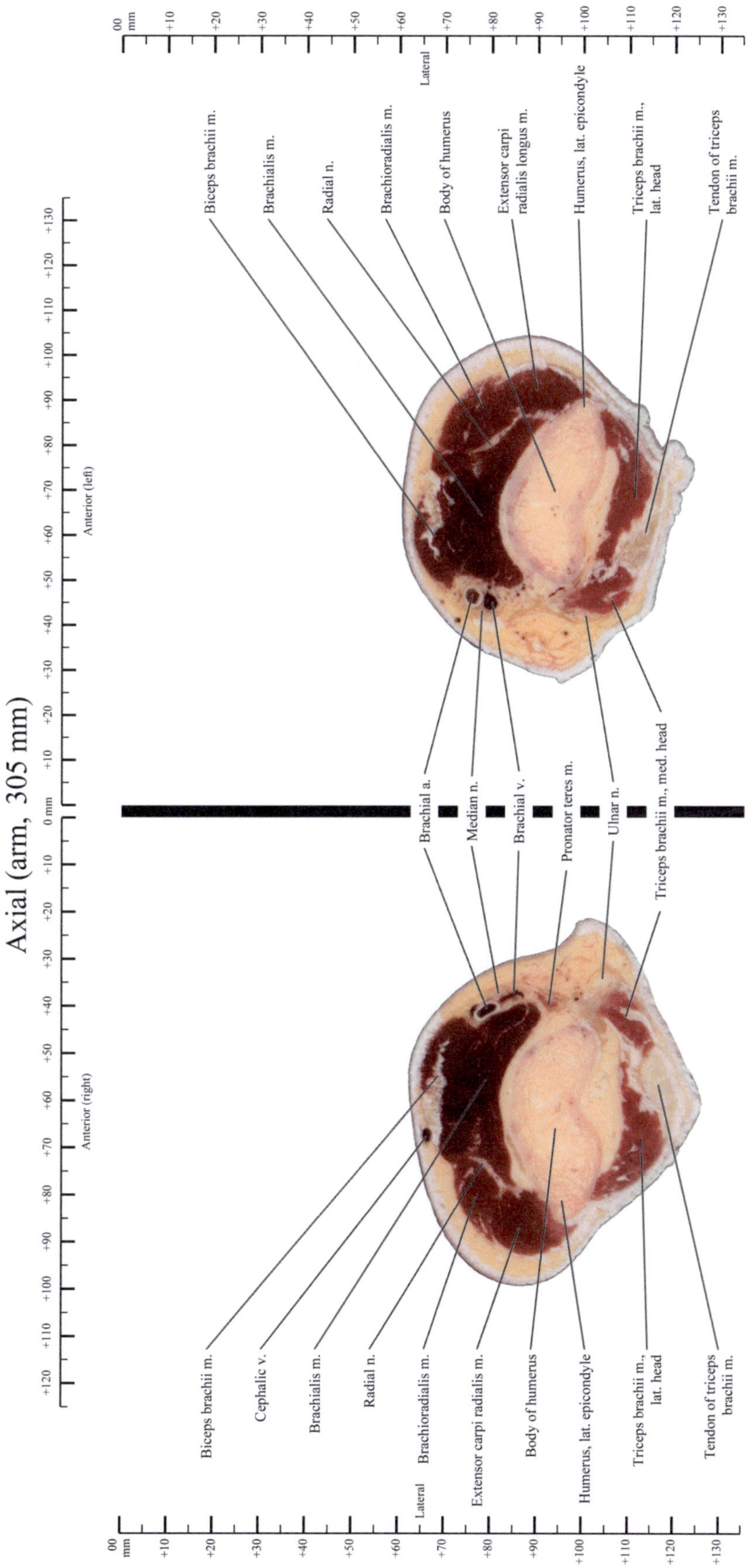
Axial (arm, 305 mm)
Anterior (left)
Anterior (right)
Lateral
Biceps brachii m.
Brachialis m.
Radial n.
Brachioradialis m.
Body of humerus
Extensor carpi radialis longus m.
Humerus, lat. epicondyle
Triceps brachii m., lat. head
Tendon of triceps brachii m.
Brachial a.
Median n.
Brachial v.
Pronator teres m.
Ulnar n.
Triceps brachii m., med. head
Biceps brachii m.
Cephalic v.
Brachialis m.
Radial n.
Brachioradialis m.
Extensor carpi radialis m.
Body of humerus
Humerus, lat. epicondyle
Triceps brachii m., lat. head
Tendon of triceps brachii m.
Lateral

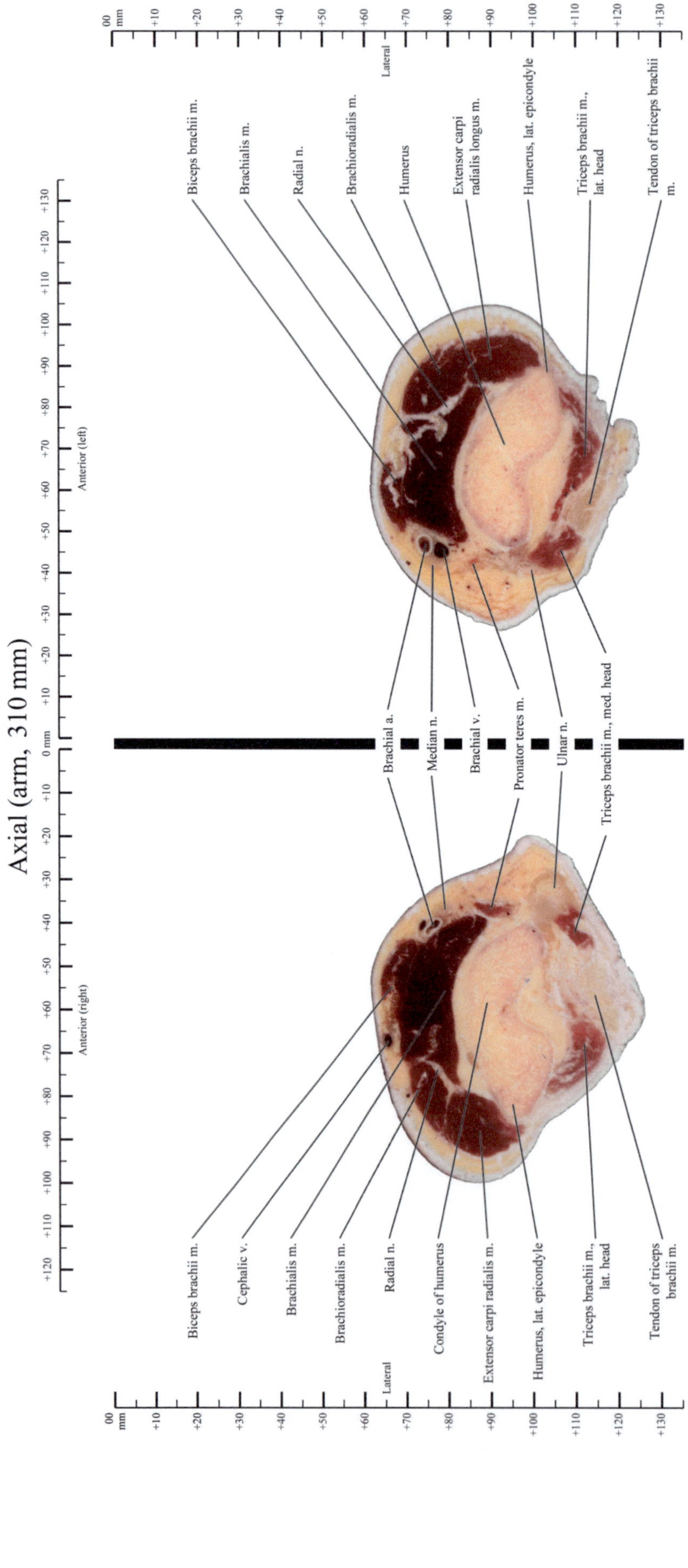
Axial (arm, 310 mm)
Anterior (left)
Anterior (right)
Lateral
Lateral
Biceps brachii m.
Brachialis m.
Radial n.
Brachioradialis m.
Humerus
Extensor carpi radialis longus m.
Humerus, lat. epicondyle
Triceps brachii m., lat. head
Tendon of triceps brachii m.
Brachial a.
Median n.
Brachial v.
Pronator teres m.
Ulnar n.
Triceps brachii m., med. head
Biceps brachii m.
Cephalic v.
Brachialis m.
Brachioradialis m.
Radial n.
Condyle of humerus
Extensor carpi radialis m.
Humerus, lat. epicondyle
Triceps brachii m., lat. head
Tendon of triceps brachii m.

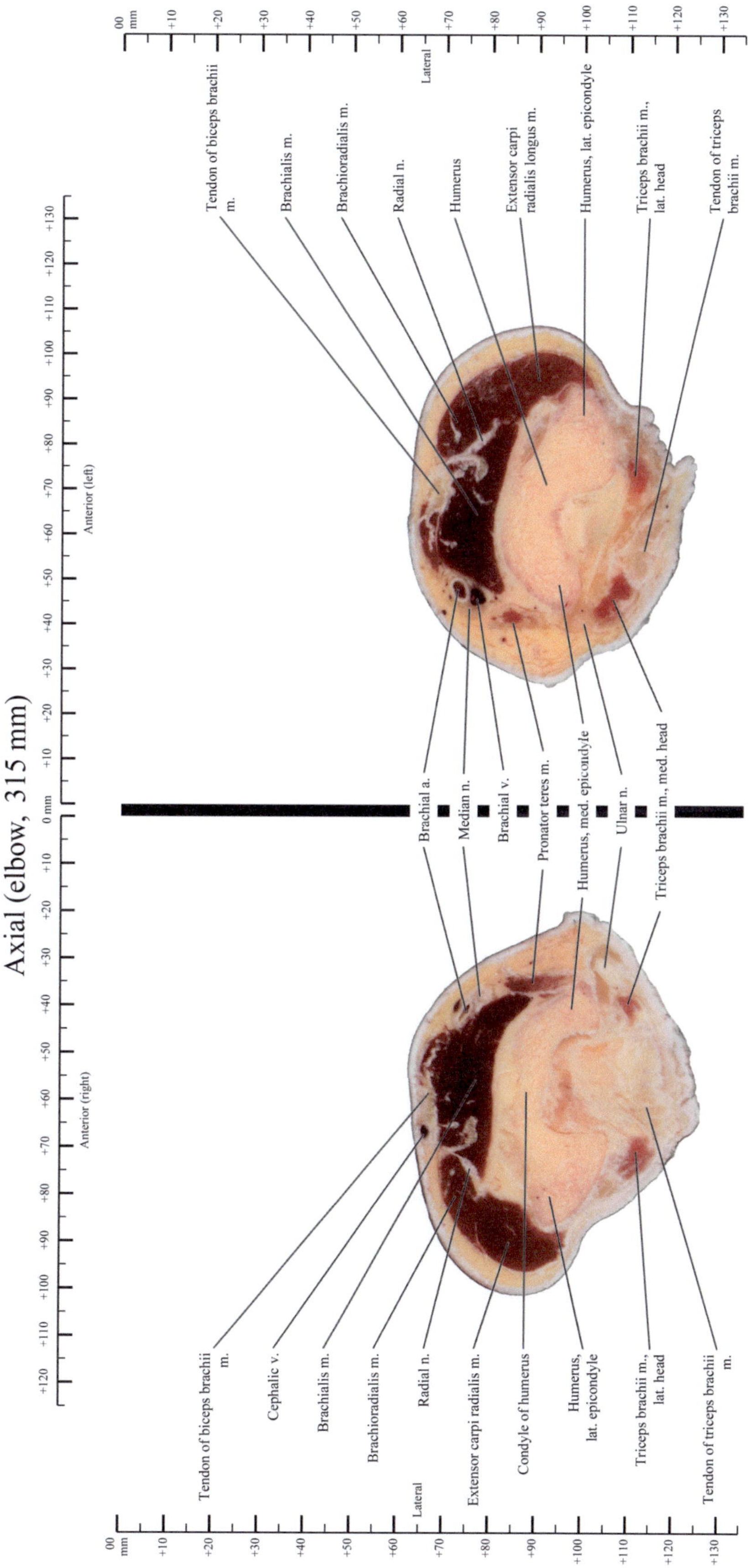
Axial (elbow, 315 mm)
Anterior (left)
Anterior (right)
Lateral
Tendon of biceps brachii m.
Brachialis m.
Brachioradialis m.
Radial n.
Humerus
Extensor carpi radialis longus m.
Humerus, lat. epicondyle
Triceps brachii m., lat. head
Tendon of triceps brachii m.
Brachial a.
Median n.
Brachial v.
Pronator teres m.
Humerus, med. epicondyle
Ulnar n.
Triceps brachii m., med. head
Cephalic v.
Extensor carpi radialis m.
Condyle of humerus
Humerus, lat. epicondyle

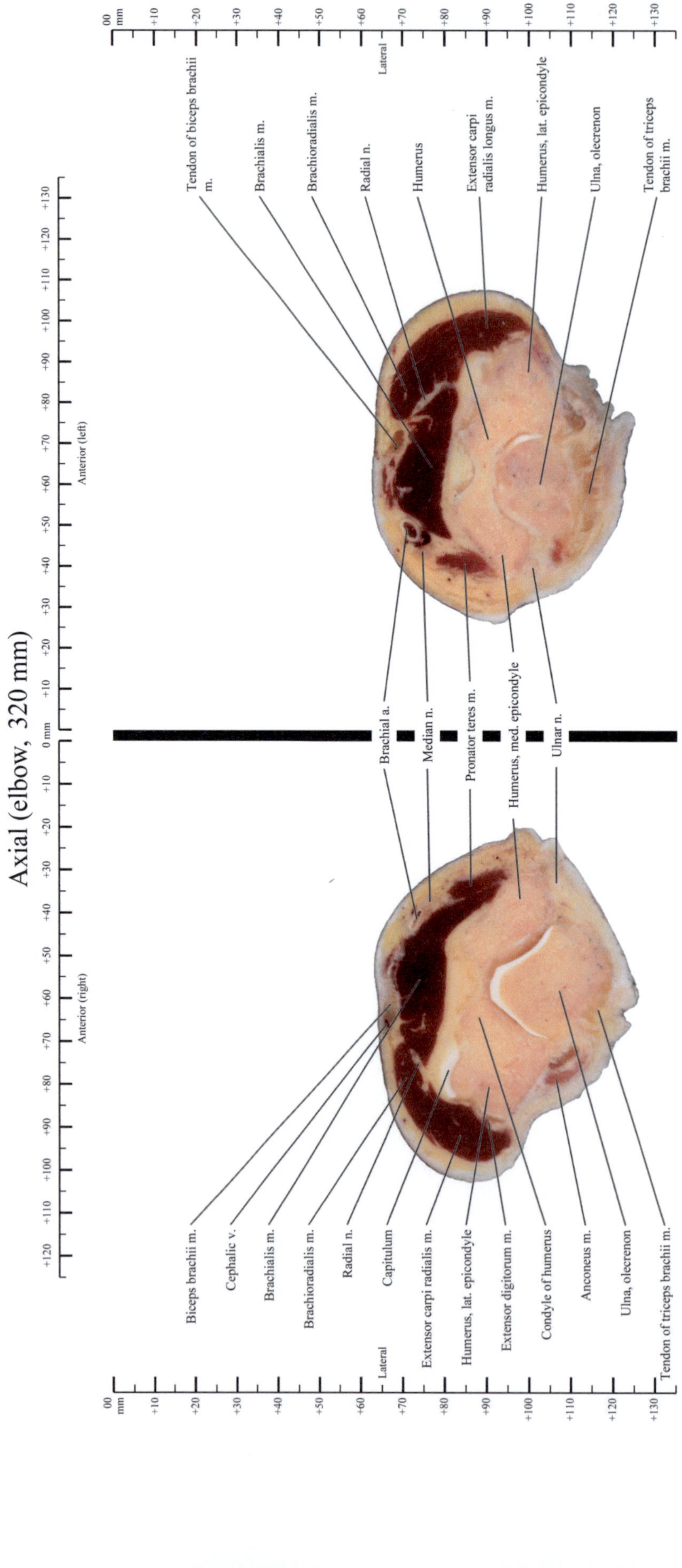
Axial (elbow, 320 mm)
Anterior (left)
Anterior (right)
Lateral
Tendon of biceps brachii m.
Brachialis m.
Brachioradialis m.
Radial n.
Humerus
Extensor carpi radialis longus m.
Humerus, lat. epicondyle
Ulna, olecrenon
Tendon of triceps brachii m.
Brachial a.
Median n.
Pronator teres m.
Humerus, med. epicondyle
Ulnar n.
Biceps brachii m.
Cephalic v.
Brachialis m.
Brachioradialis m.
Radial n.
Capitulum
Extensor carpi radialis m.
Humerus, lat. epicondyle
Extensor digitorum m.
Condyle of humerus
Anconeus m.
Ulna, olecrenon
Tendon of triceps brachii m.

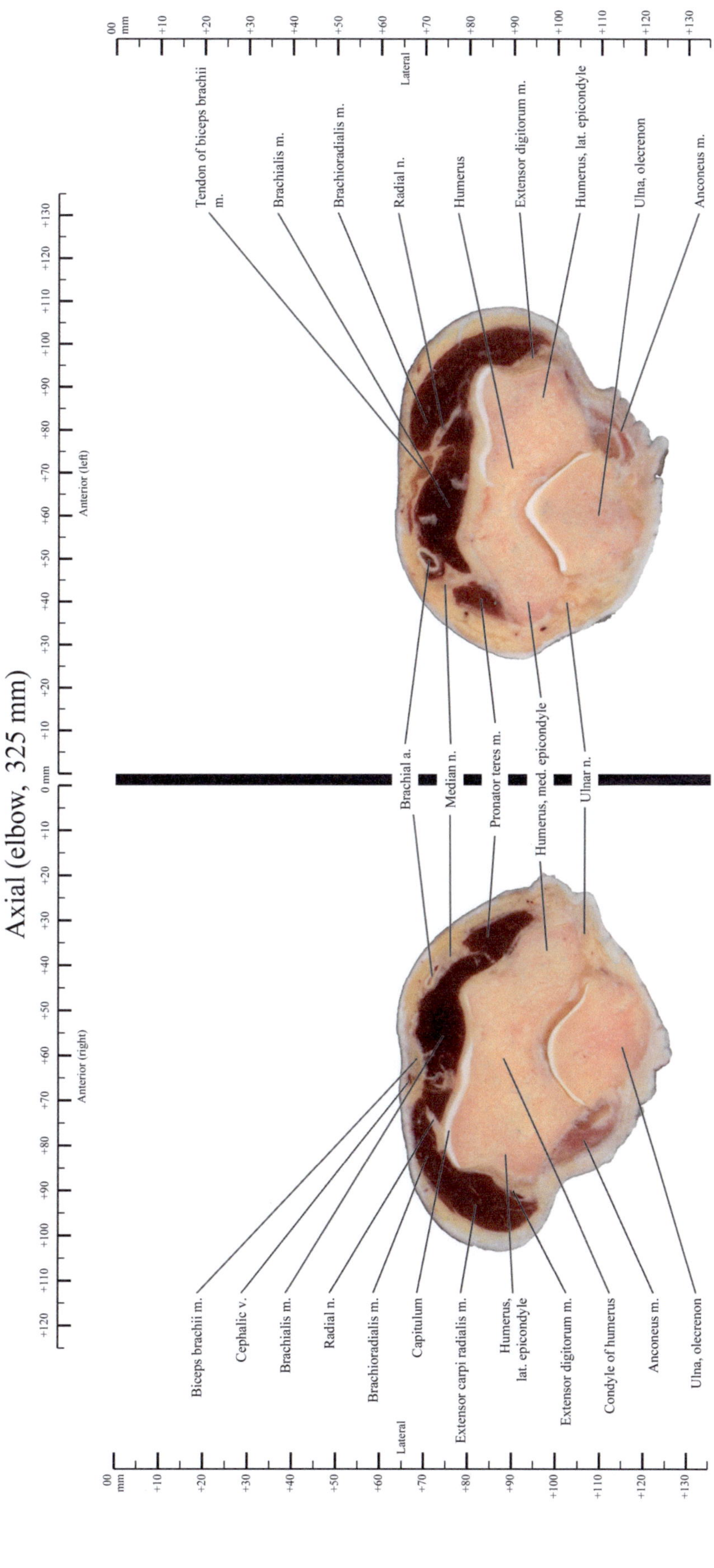
Axial (elbow, 325 mm)
Anterior (left)
Anterior (right)
Lateral
Tendon of biceps brachii m.
Brachialis m.
Brachioradialis m.
Radial n.
Humerus
Extensor digitorum m.
Humerus, lat. epicondyle
Ulna, olecrenon
Anconeus m.
Brachial a.
Median n.
Pronator teres m.
Humerus, med. epicondyle
Ulnar n.
Biceps brachii m.
Cephalic v.
Brachialis m.
Radial n.
Brachioradialis m.
Capitulum
Extensor carpi radialis m.
Humerus, lat. epicondyle
Extensor digitorum m.
Condyle of humerus
Anconeus m.
Ulna, olecrenon

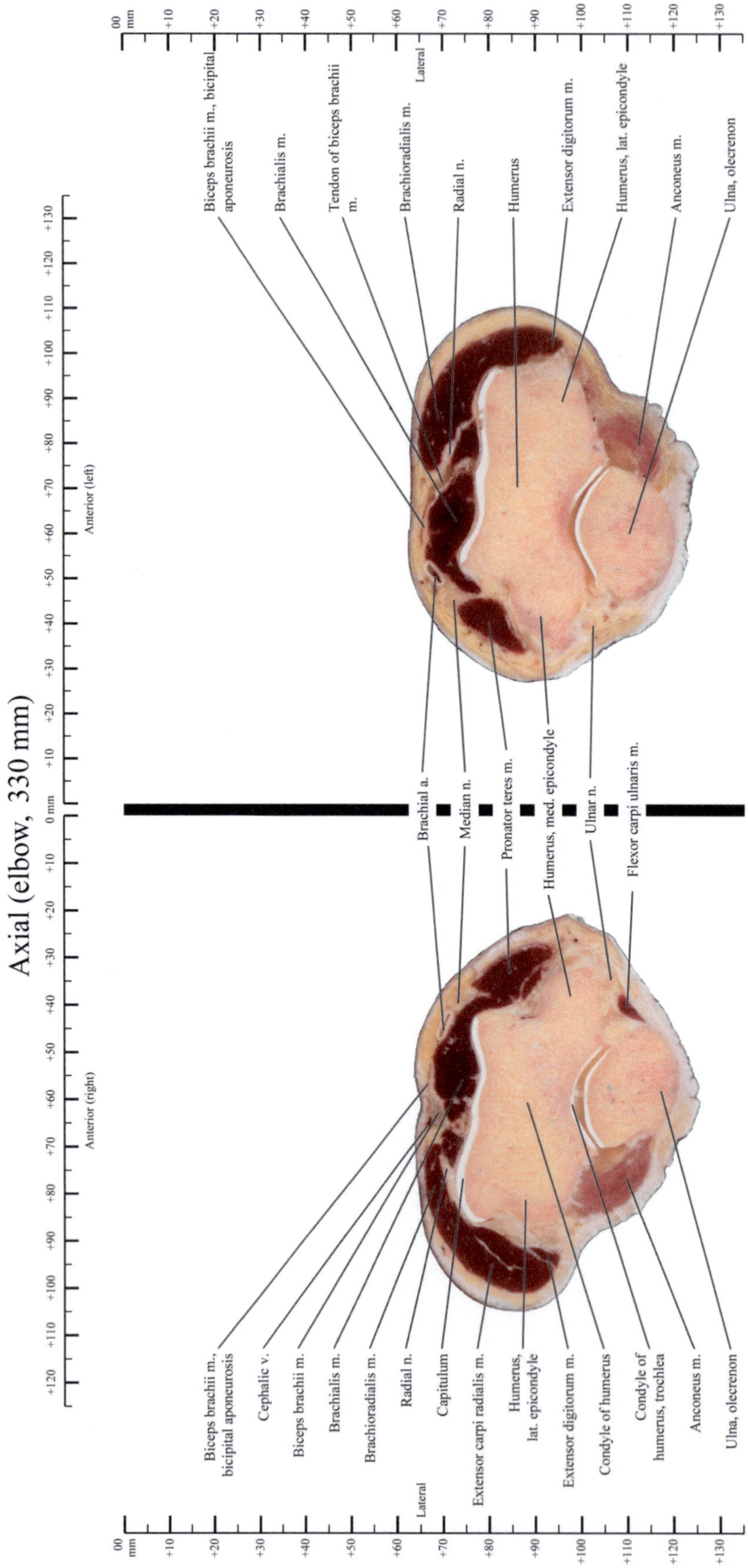
Axial (elbow, 330 mm)
Anterior (left)
Anterior (right)
Lateral
Lateral
Biceps brachii m., bicipital aponeurosis
Brachialis m.
Tendon of biceps brachii m.
Brachioradialis m.
Radial n.
Humerus
Extensor digitorum m.
Humerus, lat. epicondyle
Anconeus m.
Ulna, olecrenon
Brachial a.
Median n.
Pronator teres m.
Humerus, med. epicondyle
Ulnar n.
Flexor carpi ulnaris m.
Biceps brachii m., bicipital aponeurosis
Cephalic v.
Biceps brachii m.
Brachialis m.
Brachioradialis m.
Radial n.
Capitulum
Extensor carpi radialis m.
Humerus, lat. epicondyle
Extensor digitorum m.
Condyle of humerus
Condyle of humerus, trochlea
Anconeus m.
Ulna, olecrenon
0 mm
+10
+20
+30
+40
+50
+60
+70
+80
+90
+100
+110
+120
+130
00 mm

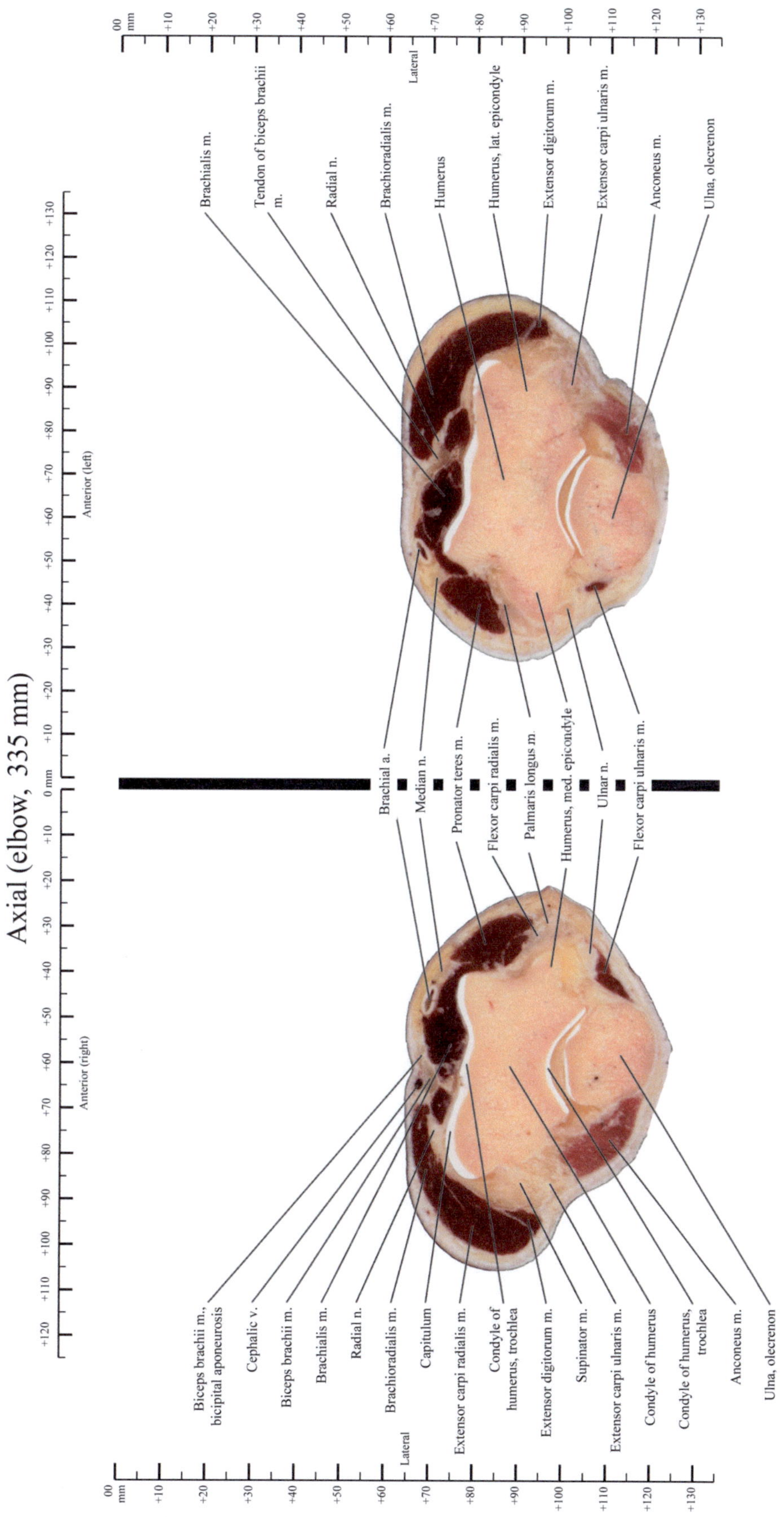

Axial (elbow, 335 mm)
Anterior (left)
Anterior (right)
Lateral
Brachialis m.
Tendon of biceps brachii m.
Radial n.
Brachioradialis m.
Humerus
Humerus, lat. epicondyle
Extensor digitorum m.
Extensor carpi ulnaris m.
Anconeus m.
Ulna, olecrenon
Brachial a.
Median n.
Pronator teres m.
Flexor carpi radialis m.
Palmaris longus m.
Humerus, med. epicondyle
Ulnar n.
Flexor carpi ulnaris m.
Biceps brachii m., bicipital aponeurosis
Cephalic v.
Biceps brachii m.
Brachialis m.
Radial n.
Brachioradialis m.
Capitulum
Extensor carpi radialis m.
Condyle of humerus, trochlea
Extensor digitorum m.
Supinator m.
Extensor carpi ulnaris m.
Condyle of humerus
Condyle of humerus, trochlea
Anconeus m.
Ulna, olecrenon

Axial (elbow, 340 mm)

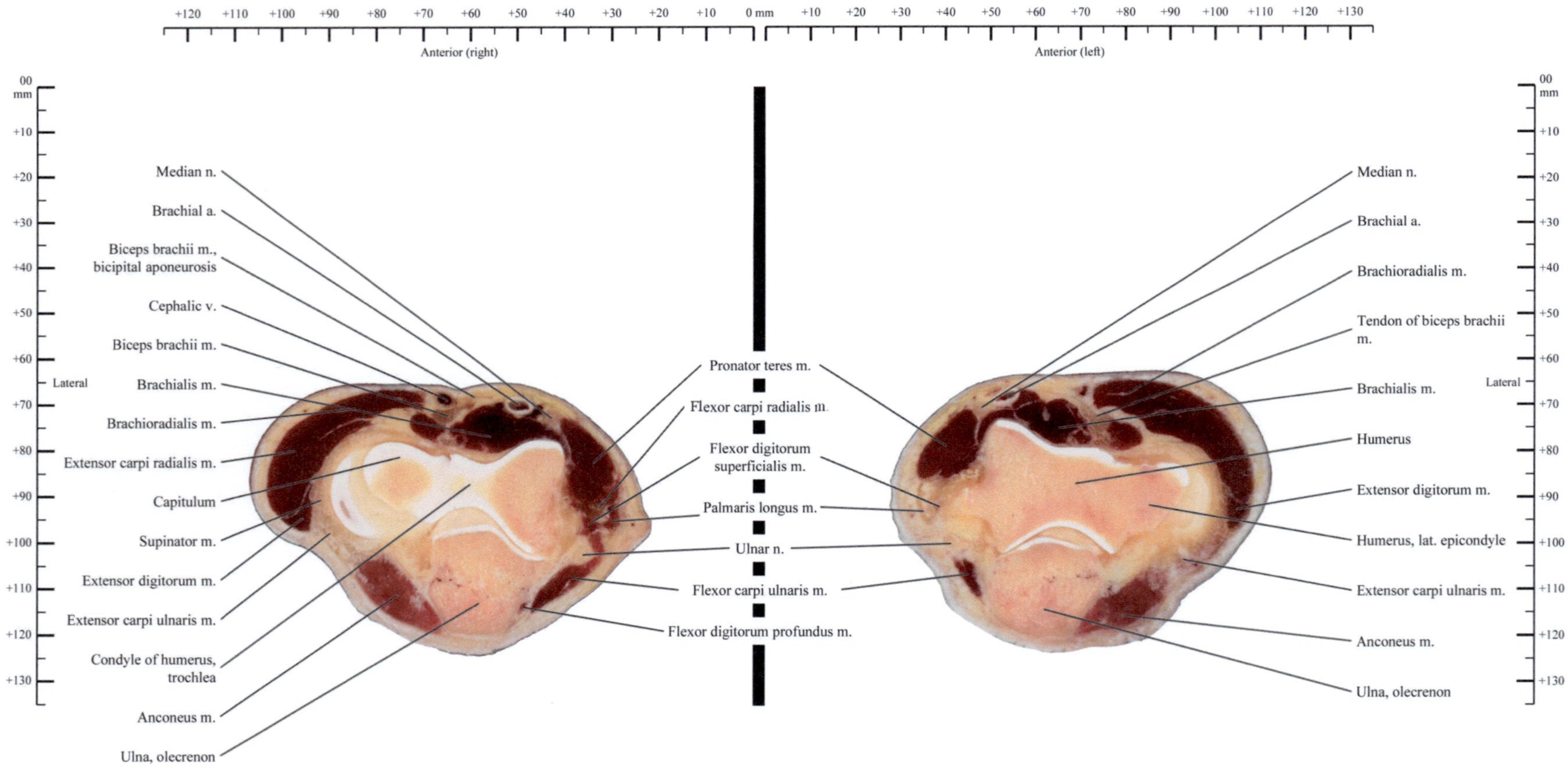

Axial (elbow, 345 mm)

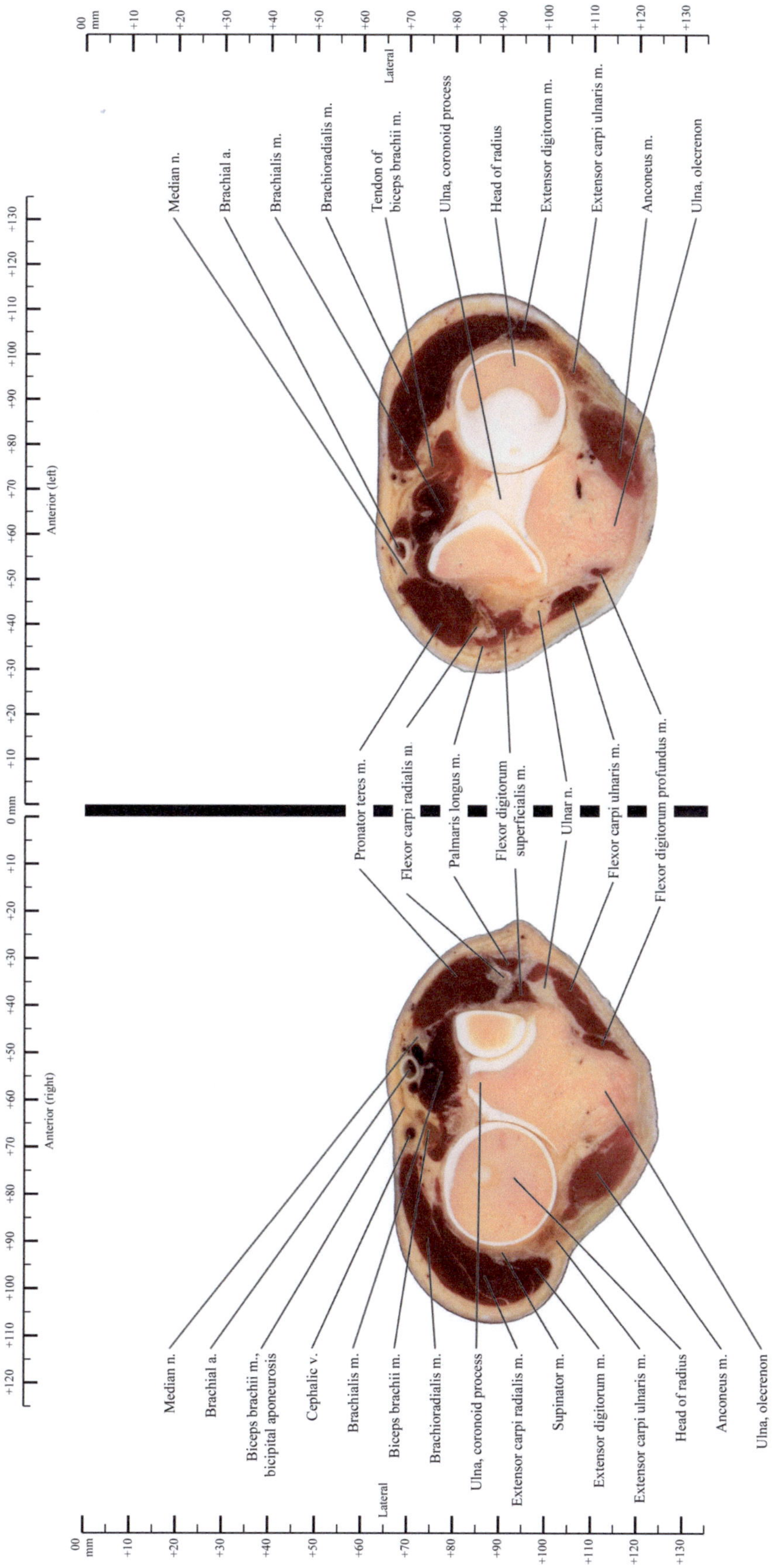

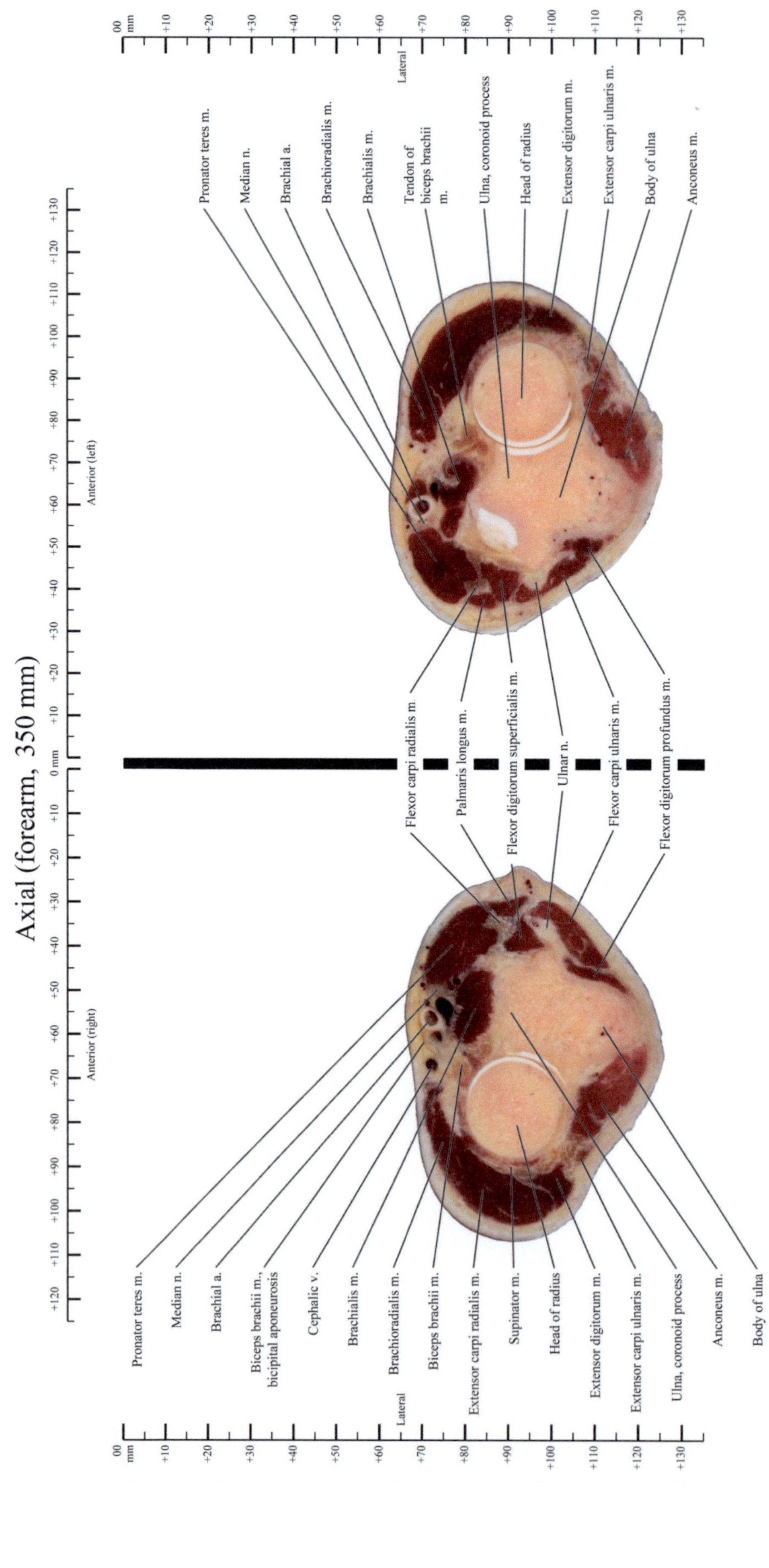
Axial (forearm, 350 mm)
Anterior (left)
Anterior (right)
Lateral
Lateral
0 mm
Pronator teres m.
Median n.
Brachial a.
Brachioradialis m.
Brachialis m.
Tendon of biceps brachii m.
Ulna, coronoid process
Head of radius
Extensor digitorum m.
Extensor carpi ulnaris m.
Body of ulna
Anconeus m.
Flexor carpi radialis m
Palmaris longus m.
Flexor digitorum superficialis m.
Ulnar n.
Flexor carpi ulnaris m.
Flexor digitorum profundus m.
Pronator teres m.
Median n.
Brachial a.
Biceps brachii m., bicipital aponeurosis
Cephalic v.
Brachialis m.
Brachioradialis m.
Biceps brachii m.
Extensor carpi radialis m.
Supinator m.
Head of radius
Extensor digitorum m.
Extensor carpi ulnaris m.
Ulna, coronoid process
Anconeus m.
Body of ulna

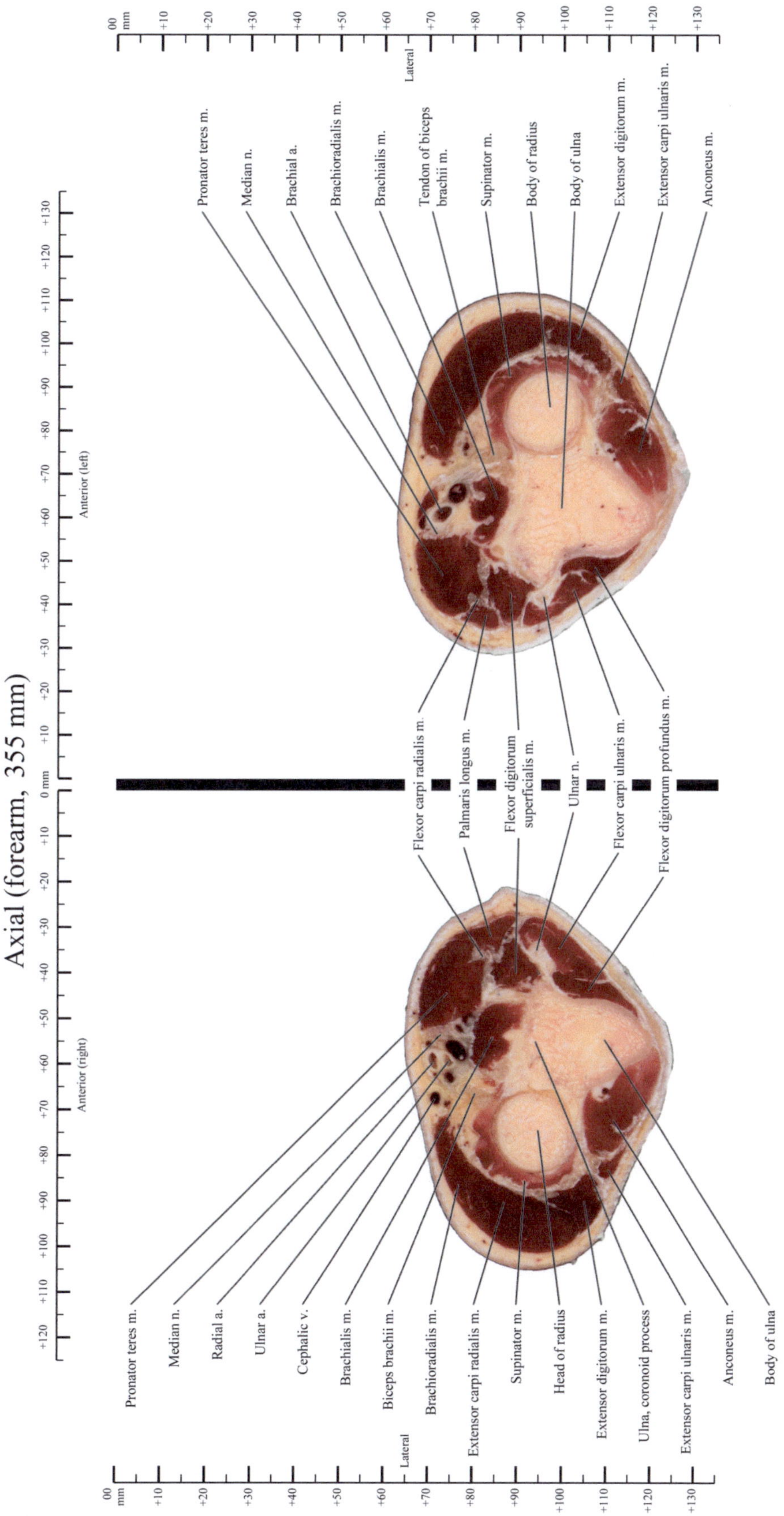
Axial (forearm, 355 mm)
Anterior (left)
Anterior (right)
Lateral
Pronator teres m.
Median n.
Brachial a.
Brachioradialis m.
Brachialis m.
Tendon of biceps brachii m.
Supinator m.
Body of radius
Body of ulna
Extensor digitorum m.
Extensor carpi ulnaris m.
Anconeus m.
Flexor carpi radialis m.
Palmaris longus m.
Flexor digitorum superficialis m.
Ulnar n.
Flexor carpi ulnaris m.
Flexor digitorum profundus m.
Pronator teres m.
Median n.
Radial a.
Ulnar a.
Cephalic v.
Brachialis m.
Biceps brachii m.
Brachioradialis m.
Extensor carpi radialis m.
Supinator m.
Head of radius
Extensor digitorum m.
Ulna, coronoid process
Extensor carpi ulnaris m.
Anconeus m.
Body of ulna

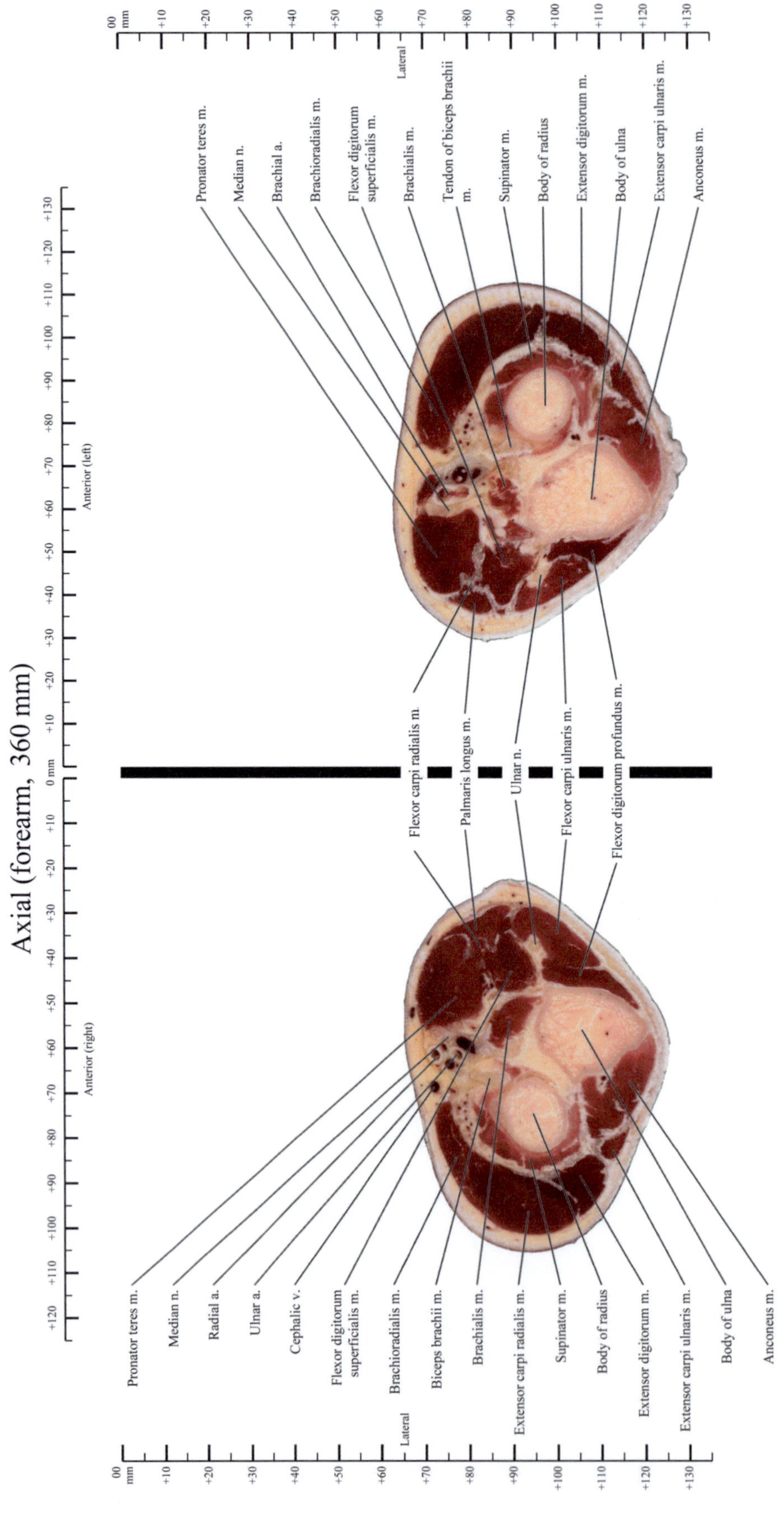
Axial (forearm, 360 mm)
Anterior (left)
Anterior (right)
Lateral
Pronator teres m.
Median n.
Brachial a.
Brachioradialis m.
Flexor digitorum superficialis m.
Brachialis m.
Tendon of biceps brachii m.
Supinator m.
Body of radius
Extensor digitorum m.
Body of ulna
Extensor carpi ulnaris m.
Anconeus m.
Flexor carpi radialis m.
Palmaris longus m.
Ulnar n.
Flexor carpi ulnaris m.
Flexor digitorum profundus m.
Radial a.
Ulnar a.
Cephalic v.
Biceps brachii m.
Extensor carpi radialis m.

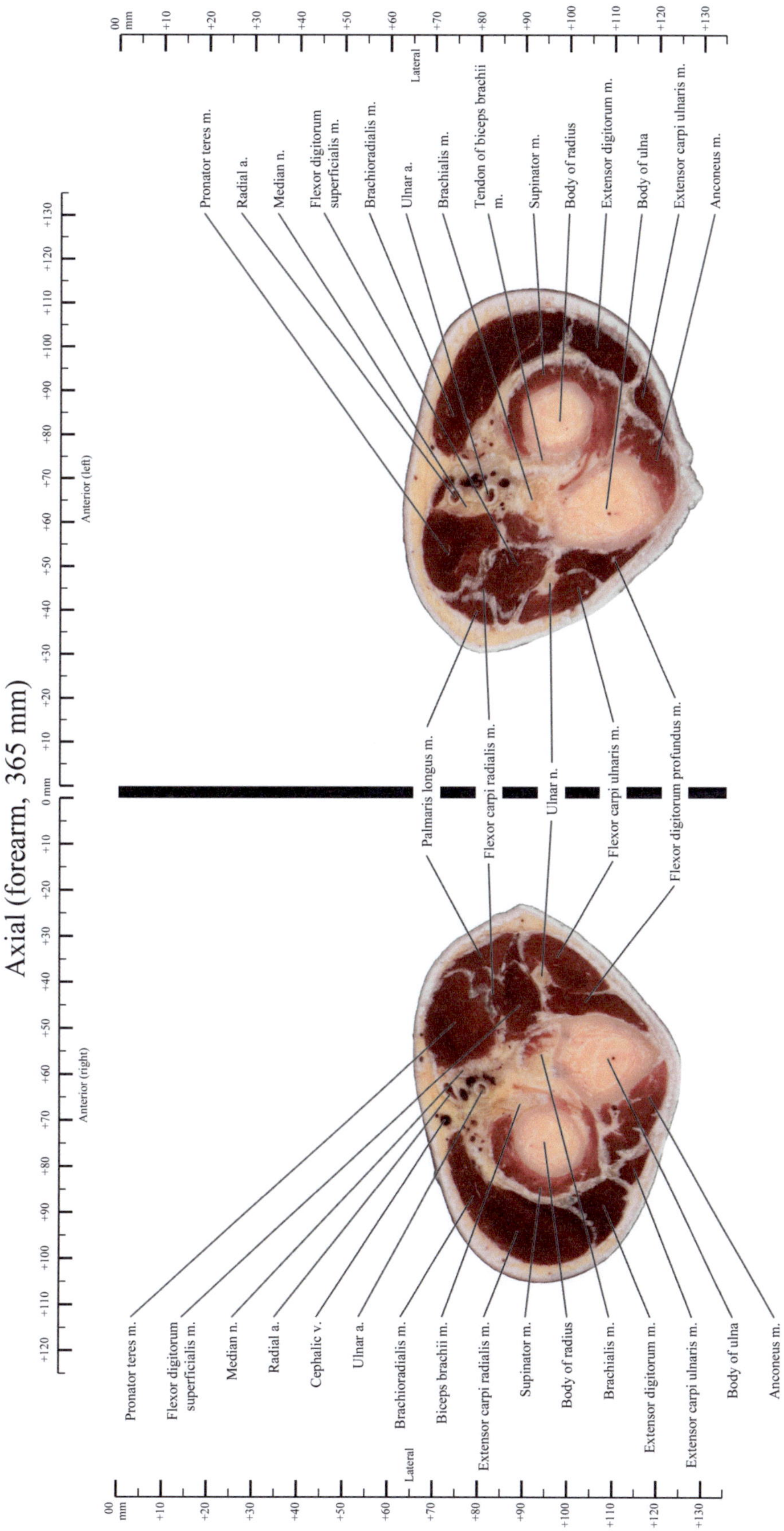
Axial (forearm, 365 mm)
Anterior (left)
Anterior (right)
Lateral
Pronator teres m.
Radial a.
Median n.
Flexor digitorum superficialis m.
Brachioradialis m.
Ulnar a.
Brachialis m.
Tendon of biceps brachii m.
Supinator m.
Body of radius
Extensor digitorum m.
Body of ulna
Extensor carpi ulnaris m.
Anconeus m.
Palmaris longus m.
Flexor carpi radialis m.
Ulnar n.
Flexor carpi ulnaris m.
Flexor digitorum profundus m.
Pronator teres m.
Flexor digitorum superficialis m.
Median n.
Radial a.
Cephalic v.
Ulnar a.
Brachioradialis m.
Biceps brachii m.
Extensor carpi radialis m.
Supinator m.
Body of radius
Brachialis m.
Extensor digitorum m.
Extensor carpi ulnaris m.
Body of ulna
Anconeus m.

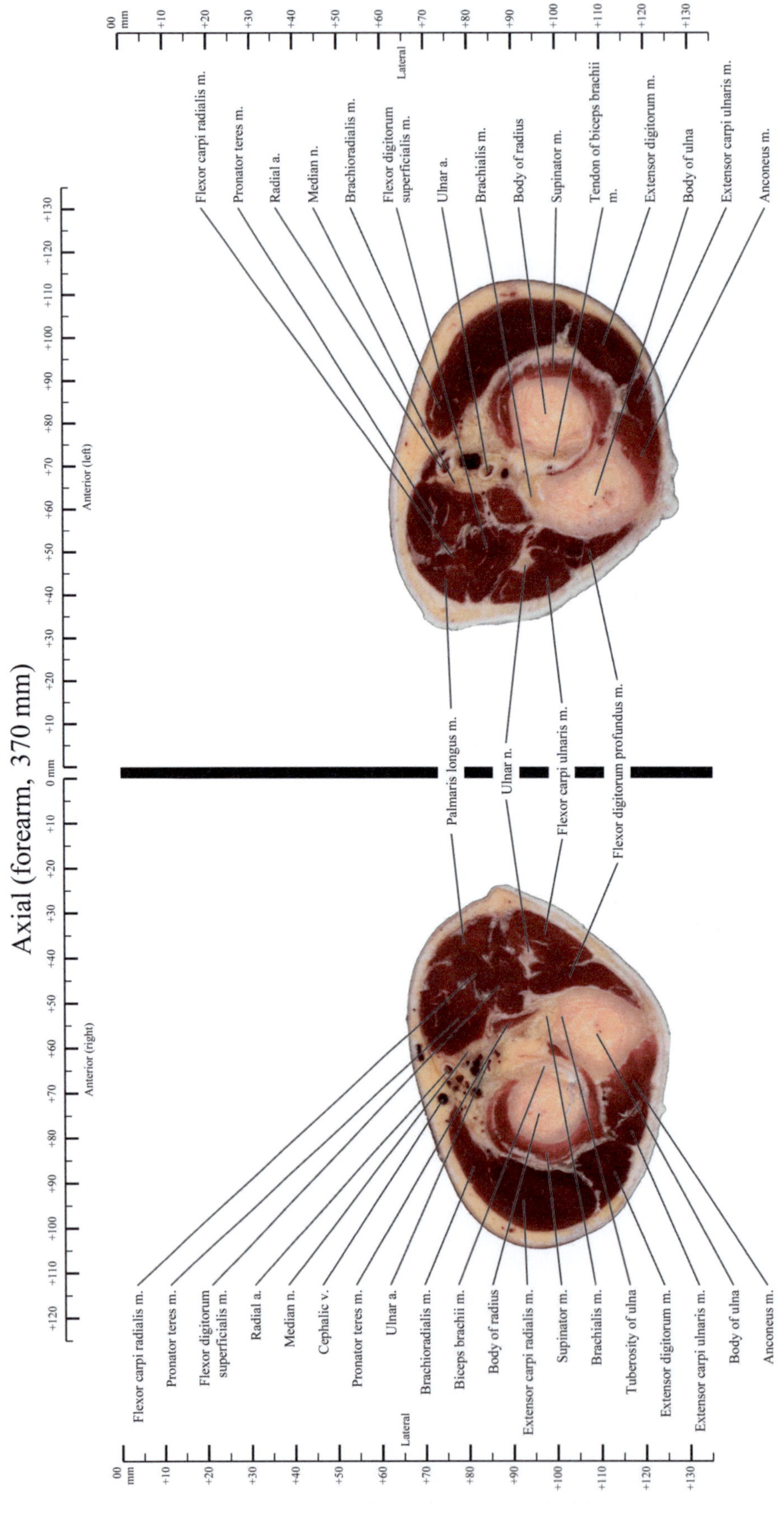
Axial (forearm, 370 mm)
Anterior (left)
Anterior (right)
Lateral
Lateral
Flexor carpi radialis m.
Pronator teres m.
Radial a.
Median n.
Brachioradialis m.
Flexor digitorum superficialis m.
Ulnar a.
Brachialis m.
Body of radius
Supinator m.
Tendon of biceps brachii m.
Extensor digitorum m.
Body of ulna
Extensor carpi ulnaris m.
Anconeus m.
Palmaris longus m.
Ulnar n.
Flexor carpi ulnaris m.
Flexor digitorum profundus m.
Flexor carpi radialis m.
Pronator teres m.
Flexor digitorum superficialis m.
Radial a.
Median n.
Cephalic v.
Pronator teres m.
Ulnar a.
Brachioradialis m.
Biceps brachii m.
Body of radius
Extensor carpi radialis m.
Supinator m.
Brachialis m.
Tuberosity of ulna
Extensor digitorum m.
Extensor carpi ulnaris m.
Body of ulna
Anconeus m.

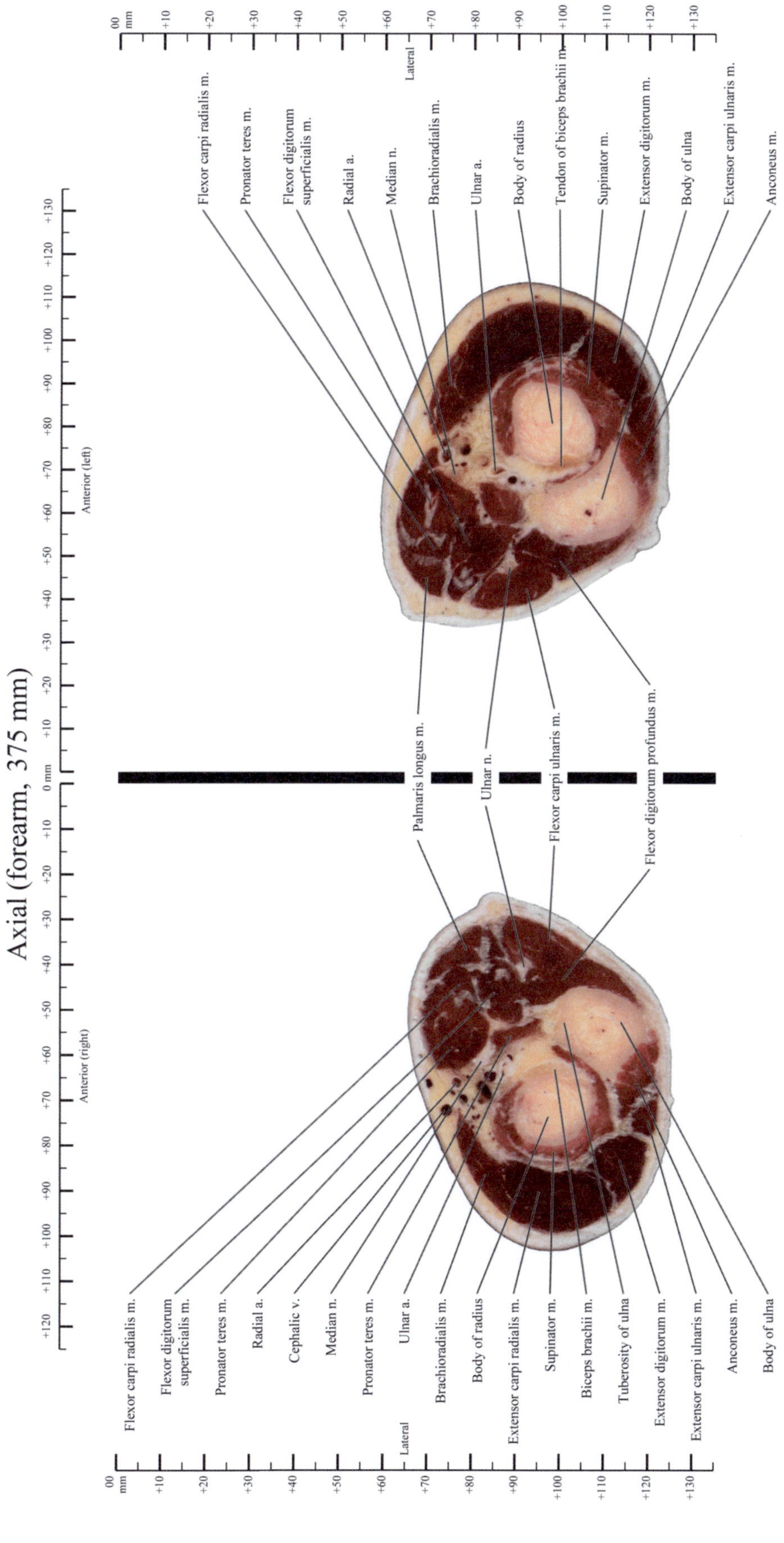

Axial (forearm, 375 mm)
Anterior (left)
Anterior (right)
Lateral
Flexor carpi radialis m.
Pronator teres m.
Flexor digitorum superficialis m.
Radial a.
Median n.
Brachioradialis m.
Ulnar a.
Body of radius
Tendon of biceps brachii m.
Supinator m.
Extensor digitorum m.
Body of ulna
Extensor carpi ulnaris m.
Anconeus m.
Palmaris longus m.
Ulnar n.
Flexor carpi ulnaris m.
Flexor digitorum profundus m.
Flexor carpi radialis m.
Flexor digitorum superficialis m.
Pronator teres m.
Radial a.
Cephalic v.
Median n.
Pronator teres m.
Ulnar a.
Brachioradialis m.
Body of radius
Extensor carpi radialis m.
Supinator m.
Biceps brachii m.
Tuberosity of ulna
Extensor digitorum m.
Extensor carpi ulnaris m.
Anconeus m.
Body of ulna

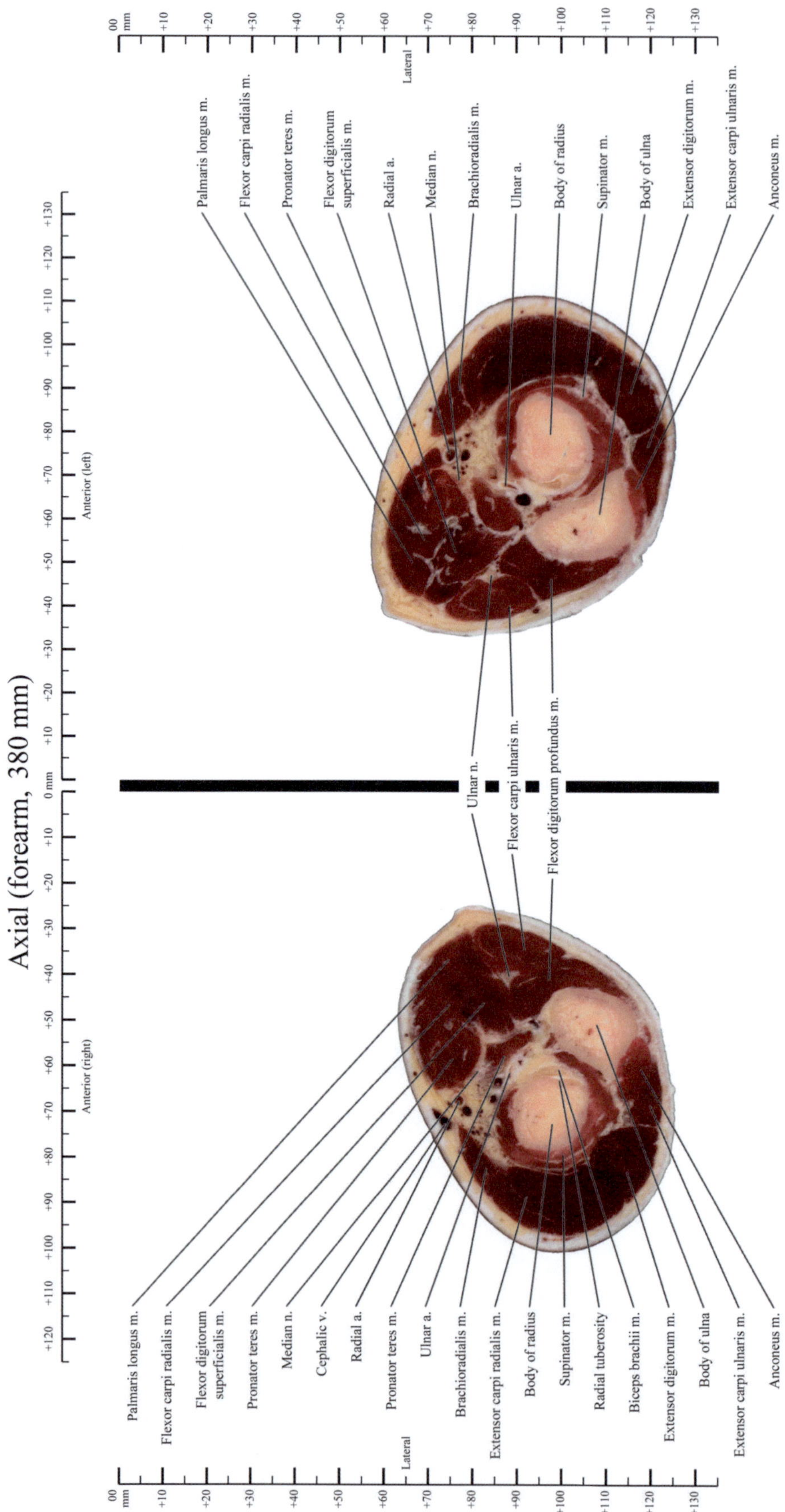
Axial (forearm, 380 mm)
Anterior (left)
Anterior (right)
Lateral
Lateral
Palmaris longus m.
Flexor carpi radialis m.
Pronator teres m.
Flexor digitorum superficialis m.
Radial a.
Median n.
Brachioradialis m.
Ulnar a.
Body of radius
Supinator m.
Body of ulna
Extensor digitorum m.
Extensor carpi ulnaris m.
Anconeus m.
Ulnar n.
Flexor carpi ulnaris m.
Flexor digitorum profundus m.
Palmaris longus m.
Flexor carpi radialis m.
Flexor digitorum superficialis m.
Pronator teres m.
Median n.
Cephalic v.
Radial a.
Pronator teres m.
Ulnar a.
Brachioradialis m.
Extensor carpi radialis m.
Body of radius
Supinator m.
Radial tuberosity
Biceps brachii m.
Extensor digitorum m.
Body of ulna
Extensor carpi ulnaris m.
Anconeus m.

Axial (forearm, 385 mm)

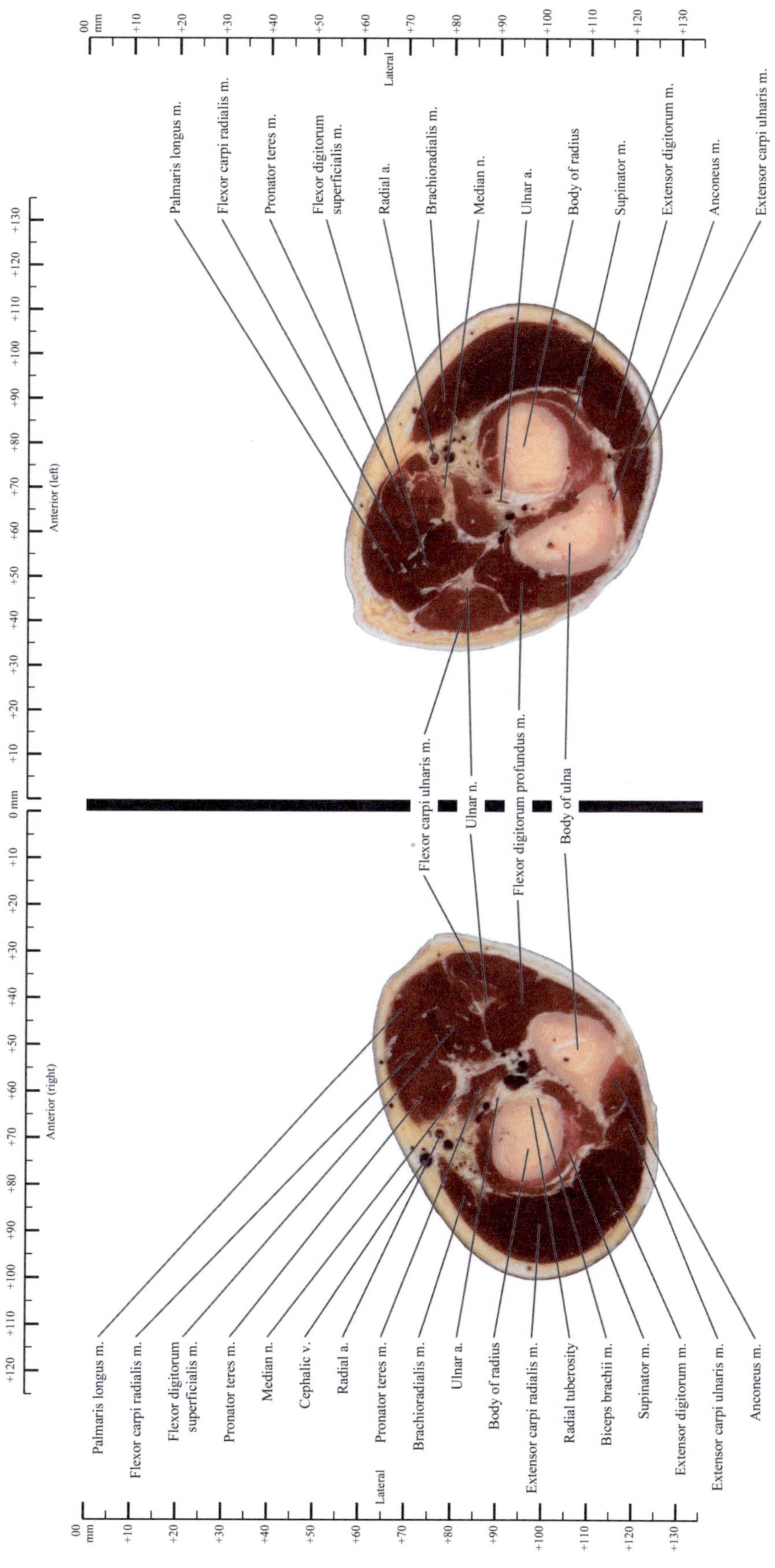

Axial (forearm, 390 mm)

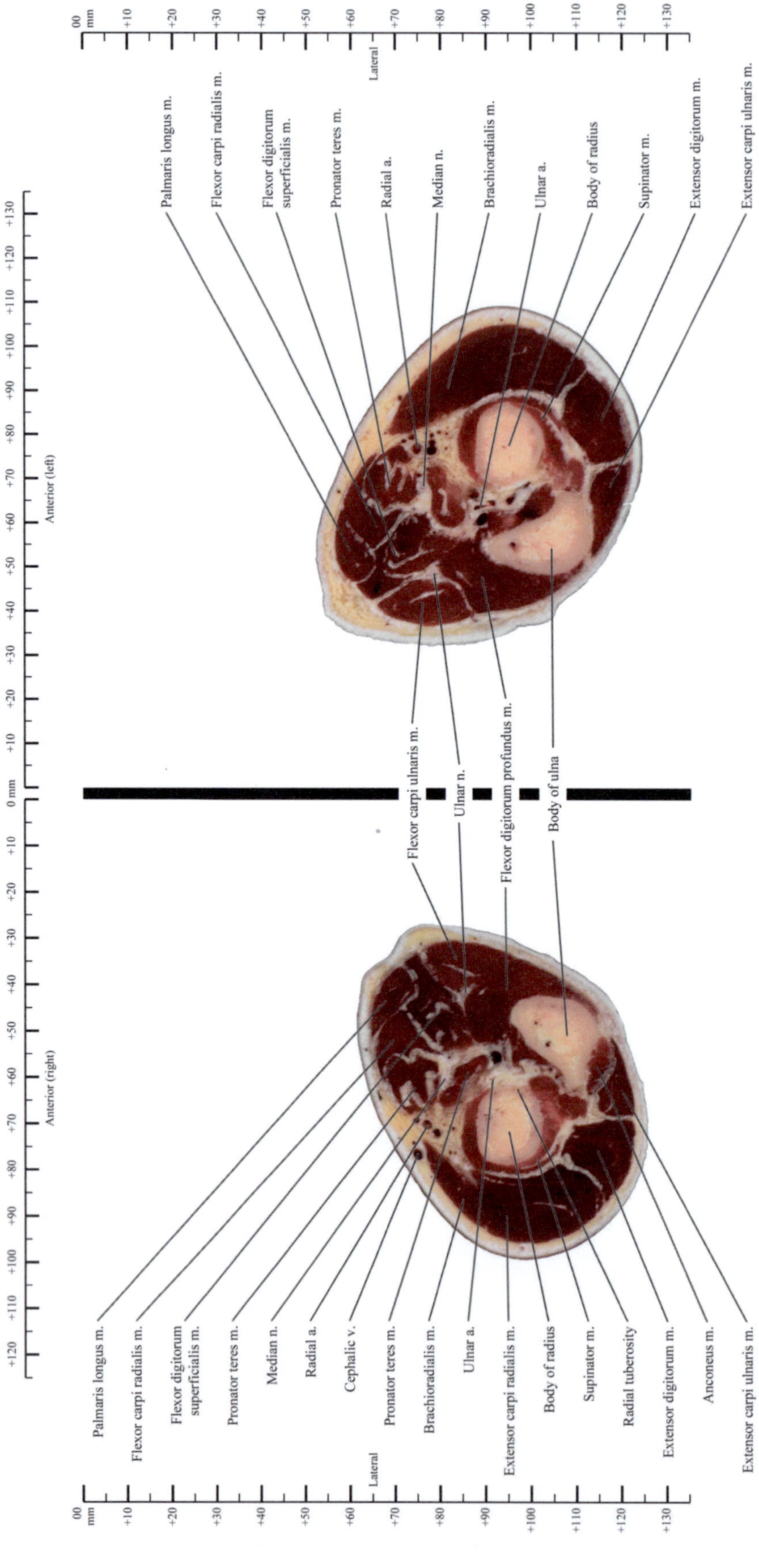

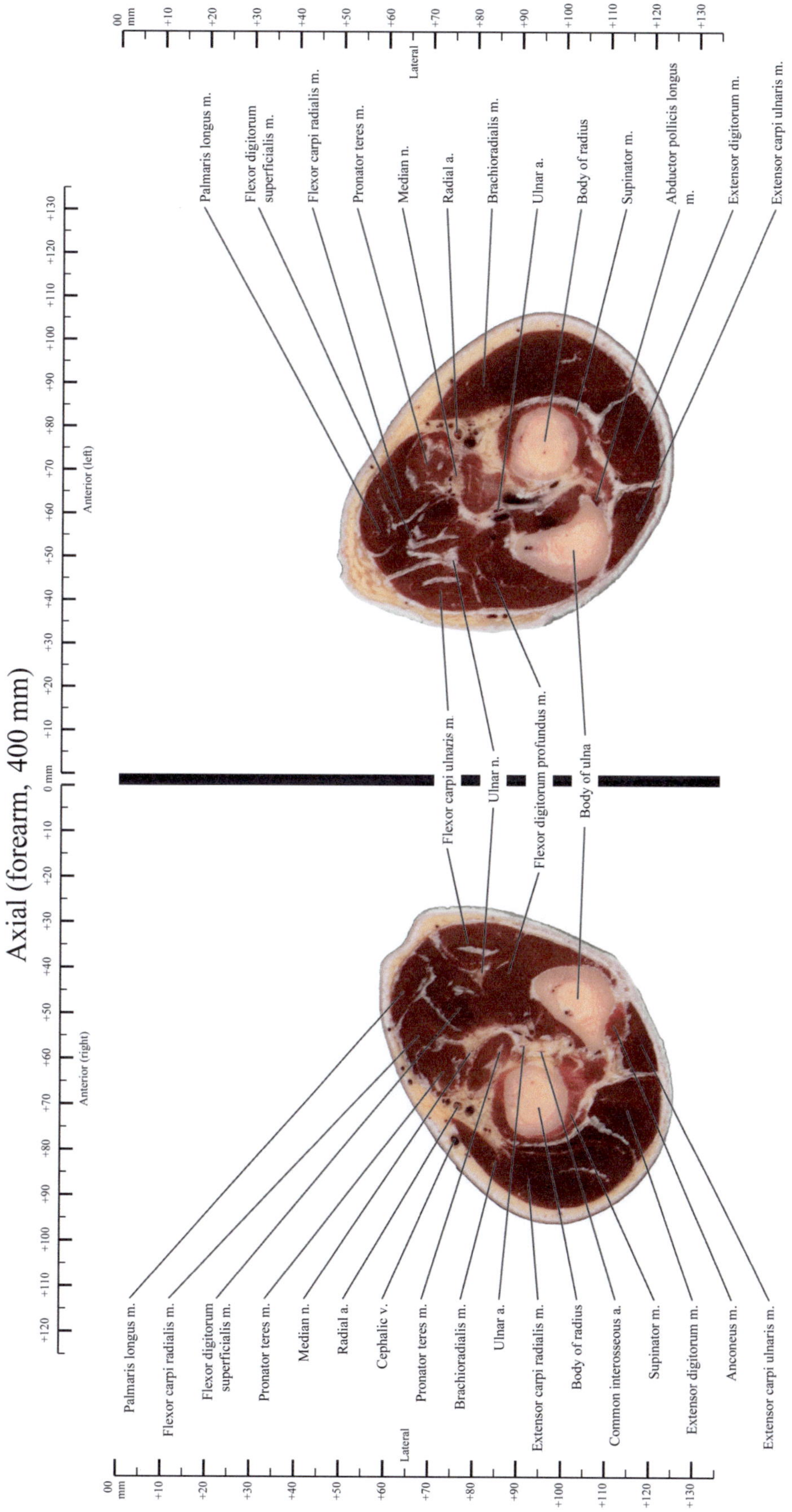

Axial (forearm, 400 mm)
Anterior (left)
Anterior (right)
Lateral
Lateral
Palmaris longus m.
Flexor digitorum superficialis m.
Flexor carpi radialis m.
Pronator teres m.
Median n.
Radial a.
Brachioradialis m.
Ulnar a.
Body of radius
Supinator m.
Abductor pollicis longus m.
Extensor digitorum m.
Extensor carpi ulnaris m.
Flexor carpi ulnaris m.
Ulnar n.
Flexor digitorum profundus m.
Body of ulna
Palmaris longus m.
Flexor carpi radialis m.
Flexor digitorum superficialis m.
Pronator teres m.
Median n.
Radial a.
Cephalic v.
Pronator teres m.
Brachioradialis m.
Ulnar a.
Extensor carpi radialis m.
Body of radius
Common interosseus a.
Supinator m.
Extensor digitorum m.
Anconeus m.
Extensor carpi ulnaris m.

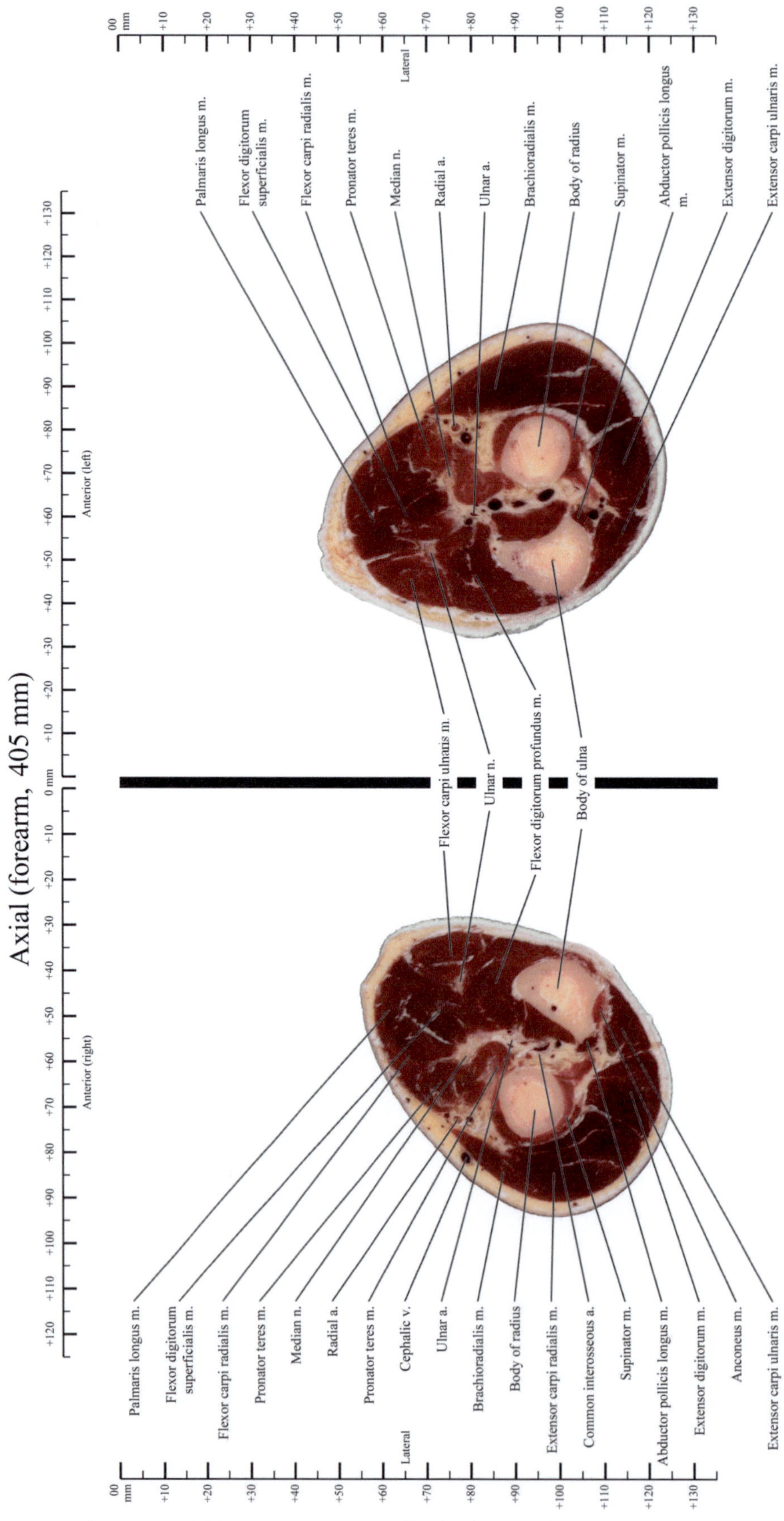
Axial (forearm, 405 mm)
Anterior (left)
Anterior (right)
Lateral
Palmaris longus m.
Flexor digitorum superficialis m.
Flexor carpi radialis m.
Pronator teres m.
Median n.
Radial a.
Ulnar a.
Brachioradialis m.
Body of radius
Supinator m.
Abductor pollicis longus m.
Extensor digitorum m.
Extensor carpi ulnaris m.
Flexor carpi ulnaris m.
Ulnar n.
Flexor digitorum profundus m.
Body of ulna
Palmaris longus m.
Flexor digitorum superficialis m.
Flexor carpi radialis m.
Pronator teres m.
Median n.
Radial a.
Pronator teres m.
Cephalic v.
Ulnar a.
Brachioradialis m.
Body of radius
Extensor carpi radialis m.
Common interosseous a.
Supinator m.
Abductor pollicis longus m.
Extensor digitorum m.
Anconeus m.
Extensor carpi ulnaris m.

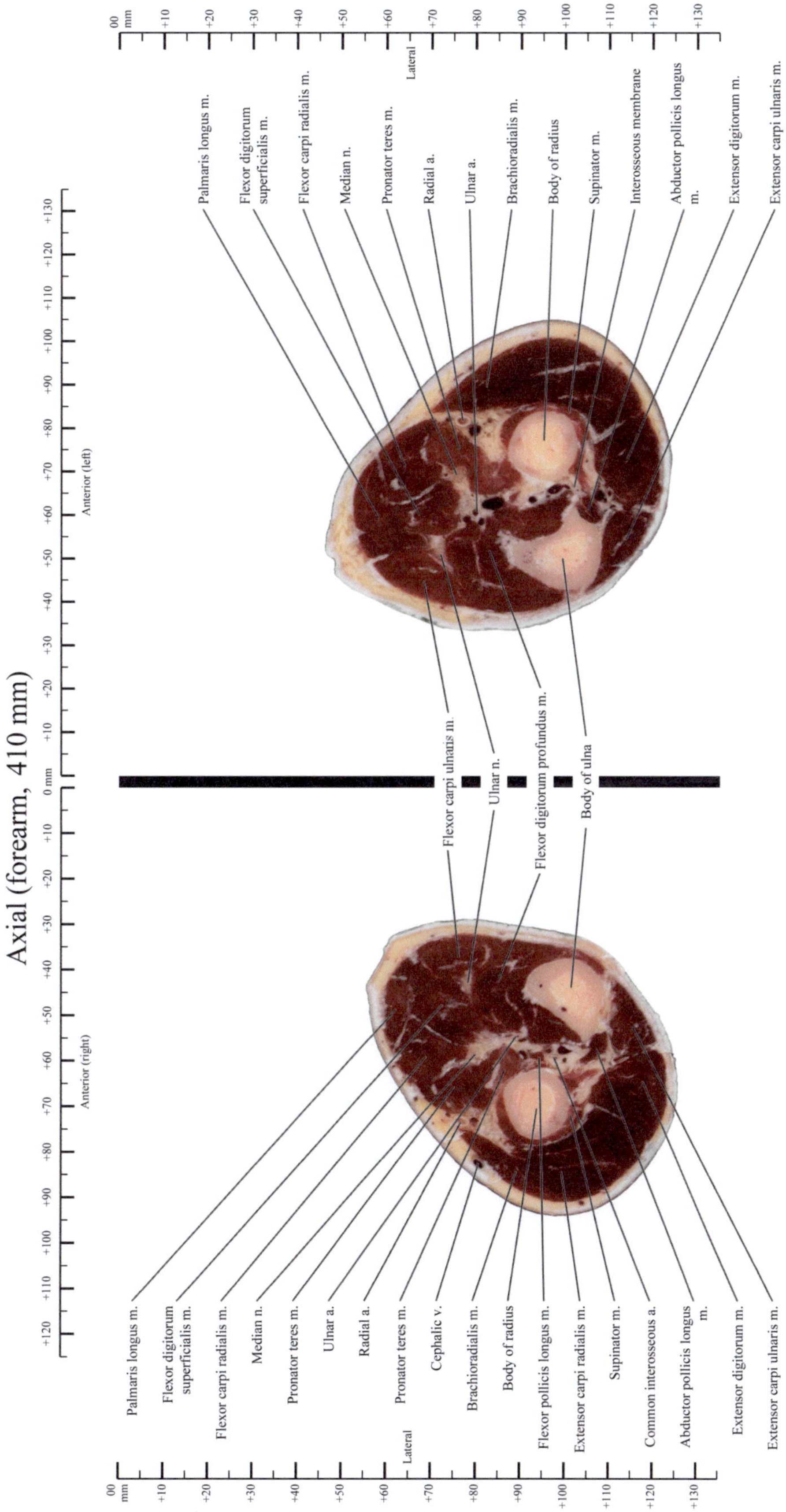
Axial (forearm, 410 mm)
Anterior (left)
Anterior (right)
Lateral
Palmaris longus m.
Flexor digitorum superficialis m.
Flexor carpi radialis m.
Median n.
Pronator teres m.
Radial a.
Ulnar a.
Brachioradialis m.
Body of radius
Supinator m.
Interosseous membrane
Abductor pollicis longus m.
Extensor digitorum m.
Extensor carpi ulnaris m.
Flexor carpi ulnaris m.
Ulnar n.
Flexor digitorum profundus m.
Body of ulna
Cephalic v.
Flexor pollicis longus m.
Extensor carpi radialis m.
Common interosseous a.

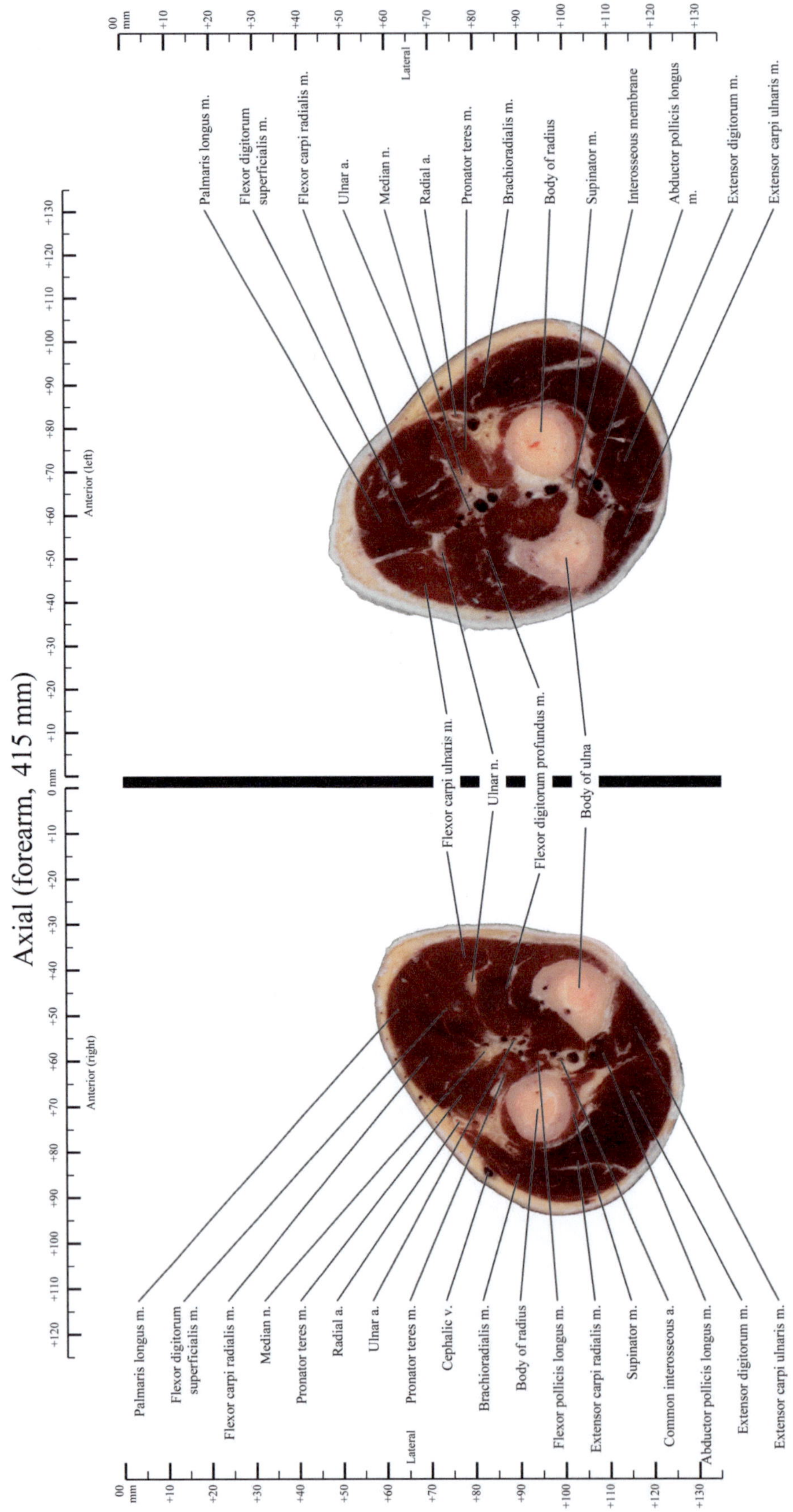
Axial (forearm, 415 mm)
Anterior (left)
Anterior (right)
Lateral
Lateral
Palmaris longus m.
Flexor digitorum superficialis m.
Flexor carpi radialis m.
Ulnar a.
Median n.
Radial a.
Pronator teres m.
Brachioradialis m.
Body of radius
Supinator m.
Interosseous membrane
Abductor pollicis longus m.
Extensor digitorum m.
Extensor carpi ulnaris m.
Flexor carpi ulnaris m.
Ulnar n.
Flexor digitorum profundus m.
Body of ulna
Palmaris longus m.
Flexor digitorum superficialis m.
Flexor carpi radialis m.
Median n.
Pronator teres m.
Radial a.
Ulnar a.
Pronator teres m.
Cephalic v.
Brachioradialis m.
Body of radius
Flexor pollicis longus m.
Extensor carpi radialis m.
Supinator m.
Common interosseous a.
Abductor pollicis longus m.
Extensor digitorum m.
Extensor carpi ulnaris m.

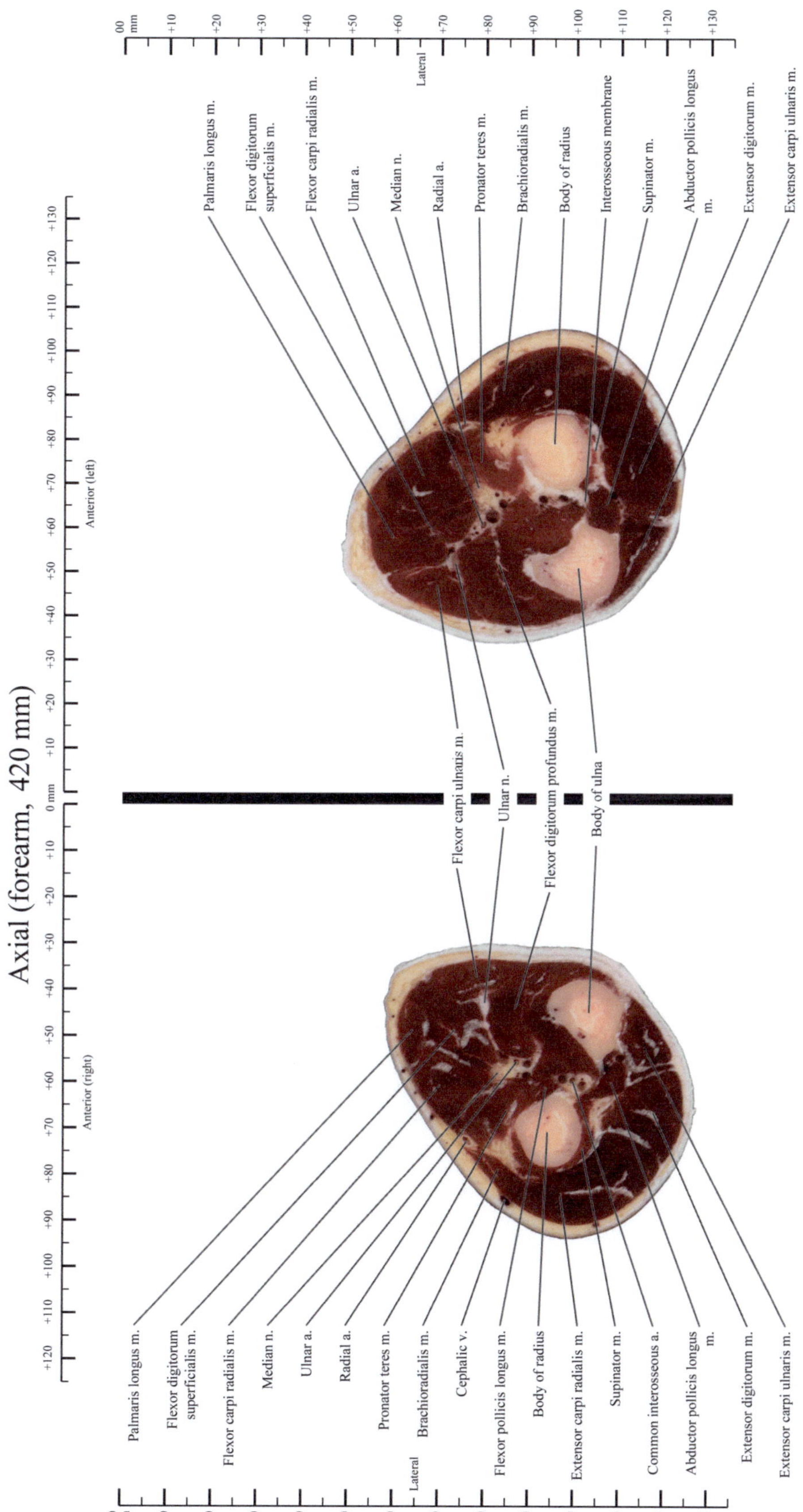
Axial (forearm, 420 mm)
Anterior (left)
Anterior (right)
Lateral
Palmaris longus m.
Flexor digitorum superficialis m.
Flexor carpi radialis m.
Ulnar a.
Median n.
Radial a.
Pronator teres m.
Brachioradialis m.
Body of radius
Interosseous membrane
Supinator m.
Abductor pollicis longus m.
Extensor digitorum m.
Extensor carpi ulnaris m.
Flexor carpi ulnaris m.
Ulnar n.
Flexor digitorum profundus m.
Body of ulna
Cephalic v.
Flexor pollicis longus m.
Extensor carpi radialis m.
Common interosseous a.

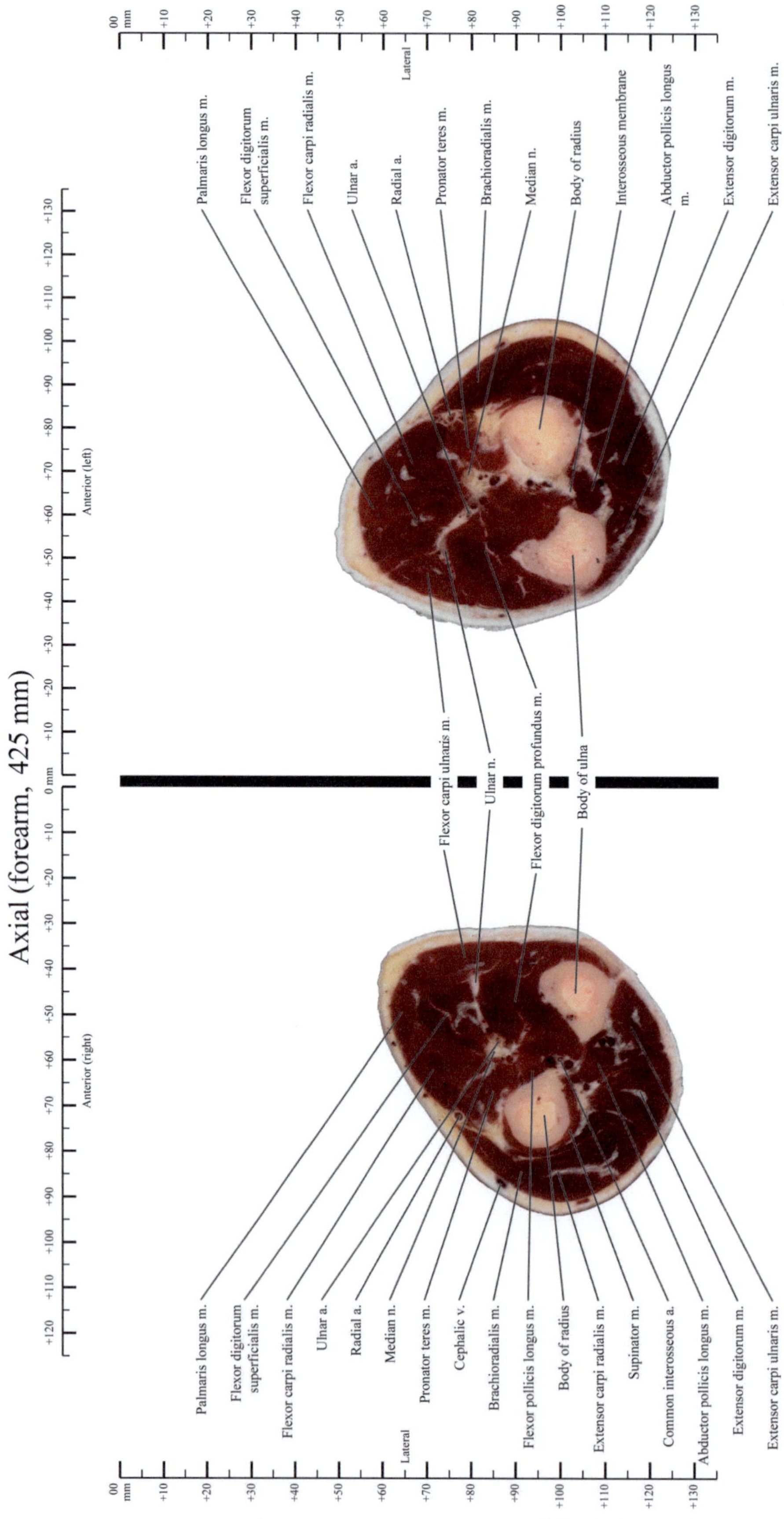
Axial (forearm, 425 mm)
Anterior (left)
Anterior (right)
Lateral
Palmaris longus m.
Flexor digitorum superficialis m.
Flexor carpi radialis m.
Ulnar a.
Radial a.
Pronator teres m.
Brachioradialis m.
Median n.
Body of radius
Interosseous membrane
Abductor pollicis longus m.
Extensor digitorum m.
Extensor carpi ulnaris m.
Flexor carpi ulnaris m.
Ulnar n.
Flexor digitorum profundus m.
Body of ulna
Lateral
Palmaris longus m.
Flexor digitorum superficialis m.
Flexor carpi radialis m.
Ulnar a.
Radial a.
Median n.
Pronator teres m.
Cephalic v.
Brachioradialis m.
Flexor pollicis longus m.
Body of radius
Extensor carpi radialis m.
Supinator m.
Common interosseous a.
Abductor pollicis longus m.
Extensor digitorum m.
Extensor carpi ulnaris m.

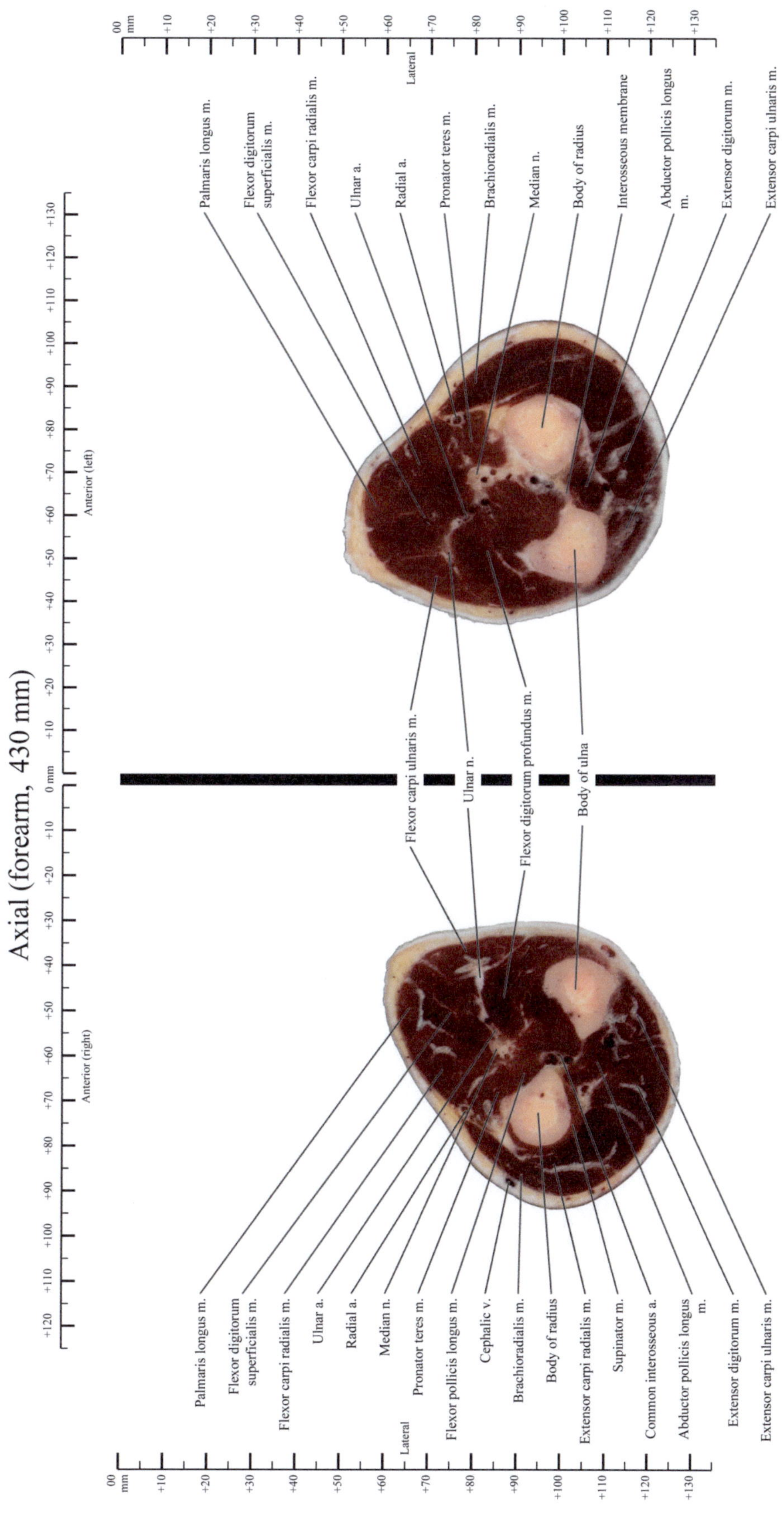
Axial (forearm, 430 mm)
Anterior (left)
Lateral
Palmaris longus m.
Flexor digitorum superficialis m.
Flexor carpi radialis m.
Ulnar a.
Radial a.
Pronator teres m.
Brachioradialis m.
Median n.
Body of radius
Interosseous membrane
Abductor pollicis longus m.
Extensor digitorum m.
Extensor carpi ulnaris m.
Flexor carpi ulnaris m.
Ulnar n.
Flexor digitorum profundus m.
Body of ulna
Anterior (right)
Palmaris longus m.
Flexor digitorum superficialis m.
Flexor carpi radialis m.
Ulnar a.
Radial a.
Median n.
Pronator teres m.
Flexor pollicis longus m.
Cephalic v.
Brachioradialis m.
Body of radius
Extensor carpi radialis m.
Supinator m.
Common interosseous a.
Abductor pollicis longus m.
Extensor digitorum m.
Extensor carpi ulnaris m.
Lateral

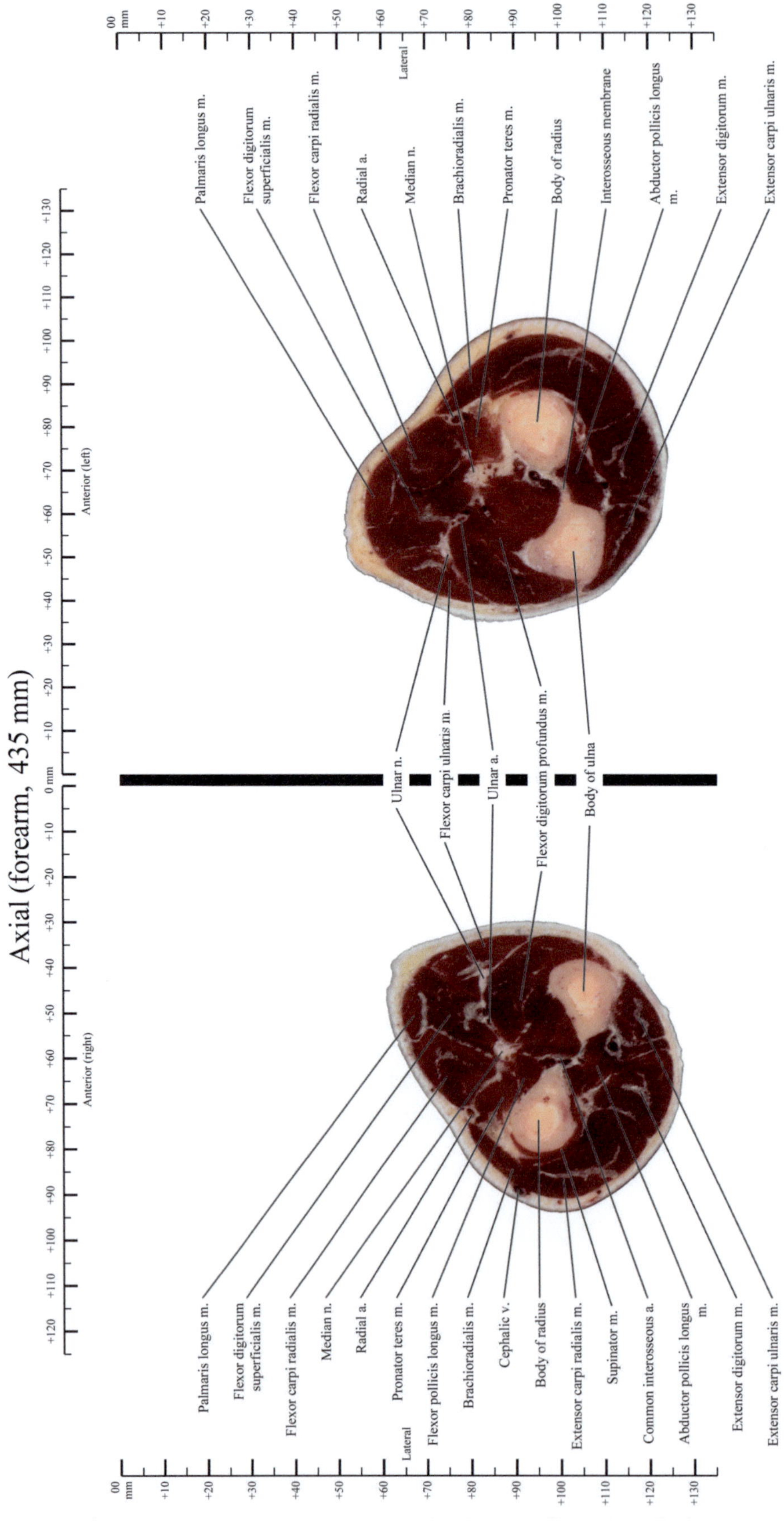
Axial (forearm, 435 mm)
Anterior (left)
Anterior (right)
Lateral
Palmaris longus m.
Flexor digitorum superficialis m.
Flexor carpi radialis m.
Radial a.
Median n.
Brachioradialis m.
Pronator teres m.
Body of radius
Interosseous membrane
Abductor pollicis longus m.
Extensor digitorum m.
Extensor carpi ulnaris m.
Ulnar n.
Flexor carpi ulnaris m.
Ulnar a.
Flexor digitorum profundus m.
Body of ulna
Palmaris longus m.
Flexor digitorum superficialis m.
Flexor carpi radialis m.
Median n.
Radial a.
Pronator teres m.
Flexor pollicis longus m.
Brachioradialis m.
Cephalic v.
Body of radius
Extensor carpi radialis m.
Supinator m.
Common interosseous a.
Abductor pollicis longus m.
Extensor digitorum m.
Extensor carpi ulnaris m.

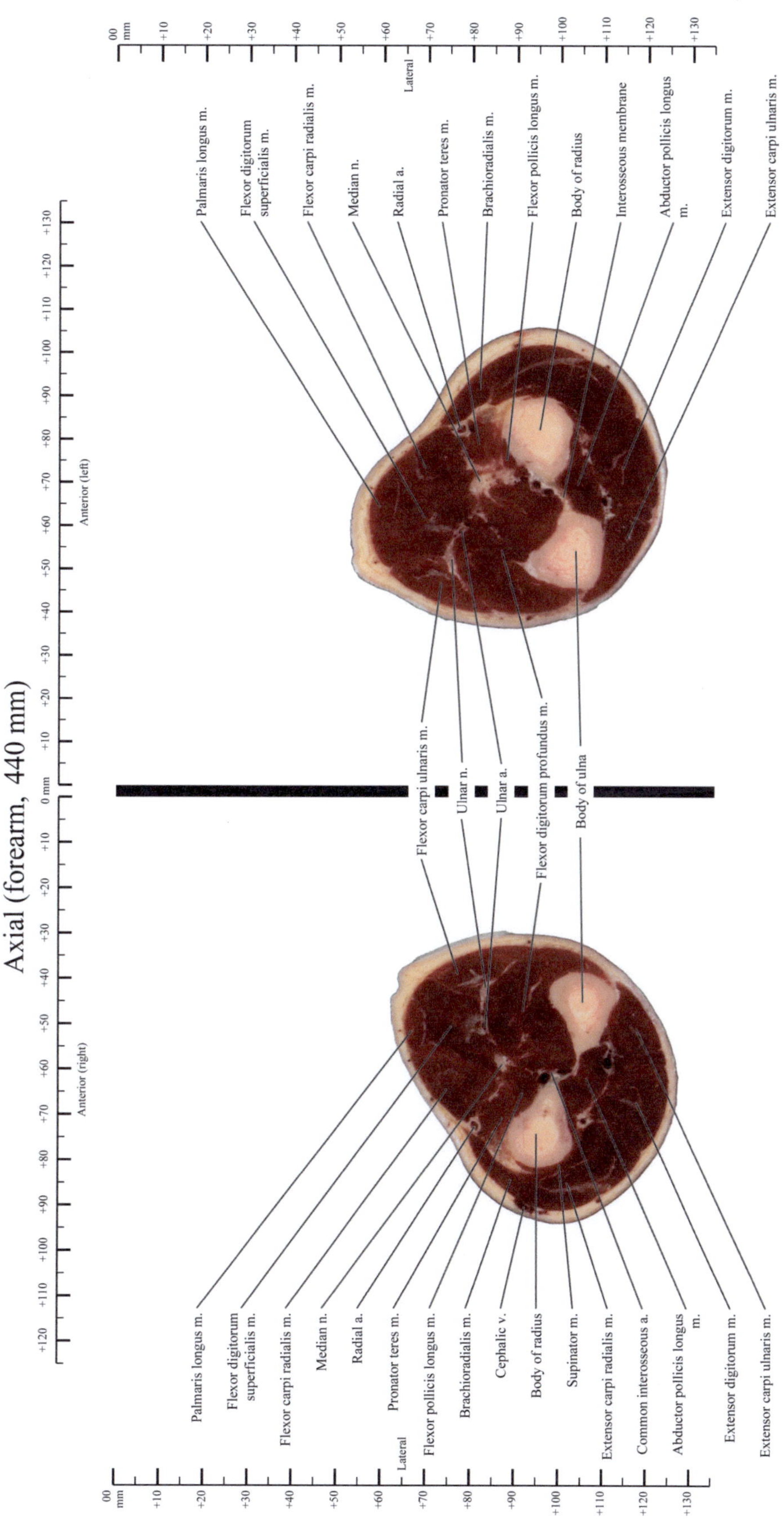
Axial (forearm, 440 mm)
Anterior (left)
Anterior (right)
Lateral
Palmaris longus m.
Flexor digitorum superficialis m.
Flexor carpi radialis m.
Median n.
Radial a.
Pronator teres m.
Brachioradialis m.
Flexor pollicis longus m.
Body of radius
Interosseous membrane
Abductor pollicis longus m.
Extensor digitorum m.
Extensor carpi ulnaris m.
Flexor carpi ulnaris m.
Ulnar n.
Ulnar a.
Flexor digitorum profundus m.
Body of ulna
Cephalic v.
Supinator m.
Extensor carpi radialis m.
Common interosseous a.
0 mm
mm

Axial (forearm, 445 mm)

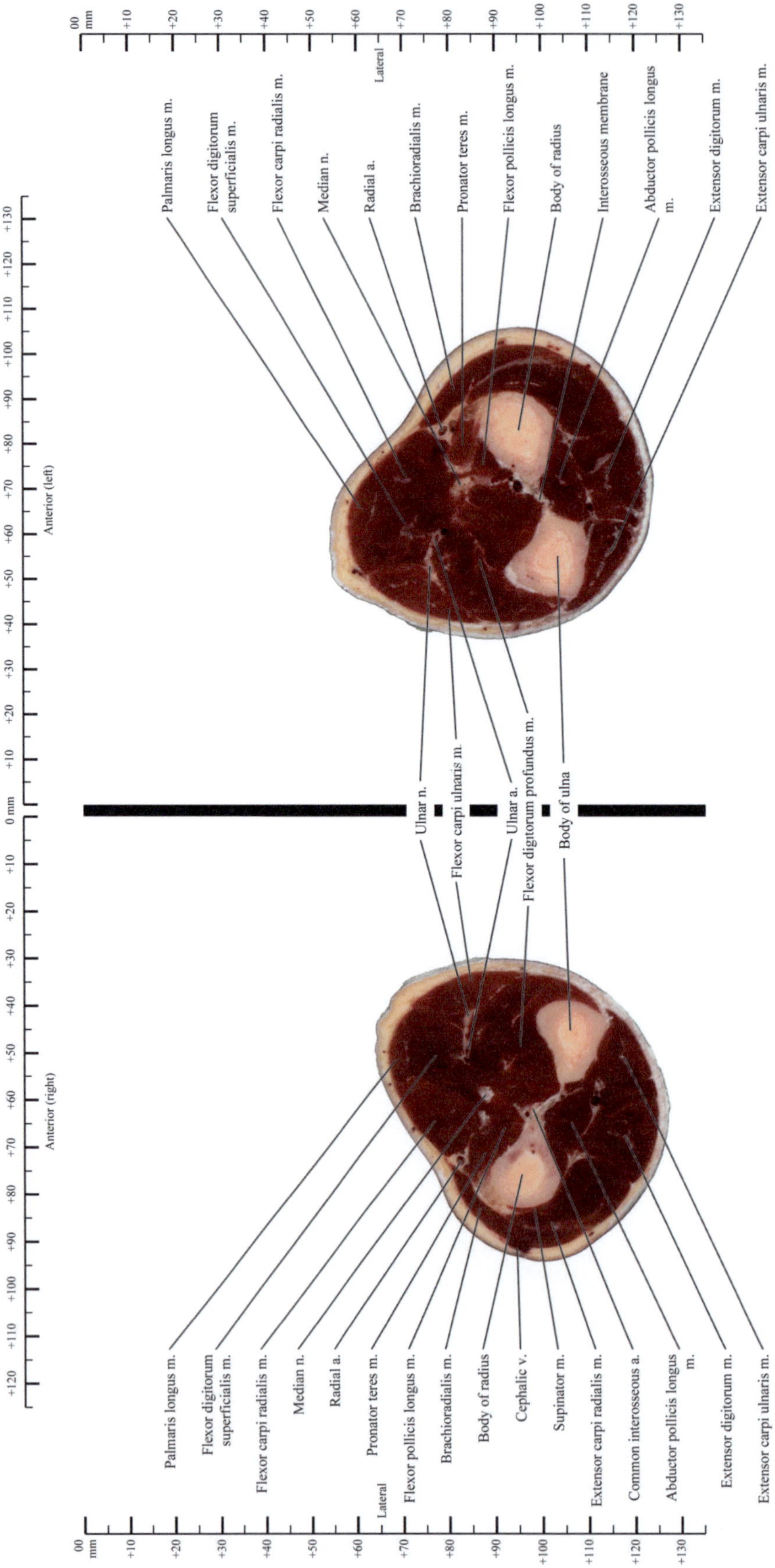

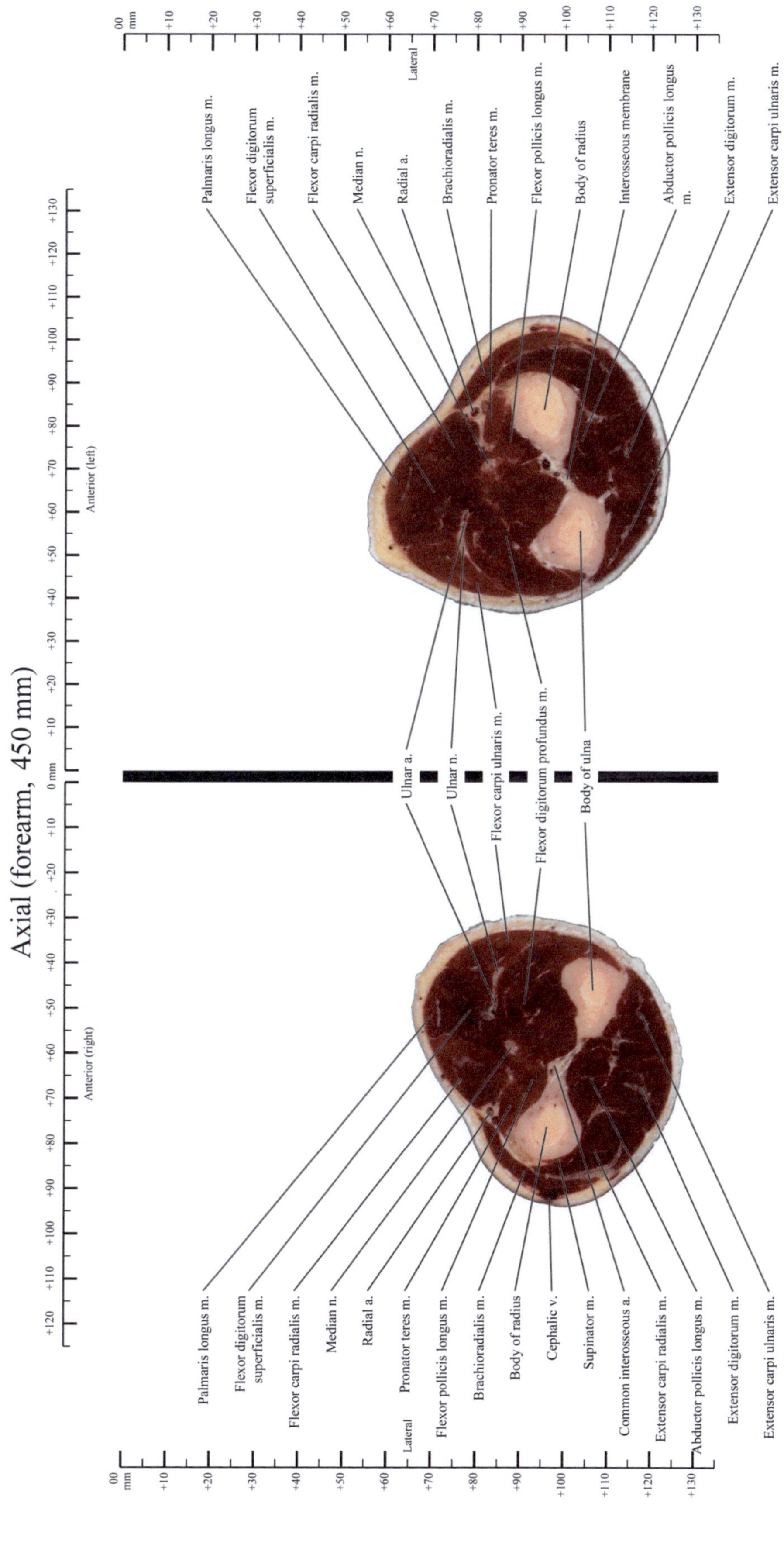

Axial (forearm, 450 mm)
Anterior (left)
Anterior (right)
Lateral
Palmaris longus m.
Flexor digitorum superficialis m.
Flexor carpi radialis m.
Median n.
Radial a.
Brachioradialis m.
Pronator teres m.
Flexor pollicis longus m.
Body of radius
Interosseous membrane
Abductor pollicis longus m.
Extensor digitorum m.
Extensor carpi ulnaris m.
Ulnar a.
Ulnar n.
Flexor carpi ulnaris m.
Flexor digitorum profundus m.
Body of ulna
Cephalic v.
Supinator m.
Common interosseous a.
Extensor carpi radialis m.

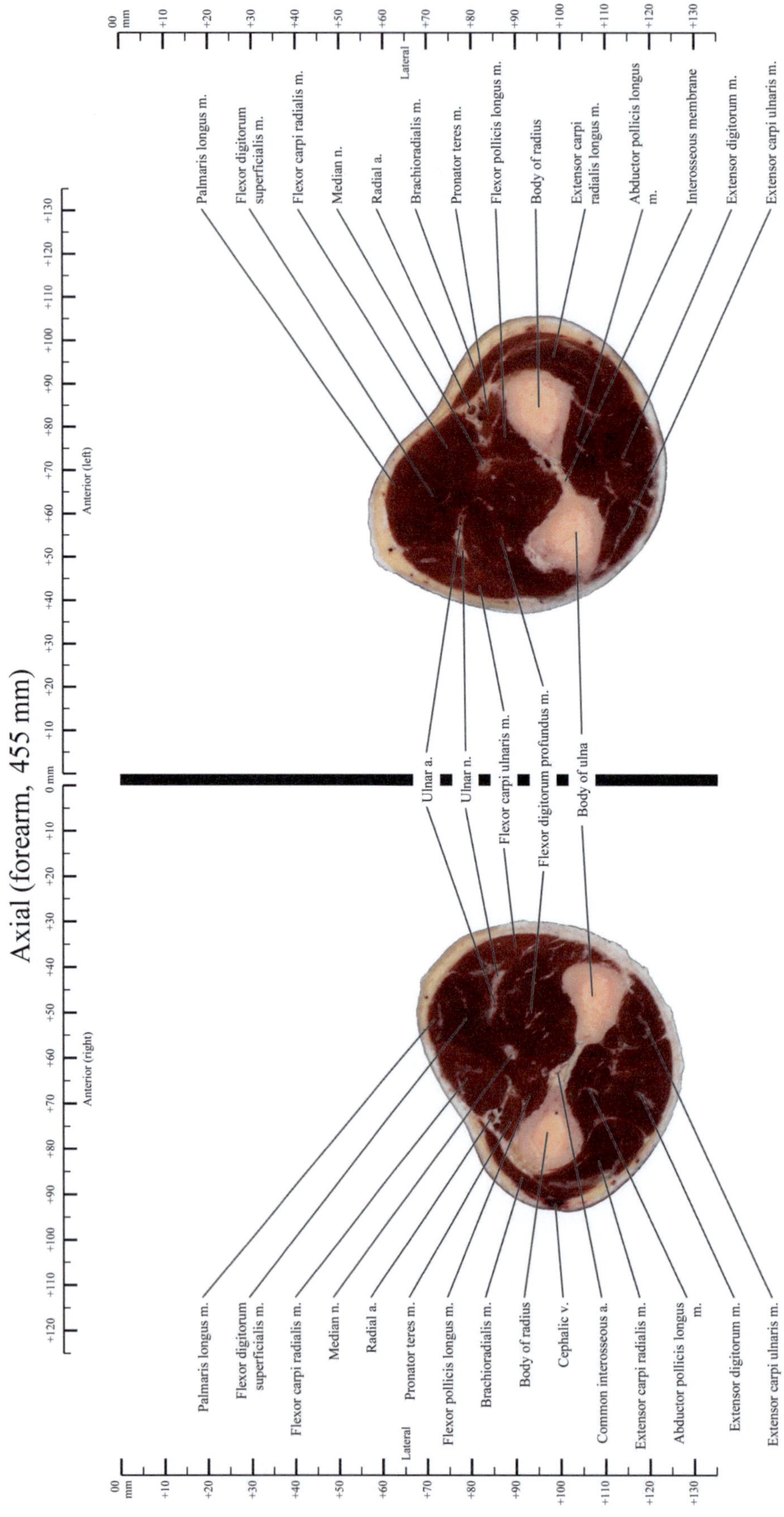
Axial (forearm, 455 mm)
Anterior (left)
Anterior (right)
Lateral
Palmaris longus m.
Flexor digitorum superficialis m.
Flexor carpi radialis m.
Median n.
Radial a.
Brachioradialis m.
Pronator teres m.
Flexor pollicis longus m.
Body of radius
Extensor carpi radialis longus m.
Abductor pollicis longus m.
Interosseous membrane
Extensor digitorum m.
Extensor carpi ulnaris m.
Ulnar a.
Ulnar n.
Flexor carpi ulnaris m.
Flexor digitorum profundus m.
Body of ulna
Palmaris longus m.
Flexor digitorum superficialis m.
Flexor carpi radialis m.
Median n.
Radial a.
Pronator teres m.
Flexor pollicis longus m.
Brachioradialis m.
Body of radius
Cephalic v.
Common interosseous a.
Extensor carpi radialis m.
Abductor pollicis longus m.
Extensor digitorum m.
Extensor carpi ulnaris m.

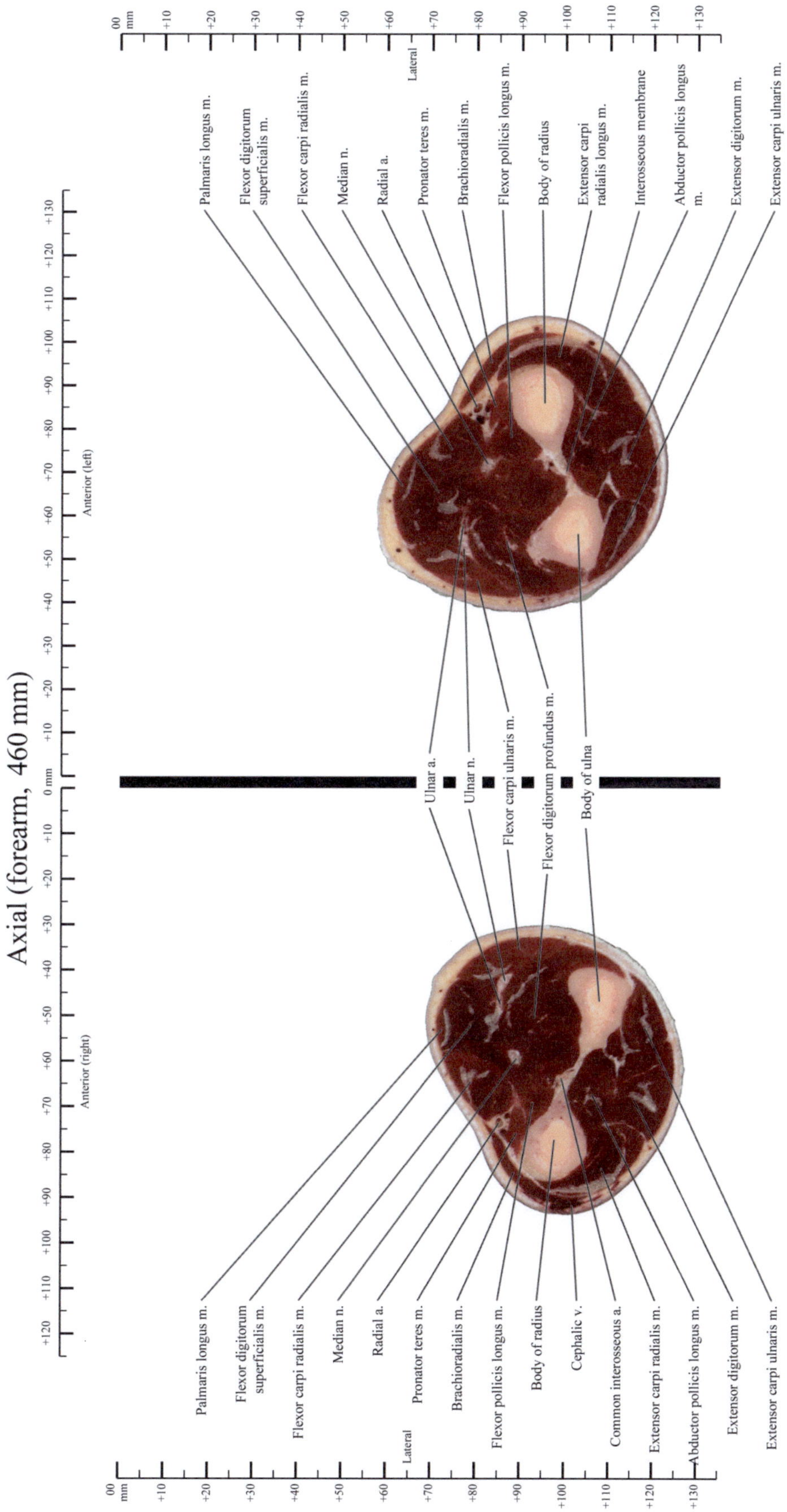
Axial (forearm, 460 mm)
Anterior (left)
Anterior (right)
Lateral
Palmaris longus m.
Flexor digitorum superficialis m.
Flexor carpi radialis m.
Median n.
Radial a.
Pronator teres m.
Brachioradialis m.
Flexor pollicis longus m.
Body of radius
Extensor carpi radialis longus m.
Interosseous membrane
Abductor pollicis longus m.
Extensor digitorum m.
Extensor carpi ulnaris m.
Ulnar a.
Ulnar n.
Flexor carpi ulnaris m.
Flexor digitorum profundus m.
Body of ulna
Palmaris longus m.
Flexor digitorum superficialis m.
Flexor carpi radialis m.
Median n.
Radial a.
Pronator teres m.
Brachioradialis m.
Flexor pollicis longus m.
Body of radius
Cephalic v.
Common interosseous a.
Extensor carpi radialis m.
Abductor pollicis longus m.
Extensor digitorum m.
Extensor carpi ulnaris m.

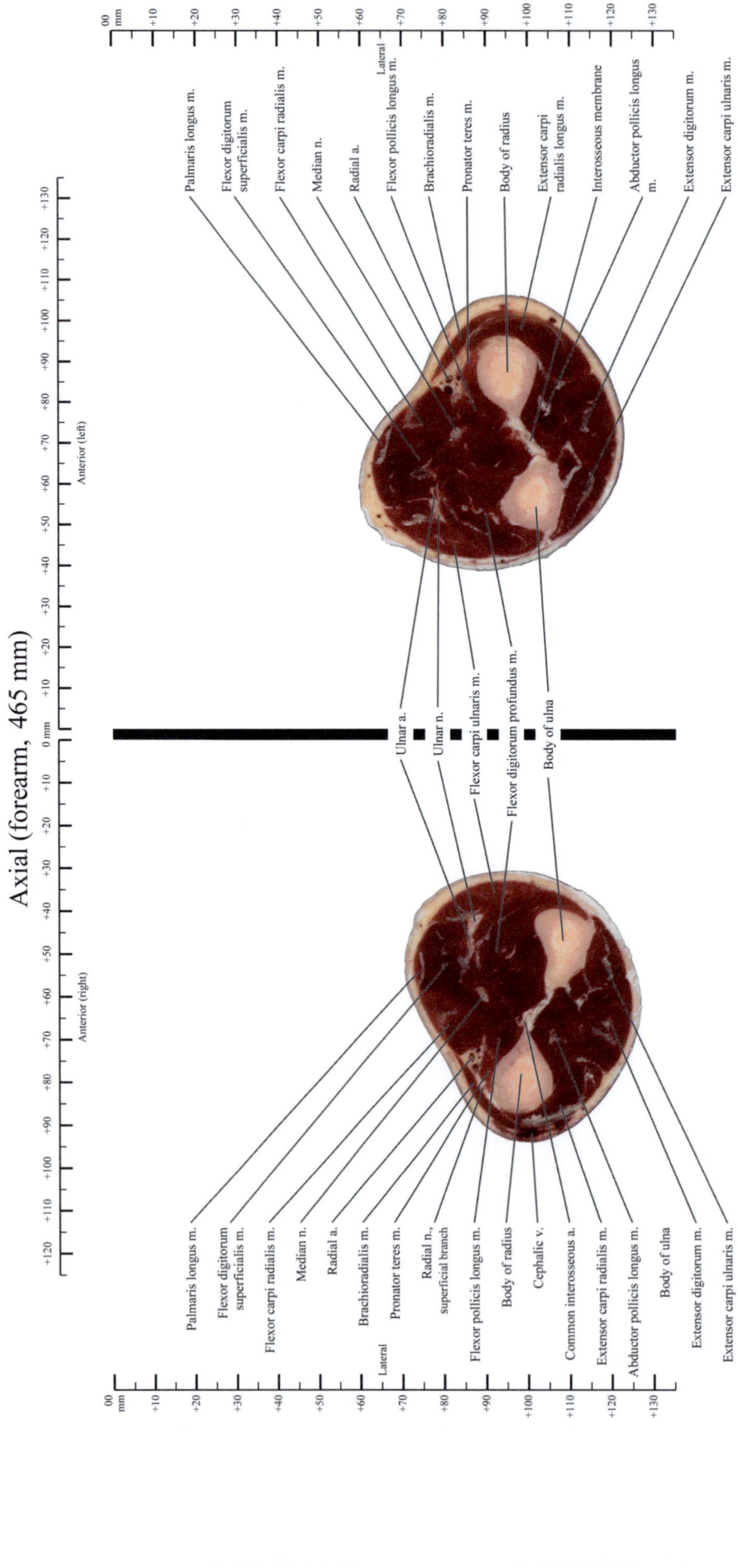
Axial (forearm, 465 mm)
Anterior (left)
Anterior (right)
Lateral
Palmaris longus m.
Flexor digitorum superficialis m.
Flexor carpi radialis m.
Median n.
Radial a.
Flexor pollicis longus m.
Brachioradialis m.
Pronator teres m.
Body of radius
Extensor carpi radialis longus m.
Interosseous membrane
Abductor pollicis longus m.
Extensor digitorum m.
Extensor carpi ulnaris m.
Ulnar a.
Ulnar n.
Flexor carpi ulnaris m.
Flexor digitorum profundus m.
Body of ulna
Radial n., superficial branch
Cephalic v.
Common interosseous a.
Extensor carpi radialis m.

Axial (forearm, 470 mm)

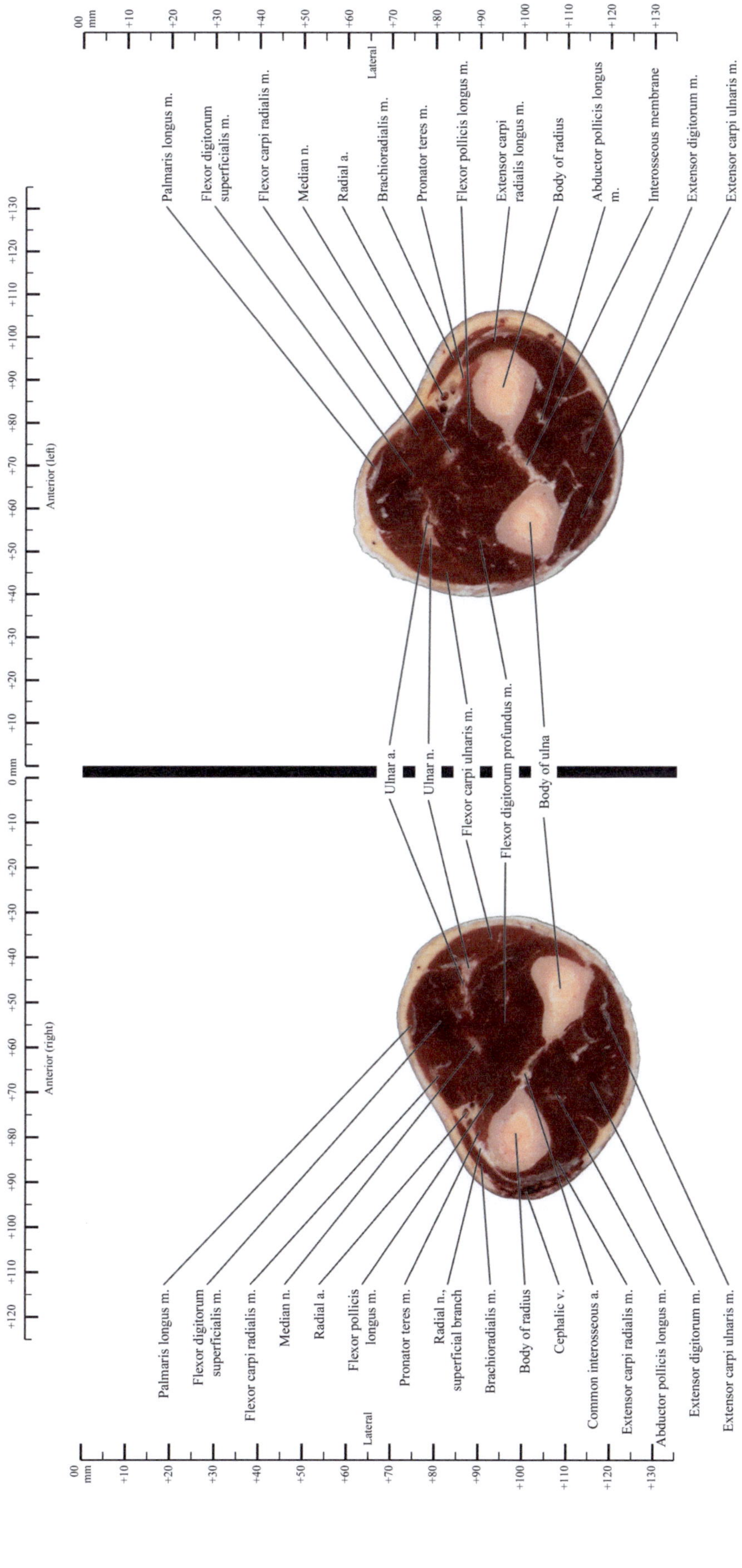

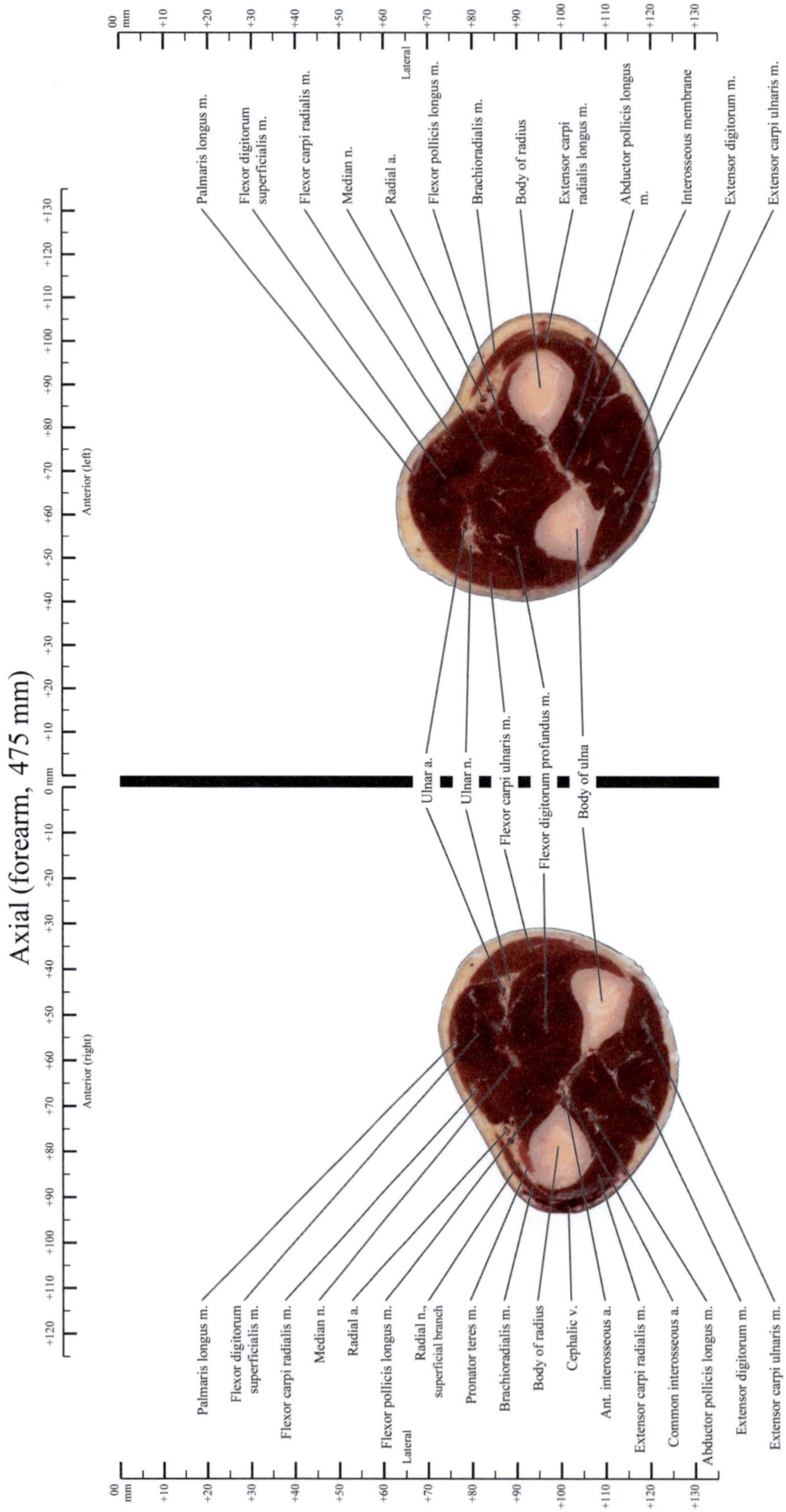
Axial (forearm, 475 mm)
Anterior (left)
Anterior (right)
Lateral
Palmaris longus m.
Flexor digitorum superficialis m.
Flexor carpi radialis m.
Median n.
Radial a.
Flexor pollicis longus m.
Brachioradialis m.
Body of radius
Extensor carpi radialis longus m.
Abductor pollicis longus m.
Interosseous membrane
Extensor digitorum m.
Extensor carpi ulnaris m.
Ulnar a.
Ulnar n.
Flexor carpi ulnaris m.
Flexor digitorum profundus m.
Body of ulna
Radial n., superficial branch
Pronator teres m.
Cephalic v.
Ant. interosseous a.
Extensor carpi radialis m.
Common interosseous a.
Abductor pollicis longus m.

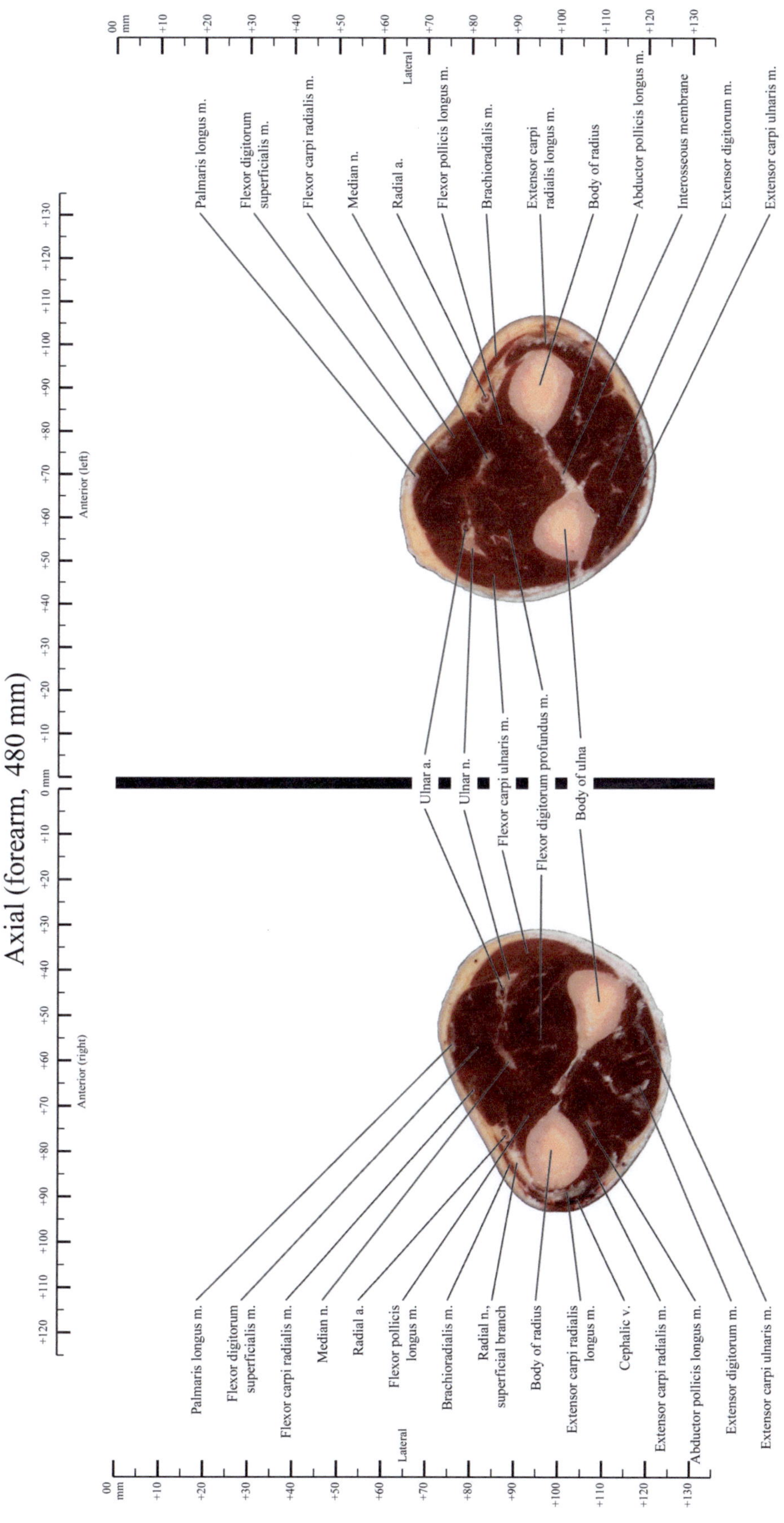

Axial (forearm, 480 mm)
Anterior (left)
Lateral
Palmaris longus m.
Flexor digitorum superficialis m.
Flexor carpi radialis m.
Median n.
Radial a.
Flexor pollicis longus m.
Brachioradialis m.
Extensor carpi radialis longus m.
Body of radius
Abductor pollicis longus m.
Interosseous membrane
Extensor digitorum m.
Extensor carpi ulnaris m.
Ulnar a.
Ulnar n.
Flexor carpi ulnaris m.
Flexor digitorum profundus m.
Body of ulna
Anterior (right)
Lateral
Palmaris longus m.
Flexor digitorum superficialis m.
Flexor carpi radialis m.
Median n.
Radial a.
Flexor pollicis longus m.
Brachioradialis m.
Radial n., superficial branch
Body of radius
Extensor carpi radialis longus m.
Cephalic v.
Extensor carpi radialis m.
Abductor pollicis longus m.
Extensor digitorum m.
Extensor carpi ulnaris m.

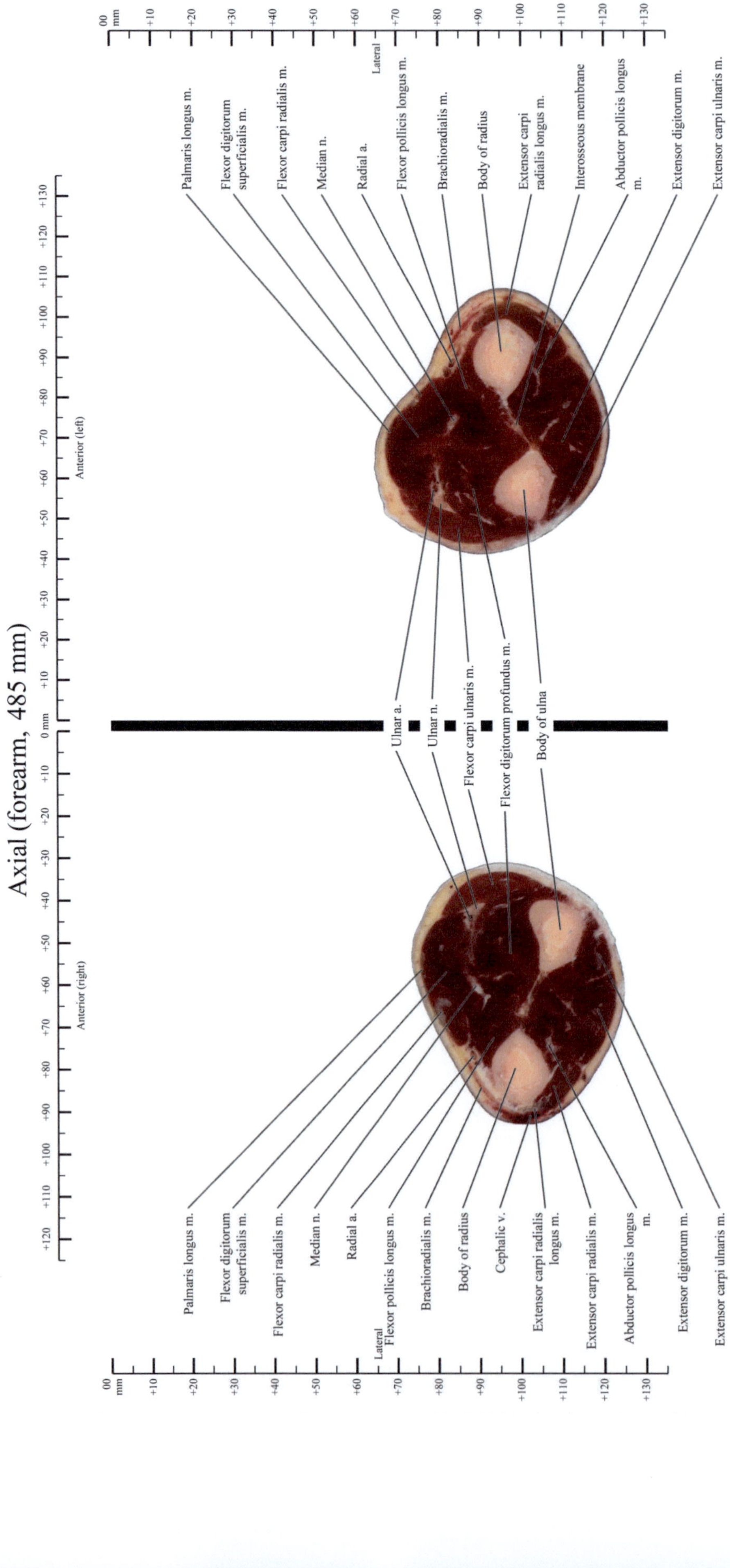
Axial (forearm, 485 mm)
Anterior (left)
Anterior (right)
Lateral
Palmaris longus m.
Flexor digitorum superficialis m.
Flexor carpi radialis m.
Median n.
Radial a.
Flexor pollicis longus m.
Brachioradialis m.
Body of radius
Extensor carpi radialis longus m.
Interosseous membrane
Abductor pollicis longus m.
Extensor digitorum m.
Extensor carpi ulnaris m.
Ulnar a.
Ulnar n.
Flexor carpi ulnaris m.
Flexor digitorum profundus m.
Body of ulna
Cephalic v.
Extensor carpi radialis m.

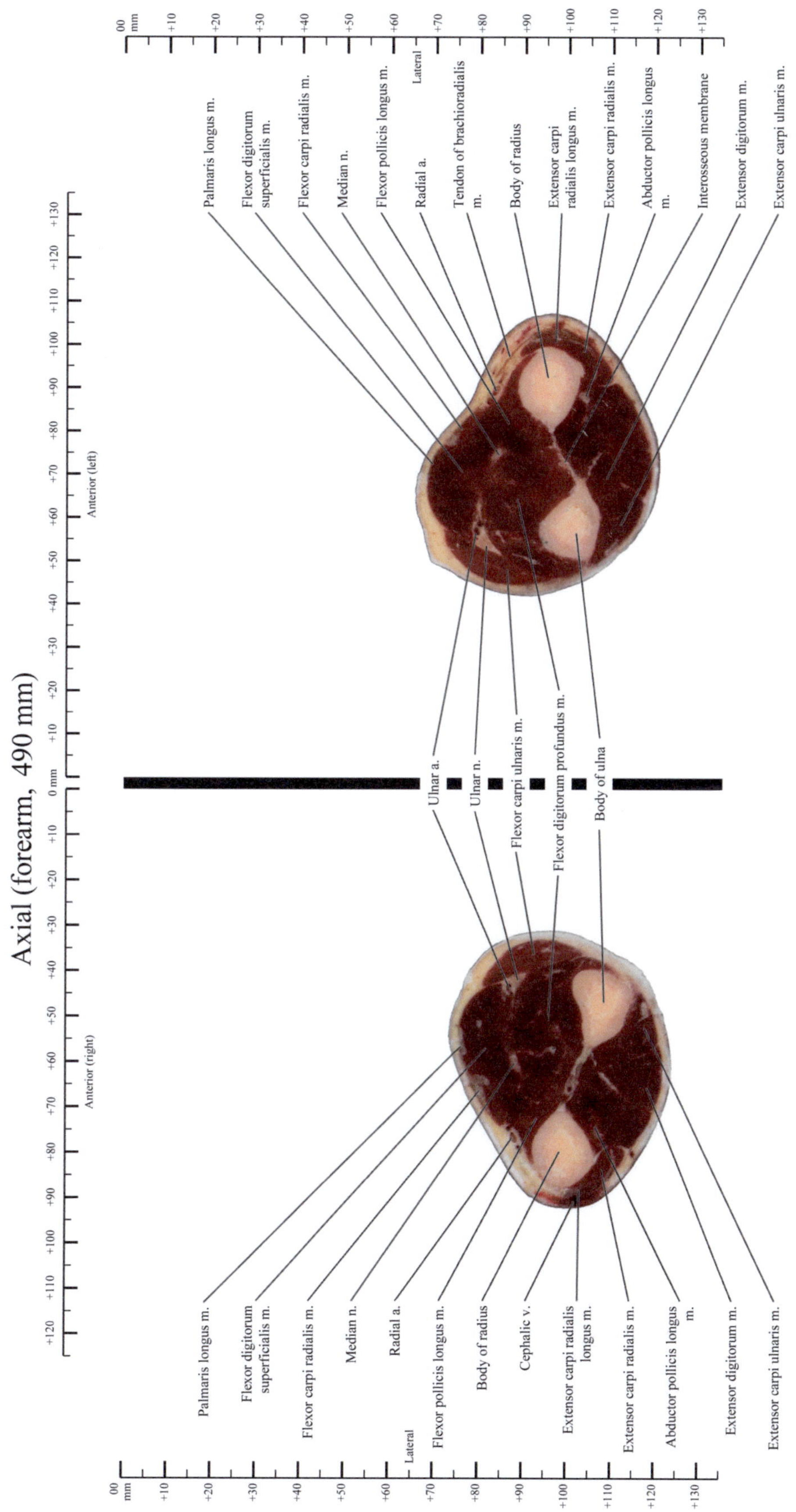
Axial (forearm, 490 mm)
Anterior (left)
Anterior (right)
Lateral
Palmaris longus m.
Flexor digitorum superficialis m.
Flexor carpi radialis m.
Median n.
Flexor pollicis longus m.
Radial a.
Tendon of brachioradialis m.
Body of radius
Extensor carpi radialis longus m.
Extensor carpi radialis m.
Abductor pollicis longus m.
Interosseous membrane
Extensor digitorum m.
Extensor carpi ulnaris m.
Ulnar a.
Ulnar n.
Flexor carpi ulnaris m.
Flexor digitorum profundus m.
Body of ulna
Cephalic v.

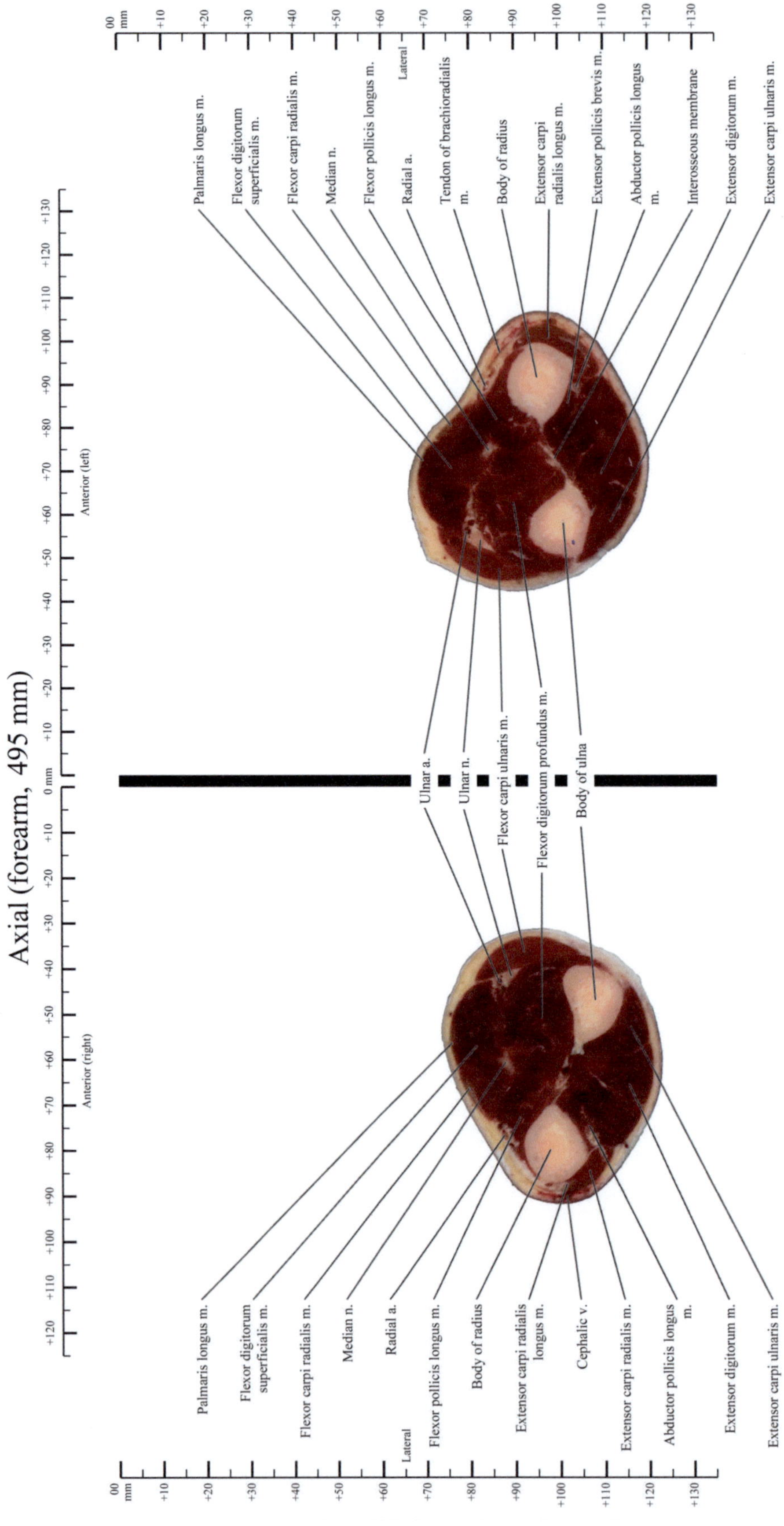
Axial (forearm, 495 mm)
Anterior (left)
Anterior (right)
Lateral
Palmaris longus m.
Flexor digitorum superficialis m.
Flexor carpi radialis m.
Median n.
Flexor pollicis longus m.
Radial a.
Tendon of brachioradialis m.
Body of radius
Extensor carpi radialis longus m.
Extensor pollicis brevis m.
Abductor pollicis longus m.
Interosseous membrane
Extensor digitorum m.
Extensor carpi ulnaris m.
Ulnar a.
Ulnar n.
Flexor carpi ulnaris m.
Flexor digitorum profundus m.
Body of ulna
Cephalic v.
Extensor carpi radialis m.

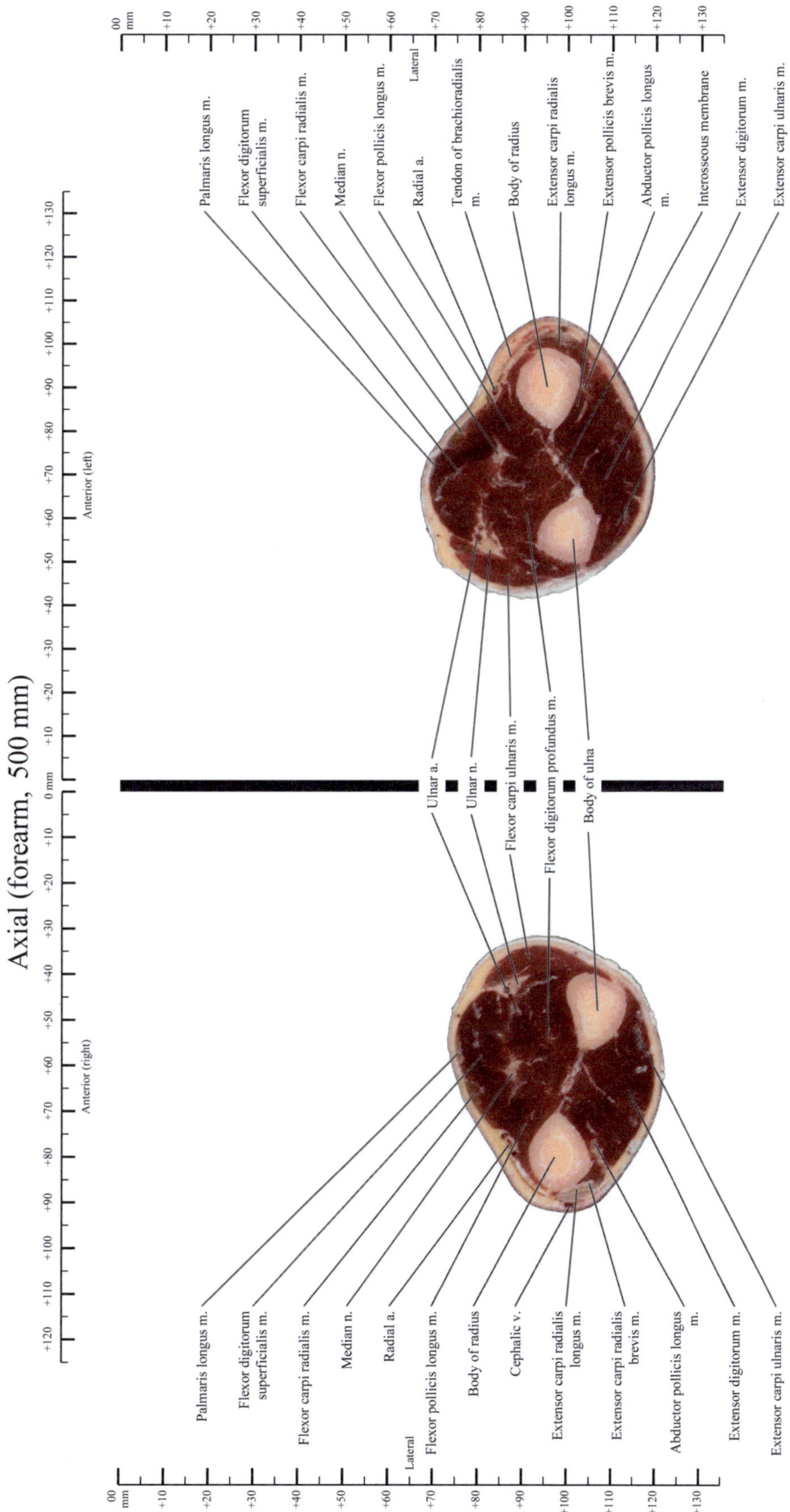
Axial (forearm, 500 mm)
Anterior (left)
Anterior (right)
Lateral
Palmaris longus m.
Flexor digitorum superficialis m.
Flexor carpi radialis m.
Median n.
Flexor pollicis longus m.
Radial a.
Tendon of brachioradialis m.
Body of radius
Extensor carpi radialis longus m.
Extensor pollicis brevis m.
Abductor pollicis longus m.
Interosseous membrane
Extensor digitorum m.
Extensor carpi ulnaris m.
Ulnar a.
Ulnar n.
Flexor carpi ulnaris m.
Flexor digitorum profundus m.
Body of ulna
Cephalic v.
Extensor carpi radialis brevis m.

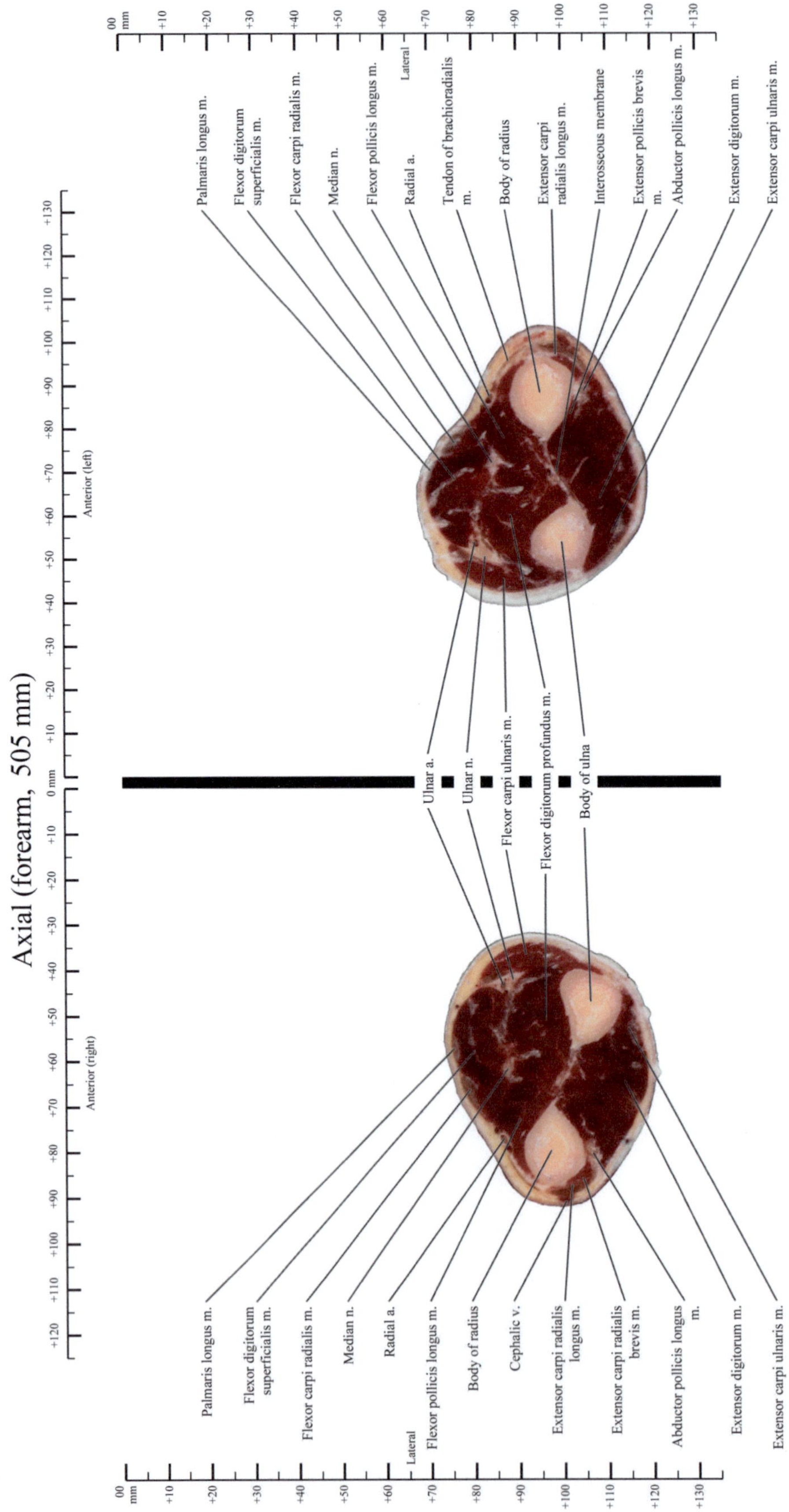
Axial (forearm, 505 mm)
Anterior (left)
Anterior (right)
Lateral
Palmaris longus m.
Flexor digitorum superficialis m.
Flexor carpi radialis m.
Median n.
Flexor pollicis longus m.
Radial a.
Tendon of brachioradialis m.
Body of radius
Extensor carpi radialis longus m.
Interosseous membrane
Extensor pollicis brevis m.
Abductor pollicis longus m.
Extensor digitorum m.
Extensor carpi ulnaris m.
Ulnar a.
Ulnar n.
Flexor carpi ulnaris m.
Flexor digitorum profundus m.
Body of ulna
Cephalic v.
Extensor carpi radialis brevis m.

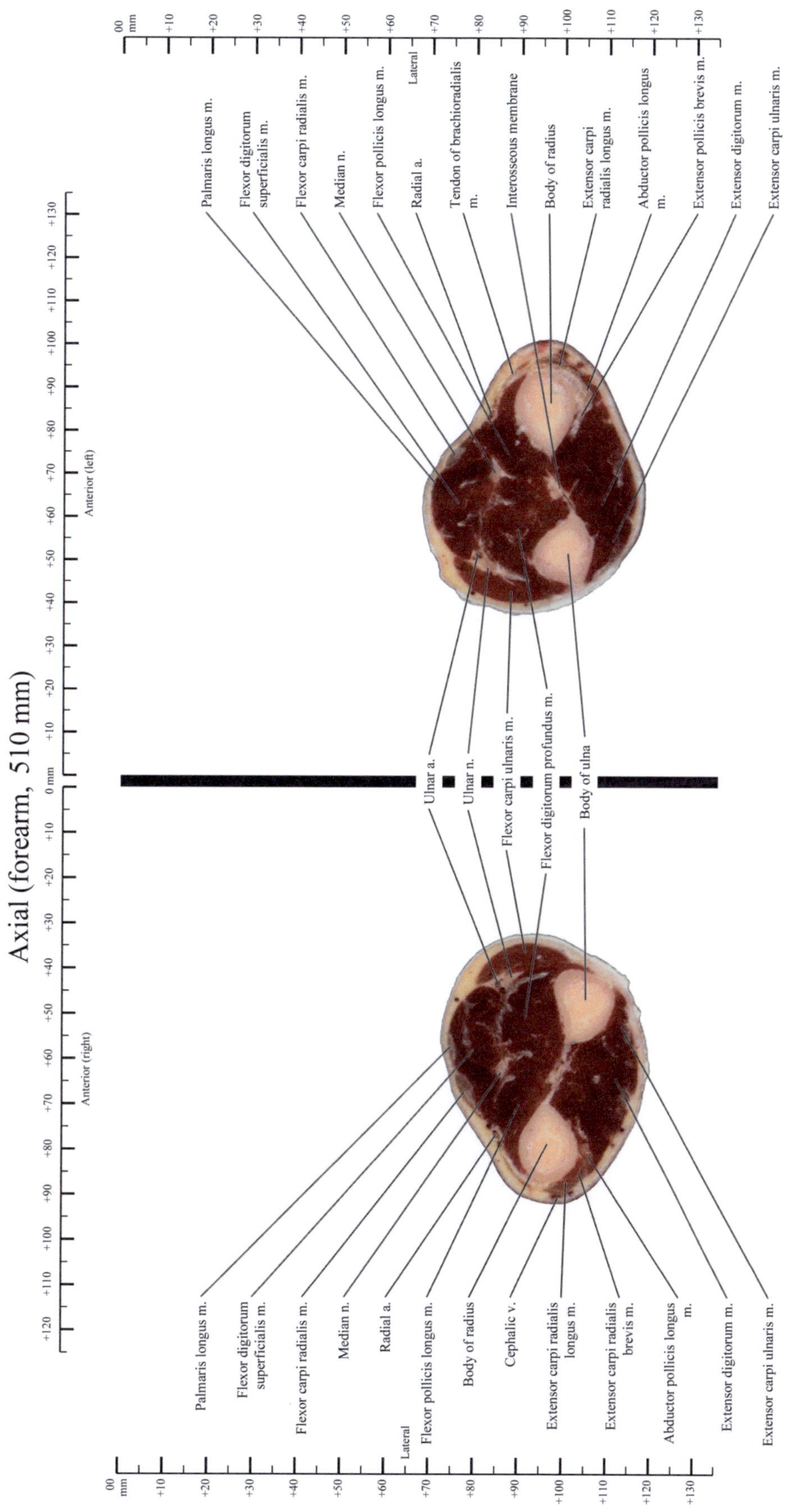
Axial (forearm, 510 mm)
Anterior (left)
Anterior (right)
Lateral
Palmaris longus m.
Flexor digitorum superficialis m.
Flexor carpi radialis m.
Median n.
Flexor pollicis longus m.
Radial a.
Tendon of brachioradialis m.
Interosseous membrane
Body of radius
Extensor carpi radialis longus m.
Abductor pollicis longus m.
Extensor pollicis brevis m.
Extensor digitorum m.
Extensor carpi ulnaris m.
Ulnar a.
Ulnar n.
Flexor carpi ulnaris m.
Flexor digitorum profundus m.
Body of ulna
Palmaris longus m.
Flexor digitorum superficialis m.
Flexor carpi radialis m.
Median n.
Radial a.
Flexor pollicis longus m.
Body of radius
Cephalic v.
Extensor carpi radialis longus m.
Extensor carpi radialis brevis m.
Abductor pollicis longus m.
Extensor digitorum m.
Extensor carpi ulnaris m.

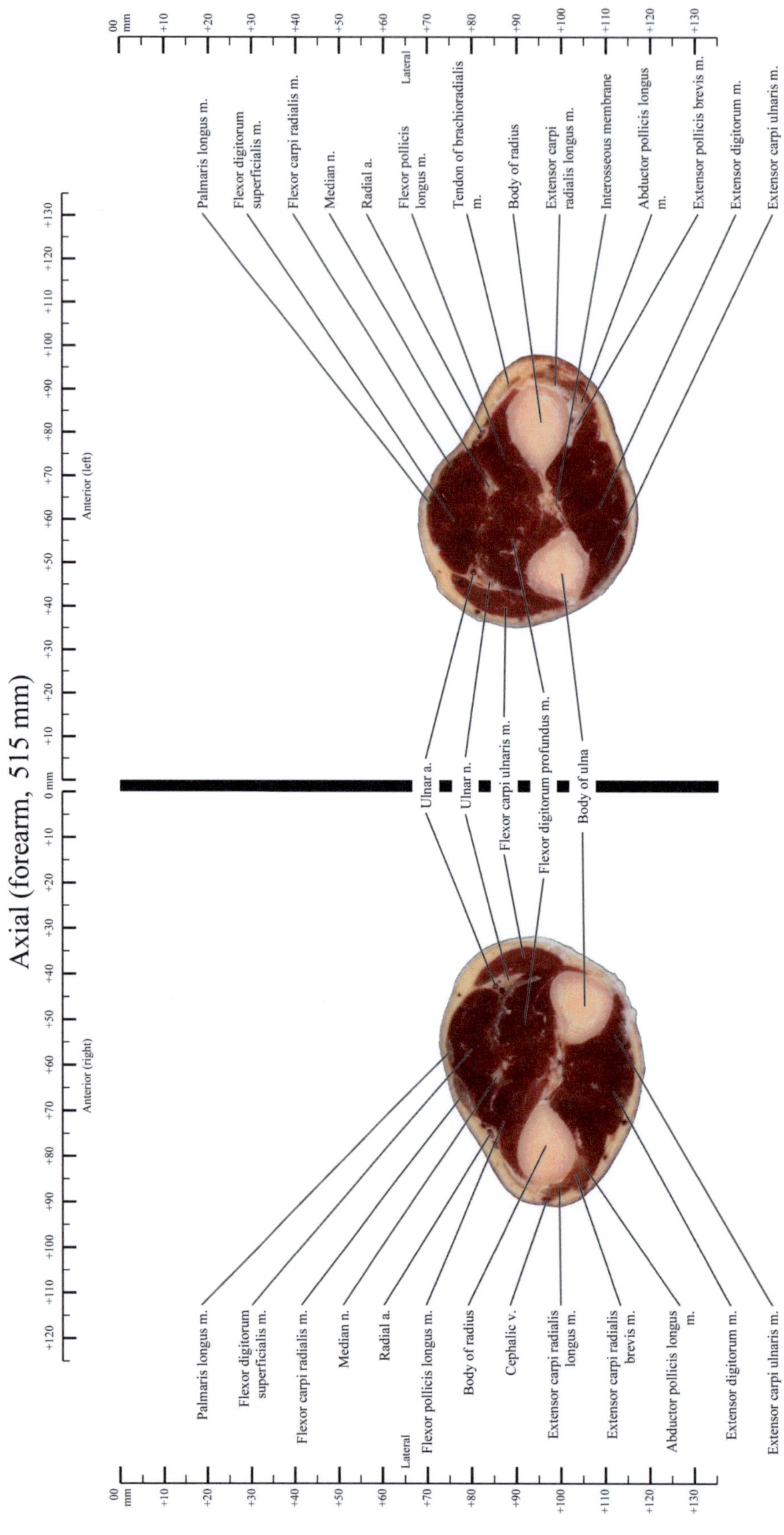
Axial (forearm, 515 mm)
Anterior (left)
Anterior (right)
Lateral
Lateral
Palmaris longus m.
Flexor digitorum superficialis m.
Flexor carpi radialis m.
Median n.
Radial a.
Flexor pollicis longus m.
Tendon of brachioradialis m.
Body of radius
Extensor carpi radialis longus m.
Interosseous membrane
Abductor pollicis longus m.
Extensor pollicis brevis m.
Extensor digitorum m.
Extensor carpi ulnaris m.
Ulnar a.
Ulnar n.
Flexor carpi ulnaris m.
Flexor digitorum profundus m.
Body of ulna
Palmaris longus m.
Flexor digitorum superficialis m.
Flexor carpi radialis m.
Median n.
Radial a.
Flexor pollicis longus m.
Body of radius
Cephalic v.
Extensor carpi radialis longus m.
Extensor carpi radialis brevis m.
Abductor pollicis longus m.
Extensor digitorum m.
Extensor carpi ulnaris m.

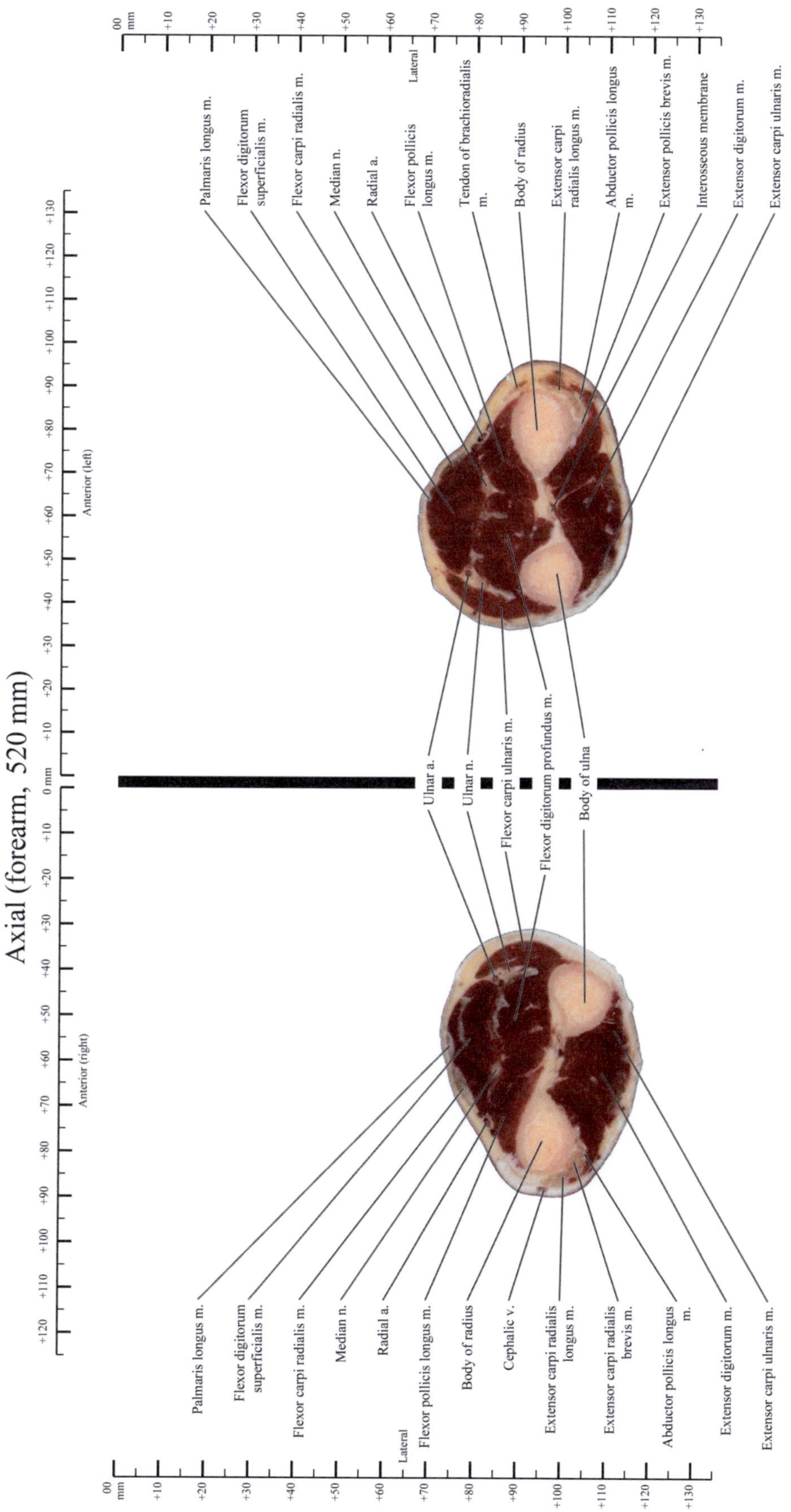
Axial (forearm, 520 mm)
Anterior (left)
Anterior (right)
Lateral
Palmaris longus m.
Flexor digitorum superficialis m.
Flexor carpi radialis m.
Median n.
Radial a.
Flexor pollicis longus m.
Tendon of brachioradialis m.
Body of radius
Extensor carpi radialis longus m.
Abductor pollicis longus m.
Extensor pollicis brevis m.
Interosseous membrane
Extensor digitorum m.
Extensor carpi ulnaris m.
Ulnar a.
Ulnar n.
Flexor carpi ulnaris m.
Flexor digitorum profundus m.
Body of ulna
Cephalic v.
Extensor carpi radialis brevis m.

Axial (forearm, 525 mm)

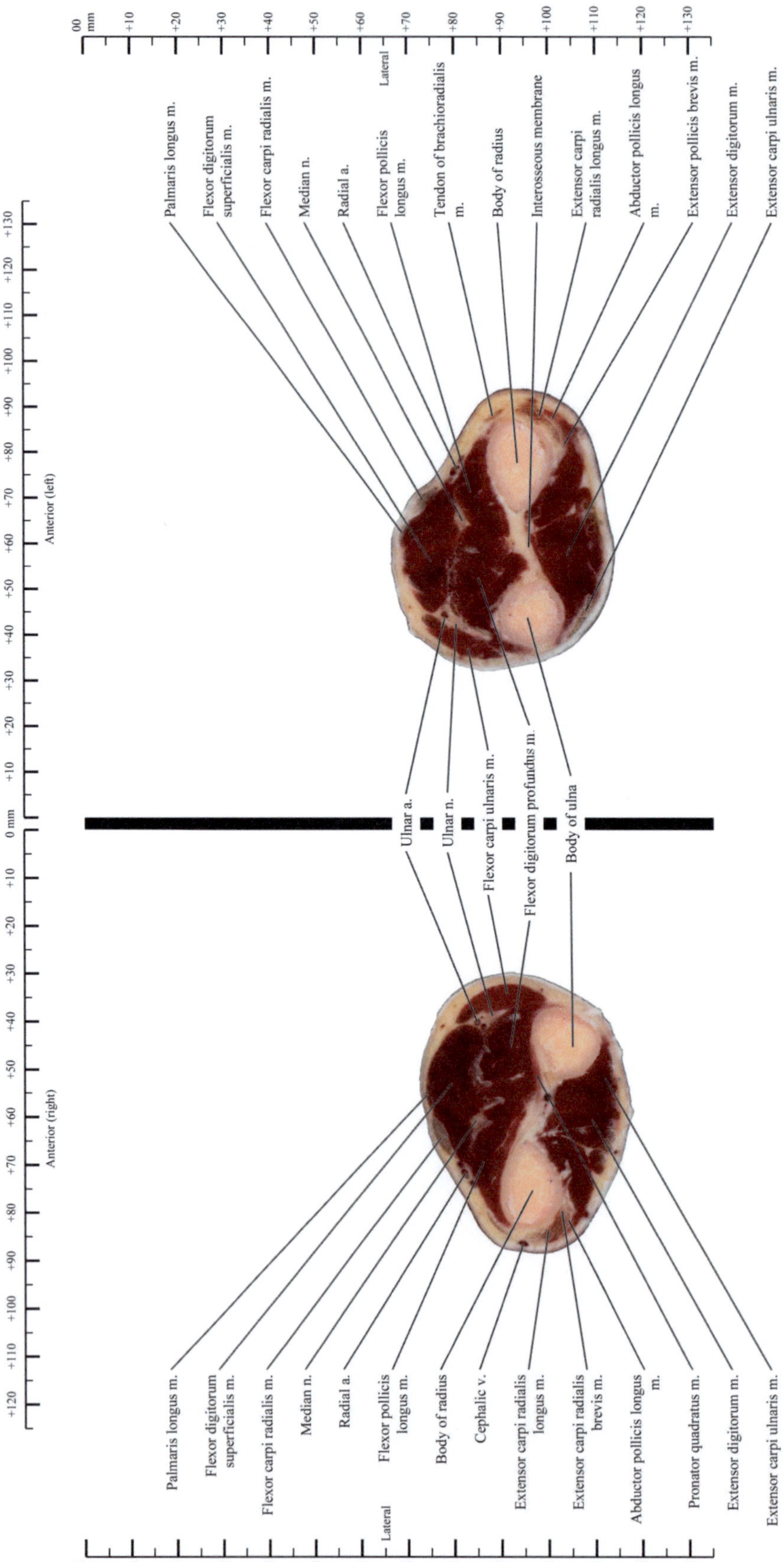

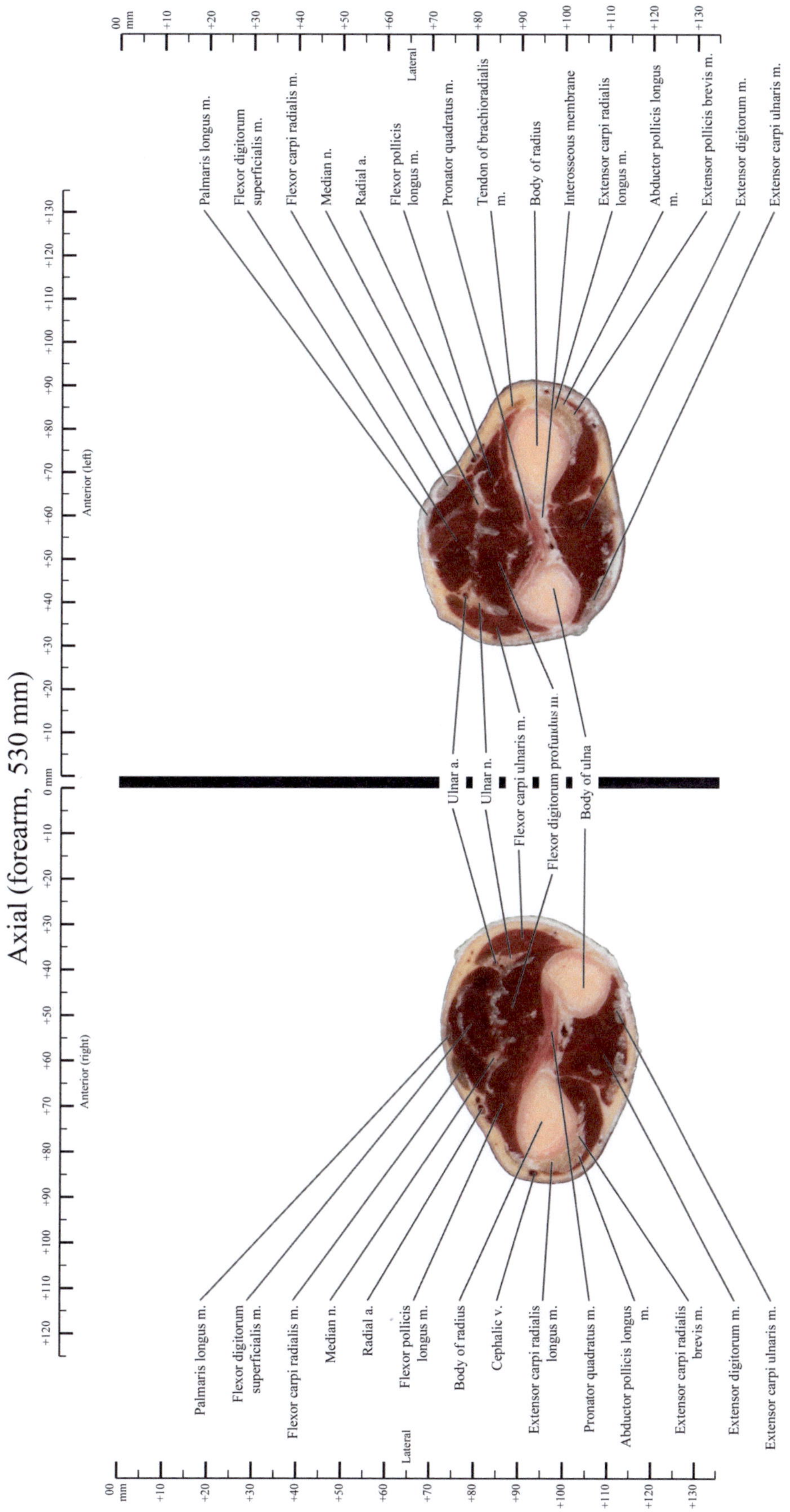
Axial (forearm, 530 mm)
Anterior (left)
Anterior (right)
Lateral
Lateral
Palmaris longus m.
Flexor digitorum superficialis m.
Flexor carpi radialis m.
Median n.
Radial a.
Flexor pollicis longus m.
Pronator quadratus m.
Tendon of brachioradialis m.
Body of radius
Interosseous membrane
Extensor carpi radialis longus m.
Abductor pollicis longus m.
Extensor pollicis brevis m.
Extensor digitorum m.
Extensor carpi ulnaris m.
Ulnar a.
Ulnar n.
Flexor carpi ulnaris m.
Flexor digitorum profundus m.
Body of ulna
Palmaris longus m.
Flexor digitorum superficialis m.
Flexor carpi radialis m.
Median n.
Radial a.
Flexor pollicis longus m.
Body of radius
Cephalic v.
Extensor carpi radialis longus m.
Pronator quadratus m.
Abductor pollicis longus m.
Extensor carpi radialis brevis m.
Extensor digitorum m.
Extensor carpi ulnaris m.

Axial (forearm, 535 mm)

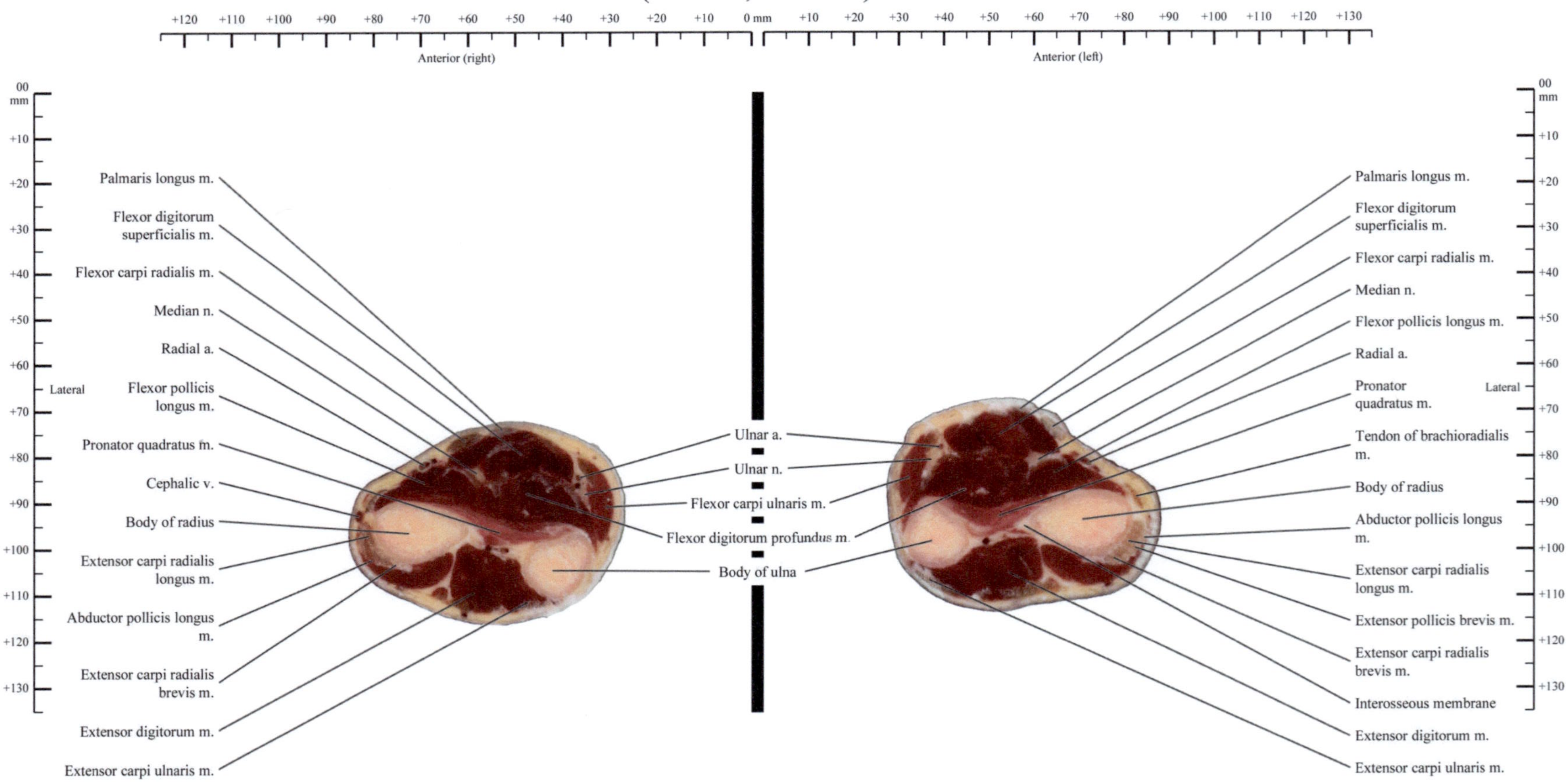

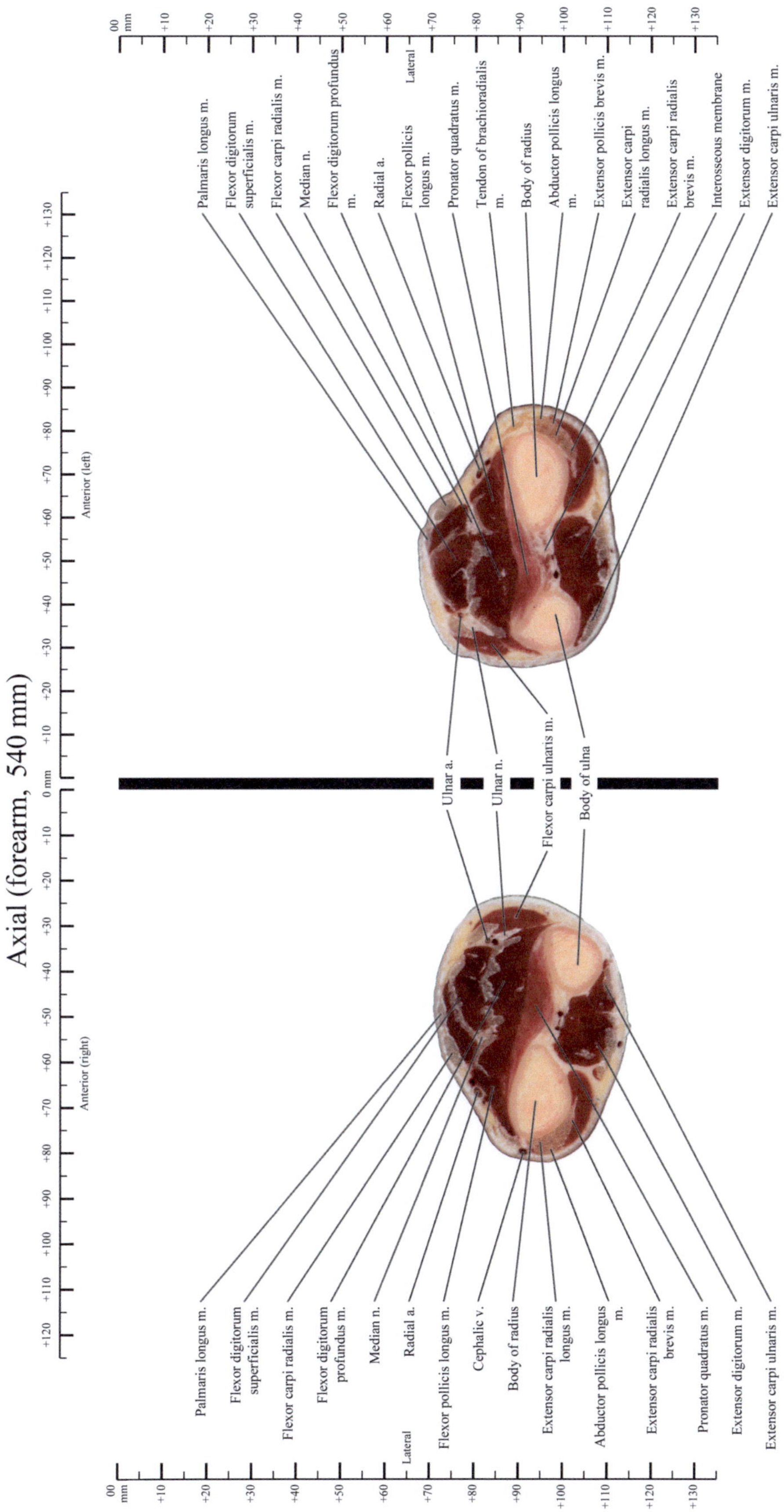
Axial (forearm, 540 mm)
Anterior (left)
Anterior (right)
Lateral
Palmaris longus m.
Flexor digitorum superficialis m.
Flexor carpi radialis m.
Median n.
Flexor digitorum profundus m.
Radial a.
Flexor pollicis longus m.
Pronator quadratus m.
Tendon of brachioradialis m.
Body of radius
Abductor pollicis longus m.
Extensor pollicis brevis m.
Extensor carpi radialis longus m.
Extensor carpi radialis brevis m.
Interosseous membrane
Extensor digitorum m.
Extensor carpi ulnaris m.
Ulnar a.
Ulnar n.
Flexor carpi ulnaris m.
Body of ulna
Cephalic v.

Axial (forearm, 545 mm)

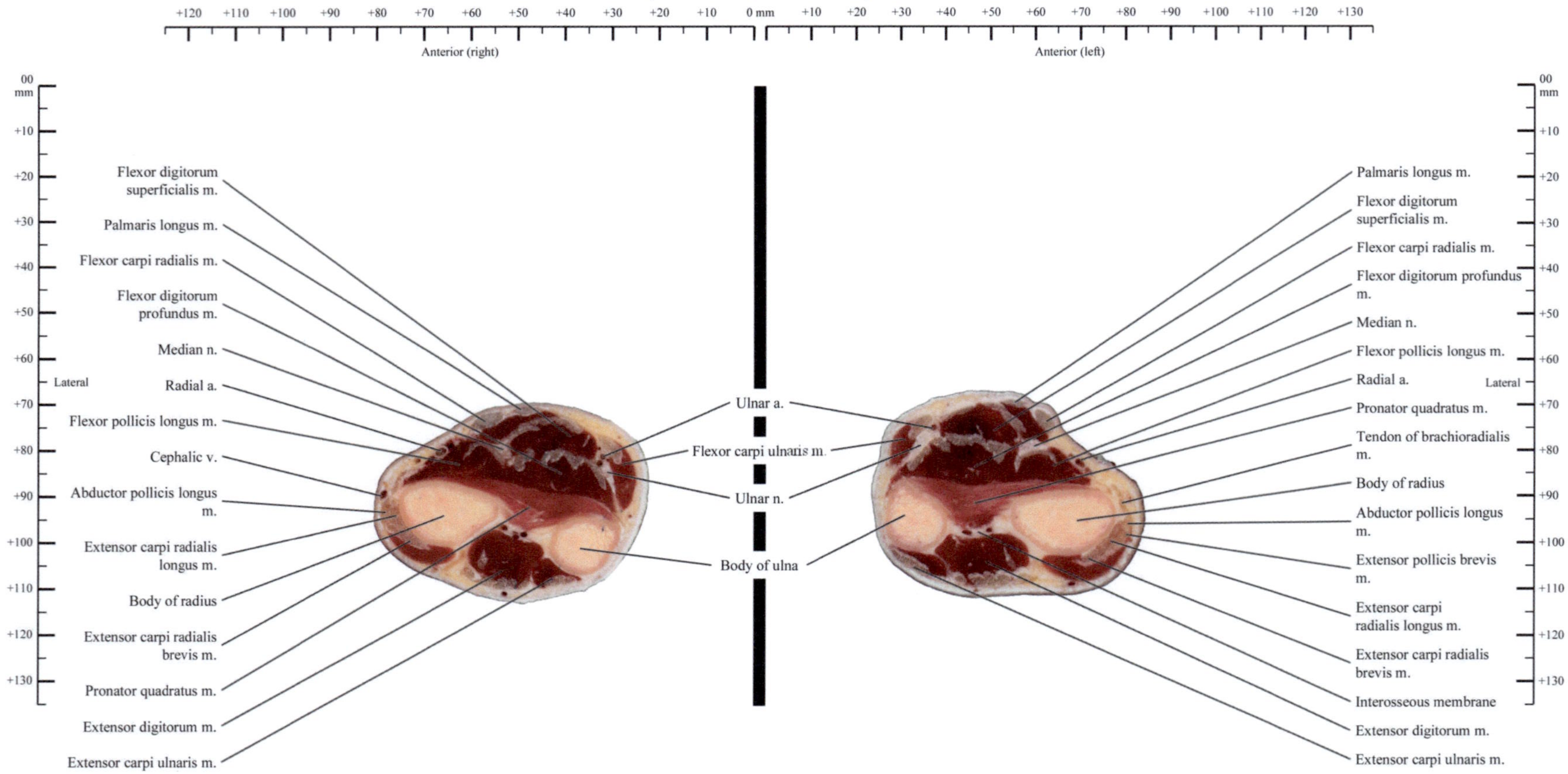

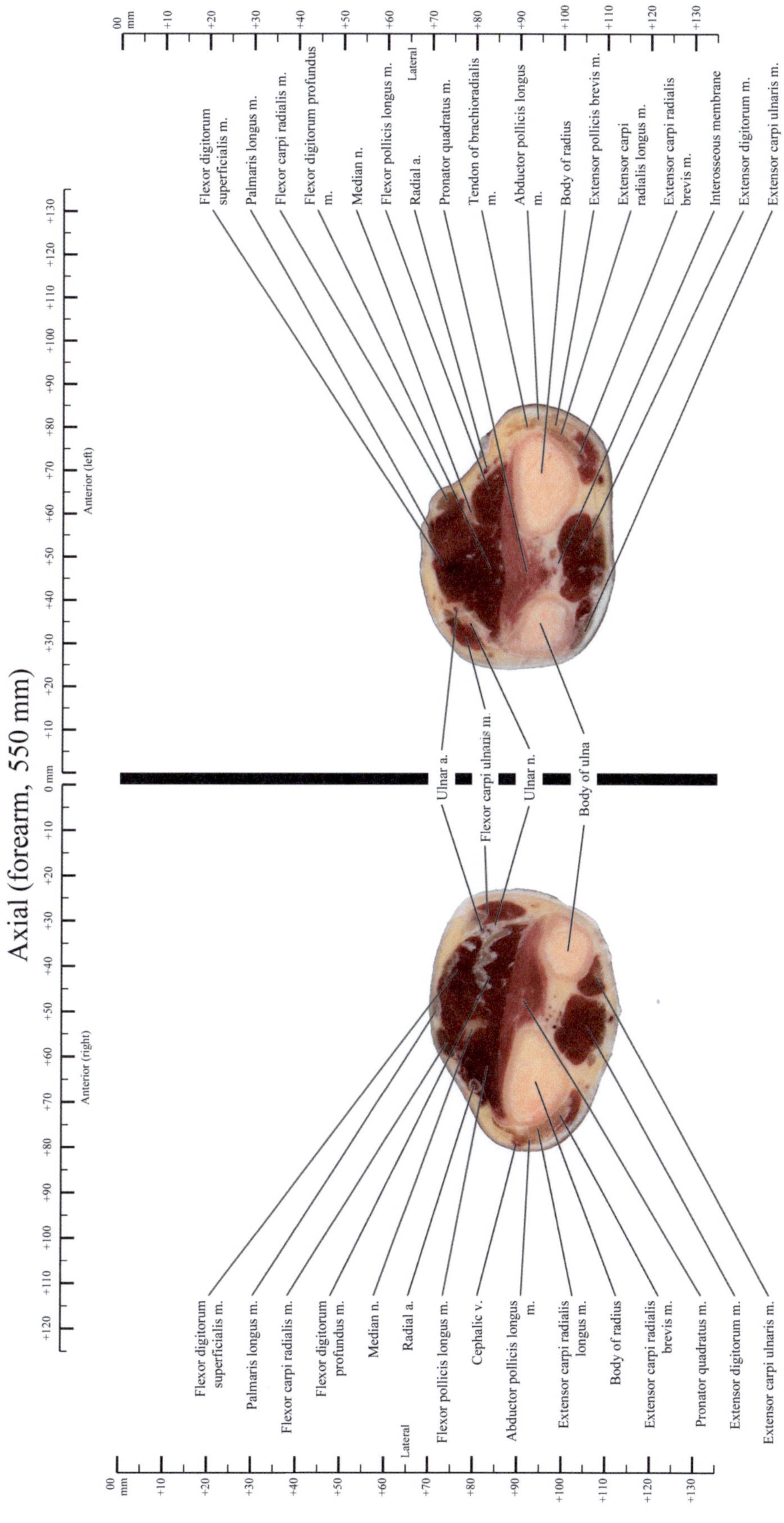
Axial (forearm, 550 mm)
Anterior (left)
Anterior (right)
Lateral
Flexor digitorum superficialis m.
Palmaris longus m.
Flexor carpi radialis m.
Flexor digitorum profundus m.
Median n.
Flexor pollicis longus m.
Radial a.
Pronator quadratus m.
Tendon of brachioradialis m.
Abductor pollicis longus m.
Body of radius
Extensor pollicis brevis m.
Extensor carpi radialis longus m.
Extensor carpi radialis brevis m.
Interosseous membrane
Extensor digitorum m.
Extensor carpi ulnaris m.
Ulnar a.
Flexor carpi ulnaris m.
Ulnar n.
Body of ulna
Lateral
Flexor digitorum superficialis m.
Palmaris longus m.
Flexor carpi radialis m.
Flexor digitorum profundus m.
Median n.
Radial a.
Flexor pollicis longus m.
Cephalic v.
Abductor pollicis longus m.
Extensor carpi radialis longus m.
Body of radius
Extensor carpi radialis brevis m.
Pronator quadratus m.
Extensor digitorum m.
Extensor carpi ulnaris m.

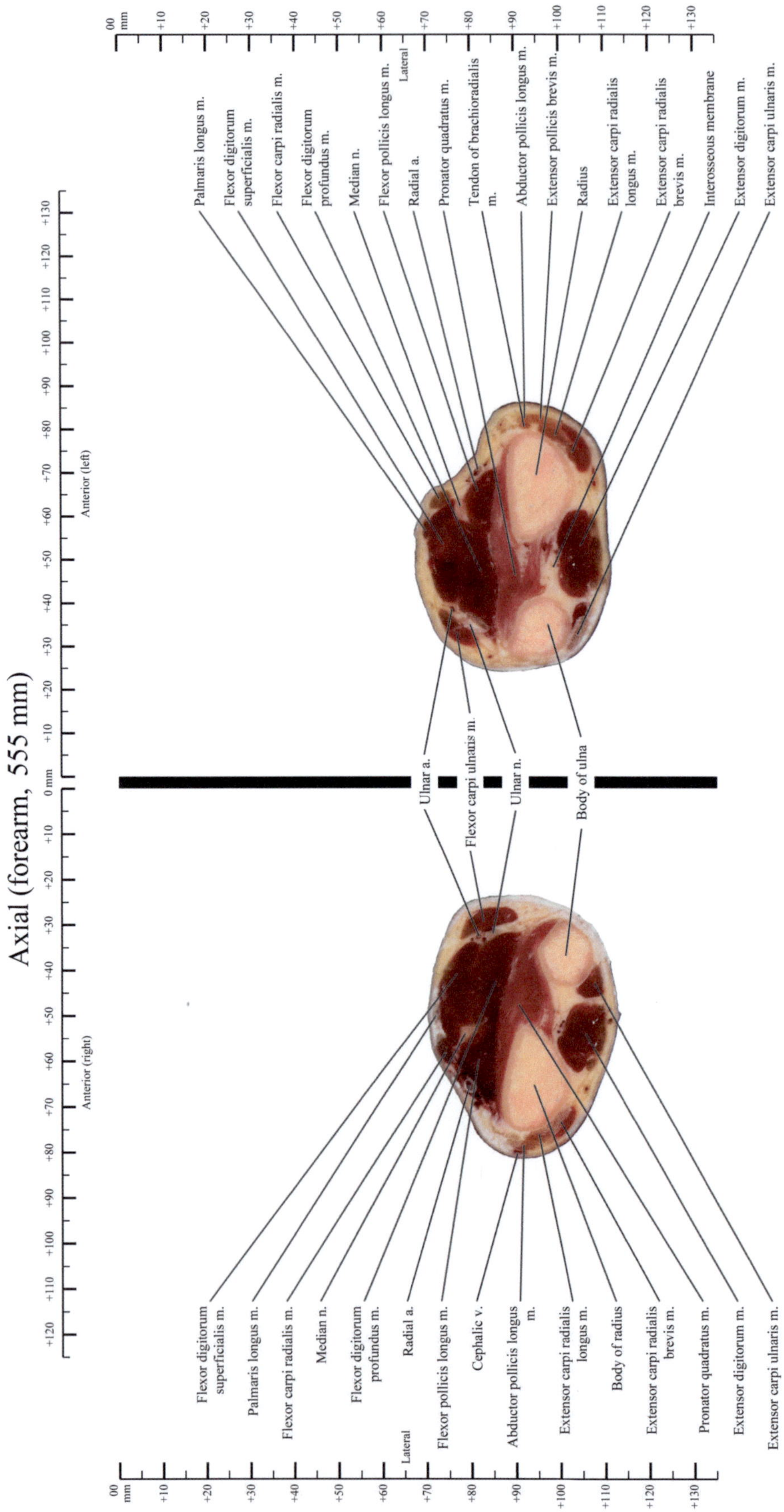
Axial (forearm, 555 mm)
Anterior (left)
Anterior (right)
Lateral
Palmaris longus m.
Flexor digitorum superficialis m.
Flexor carpi radialis m.
Flexor digitorum profundus m.
Median n.
Flexor pollicis longus m.
Radial a.
Pronator quadratus m.
Tendon of brachioradialis m.
Abductor pollicis longus m.
Extensor pollicis brevis m.
Radius
Extensor carpi radialis longus m.
Extensor carpi radialis brevis m.
Interosseous membrane
Extensor digitorum m.
Extensor carpi ulnaris m.
Ulnar a.
Flexor carpi ulnaris m.
Ulnar n.
Body of ulna
Cephalic v.
Body of radius
Flexor digitorum superficialis m.
Palmaris longus m.
Flexor carpi radialis m.
Median n.
Flexor digitorum profundus m.
Radial a.
Flexor pollicis longus m.
Abductor pollicis longus m.
Extensor carpi radialis longus m.
Extensor carpi radialis brevis m.
Pronator quadratus m.
Extensor digitorum m.
Extensor carpi ulnaris m.

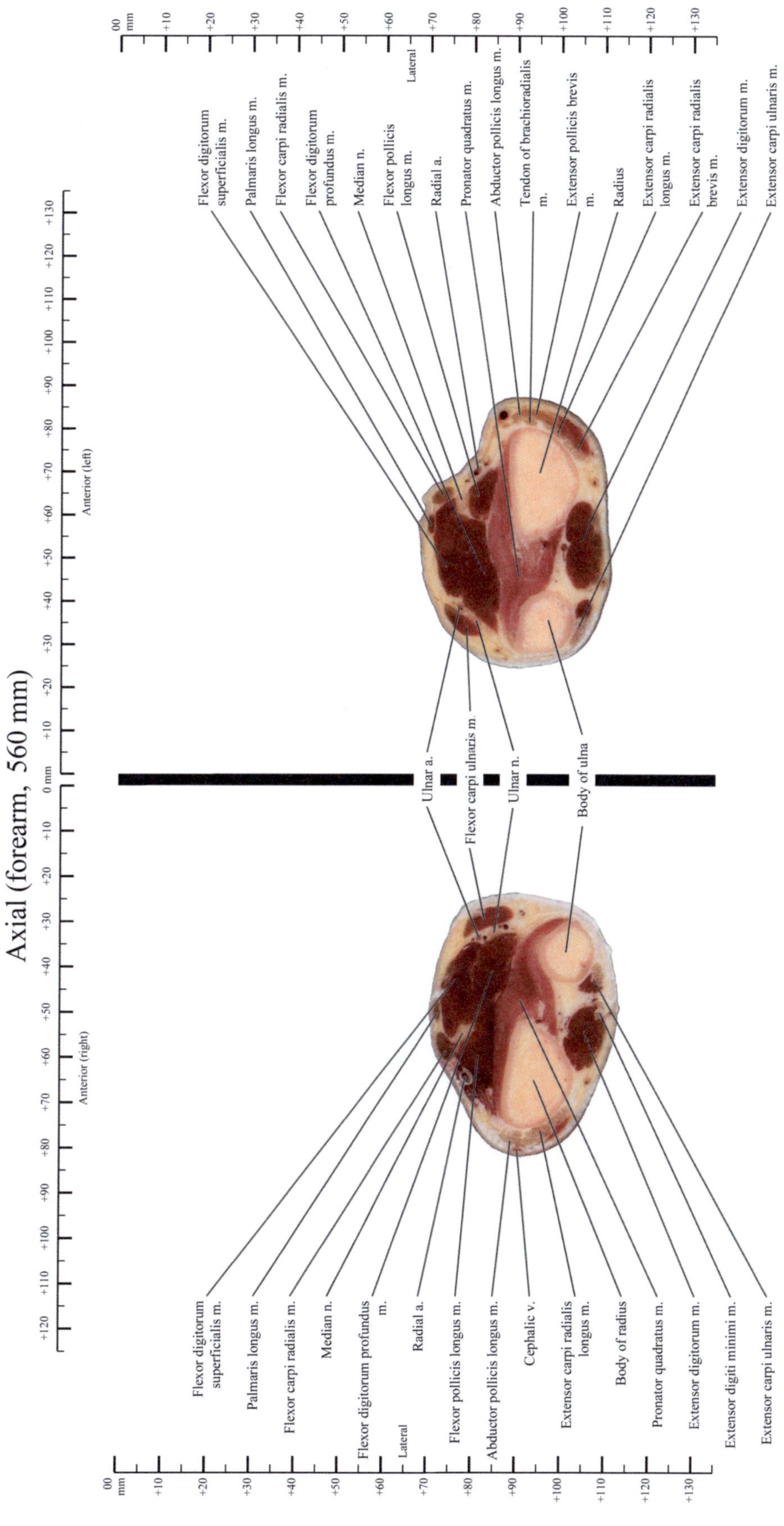
Axial (forearm, 560 mm)
Anterior (left)
Anterior (right)
Lateral
Flexor digitorum superficialis m.
Palmaris longus m.
Flexor carpi radialis m.
Flexor digitorum profundus m.
Median n.
Flexor pollicis longus m.
Radial a.
Pronator quadratus m.
Abductor pollicis longus m.
Tendon of brachioradialis m.
Extensor pollicis brevis m.
Radius
Extensor carpi radialis longus m.
Extensor carpi radialis brevis m.
Extensor digitorum m.
Extensor carpi ulnaris m.
Ulnar a.
Flexor carpi ulnaris m.
Ulnar n.
Body of ulna
Flexor digitorum superficialis m.
Palmaris longus m.
Flexor carpi radialis m.
Median n.
Flexor digitorum profundus m.
Lateral
Radial a.
Flexor pollicis longus m.
Abductor pollicis longus m.
Cephalic v.
Extensor carpi radialis longus m.
Body of radius
Pronator quadratus m.
Extensor digitorum m.
Extensor digiti minimi m.
Extensor carpi ulnaris m.

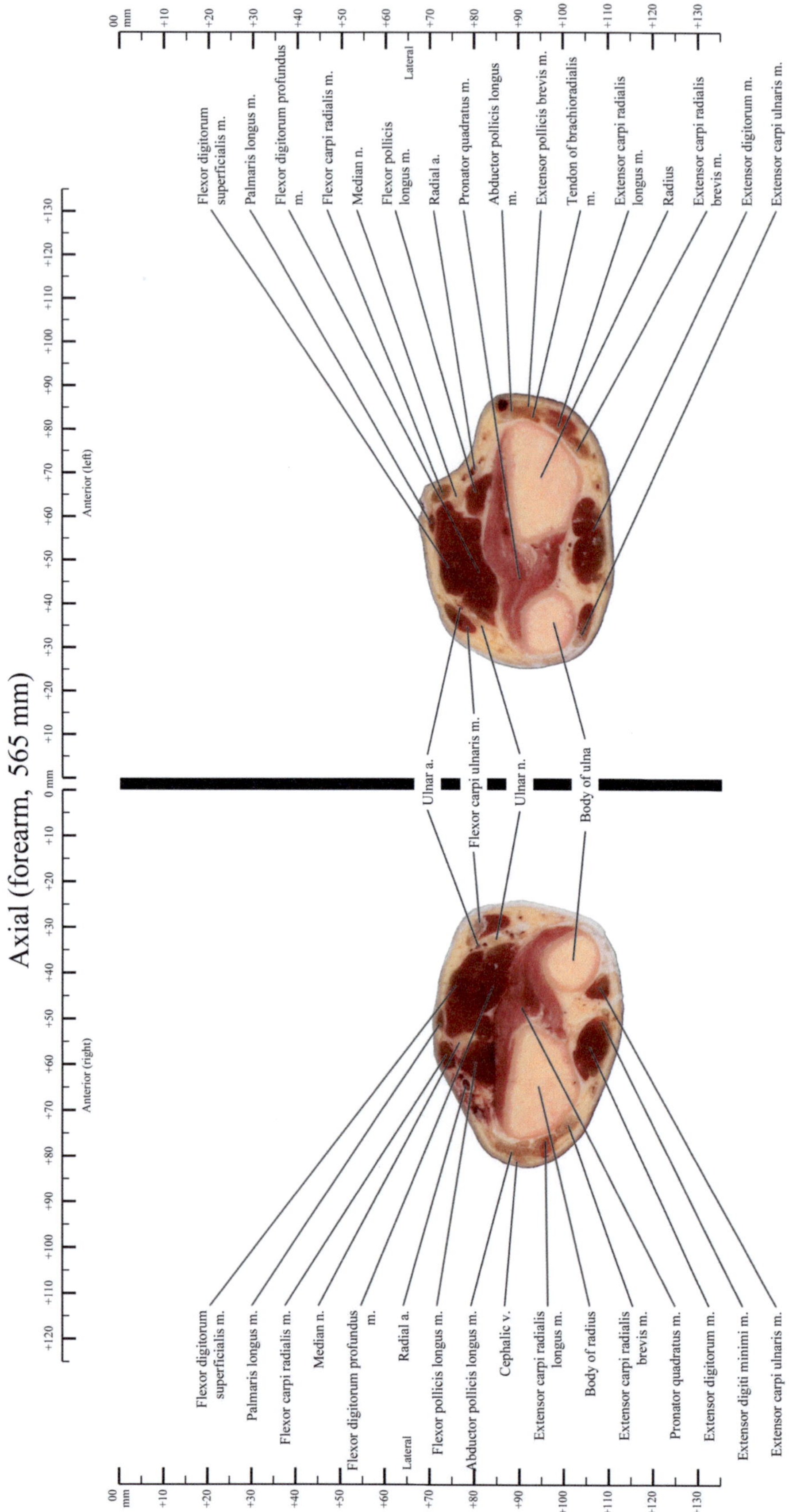
Axial (forearm, 565 mm)
Anterior (left)
Anterior (right)
Lateral
Flexor digitorum superficialis m.
Palmaris longus m.
Flexor digitorum profundus m.
Flexor carpi radialis m.
Median n.
Flexor pollicis longus m.
Radial a.
Pronator quadratus m.
Abductor pollicis longus m.
Extensor pollicis brevis m.
Tendon of brachioradialis m.
Extensor carpi radialis longus m.
Radius
Extensor carpi radialis brevis m.
Extensor digitorum m.
Extensor carpi ulnaris m.
Ulnar a.
Flexor carpi ulnaris m.
Ulnar n.
Body of ulna
Flexor digitorum superficialis m.
Palmaris longus m.
Flexor carpi radialis m.
Median n.
Flexor digitorum profundus m.
Radial a.
Flexor pollicis longus m.
Abductor pollicis longus m.
Cephalic v.
Extensor carpi radialis longus m.
Body of radius
Extensor carpi radialis brevis m.
Pronator quadratus m.
Extensor digitorum m.
Extensor digiti minimi m.
Extensor carpi ulnaris m.

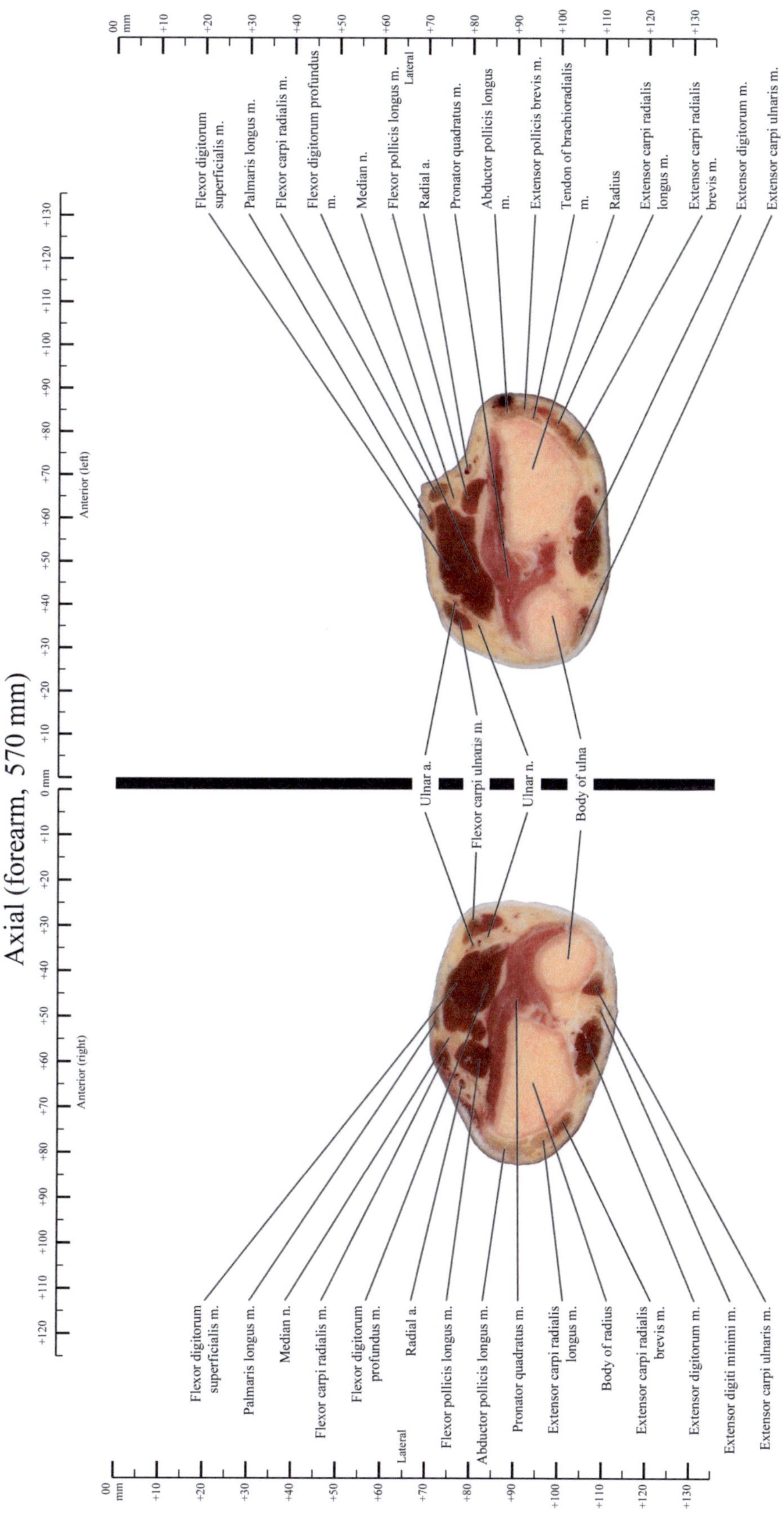
Axial (forearm, 570 mm)
Anterior (left)
Anterior (right)
Lateral
Flexor digitorum superficialis m.
Palmaris longus m.
Flexor carpi radialis m.
Flexor digitorum profundus m.
Median n.
Flexor pollicis longus m.
Radial a.
Pronator quadratus m.
Abductor pollicis longus m.
Extensor pollicis brevis m.
Tendon of brachioradialis m.
Radius
Extensor carpi radialis longus m.
Extensor carpi radialis brevis m.
Extensor digitorum m.
Extensor carpi ulnaris m.
Ulnar a.
Flexor carpi ulnaris m.
Ulnar n.
Body of ulna
Body of radius
Extensor digiti minimi m.

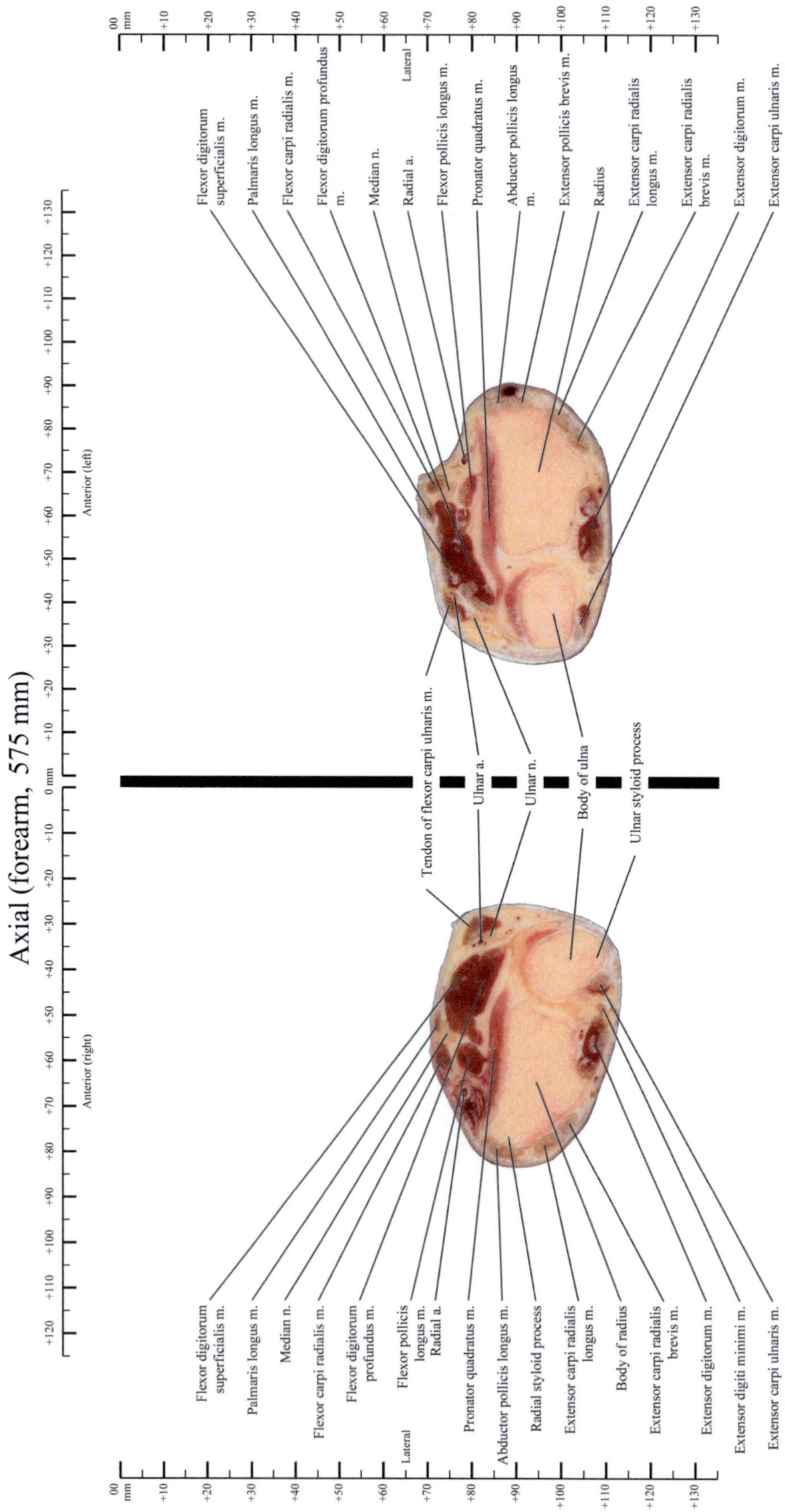
Axial (forearm, 575 mm)
Anterior (left)
Anterior (right)
Lateral
Flexor digitorum superficialis m.
Palmaris longus m.
Flexor carpi radialis m.
Flexor digitorum profundus m.
Median n.
Radial a.
Flexor pollicis longus m.
Pronator quadratus m.
Abductor pollicis longus m.
Extensor pollicis brevis m.
Radius
Extensor carpi radialis longus m.
Extensor carpi radialis brevis m.
Extensor digitorum m.
Extensor carpi ulnaris m.
Tendon of flexor carpi ulnaris m.
Ulnar a.
Ulnar n.
Body of ulna
Ulnar styloid process
Flexor digitorum superficialis m.
Palmaris longus m.
Median n.
Flexor carpi radialis m.
Flexor digitorum profundus m.
Flexor pollicis longus m.
Radial a.
Pronator quadratus m.
Abductor pollicis longus m.
Radial styloid process
Extensor carpi radialis longus m.
Body of radius
Extensor carpi radialis brevis m.
Extensor digitorum m.
Extensor digiti minimi m.
Extensor carpi ulnaris m.

Axial (forearm, 580 mm)

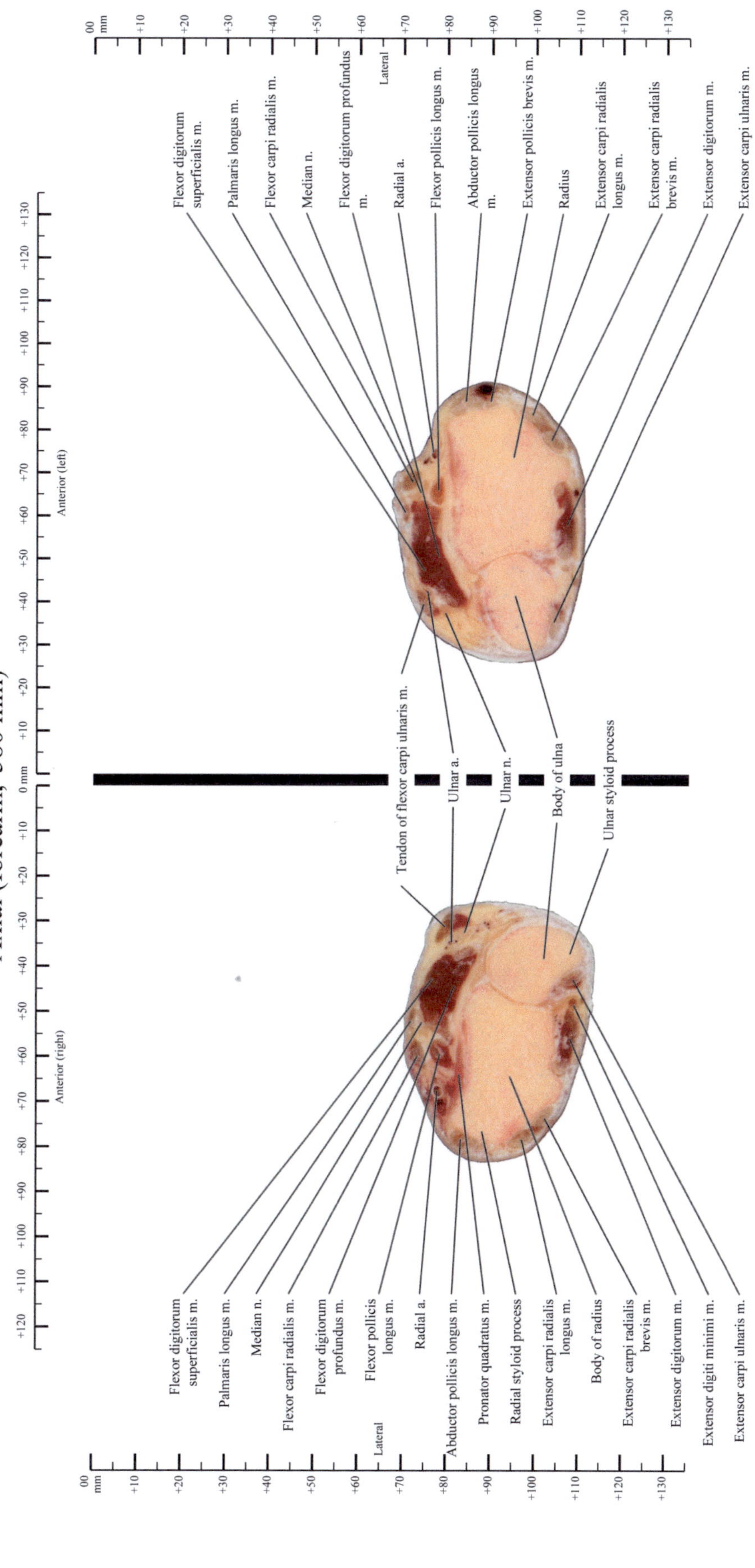

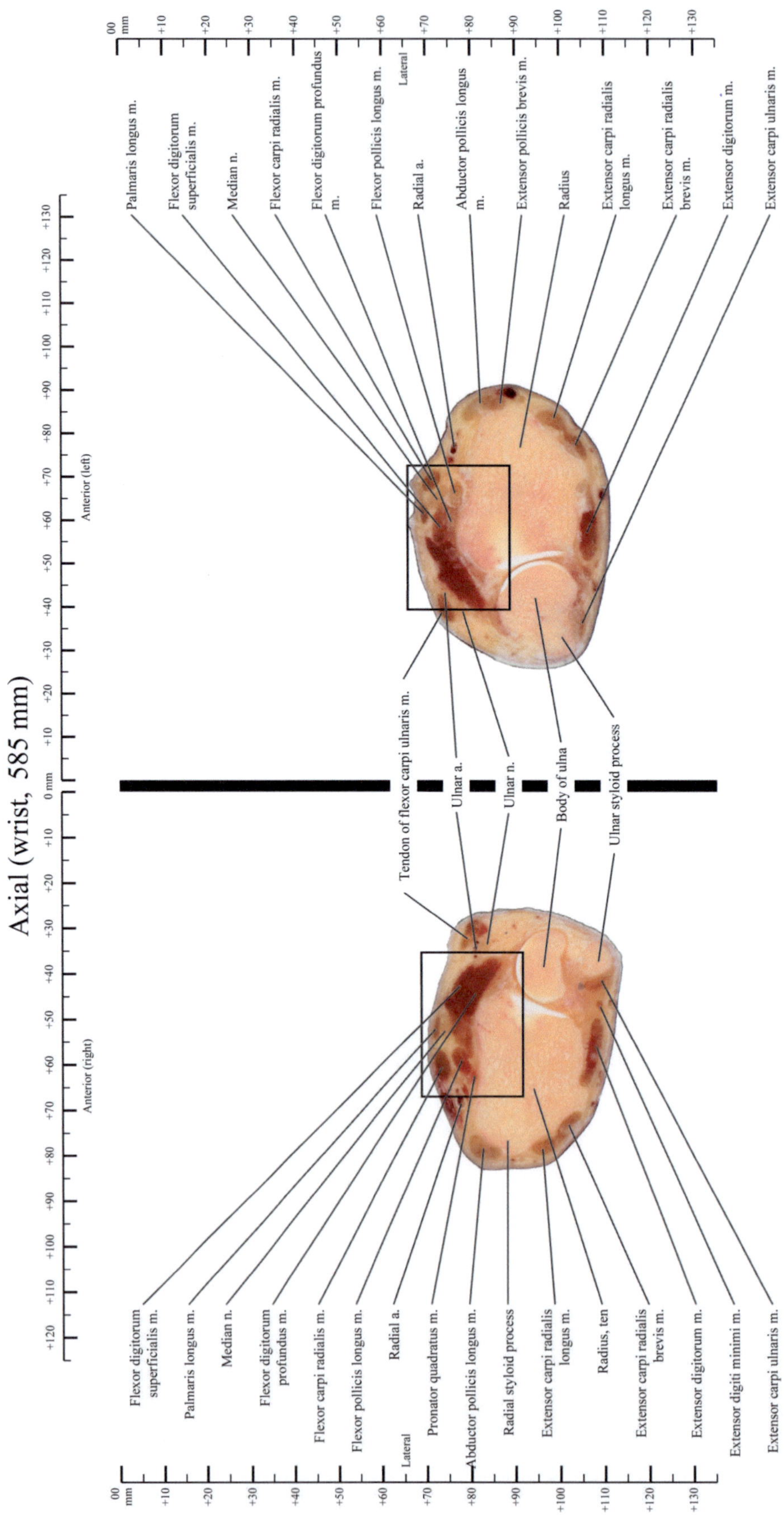
Axial (wrist, 585 mm)
Anterior (left)
Anterior (right)
Lateral
Palmaris longus m.
Flexor digitorum superficialis m.
Median n.
Flexor carpi radialis m.
Flexor digitorum profundus m.
Flexor pollicis longus m.
Radial a.
Abductor pollicis longus m.
Extensor pollicis brevis m.
Radius
Extensor carpi radialis longus m.
Extensor carpi radialis brevis m.
Extensor digitorum m.
Extensor carpi ulnaris m.
Tendon of flexor carpi ulnaris m.
Ulnar a.
Ulnar n.
Body of ulna
Ulnar styloid process
Flexor digitorum superficialis m.
Palmaris longus m.
Median n.
Flexor digitorum profundus m.
Flexor carpi radialis m.
Flexor pollicis longus m.
Radial a.
Pronator quadratus m.
Abductor pollicis longus m.
Radial styloid process
Extensor carpi radialis longus m.
Radius, ten
Extensor carpi radialis brevis m.
Extensor digitorum m.
Extensor digiti minimi m.
Extensor carpi ulnaris m.

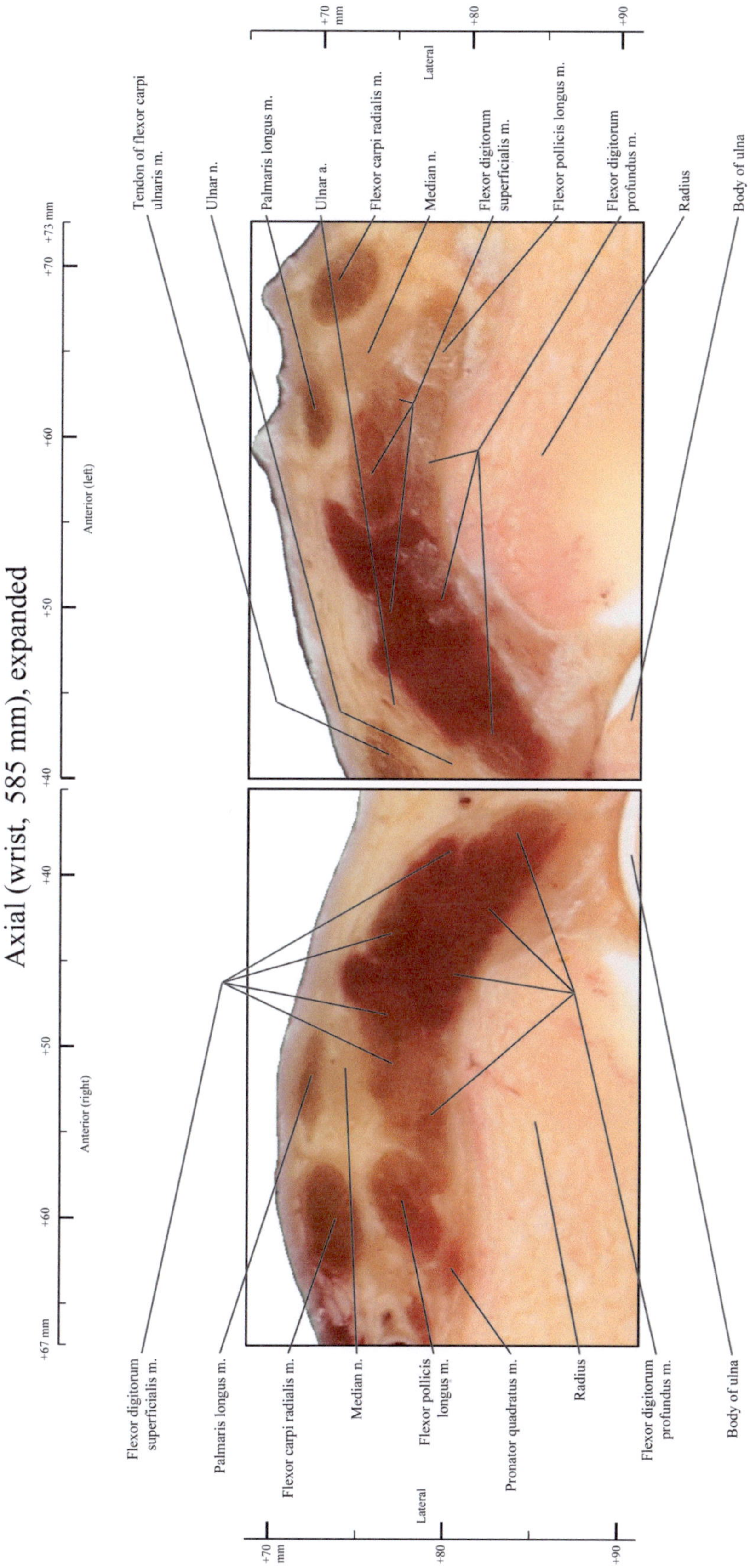

Axial (wrist, 585 mm), expanded

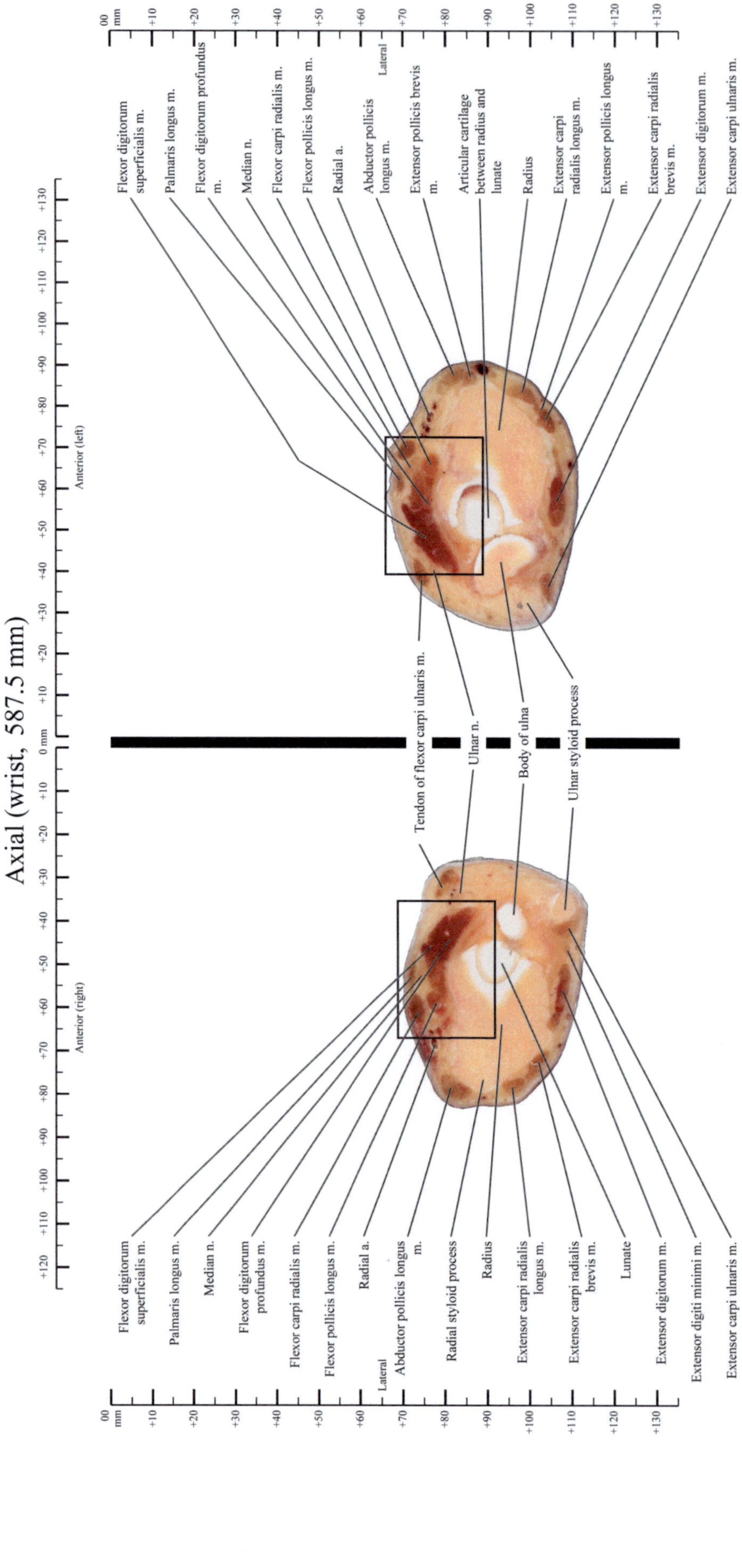
Axial (wrist, 587.5 mm)
Anterior (left)
Anterior (right)
Lateral
Flexor digitorum superficialis m.
Palmaris longus m.
Flexor digitorum profundus m.
Median n.
Flexor carpi radialis m.
Flexor pollicis longus m.
Radial a.
Abductor pollicis longus m.
Extensor pollicis brevis m.
Articular cartilage between radius and lunate
Radius
Extensor carpi radialis longus m.
Extensor pollicis longus m.
Extensor carpi radialis brevis m.
Extensor digitorum m.
Extensor carpi ulnaris m.
Tendon of flexor carpi ulnaris m.
Ulnar n.
Body of ulna
Ulnar styloid process
Radial styloid process
Lunate
Extensor digiti minimi m.

Axial (wrist, 587.5 mm), expanded

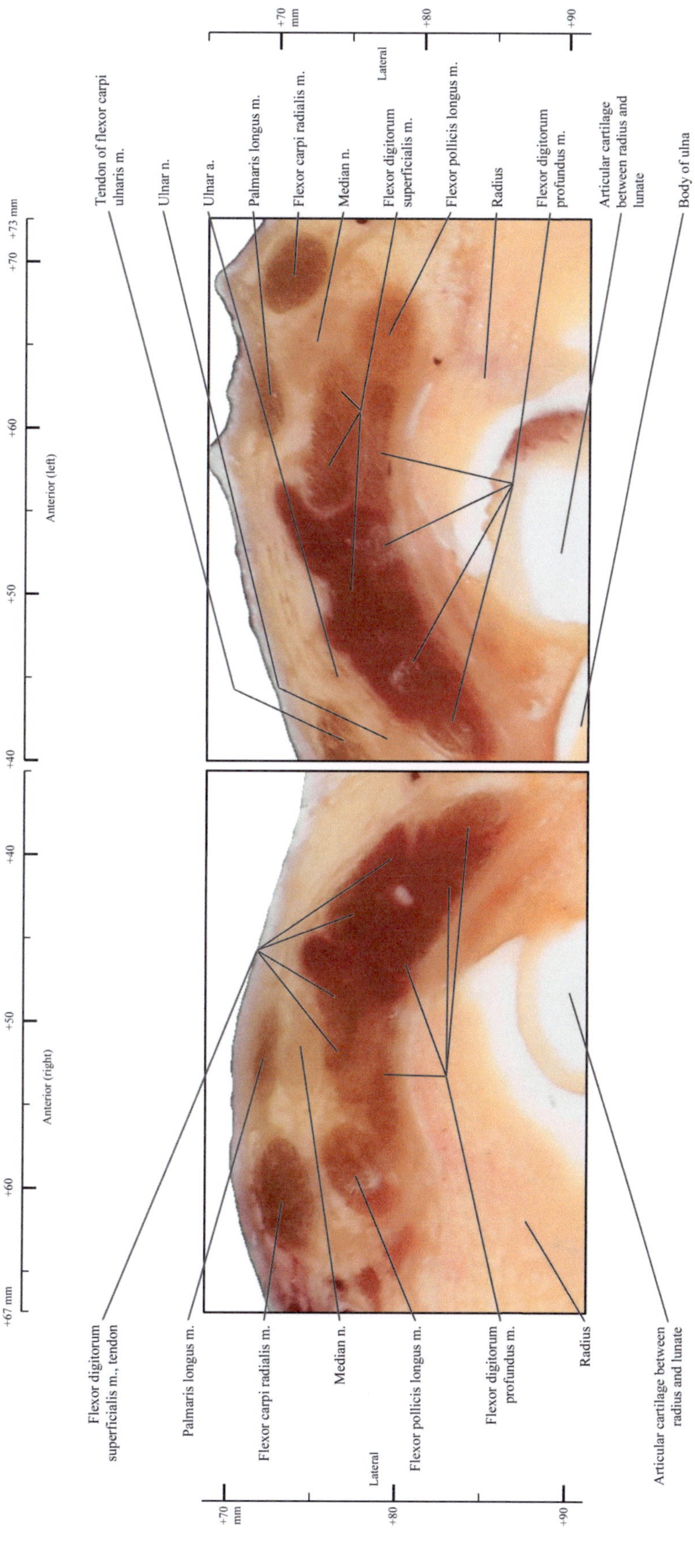
Anterior (left)
+40
+50
+60
+70
+73 mm
Tendon of flexor carpi ulnaris m.
Ulnar n.
Ulnar a.
Palmaris longus m.
Flexor carpi radialis m.
Median n.
Flexor digitorum superficialis m.
Flexor pollicis longus m.
Radius
Flexor digitorum profundus m.
Articular cartilage between radius and lunate
Body of ulna
Lateral
+70 mm
+80
+90
Anterior (right)
+67 mm
+60
+50
+40
Flexor digitorum superficialis m., tendon
Palmaris longus m.
Flexor carpi radialis m.
Median n.
Flexor pollicis longus m.
Flexor digitorum profundus m.
Radius
Articular cartilage between radius and lunate
Lateral
+70 mm
+80
+90

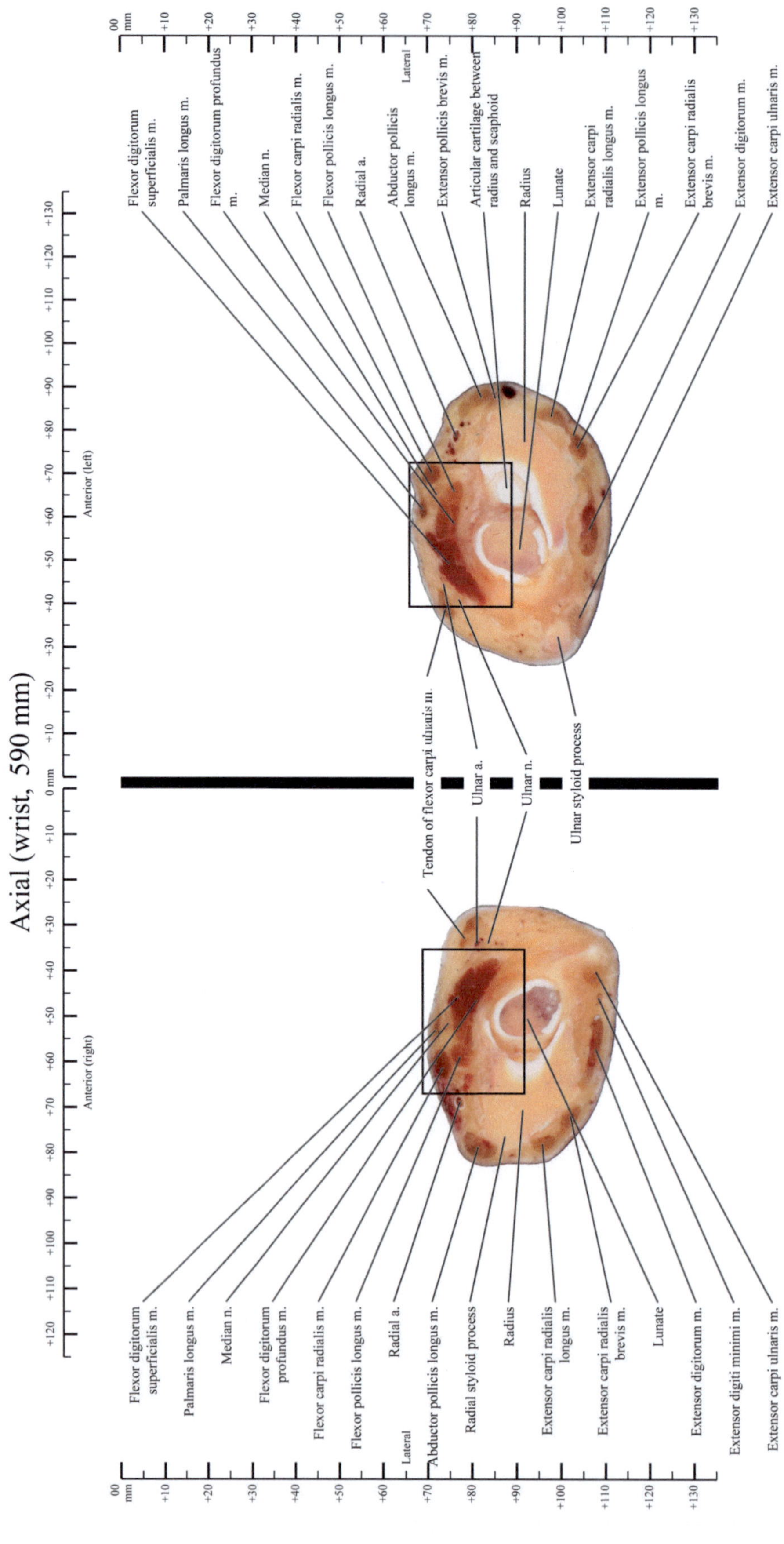
Axial (wrist, 590 mm)
Anterior (left)
Anterior (right)
Lateral
Flexor digitorum superficialis m.
Palmaris longus m.
Flexor digitorum profundus m.
Median n.
Flexor carpi radialis m.
Flexor pollicis longus m.
Radial a.
Abductor pollicis longus m.
Extensor pollicis brevis m.
Articular cartilage between radius and scaphoid
Radius
Lunate
Extensor carpi radialis longus m.
Extensor pollicis longus m.
Extensor carpi radialis brevis m.
Extensor digitorum m.
Extensor carpi ulnaris m.
Tendon of flexor carpi ulnaris m.
Ulnar a.
Ulnar n.
Ulnar styloid process
Radial styloid process
Extensor digiti minimi m.

Axial (wrist, 590 mm), expanded

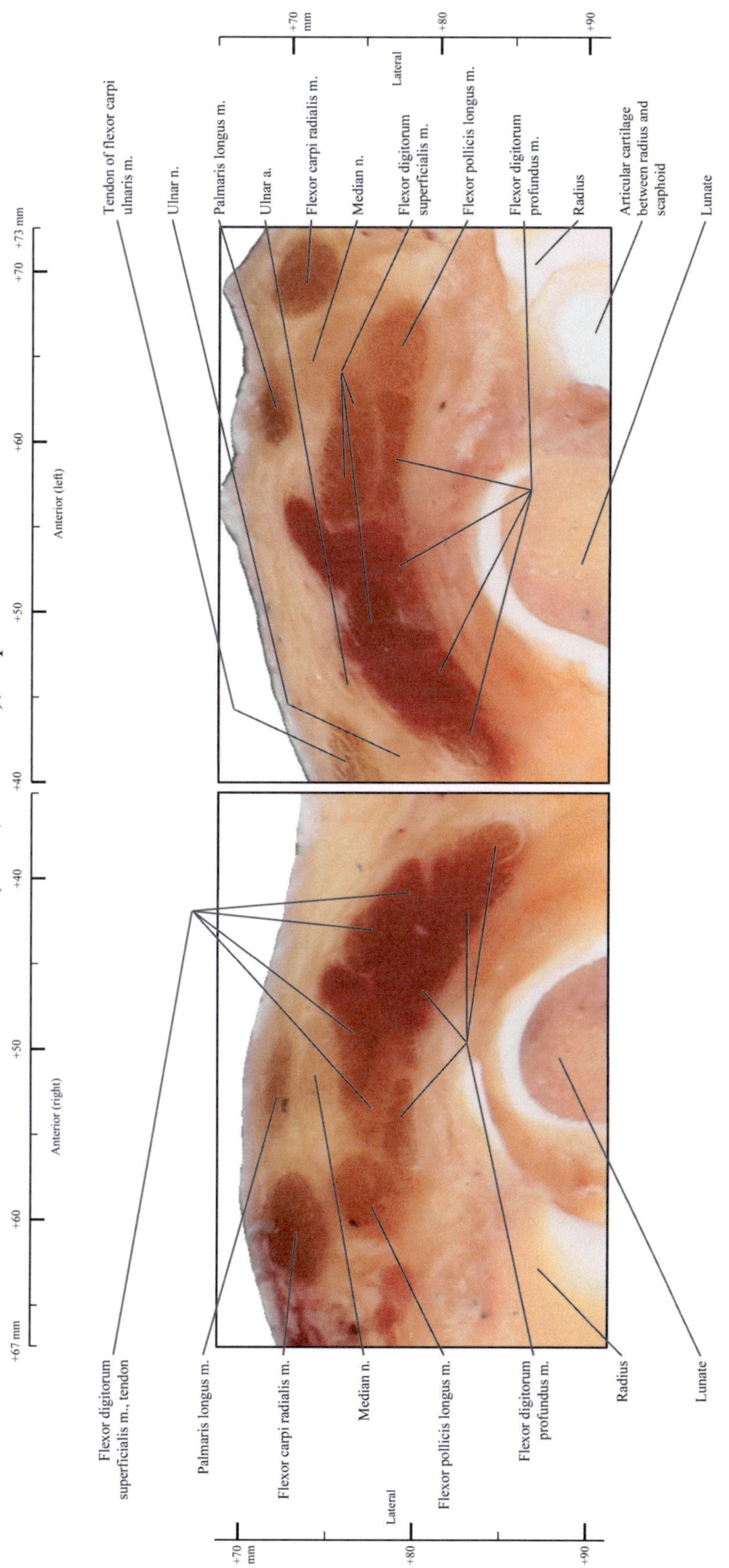

Axial (wrist, 592.5 mm)

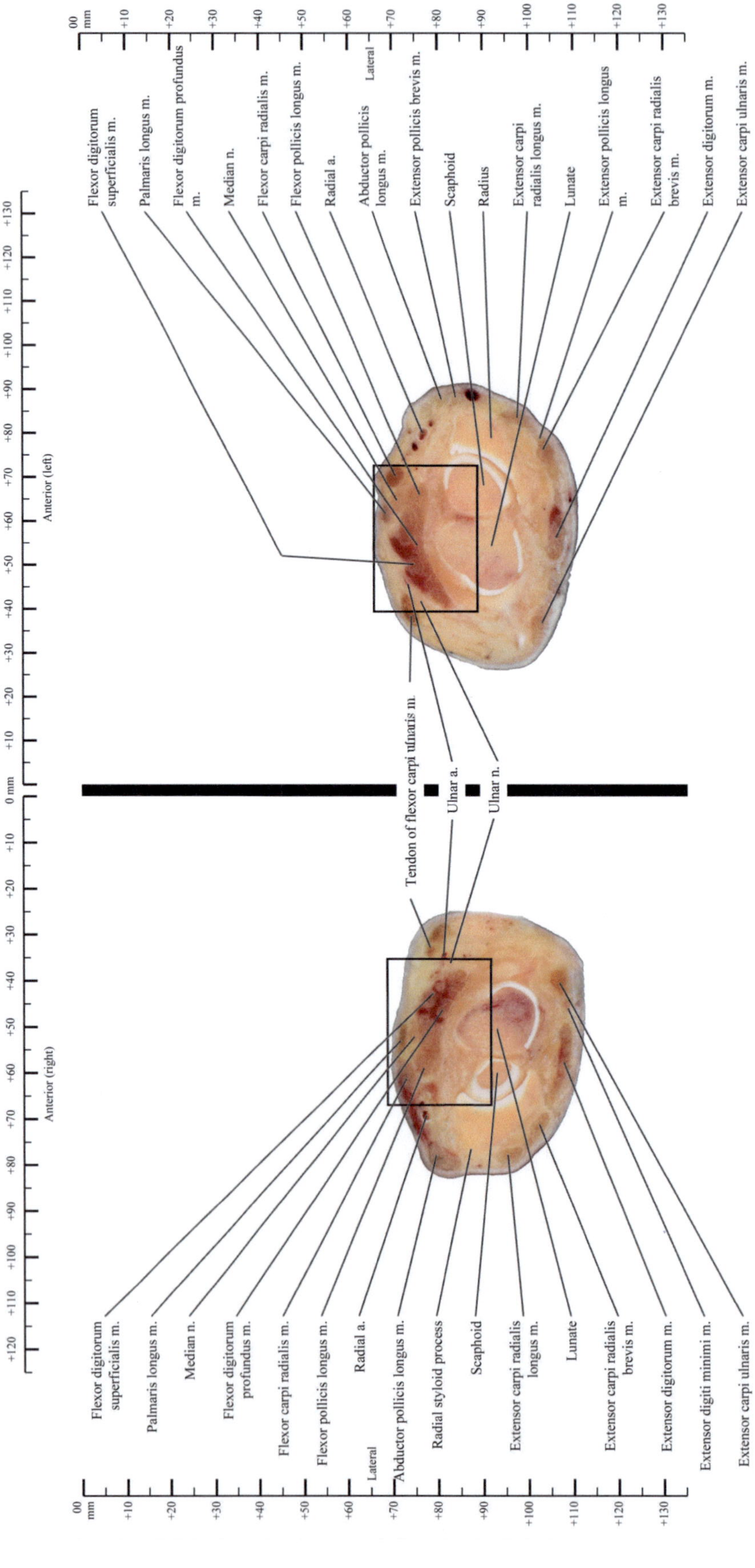

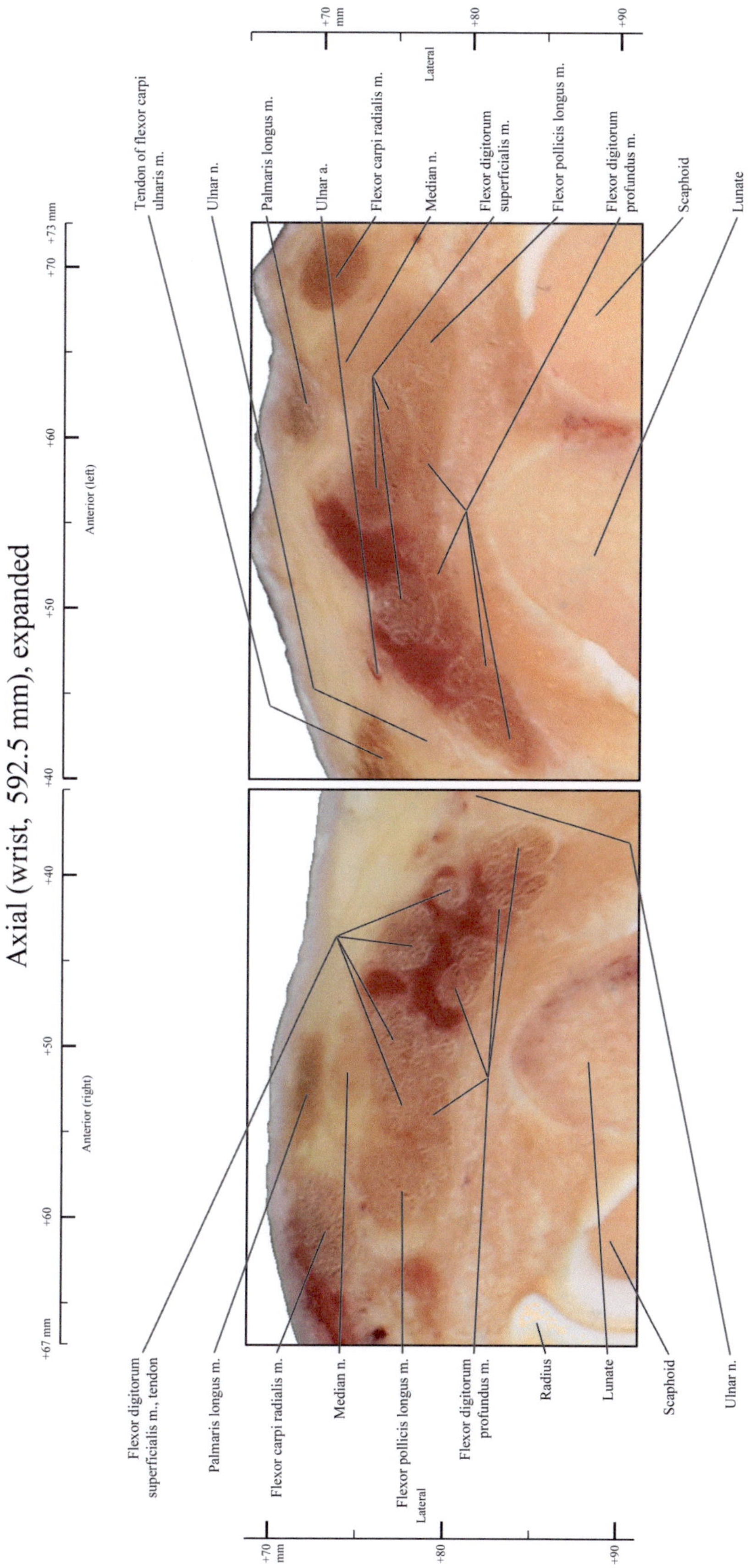
Axial (wrist, 592.5 mm), expanded
Anterior (left)
Anterior (right)
Lateral
Tendon of flexor carpi ulnaris m.
Ulnar n.
Palmaris longus m.
Ulnar a.
Flexor carpi radialis m.
Median n.
Flexor digitorum superficialis m.
Flexor pollicis longus m.
Flexor digitorum profundus m.
Scaphoid
Lunate
Flexor digitorum superficialis m., tendon
Palmaris longus m.
Flexor carpi radialis m.
Median n.
Flexor pollicis longus m.
Flexor digitorum profundus m.
Radius
Lunate
Scaphoid
Ulnar n.
+40
+50
+60
+70
+73 mm
+67 mm
+70 mm
+80
+90

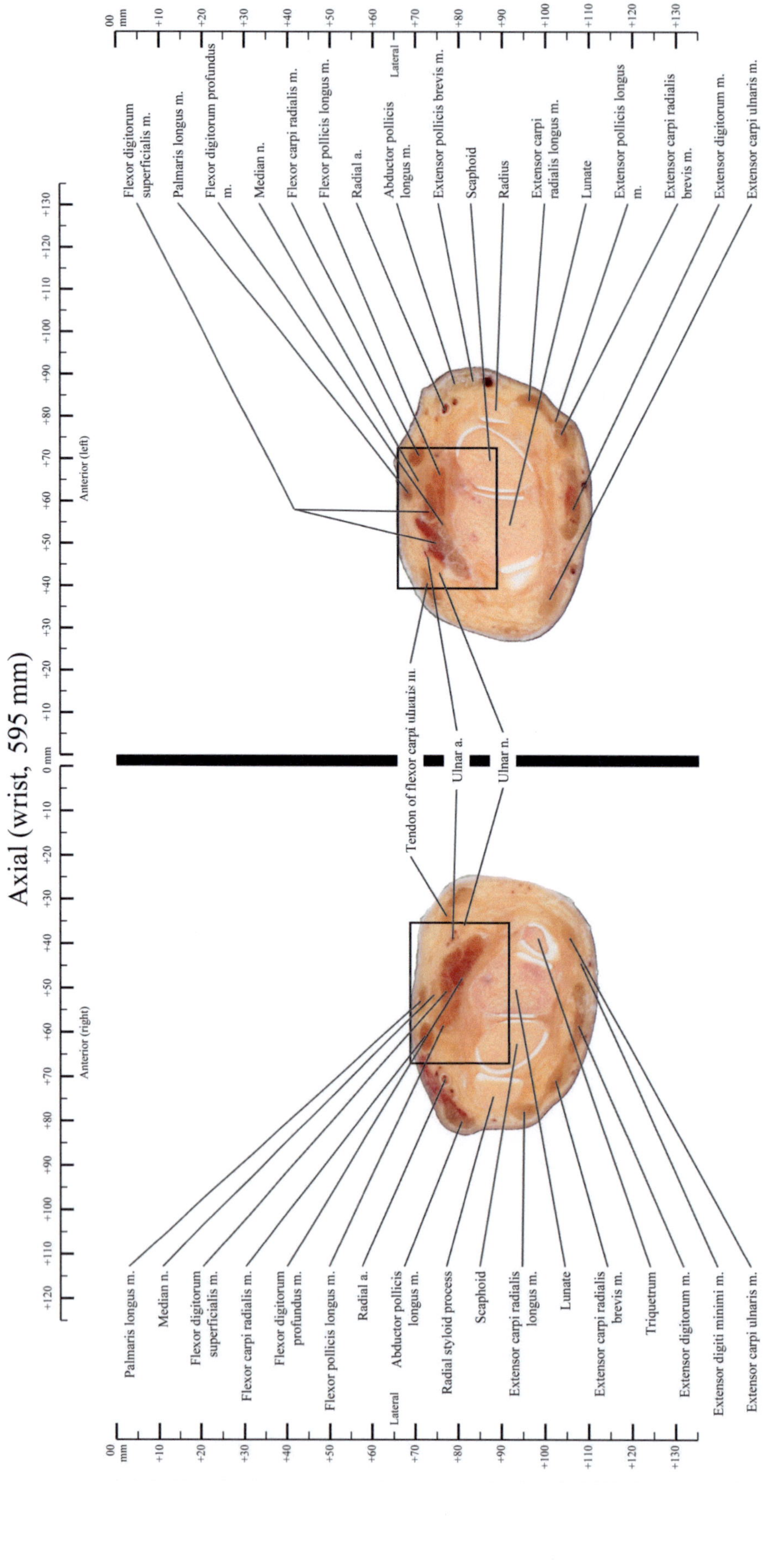
Axial (wrist, 595 mm)
Anterior (left)
Anterior (right)
Lateral
Flexor digitorum superficialis m.
Palmaris longus m.
Flexor digitorum profundus m.
Median n.
Flexor carpi radialis m.
Flexor pollicis longus m.
Radial a.
Abductor pollicis longus m.
Extensor pollicis brevis m.
Scaphoid
Radius
Extensor carpi radialis longus m.
Lunate
Extensor pollicis longus m.
Extensor carpi radialis brevis m.
Extensor digitorum m.
Extensor carpi ulnaris m.
Tendon of flexor carpi ulnaris m.
Ulnar a.
Ulnar n.
Palmaris longus m.
Median n.
Flexor digitorum superficialis m.
Flexor carpi radialis m.
Flexor digitorum profundus m.
Flexor pollicis longus m.
Radial a.
Abductor pollicis longus m.
Radial styloid process
Scaphoid
Extensor carpi radialis longus m.
Lunate
Extensor carpi radialis brevis m.
Triquetrum
Extensor digitorum m.
Extensor digiti minimi m.
Extensor carpi ulnaris m.

Axial (wrist, 595 mm), expanded

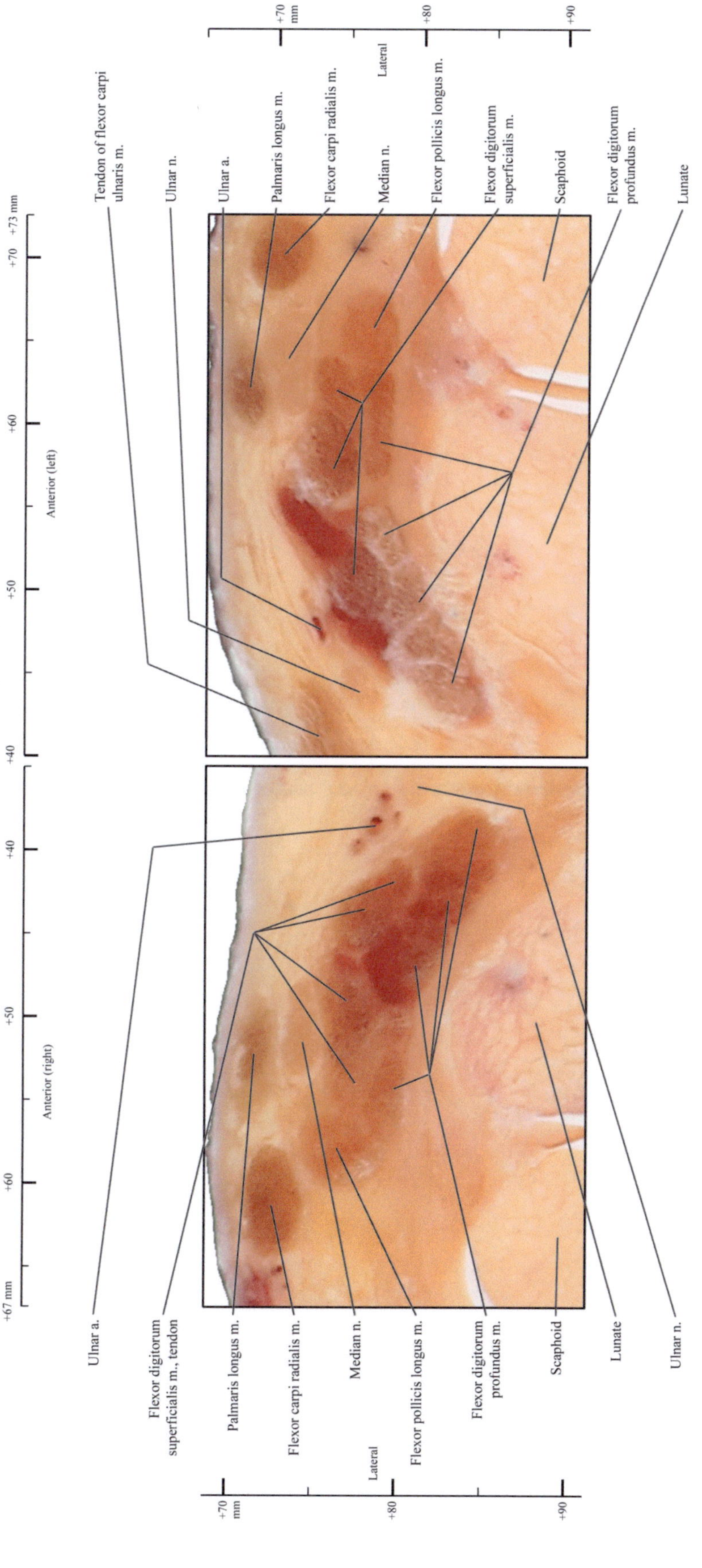

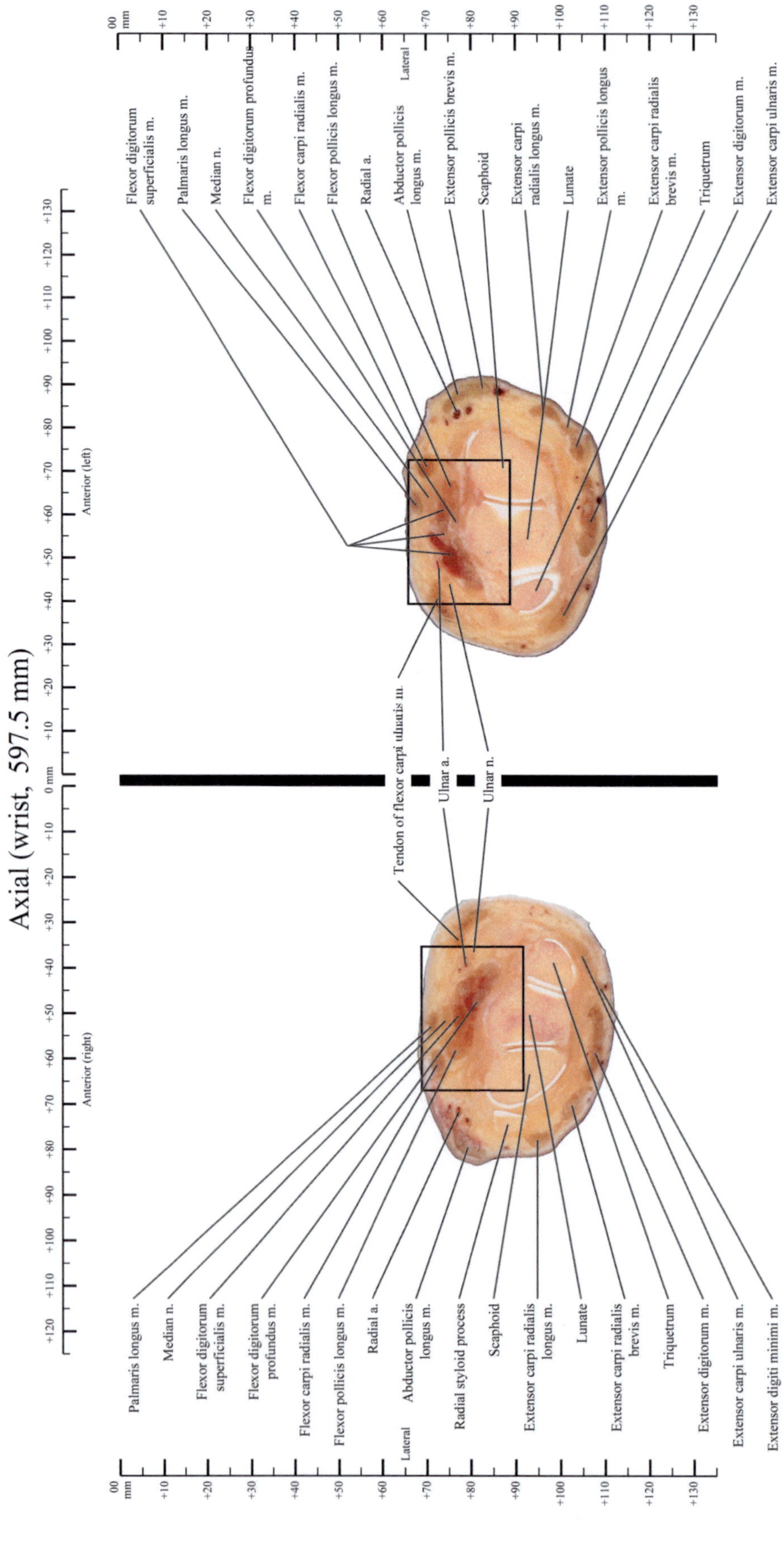
Axial (wrist, 597.5 mm)
Anterior (left)
Anterior (right)
Lateral
Lateral
Flexor digitorum superficialis m.
Palmaris longus m.
Median n.
Flexor digitorum profundus m.
Flexor carpi radialis m.
Flexor pollicis longus m.
Radial a.
Abductor pollicis longus m.
Extensor pollicis brevis m.
Scaphoid
Extensor carpi radialis longus m.
Lunate
Extensor pollicis longus m.
Extensor carpi radialis brevis m.
Triquetrum
Extensor digitorum m.
Extensor carpi ulnaris m.
Tendon of flexor carpi ulnaris m.
Ulnar a.
Ulnar n.
Palmaris longus m.
Median n.
Flexor digitorum superficialis m.
Flexor digitorum profundus m.
Flexor carpi radialis m.
Flexor pollicis longus m.
Radial a.
Abductor pollicis longus m.
Radial styloid process
Scaphoid
Extensor carpi radialis longus m.
Lunate
Extensor carpi radialis brevis m.
Triquetrum
Extensor digitorum m.
Extensor carpi ulnaris m.
Extensor digiti minimi m.

Axial (wrist, 597.5 mm), expanded

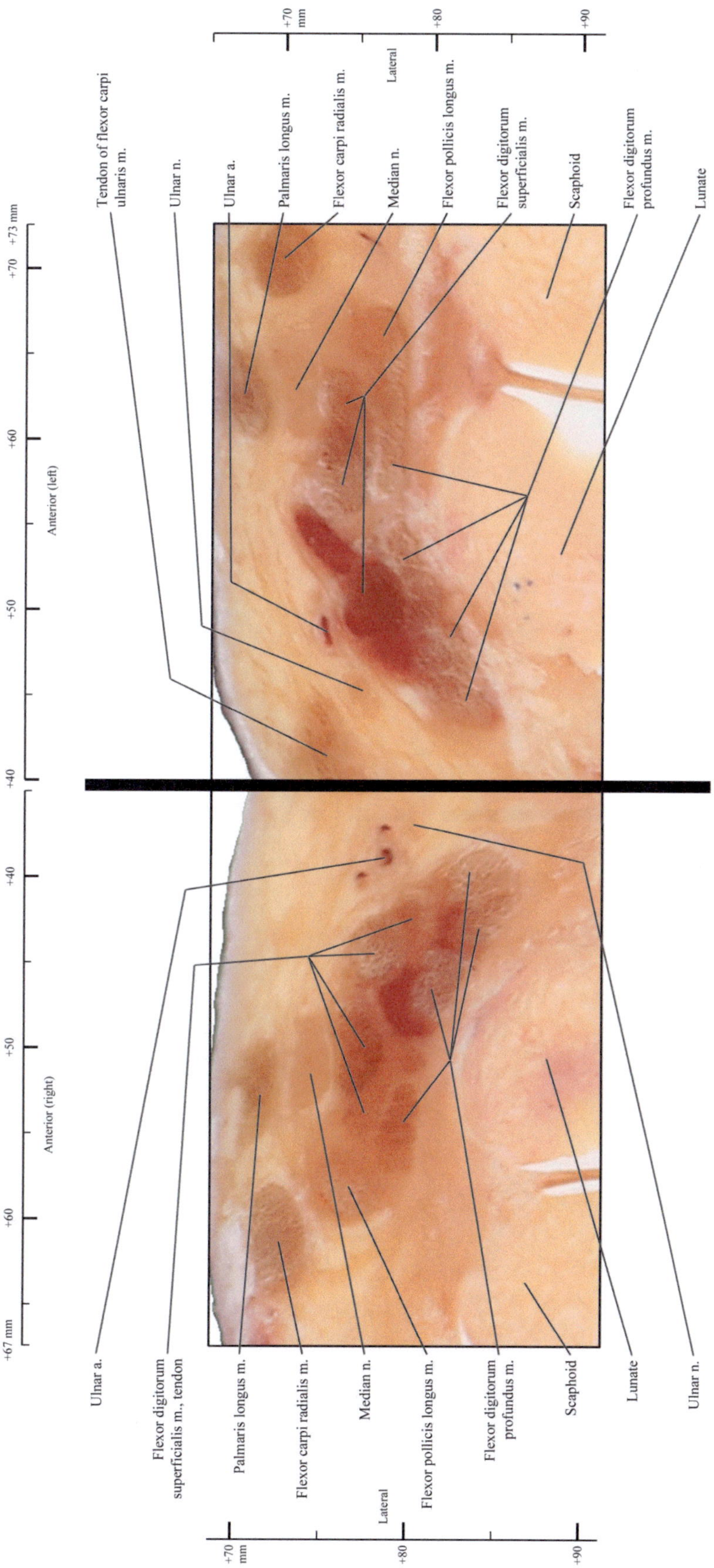

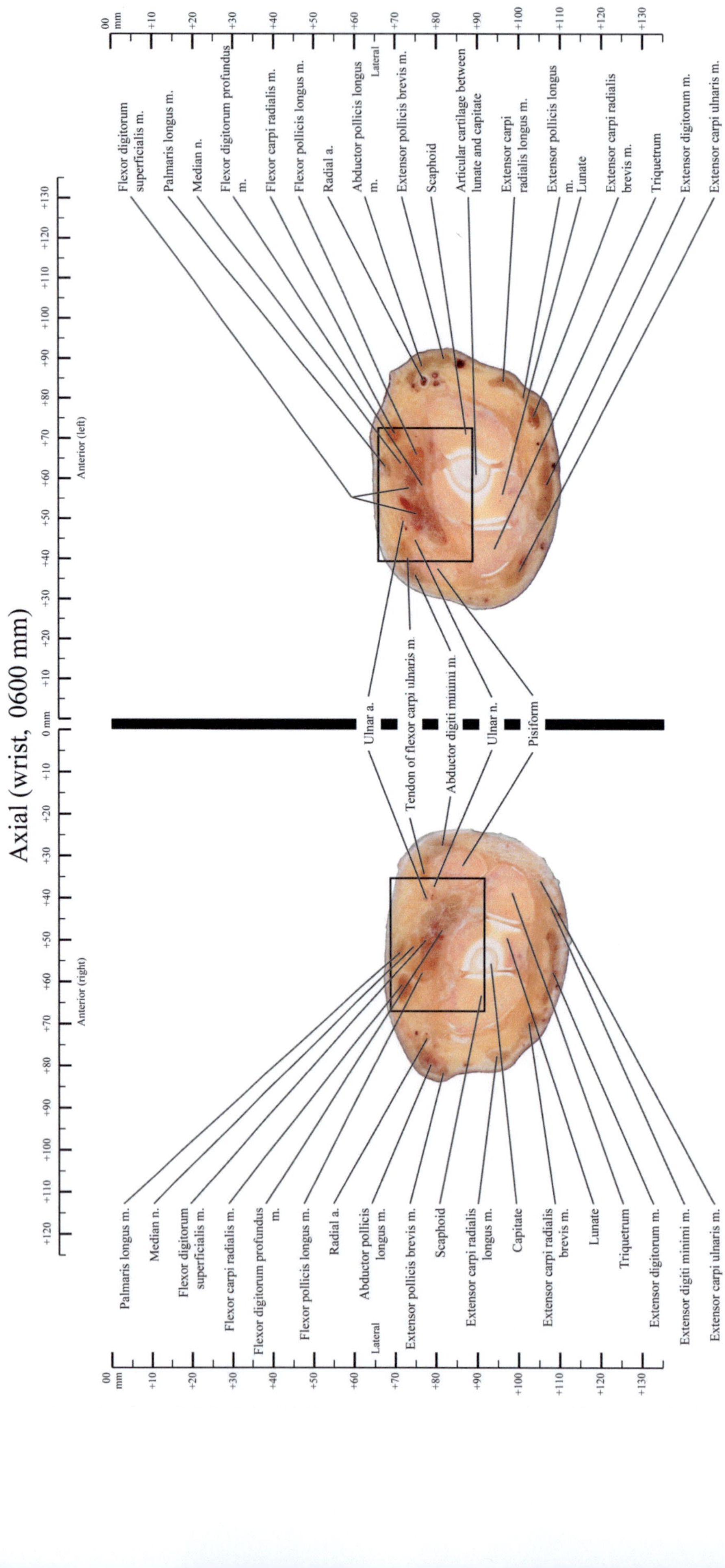
Axial (wrist, 0600 mm)
Anterior (left)
Anterior (right)
Flexor digitorum superficialis m.
Palmaris longus m.
Median n.
Flexor digitorum profundus m.
Flexor carpi radialis m.
Flexor pollicis longus m.
Radial a.
Abductor pollicis longus m.
Lateral
Extensor pollicis brevis m.
Scaphoid
Articular cartilage between lunate and capitate
Extensor carpi radialis longus m.
Extensor pollicis longus m.
Lunate
Extensor carpi radialis brevis m.
Triquetrum
Extensor digitorum m.
Extensor carpi ulnaris m.
Ulnar a.
Tendon of flexor carpi ulnaris m.
Abductor digiti minimi m.
Ulnar n.
Pisiform
Palmaris longus m.
Median n.
Flexor digitorum superficialis m.
Flexor carpi radialis m.
Flexor digitorum profundus m.
Flexor pollicis longus m.
Radial a.
Abductor pollicis longus m.
Lateral
Extensor pollicis brevis m.
Scaphoid
Extensor carpi radialis longus m.
Capitate
Extensor carpi radialis brevis m.
Lunate
Triquetrum
Extensor digitorum m.
Extensor digiti minimi m.
Extensor carpi ulnaris m.

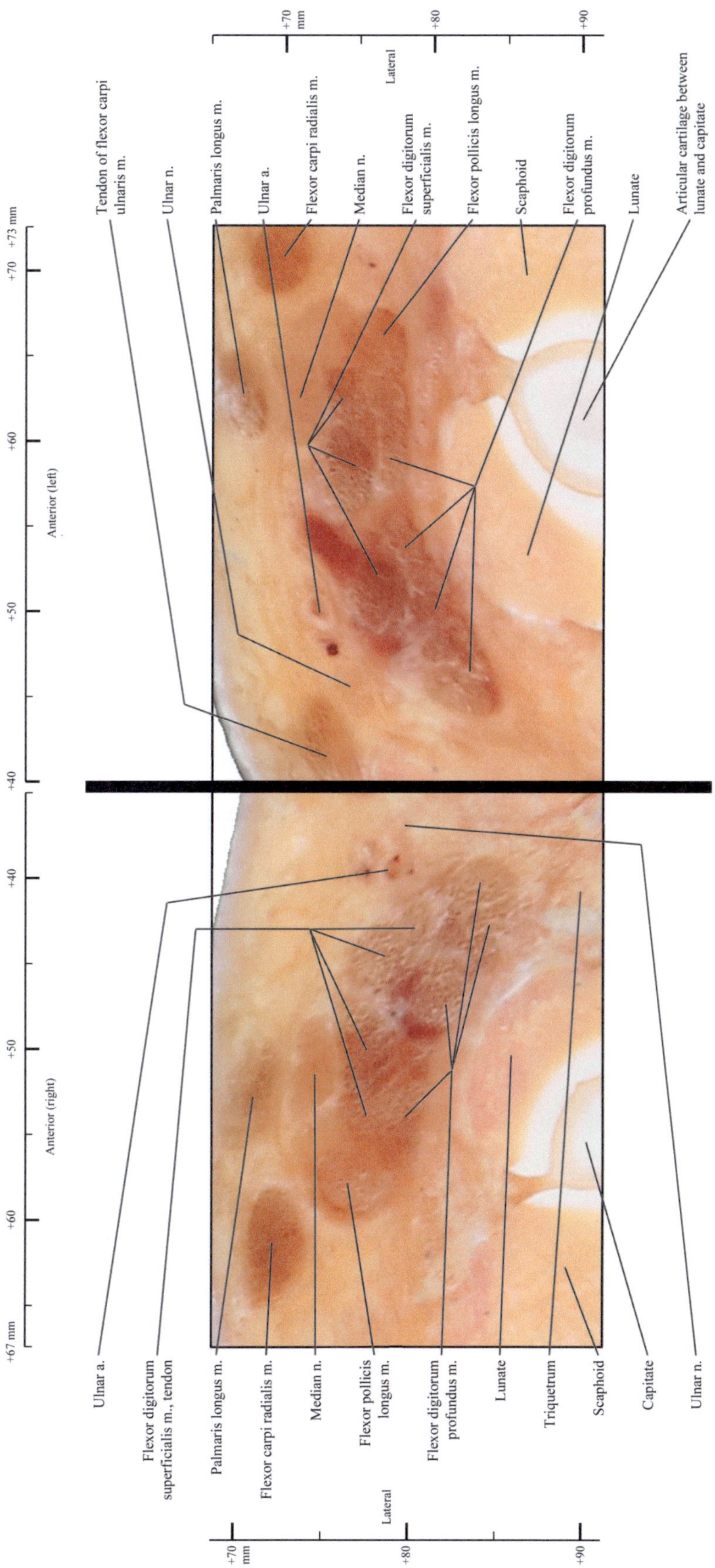
Tendon of flexor carpi ulnaris m.
Ulnar n.
Palmaris longus m.
Ulnar a.
Flexor carpi radialis m.
Median n.
Flexor digitorum superficialis m.
Flexor pollicis longus m.
Scaphoid
Flexor digitorum profundus m.
Lunate
Articular cartilage between lunate and capitate
Lateral
Anterior (left)
+70 +73 mm
+60
+50
+40
+70 mm
+80
+90
Ulnar a.
Flexor digitorum superficialis m., tendon
Palmaris longus m.
Flexor carpi radialis m.
Median n.
Flexor pollicis longus m.
Flexor digitorum profundus m.
Lunate
Triquetrum
Scaphoid
Capitate
Ulnar n.
Lateral
Anterior (right)
+67 mm
+60
+50
+40
+70 mm
+80
+90

Axial (wrist, 0600 mm), expanded

Axial (wrist, 0602.5 mm)

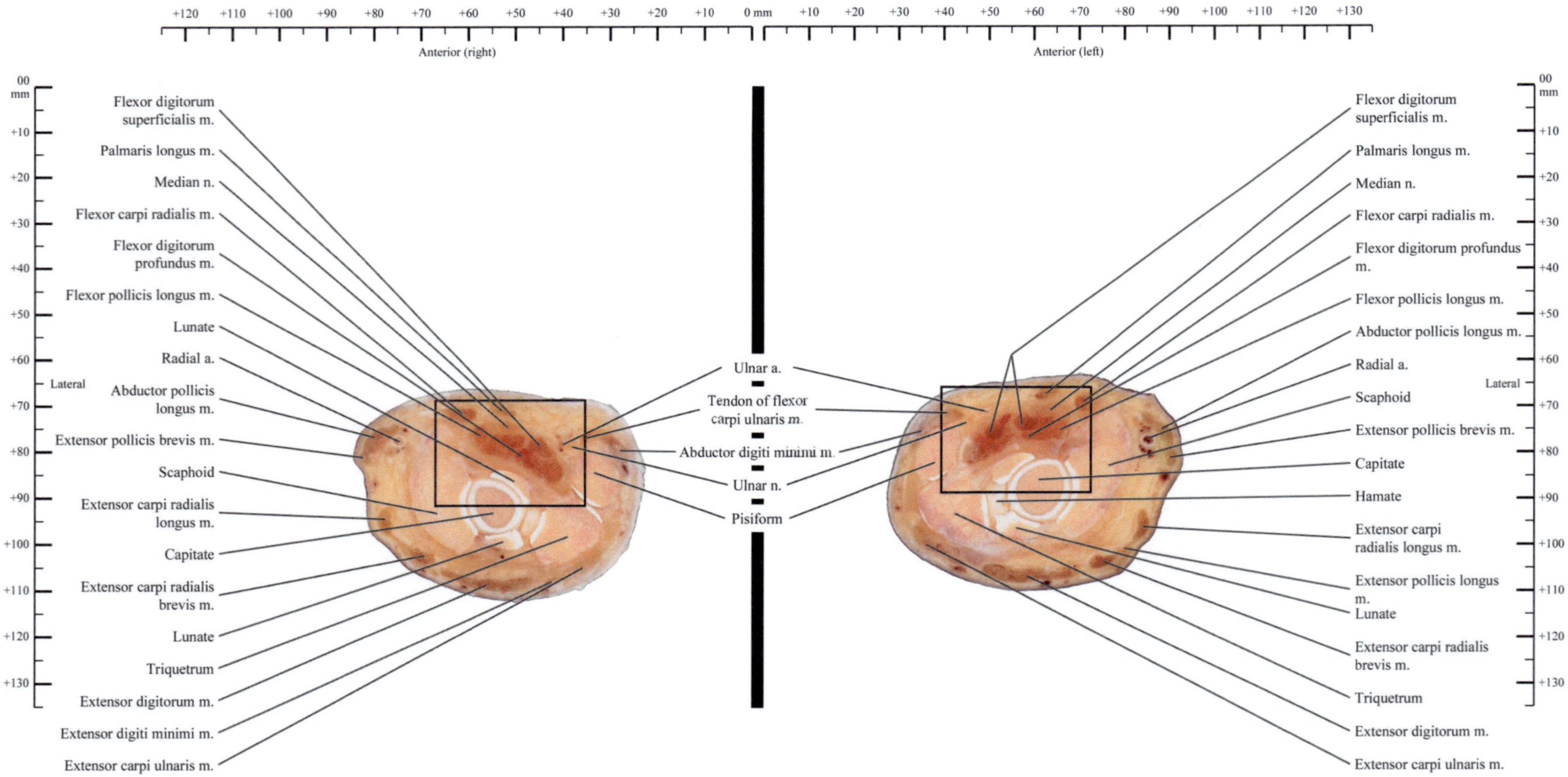

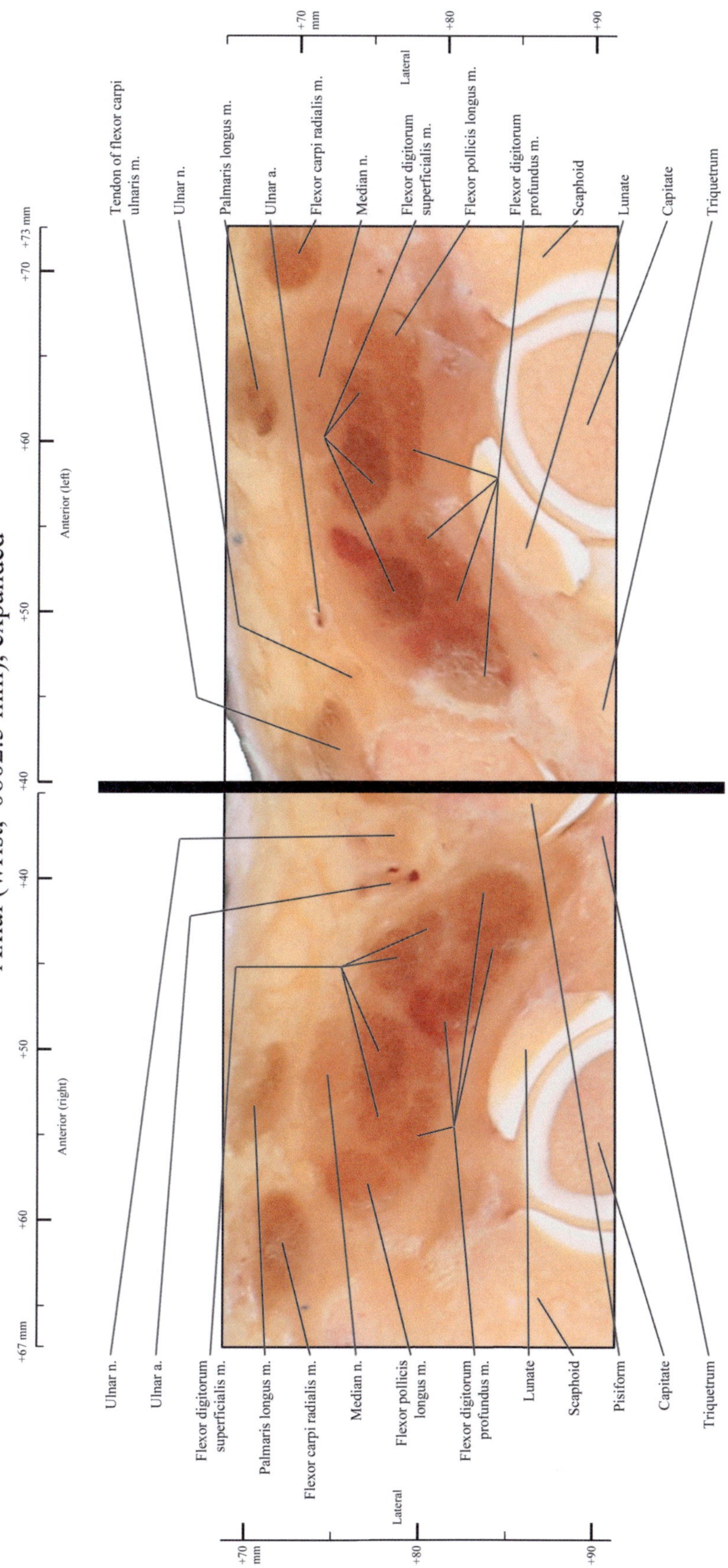

Axial (wrist, 0602.5 mm), expanded

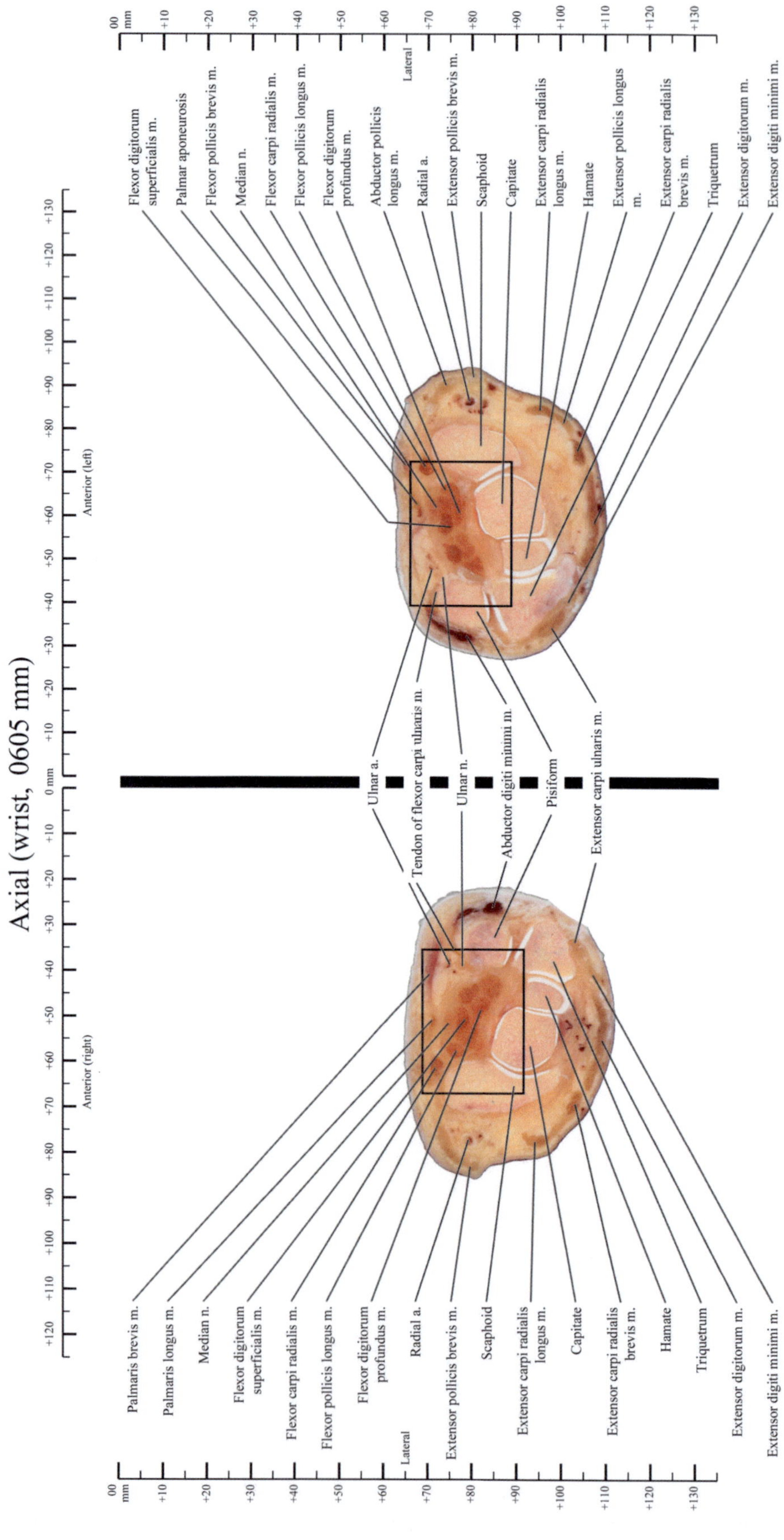
Axial (wrist, 0605 mm)
Anterior (left)
Anterior (right)
Lateral
Flexor digitorum superficialis m.
Palmar aponeurosis
Flexor pollicis brevis m.
Median n.
Flexor carpi radialis m.
Flexor pollicis longus m.
Flexor digitorum profundus m.
Abductor pollicis longus m.
Radial a.
Extensor pollicis brevis m.
Scaphoid
Capitate
Extensor carpi radialis longus m.
Hamate
Extensor pollicis longus m.
Extensor carpi radialis brevis m.
Triquetrum
Extensor digitorum m.
Extensor digiti minimi m.
Ulnar a.
Tendon of flexor carpi ulnaris m.
Ulnar n.
Abductor digiti minimi m.
Pisiform
Extensor carpi ulnaris m.
Palmaris brevis m.
Palmaris longus m.
Median n.
Flexor digitorum superficialis m.
Flexor carpi radialis m.
Flexor pollicis longus m.
Flexor digitorum profundus m.
Radial a.
Extensor pollicis brevis m.
Scaphoid
Extensor carpi radialis longus m.
Capitate
Extensor carpi radialis brevis m.
Hamate
Triquetrum
Extensor digitorum m.
Extensor digiti minimi m.

Axial (wrist, 605 mm), expanded

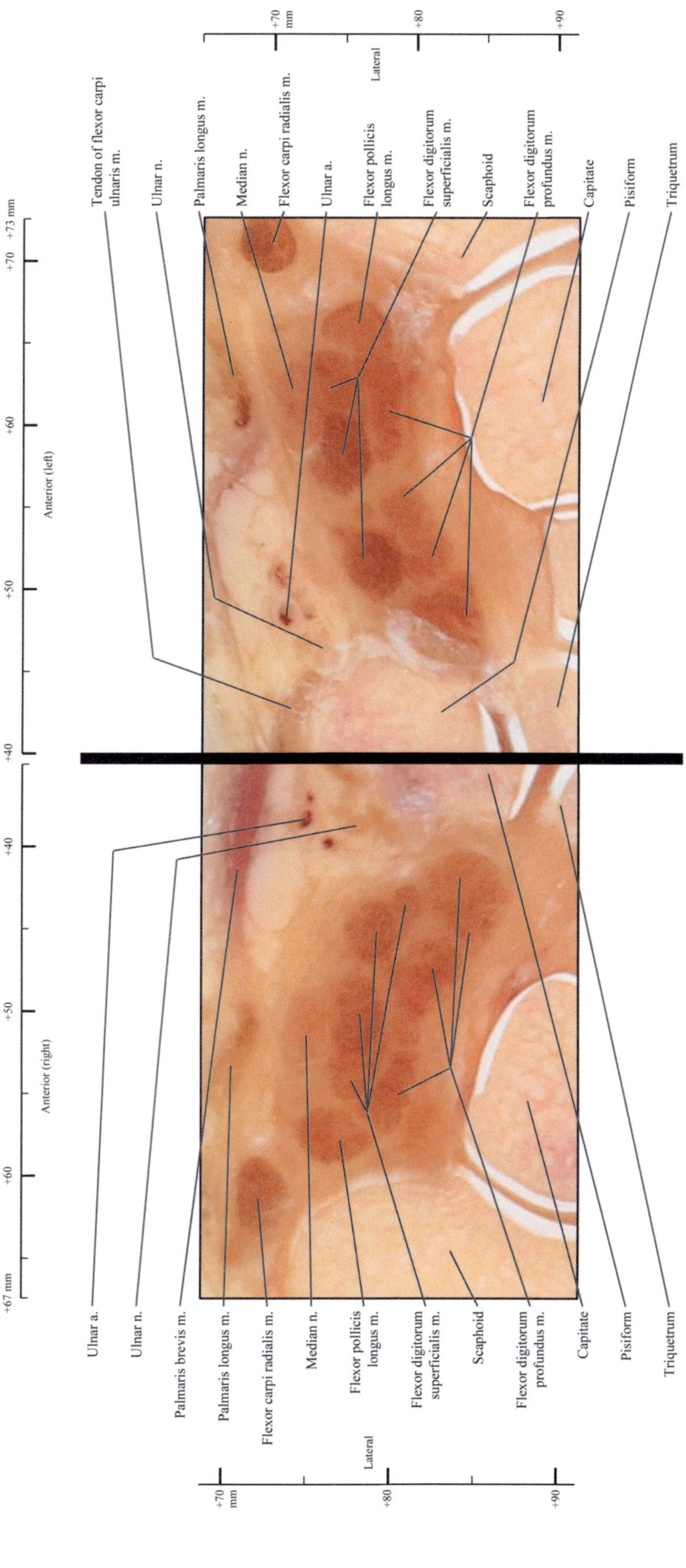
Tendon of flexor carpi ulnaris m.
Ulnar n.
Palmaris longus m.
Median n.
Flexor carpi radialis m.
Ulnar a.
Flexor pollicis longus m.
Flexor digitorum superficialis m.
Scaphoid
Flexor digitorum profundus m.
Capitate
Pisiform
Triquetrum
Lateral
+70 mm
+80
+90
+70 +73 mm
+60
+50
+40
Anterior (left)
Anterior (right)
+67 mm
Ulnar a.
Ulnar n.
Palmaris brevis m.
Palmaris longus m.
Flexor carpi radialis m.
Median n.
Flexor pollicis longus m.
Flexor digitorum superficialis m.
Scaphoid
Flexor digitorum profundus m.
Capitate
Pisiform
Triquetrum

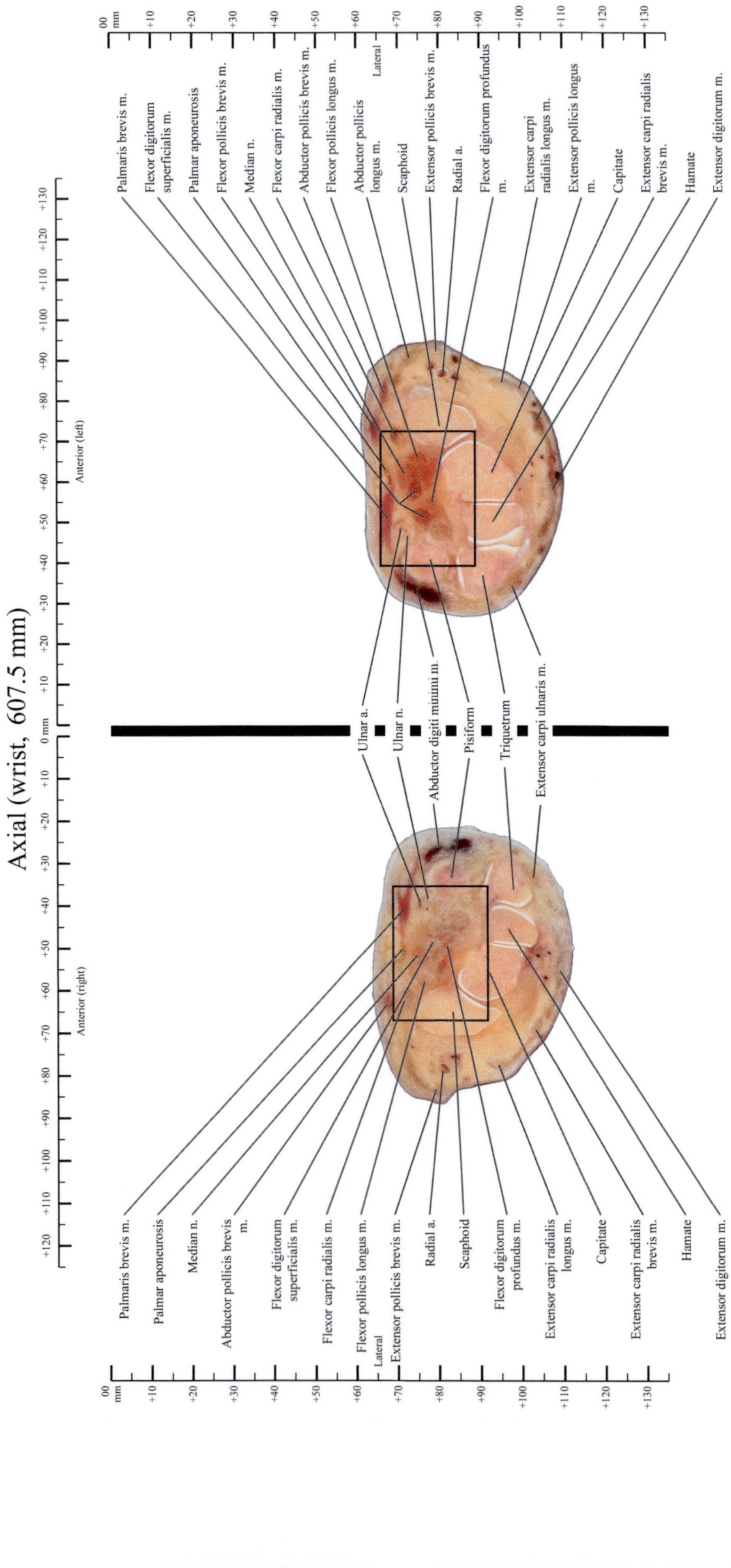
Axial (wrist, 607.5 mm)
Anterior (left)
Anterior (right)
Lateral
Palmaris brevis m.
Flexor digitorum superficialis m.
Palmar aponeurosis
Flexor pollicis brevis m.
Median n.
Flexor carpi radialis m.
Abductor pollicis brevis m.
Flexor pollicis longus m.
Abductor pollicis longus m.
Scaphoid
Extensor pollicis brevis m.
Radial a.
Flexor digitorum profundus m.
Extensor carpi radialis longus m.
Extensor pollicis longus m.
Capitate
Extensor carpi radialis brevis m.
Hamate
Extensor digitorum m.
Ulnar a.
Ulnar n.
Abductor digiti minimi m.
Pisiform
Triquetrum
Extensor carpi ulnaris m.

Axial (wrist, 607.5 mm), expanded

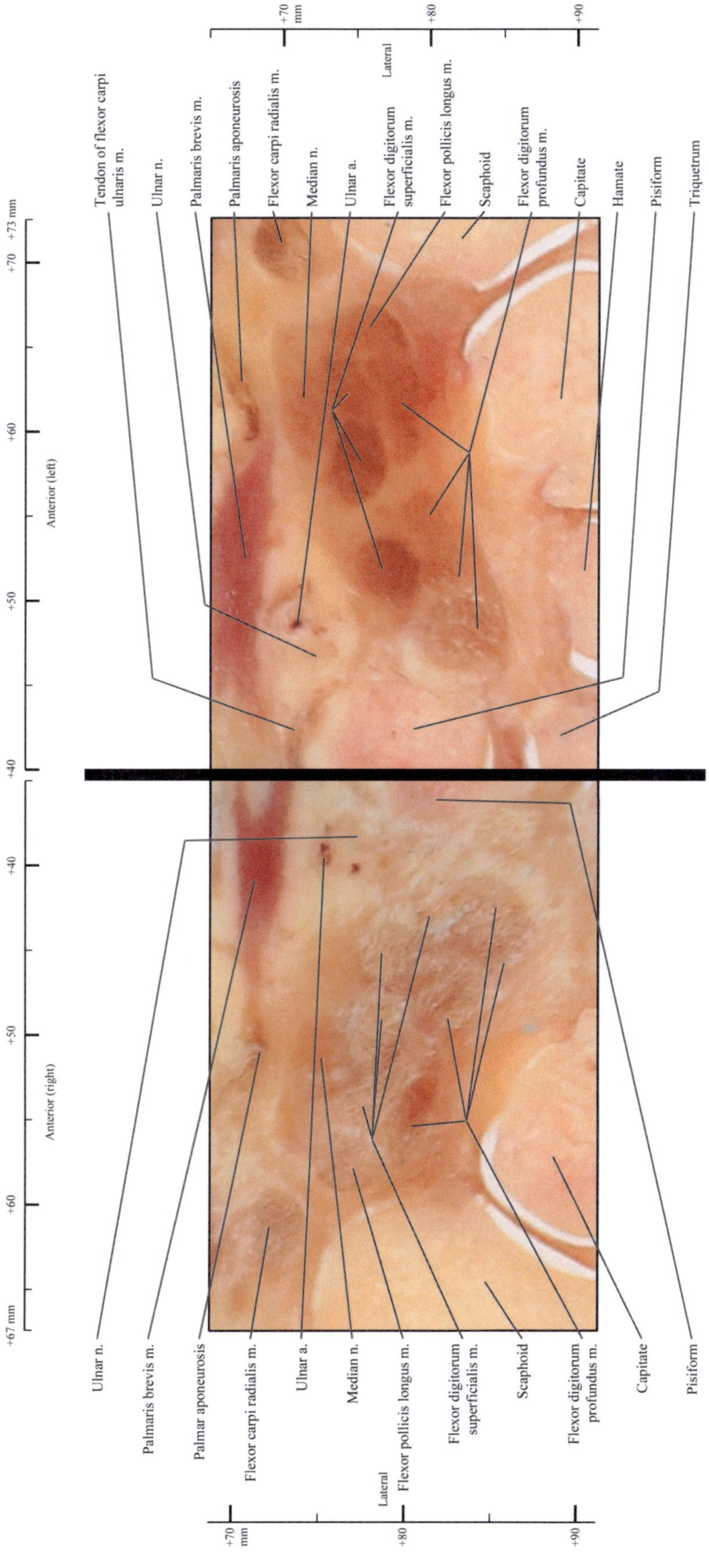

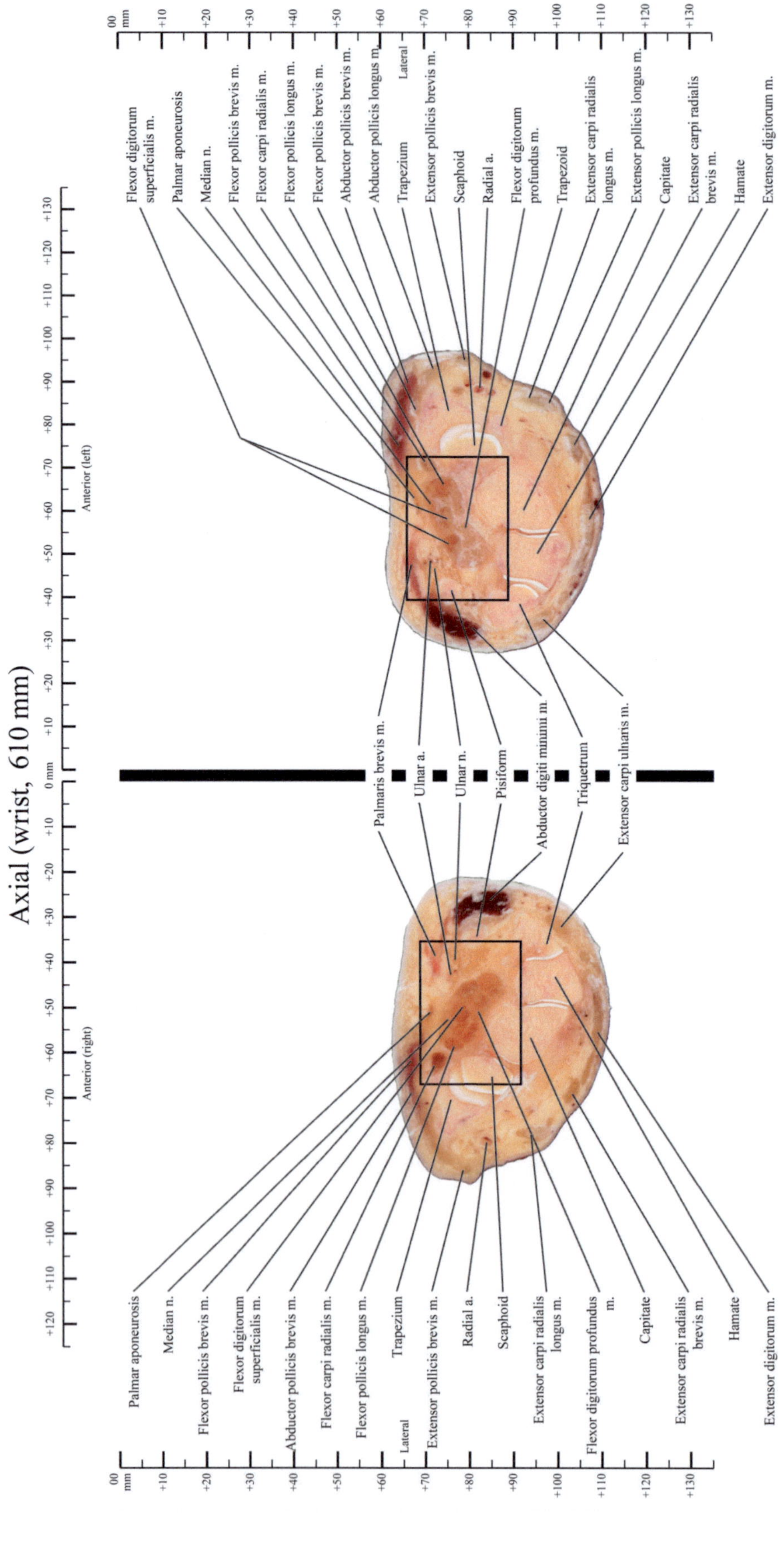
Axial (wrist, 610 mm)
Anterior (left)
Anterior (right)
Lateral
Flexor digitorum superficialis m.
Palmar aponeurosis
Median n.
Flexor pollicis brevis m.
Flexor carpi radialis m.
Flexor pollicis longus m.
Flexor pollicis brevis m.
Abductor pollicis brevis m.
Abductor pollicis longus m.
Trapezium
Extensor pollicis brevis m.
Scaphoid
Radial a.
Flexor digitorum profundus m.
Trapezoid
Extensor carpi radialis longus m.
Extensor pollicis longus m.
Capitate
Extensor carpi radialis brevis m.
Hamate
Extensor digitorum m.
Palmaris brevis m.
Ulnar a.
Ulnar n.
Pisiform
Abductor digiti minimi m.
Triquetrum
Extensor carpi ulnaris m.
Palmar aponeurosis
Median n.
Flexor pollicis brevis m.
Flexor digitorum superficialis m.
Abductor pollicis brevis m.
Flexor carpi radialis m.
Flexor pollicis longus m.
Trapezium
Extensor pollicis brevis m.
Radial a.
Scaphoid
Extensor carpi radialis longus m.
Flexor digitorum profundus m.
Capitate
Extensor carpi radialis brevis m.
Hamate
Extensor digitorum m.

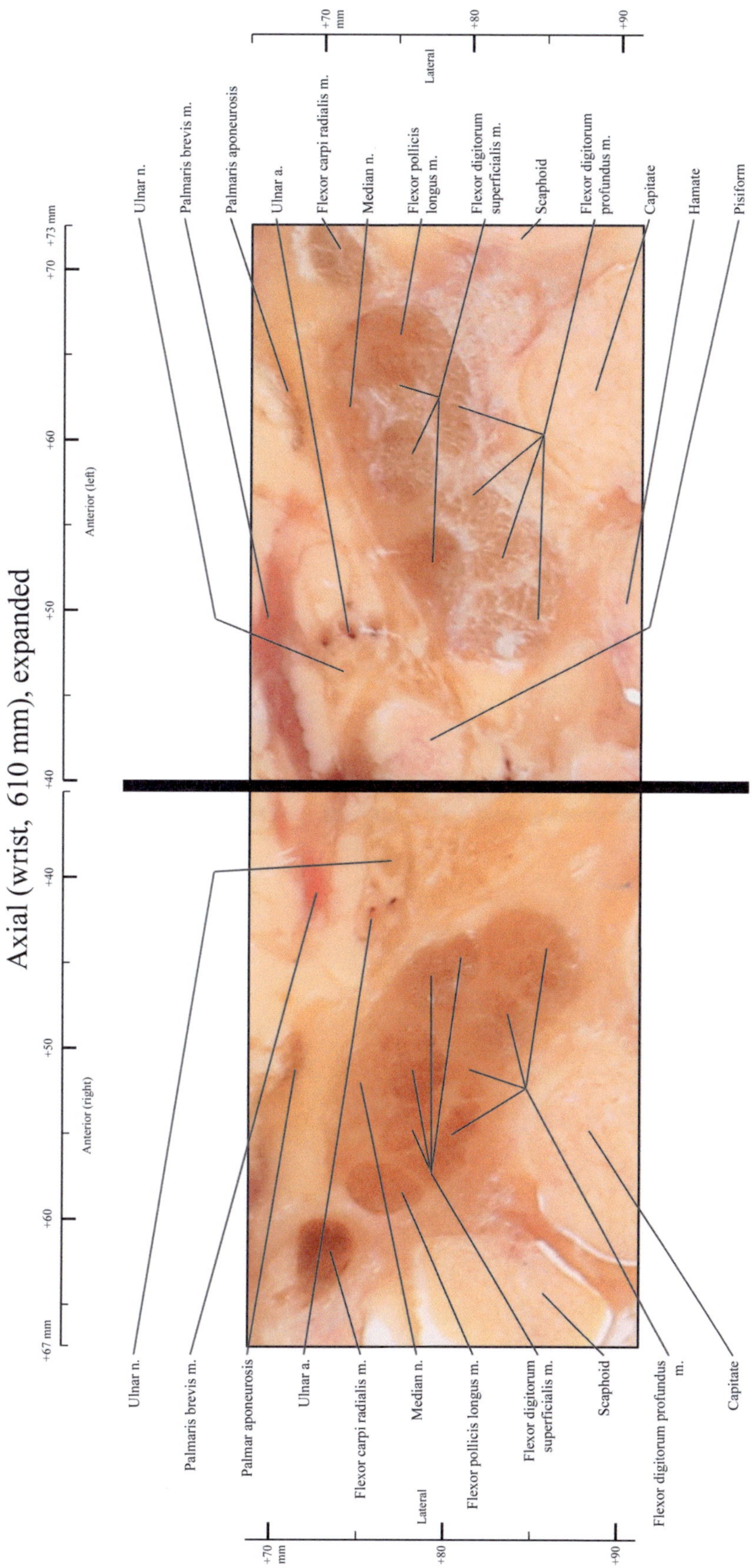
Axial (wrist, 610 mm), expanded
Anterior (left)
+40
+50
+60
+70 +73 mm
Anterior (right)
+40
+50
+60
+67 mm
Lateral
+70 mm
+80
+90
Ulnar n.
Palmaris brevis m.
Palmaris aponeurosis
Ulnar a.
Flexor carpi radialis m.
Median n.
Flexor pollicis longus m.
Flexor digitorum superficialis m.
Scaphoid
Flexor digitorum profundus m.
Capitate
Hamate
Pisiform
Ulnar n.
Palmaris brevis m.
Palmar aponeurosis
Ulnar a.
Flexor carpi radialis m.
Median n.
Flexor pollicis longus m.
Flexor digitorum superficialis m.
Scaphoid
Flexor digitorum profundus m.
Capitate
Lateral
+70 mm
+80
+90

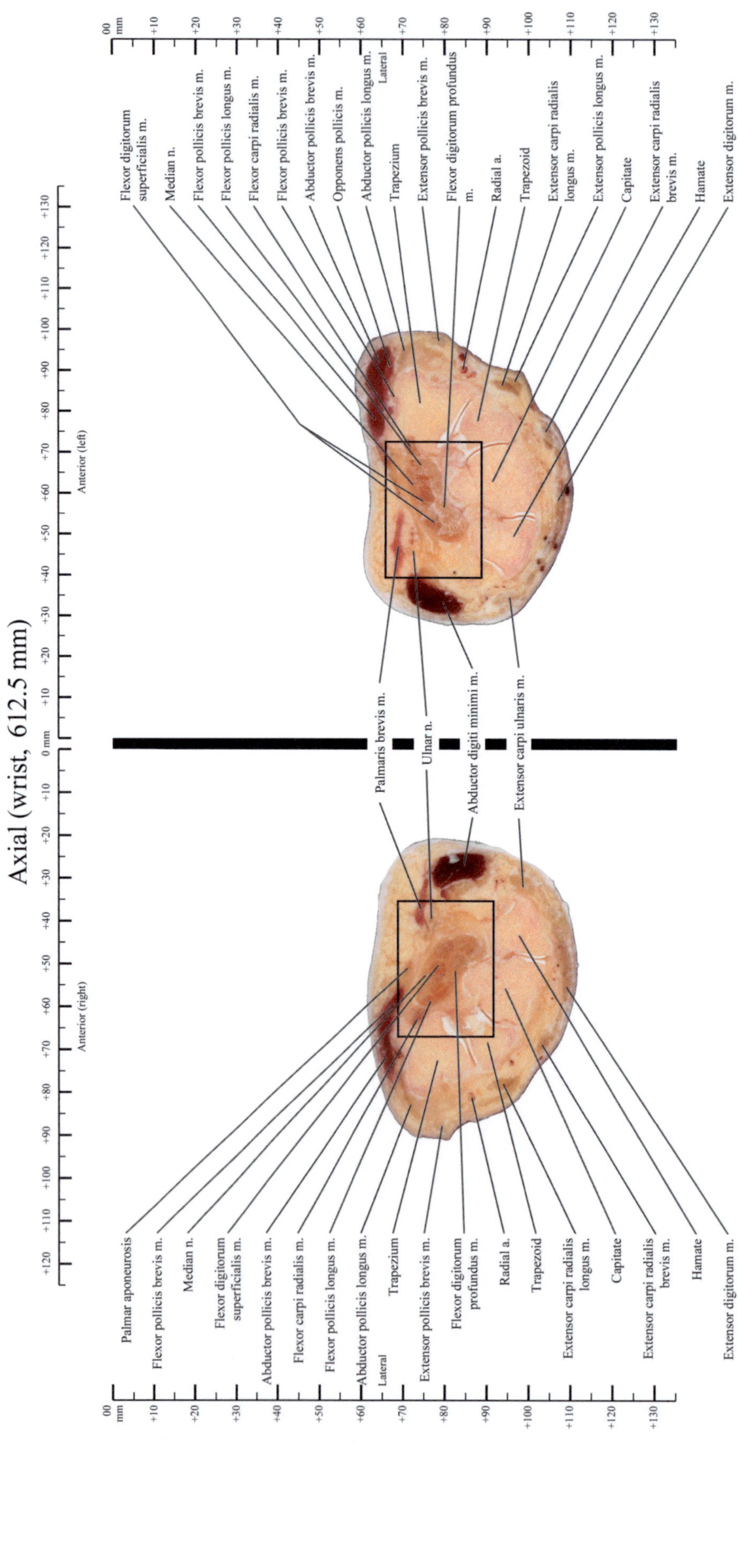
Axial (wrist, 612.5 mm)
Anterior (left)
Anterior (right)
Flexor digitorum superficialis m.
Median n.
Flexor pollicis brevis m.
Flexor pollicis longus m.
Flexor carpi radialis m.
Flexor pollicis brevis m.
Abductor pollicis brevis m.
Opponens pollicis m.
Abductor pollicis longus m.
Lateral
Trapezium
Extensor pollicis brevis m.
Flexor digitorum profundus m.
Radial a.
Trapezoid
Extensor carpi radialis longus m.
Extensor pollicis longus m.
Capitate
Extensor carpi radialis brevis m.
Hamate
Extensor digitorum m.
Palmaris brevis m.
Ulnar n.
Abductor digiti minimi m.
Extensor carpi ulnaris m.
Palmar aponeurosis
Flexor pollicis brevis m.
Median n.
Flexor digitorum superficialis m.
Abductor pollicis brevis m.
Flexor carpi radialis m.
Flexor pollicis longus m.
Abductor pollicis longus m.
Lateral
Trapezium
Extensor pollicis brevis m.
Flexor digitorum profundus m.
Radial a.
Trapezoid
Extensor carpi radialis longus m.
Capitate
Extensor carpi radialis brevis m.
Hamate
Extensor digitorum m.

Axial (wrist, 612.5 mm), expanded

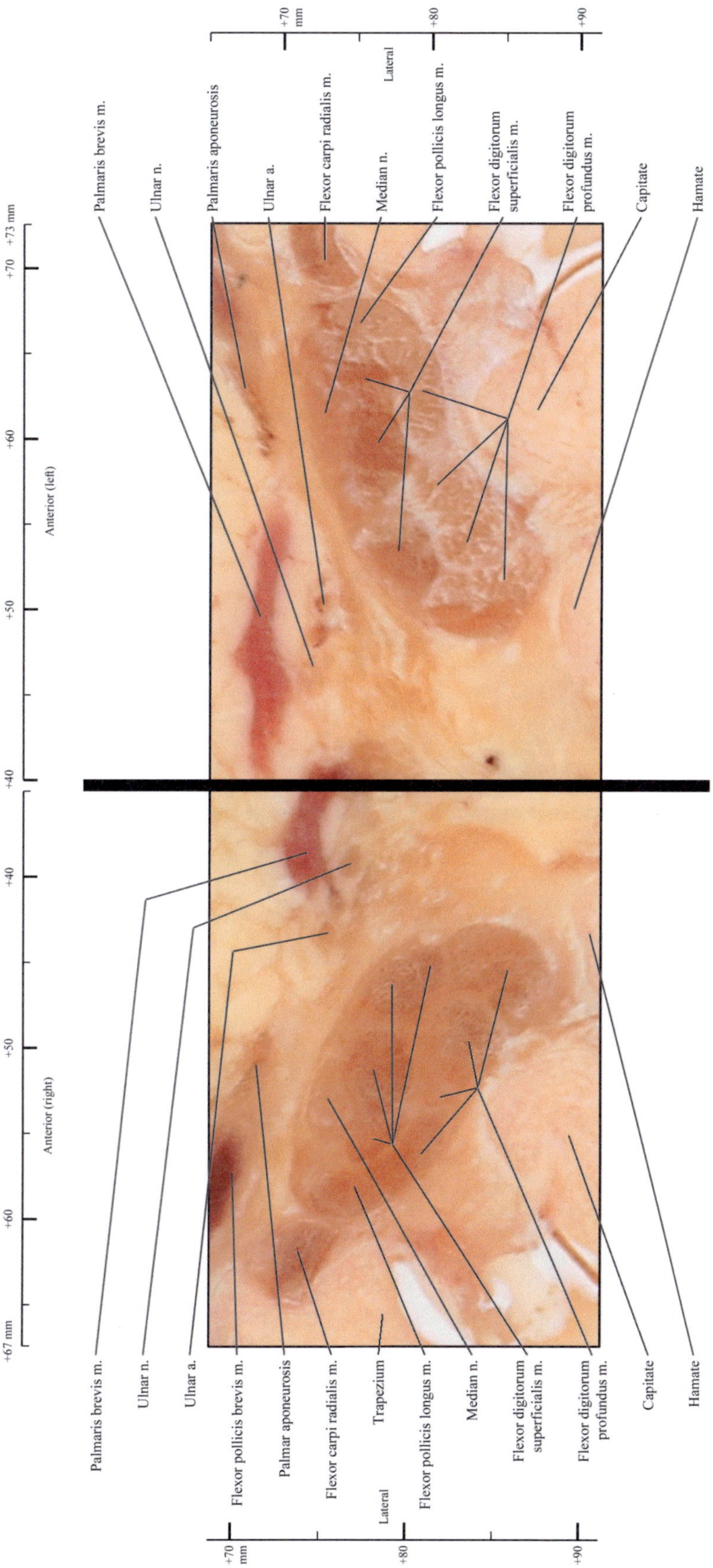

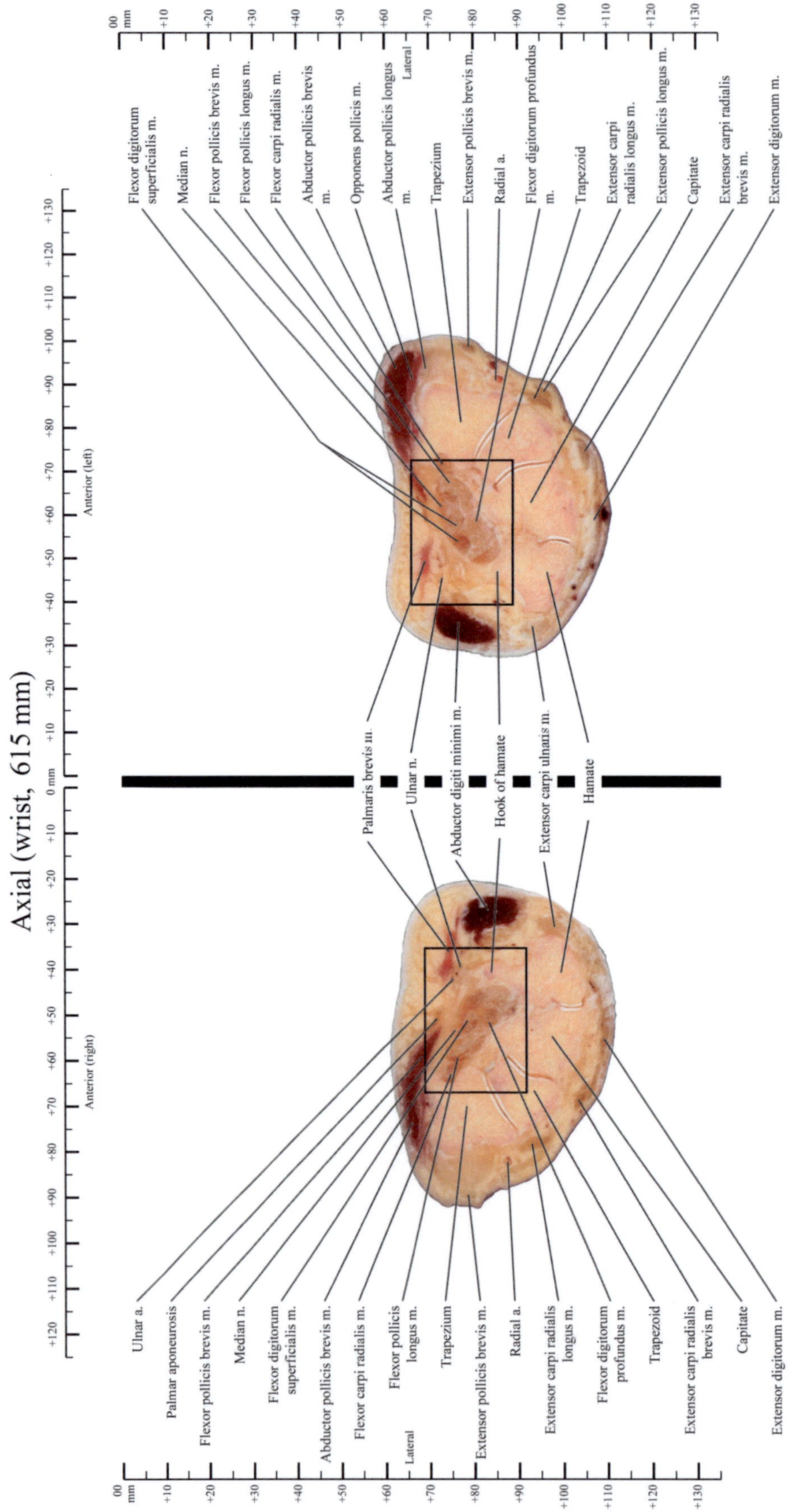
Axial (wrist, 615 mm)
Anterior (left)
Anterior (right)
Lateral
Flexor digitorum superficialis m.
Median n.
Flexor pollicis brevis m.
Flexor pollicis longus m.
Flexor carpi radialis m.
Abductor pollicis brevis m.
Opponens pollicis m.
Abductor pollicis longus m.
Trapezium
Extensor pollicis brevis m.
Radial a.
Flexor digitorum profundus m.
Trapezoid
Extensor carpi radialis longus m.
Extensor pollicis longus m.
Capitate
Extensor carpi radialis brevis m.
Extensor digitorum m.
Palmaris brevis m.
Ulnar n.
Abductor digiti minimi m.
Hook of hamate
Extensor carpi ulnaris m.
Hamate
Ulnar a.
Palmar aponeurosis
Flexor pollicis brevis m.
Median n.
Flexor digitorum superficialis m.
Abductor pollicis brevis m.
Flexor carpi radialis m.
Flexor pollicis longus m.
Trapezium
Extensor pollicis brevis m.
Radial a.
Extensor carpi radialis longus m.
Flexor digitorum profundus m.
Trapezoid
Extensor carpi radialis brevis m.
Capitate
Extensor digitorum m.

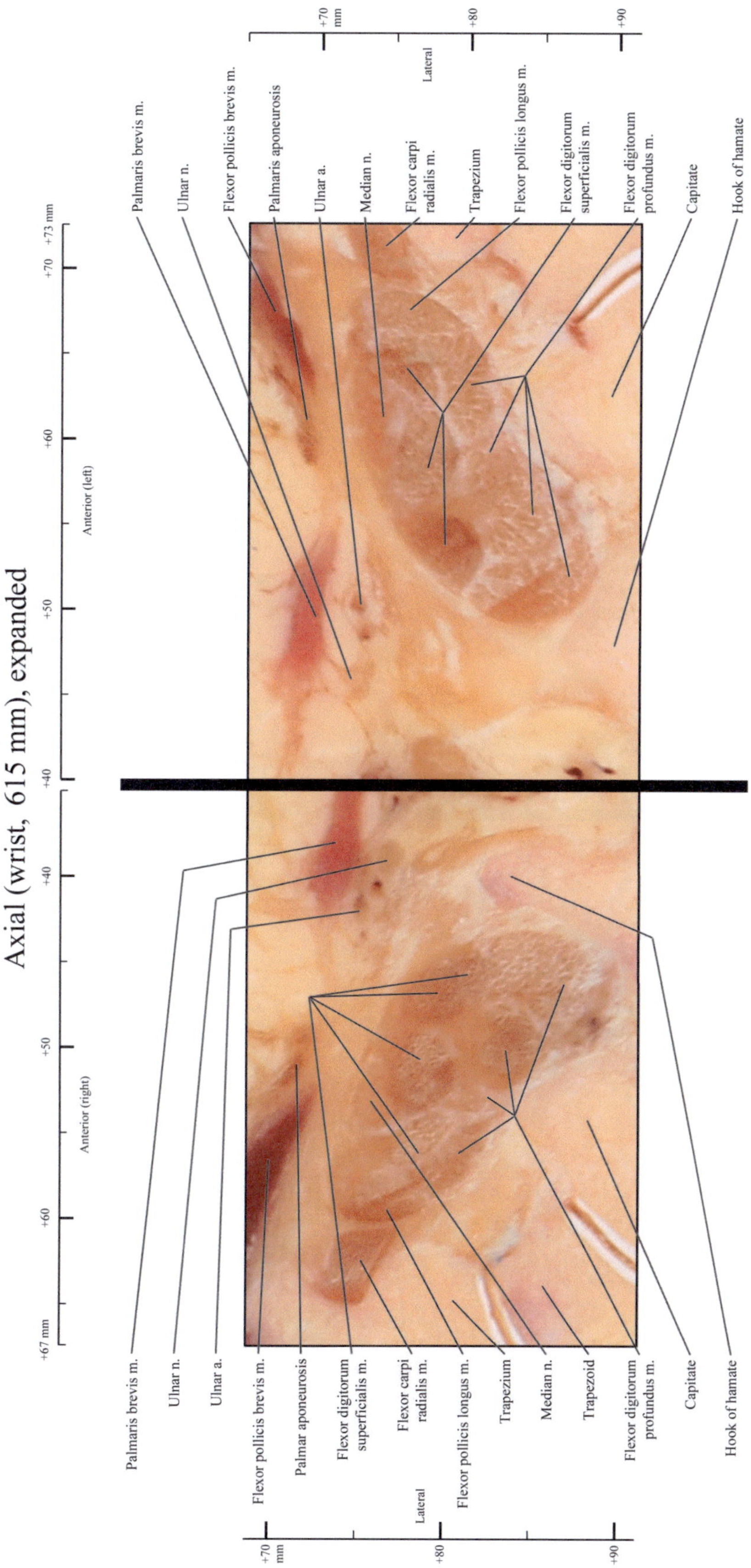

Axial (wrist, 615 mm), expanded
Anterior (left)
+70 +73 mm
+60
+50
+40
Anterior (right)
+40
+50
+60
+67 mm
Lateral
+70 mm
+80
+90
Palmaris brevis m.
Ulnar n.
Flexor pollicis brevis m.
Palmaris aponeurosis
Ulnar a.
Median n.
Flexor carpi radialis m.
Trapezium
Flexor pollicis longus m.
Flexor digitorum superficialis m.
Flexor digitorum profundus m.
Capitate
Hook of hamate
Palmaris brevis m.
Ulnar n.
Ulnar a.
Flexor pollicis brevis m.
Palmar aponeurosis
Flexor digitorum superficialis m.
Flexor carpi radialis m.
Flexor pollicis longus m.
Trapezium
Median n.
Trapezoid
Flexor digitorum profundus m.
Capitate
Hook of hamate
Lateral
+70 mm
+80
+90

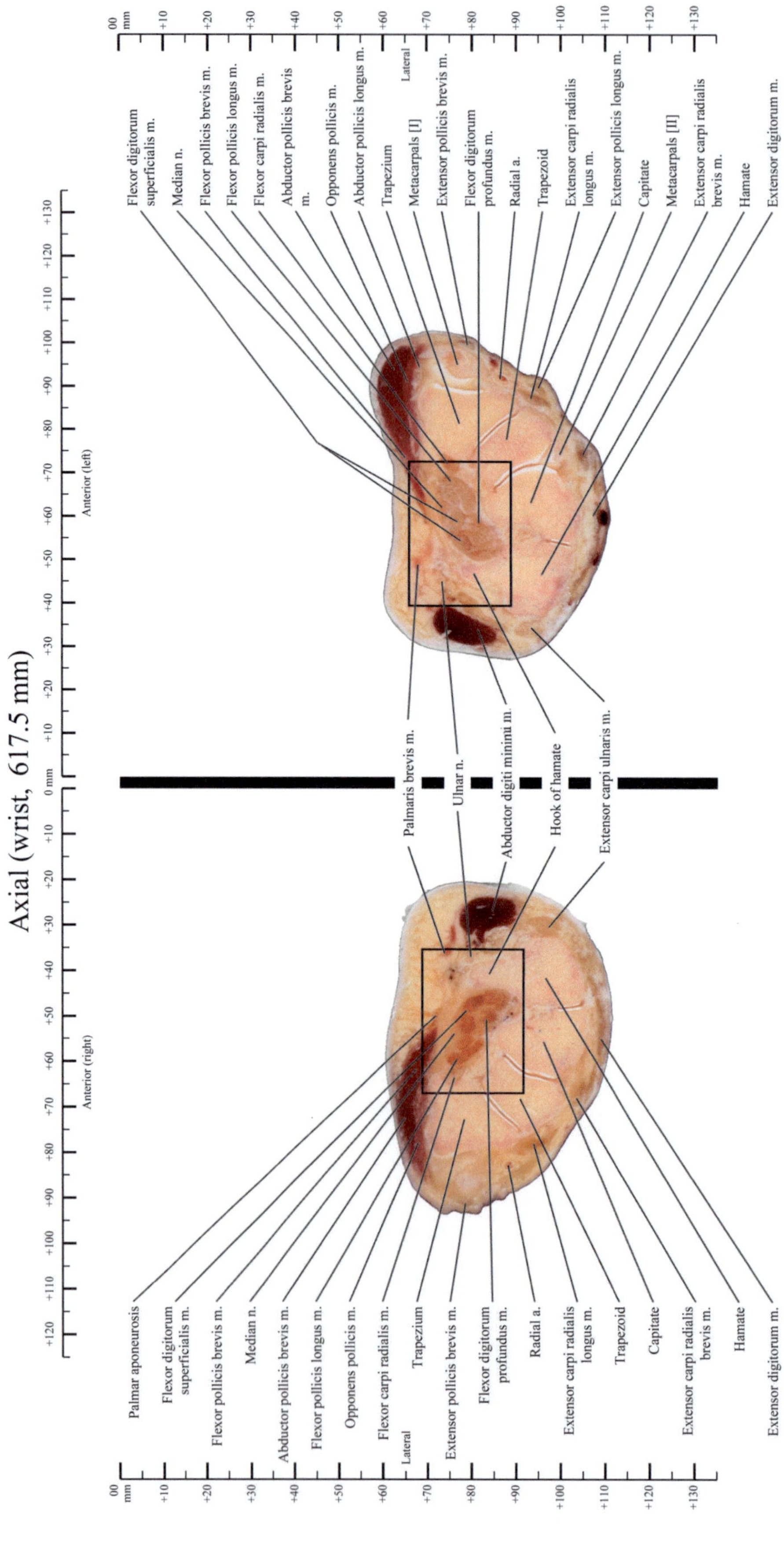
Axial (wrist, 617.5 mm)
Anterior (left)
Anterior (right)
Lateral
Flexor digitorum superficialis m.
Median n.
Flexor pollicis brevis m.
Flexor pollicis longus m.
Flexor carpi radialis m.
Abductor pollicis brevis m.
Opponens pollicis m.
Abductor pollicis longus m.
Trapezium
Metacarpals [I]
Extensor pollicis brevis m.
Flexor digitorum profundus m.
Radial a.
Trapezoid
Extensor carpi radialis longus m.
Extensor pollicis longus m.
Capitate
Metacarpals [II]
Extensor carpi radialis brevis m.
Hamate
Extensor digitorum m.
Palmaris brevis m.
Ulnar n.
Abductor digiti minimi m.
Hook of hamate
Extensor carpi ulnaris m.
Palmar aponeurosis
Flexor digitorum superficialis m.
Flexor pollicis brevis m.
Median n.
Abductor pollicis brevis m.
Flexor pollicis longus m.
Opponens pollicis m.
Flexor carpi radialis m.
Trapezium
Extensor pollicis brevis m.
Flexor digitorum profundus m.
Radial a.
Extensor carpi radialis longus m.
Trapezoid
Capitate
Extensor carpi radialis brevis m.
Hamate
Extensor digitorum m.

Axial (wrist, 617.5 mm), expanded

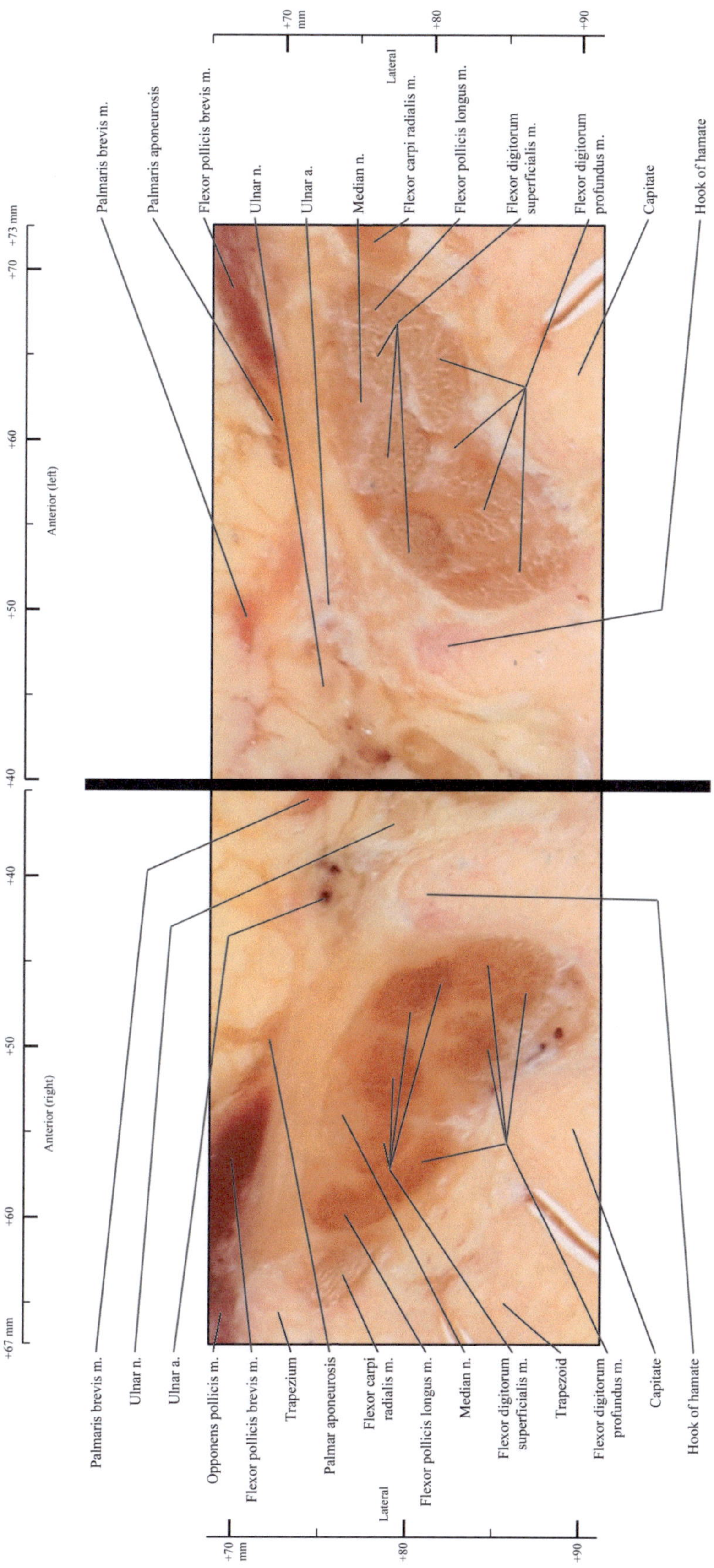

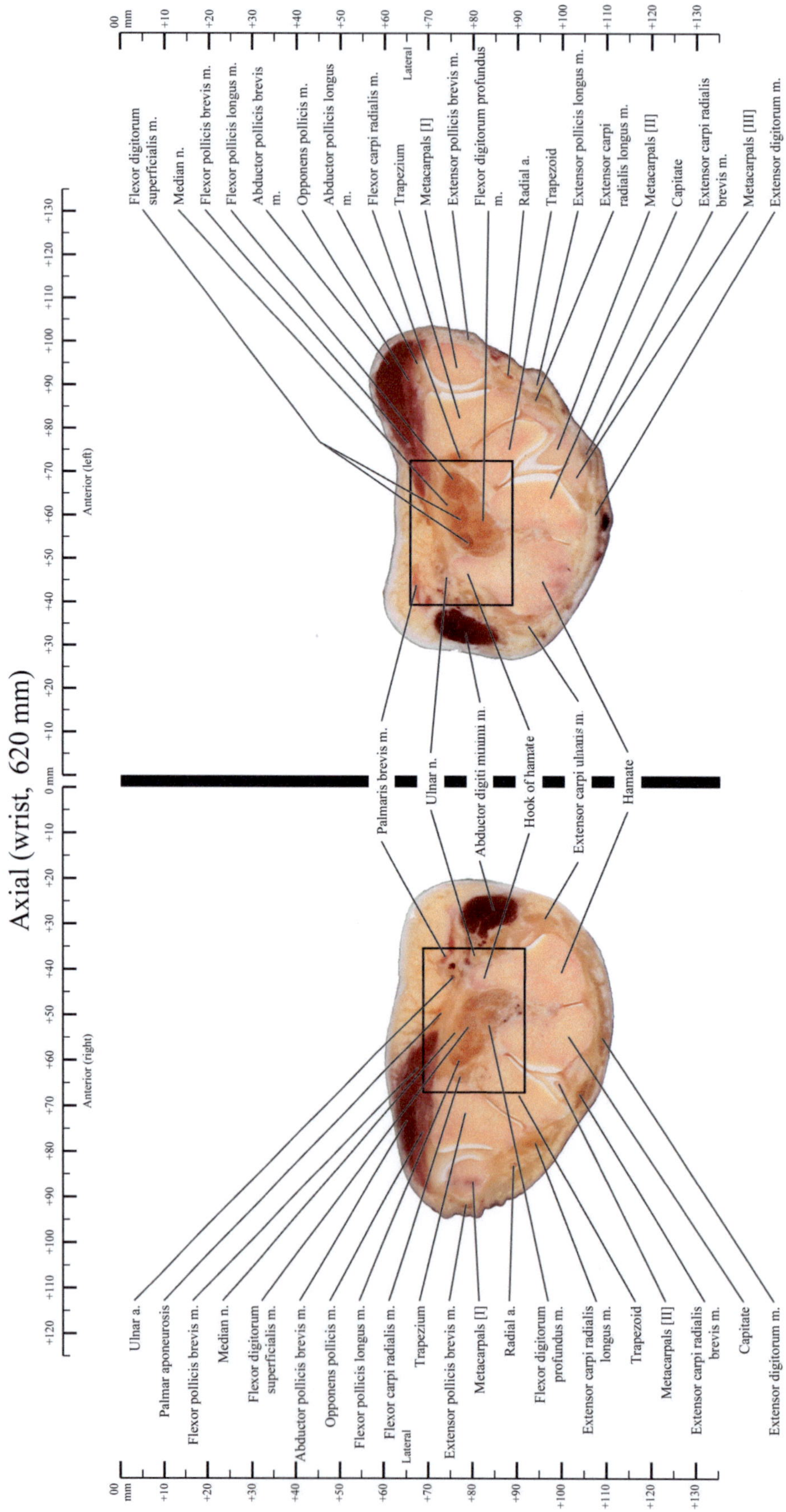
Axial (wrist, 620 mm)
Anterior (left)
Anterior (right)
Lateral
Flexor digitorum superficialis m.
Median n.
Flexor pollicis brevis m.
Flexor pollicis longus m.
Abductor pollicis brevis m.
Opponens pollicis m.
Abductor pollicis longus m.
Flexor carpi radialis m.
Trapezium
Metacarpals [I]
Extensor pollicis brevis m.
Flexor digitorum profundus m.
Radial a.
Trapezoid
Extensor pollicis longus m.
Extensor carpi radialis longus m.
Metacarpals [II]
Capitate
Extensor carpi radialis brevis m.
Metacarpals [III]
Extensor digitorum m.
Palmaris brevis m.
Ulnar n.
Abductor digiti minimi m.
Hook of hamate
Extensor carpi ulnaris m.
Hamate
Ulnar a.
Palmar aponeurosis
Flexor pollicis brevis m.
Median n.
Flexor digitorum superficialis m.
Abductor pollicis brevis m.
Opponens pollicis m.
Flexor pollicis longus m.
Flexor carpi radialis m.
Trapezium
Extensor pollicis brevis m.
Metacarpals [I]
Radial a.
Flexor digitorum profundus m.
Extensor carpi radialis longus m.
Trapezoid
Metacarpals [II]
Extensor carpi radialis brevis m.
Capitate
Extensor digitorum m.

Axial (wrist, 620 mm), expanded

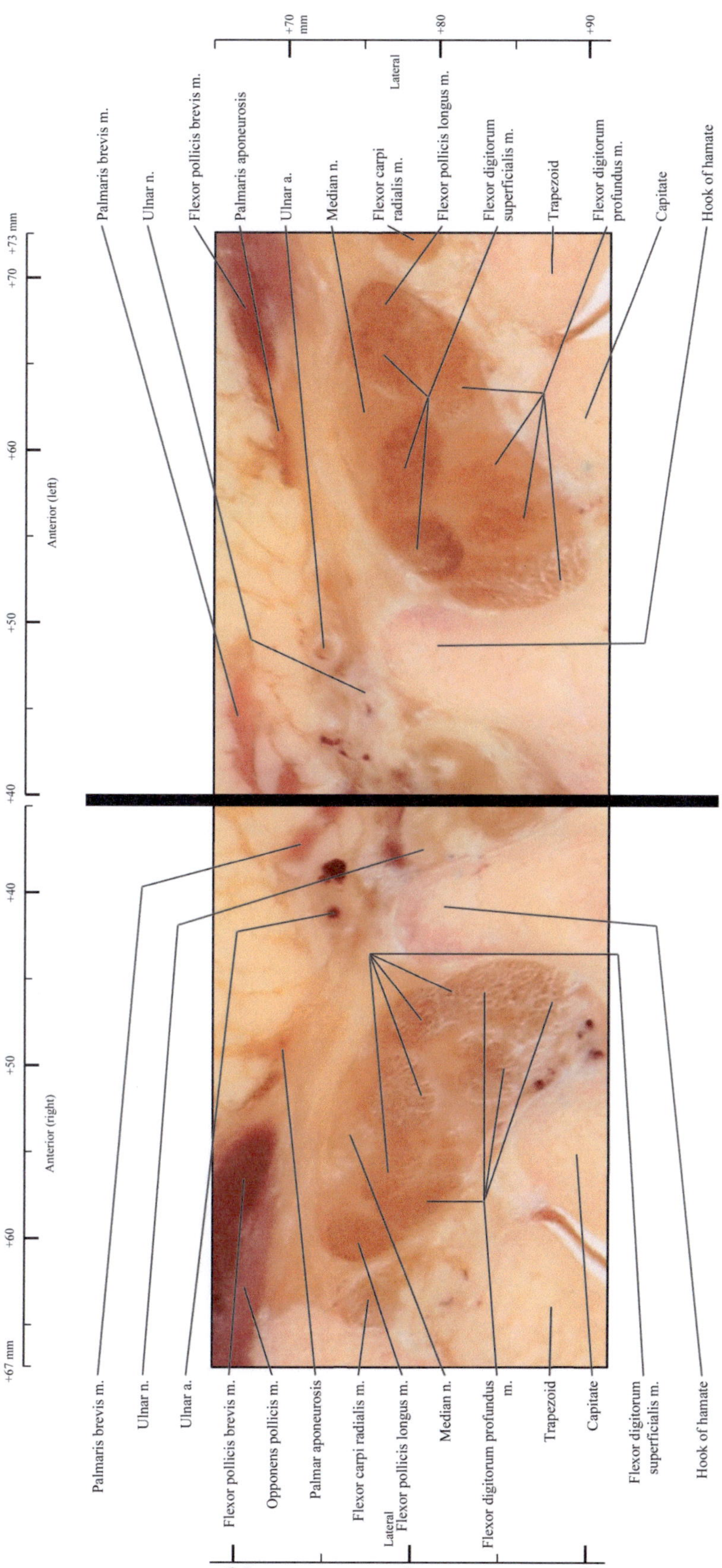

Anterior (left)
+73 mm
+70
+60
+50
+40
Palmaris brevis m.
Ulnar n.
Flexor pollicis brevis m.
Palmaris aponeurosis
Ulnar a.
Median n.
Flexor carpi radialis m.
Flexor pollicis longus m.
Flexor digitorum superficialis m.
Trapezoid
Flexor digitorum profundus m.
Capitate
Hook of hamate
Lateral
+70 mm
+80
+90
Anterior (right)
+67 mm
Palmaris brevis m.
Ulnar n.
Ulnar a.
Flexor pollicis brevis m.
Opponens pollicis m.
Palmar aponeurosis
Flexor carpi radialis m.
Flexor pollicis longus m.
Median n.
Flexor digitorum profundus m.
Trapezoid
Capitate
Flexor digitorum superficialis m.
Hook of hamate

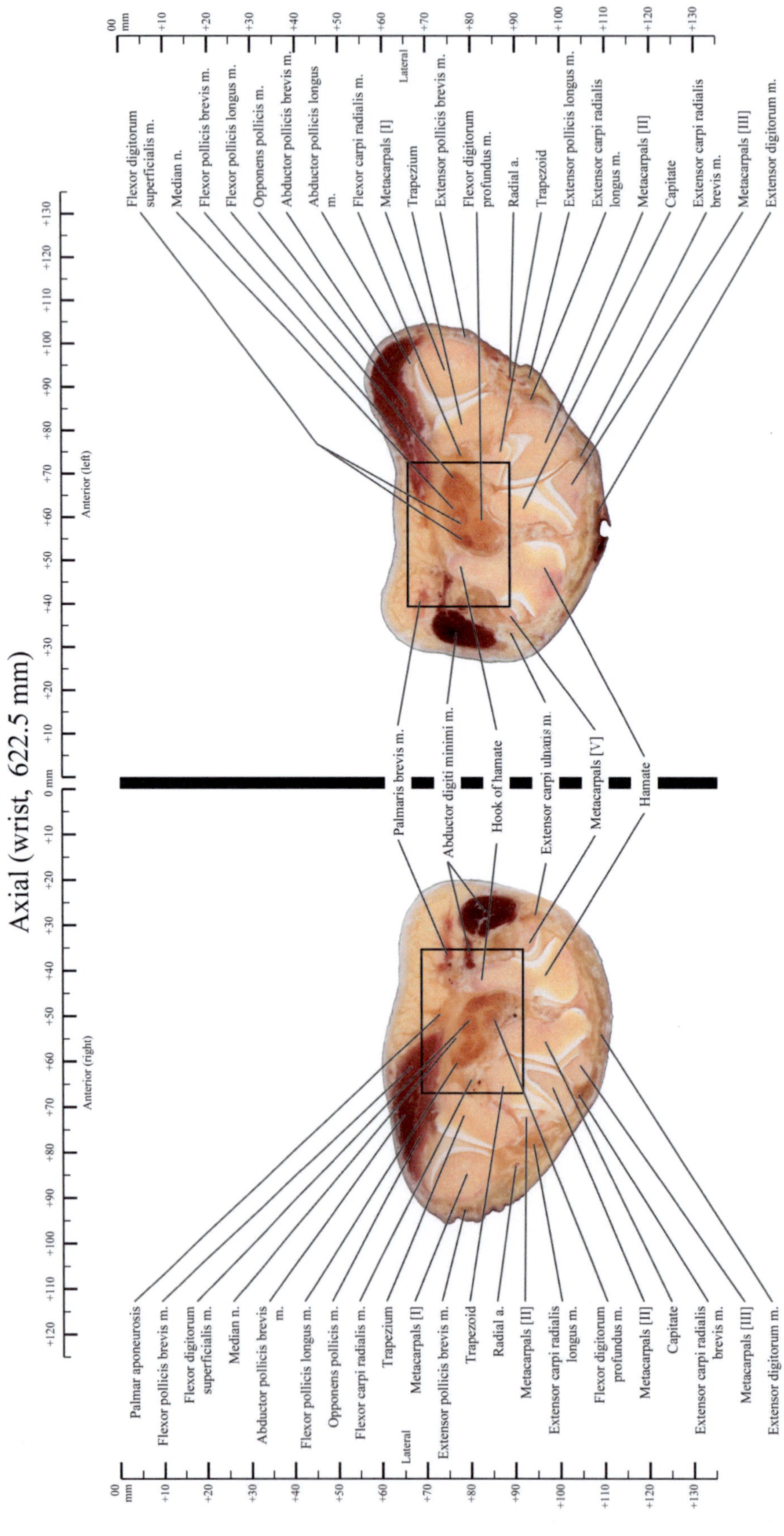
Axial (wrist, 622.5 mm)
Anterior (left)
Anterior (right)
Lateral
Flexor digitorum superficialis m.
Median n.
Flexor pollicis brevis m.
Flexor pollicis longus m.
Opponens pollicis m.
Abductor pollicis brevis m.
Abductor pollicis longus m.
Flexor carpi radialis m.
Metacarpals [I]
Trapezium
Extensor pollicis brevis m.
Flexor digitorum profundus m.
Radial a.
Trapezoid
Extensor pollicis longus m.
Extensor carpi radialis longus m.
Metacarpals [II]
Capitate
Extensor carpi radialis brevis m.
Metacarpals [III]
Extensor digitorum m.
Palmaris brevis m.
Abductor digiti minimi m.
Hook of hamate
Extensor carpi ulnaris m.
Metacarpals [V]
Hamate
Palmar aponeurosis
Flexor pollicis brevis m.
Flexor digitorum superficialis m.
Median n.
Abductor pollicis brevis m.
Flexor pollicis longus m.
Opponens pollicis m.
Flexor carpi radialis m.
Trapezium
Metacarpals [I]
Extensor pollicis brevis m.
Trapezoid
Radial a.
Metacarpals [II]
Extensor carpi radialis longus m.
Flexor digitorum profundus m.
Metacarpals [II]
Capitate
Extensor carpi radialis brevis m.
Metacarpals [III]
Extensor digitorum m.

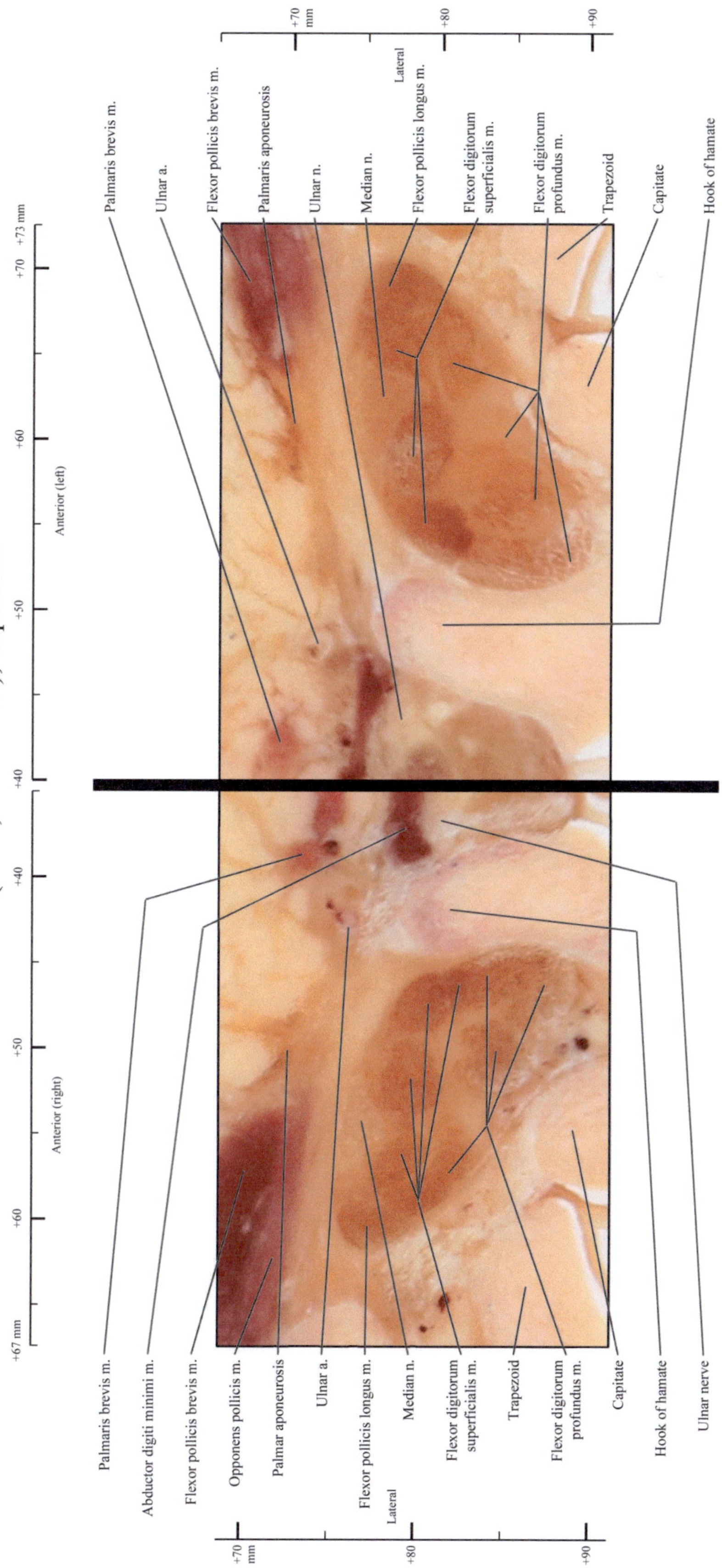
Palmaris brevis m.
Ulnar a.
Flexor pollicis brevis m.
Palmaris aponeurosis
Ulnar n.
Median n.
Lateral
Flexor pollicis longus m.
Flexor digitorum superficialis m.
Flexor digitorum profundus m.
Trapezoid
Capitate
Hook of hamate
+70 mm
+80
+90
+70 +73 mm
+60
Anterior (left)
+50
+40
+40
+50
Anterior (right)
+60
+67 mm
Palmaris brevis m.
Abductor digiti minimi m.
Flexor pollicis brevis m.
Opponens pollicis m.
Palmar aponeurosis
Ulnar a.
Flexor pollicis longus m.
Lateral
Median n.
Flexor digitorum superficialis m.
Trapezoid
Flexor digitorum profundus m.
Capitate
Hook of hamate
Ulnar nerve
+70 mm
+80
+90

Axial (wrist, 622.5 mm), expanded

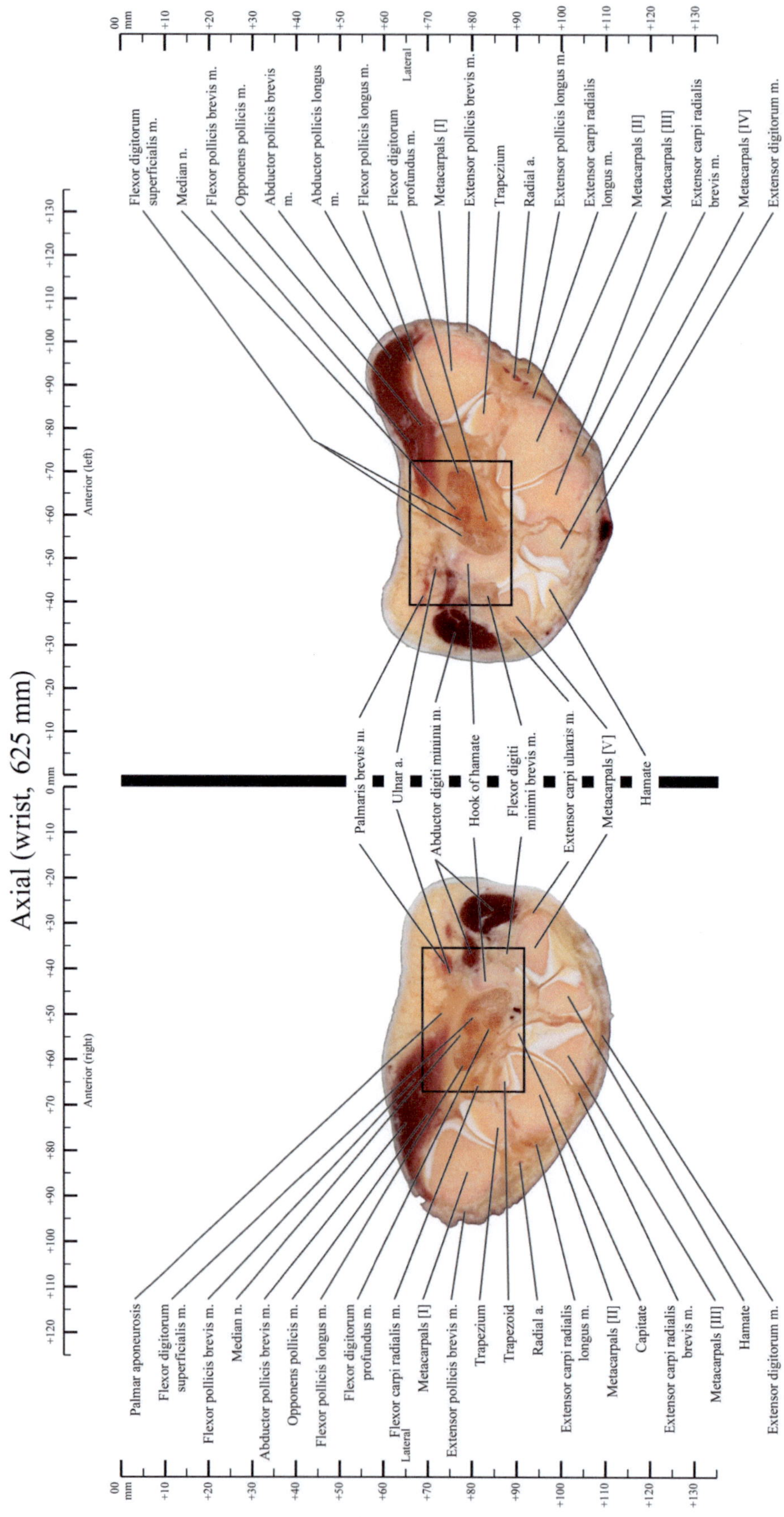

Axial (wrist, 625 mm)
Anterior (left)
Anterior (right)
Lateral
Flexor digitorum superficialis m.
Median n.
Flexor pollicis brevis m.
Opponens pollicis m.
Abductor pollicis brevis m.
Abductor pollicis longus m.
Flexor pollicis longus m.
Flexor digitorum profundus m.
Metacarpals [I]
Extensor pollicis brevis m.
Trapezium
Radial a.
Extensor pollicis longus m.
Extensor carpi radialis longus m.
Metacarpals [II]
Metacarpals [III]
Extensor carpi radialis brevis m.
Metacarpals [IV]
Extensor digitorum m.
Palmaris brevis m.
Ulnar a.
Abductor digiti minimi m.
Hook of hamate
Flexor digiti minimi brevis m.
Extensor carpi ulnaris m.
Metacarpals [V]
Hamate
Palmar aponeurosis
Flexor digitorum superficialis m.
Flexor pollicis brevis m.
Median n.
Abductor pollicis brevis m.
Opponens pollicis m.
Flexor pollicis longus m.
Flexor digitorum profundus m.
Flexor carpi radialis m.
Metacarpals [I]
Extensor pollicis brevis m.
Trapezium
Trapezoid
Radial a.
Extensor carpi radialis longus m.
Metacarpals [II]
Capitate
Extensor carpi radialis brevis m.
Metacarpals [III]
Hamate
Extensor digitorum m.

Axial (wrist, 625 mm), expanded

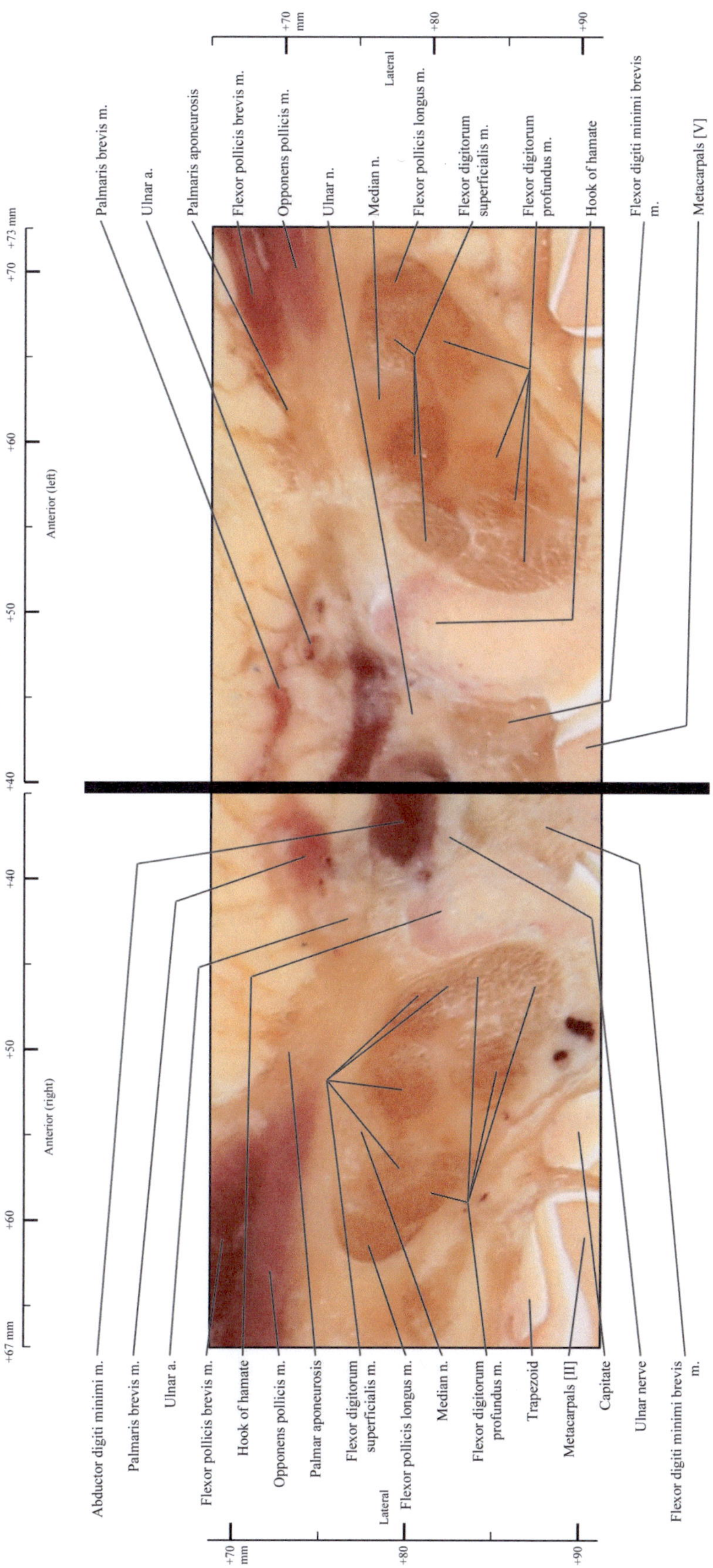
Palmaris brevis m.
Ulnar a.
Palmaris aponeurosis
Flexor pollicis brevis m.
Opponens pollicis m.
Ulnar n.
Median n.
Flexor pollicis longus m.
Flexor digitorum superficialis m.
Flexor digitorum profundus m.
Hook of hamate
Flexor digiti minimi brevis m.
Metacarpals [V]
Lateral
+70 mm
+80
+90
Anterior (left)
+73 mm
+70
+60
+50
+40
Abductor digiti minimi m.
Palmaris brevis m.
Ulnar a.
Flexor pollicis brevis m.
Hook of hamate
Opponens pollicis m.
Palmar aponeurosis
Flexor digitorum superficialis m.
Flexor pollicis longus m.
Median n.
Flexor digitorum profundus m.
Trapezoid
Metacarpals [II]
Capitate
Ulnar nerve
Flexor digiti minimi brevis m.
Lateral
+70 mm
+80
+90
Anterior (right)
+40
+50
+60
+67 mm

Axial (wrist, 627.5 mm)

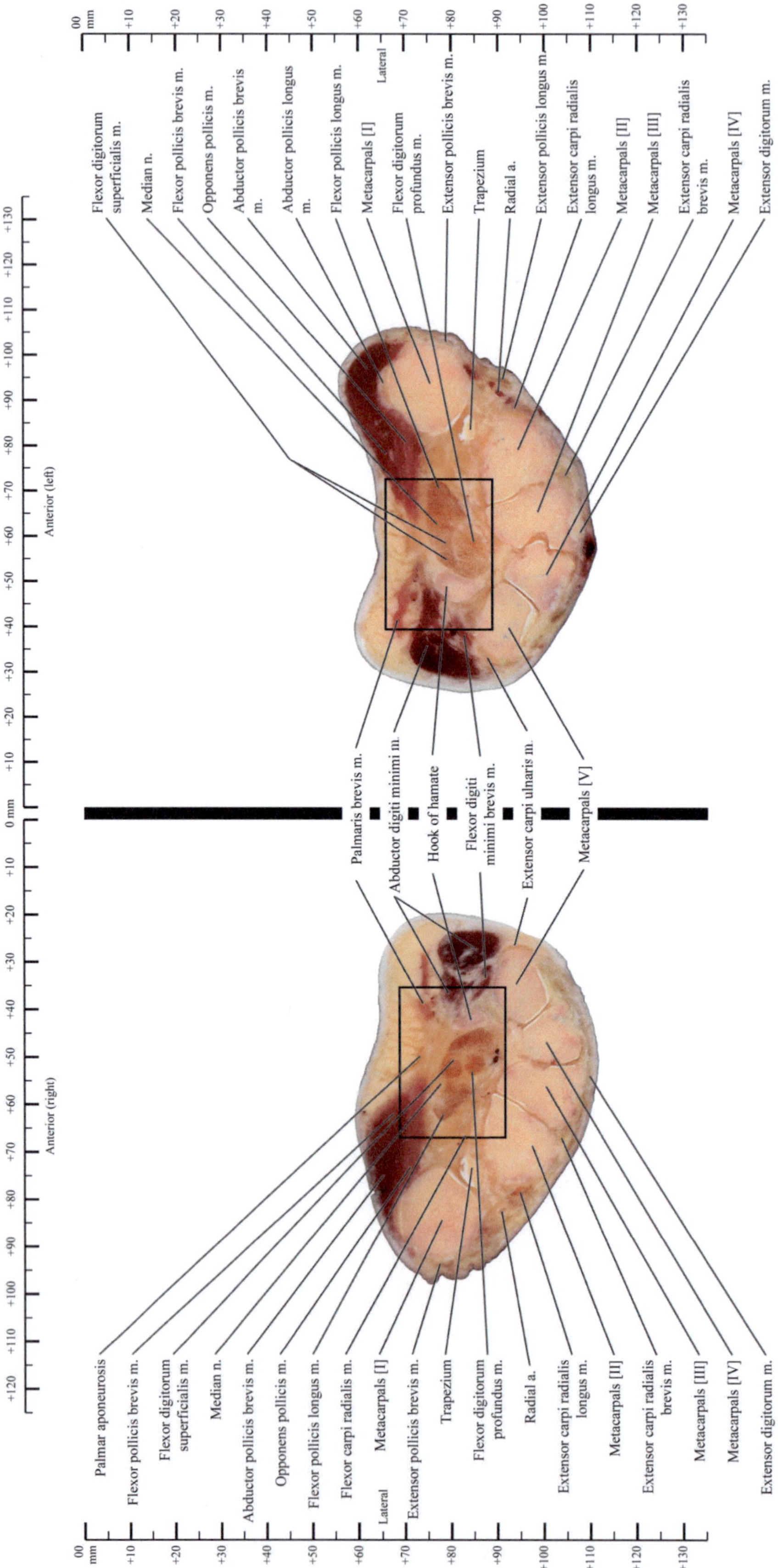

Axial (wrist, 627.5 mm), expanded

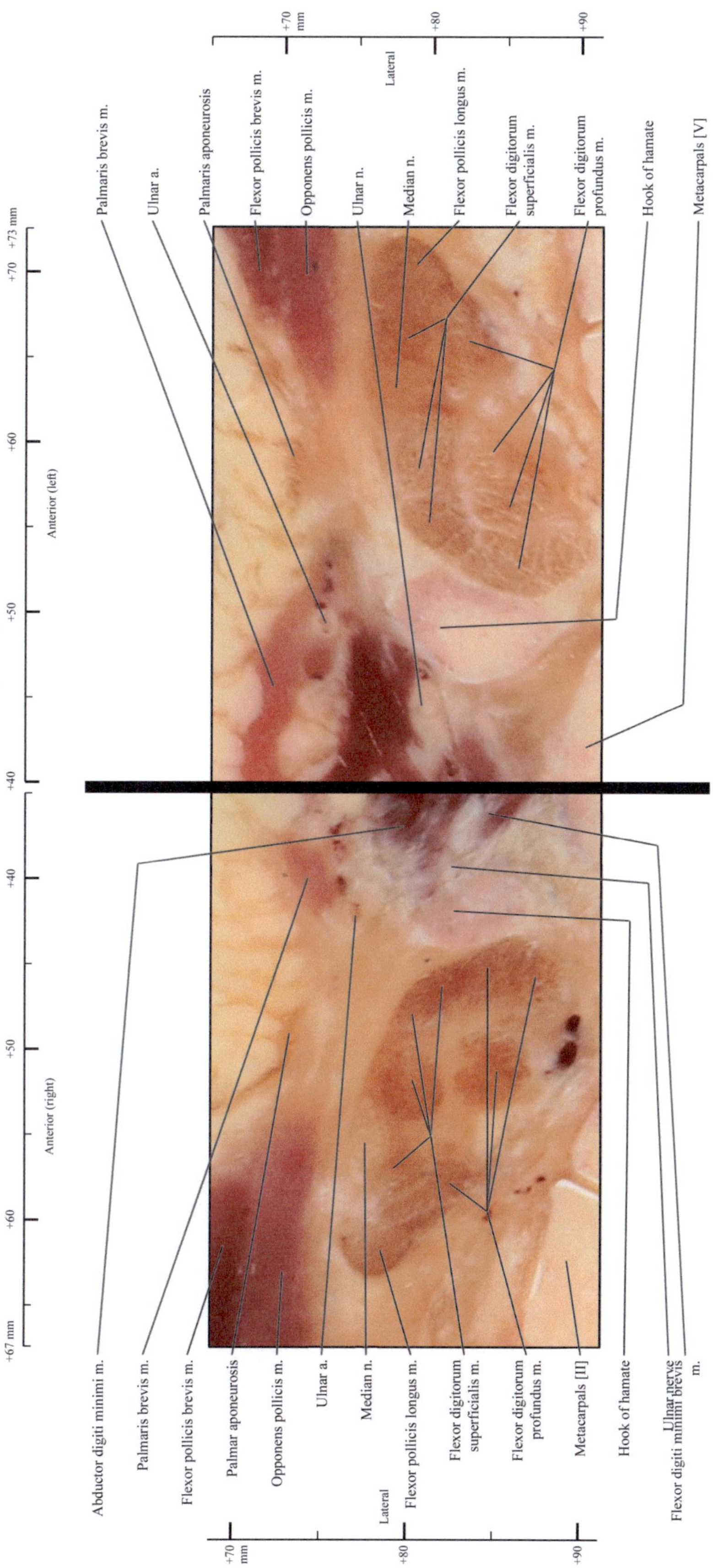

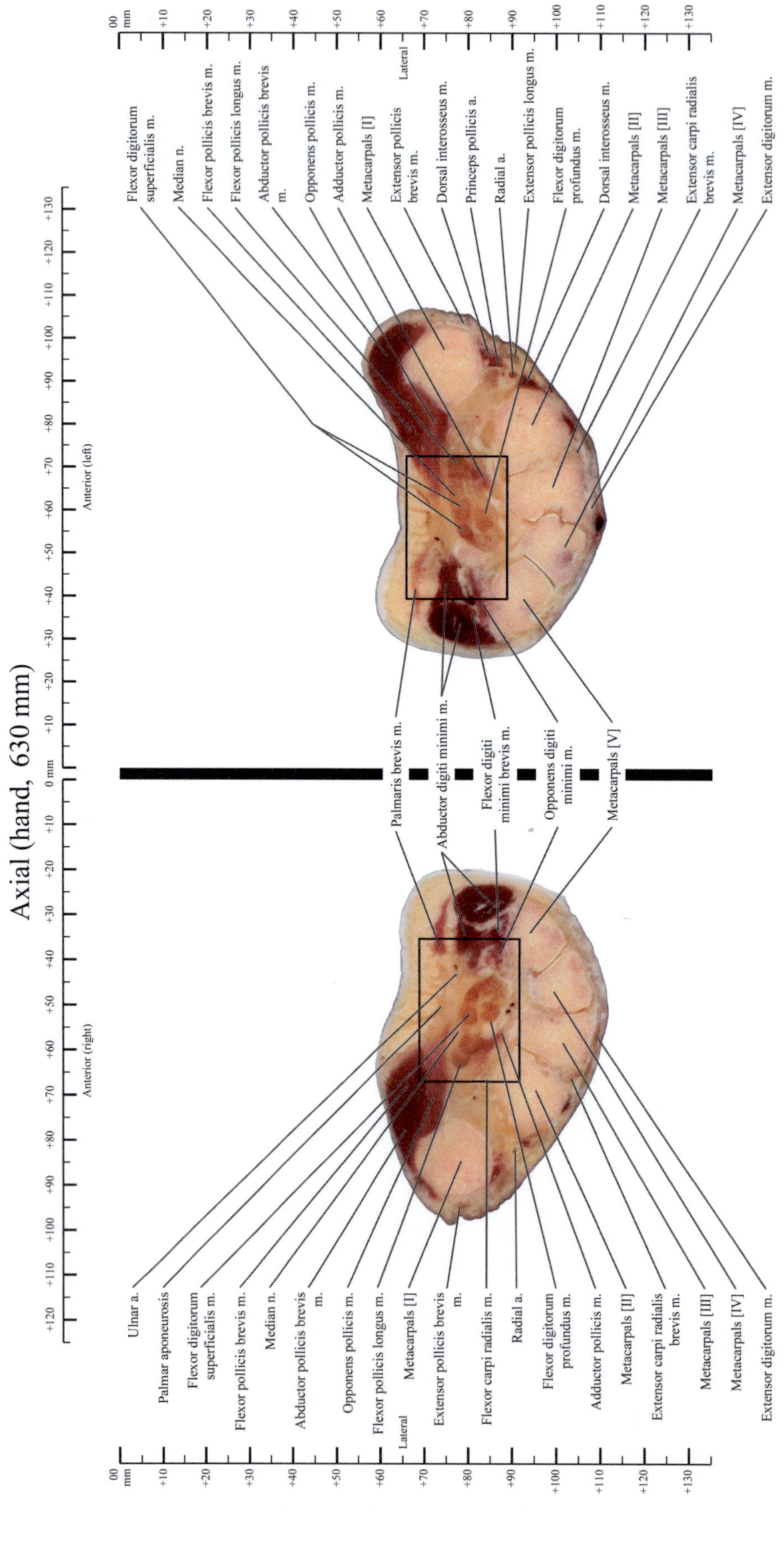
Axial (hand, 630 mm)
Anterior (left)
Anterior (right)
Lateral
Flexor digitorum superficialis m.
Median n.
Flexor pollicis brevis m.
Flexor pollicis longus m.
Abductor pollicis brevis m.
Opponens pollicis m.
Adductor pollicis m.
Metacarpals [I]
Extensor pollicis brevis m.
Dorsal interosseus m.
Princeps pollicis a.
Radial a.
Extensor pollicis longus m.
Flexor digitorum profundus m.
Dorsal interosseus m.
Metacarpals [II]
Metacarpals [III]
Extensor carpi radialis brevis m.
Metacarpals [IV]
Extensor digitorum m.
Palmaris brevis m.
Abductor digiti minimi m.
Flexor digiti minimi brevis m.
Opponens digiti minimi m.
Metacarpals [V]
Ulnar a.
Palmar aponeurosis
Flexor digitorum superficialis m.
Flexor pollicis brevis m.
Median n.
Abductor pollicis brevis m.
Opponens pollicis m.
Flexor pollicis longus m.
Metacarpals [I]
Extensor pollicis brevis m.
Flexor carpi radialis m.
Radial a.
Flexor digitorum profundus m.
Adductor pollicis m.
Metacarpals [II]
Extensor carpi radialis brevis m.
Metacarpals [III]
Metacarpals [IV]
Extensor digitorum m.

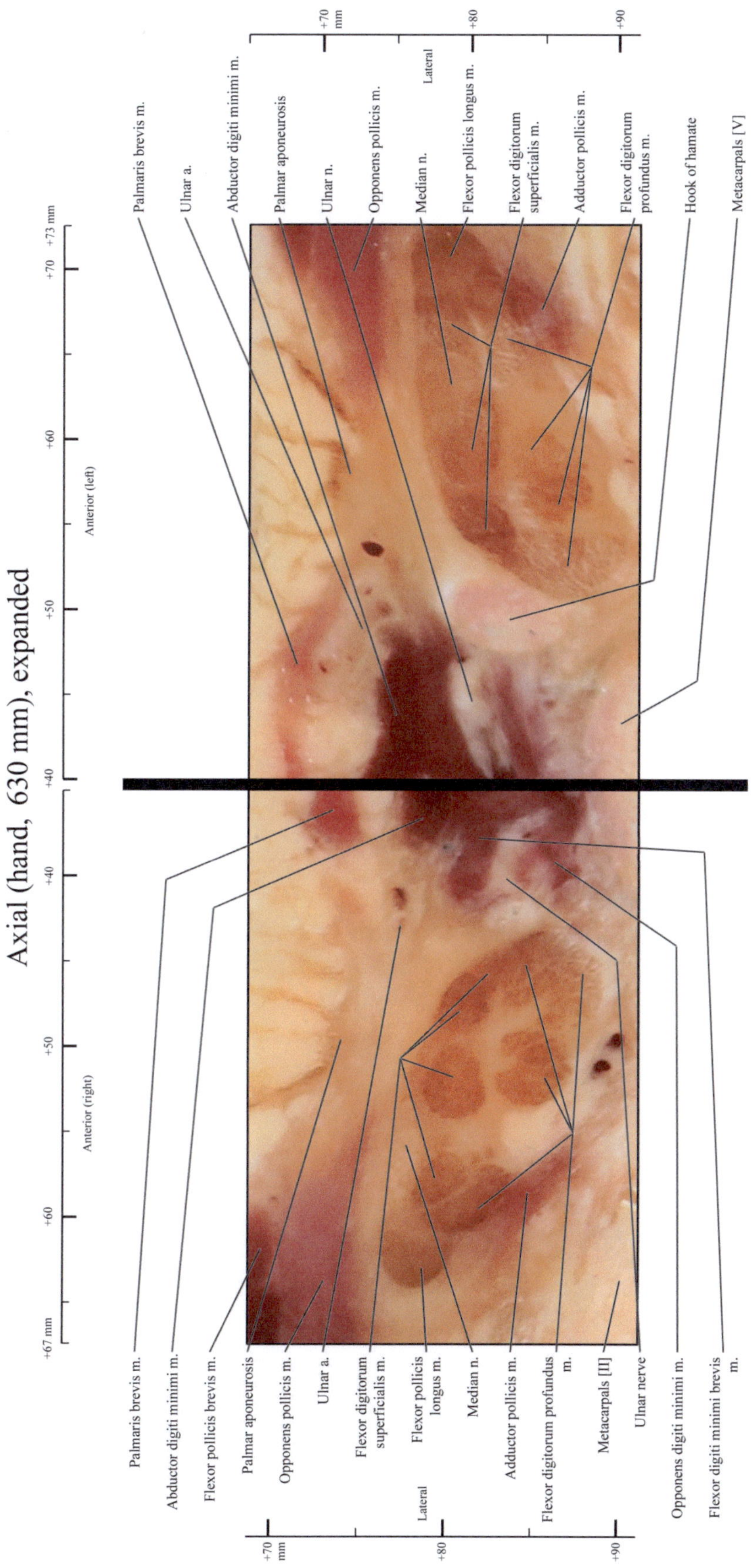

Axial (hand, 630 mm), expanded
Anterior (left)
Anterior (right)
Lateral
Lateral
+70 +73 mm
+60
+50
+40
+40
+50
+60
+67 mm
+70 mm
+80
+90
+70 mm
+80
+90
Palmaris brevis m.
Ulnar a.
Abductor digiti minimi m.
Palmar aponeurosis
Ulnar n.
Opponens pollicis m.
Median n.
Flexor pollicis longus m.
Flexor digitorum superficialis m.
Adductor pollicis m.
Flexor digitorum profundus m.
Hook of hamate
Metacarpals [V]
Palmaris brevis m.
Abductor digiti minimi m.
Flexor pollicis brevis m.
Palmar aponeurosis
Opponens pollicis m.
Ulnar a.
Flexor digitorum superficialis m.
Flexor pollicis longus m.
Median n.
Adductor pollicis m.
Flexor digitorum profundus m.
Metacarpals [II]
Ulnar nerve
Opponens digiti minimi m.
Flexor digiti minimi brevis m.

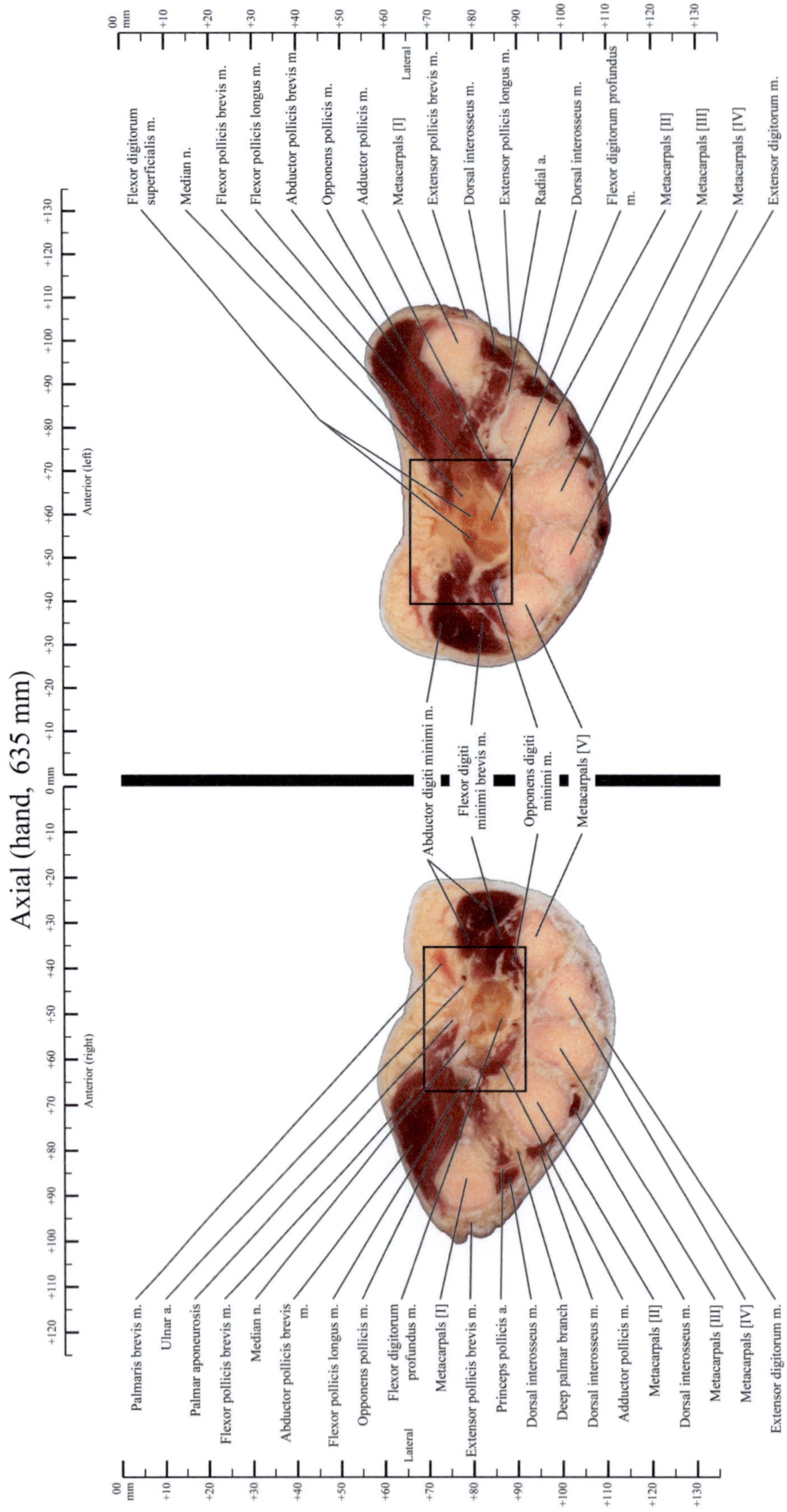

Axial (hand, 635 mm)
Anterior (left)
Anterior (right)
Lateral
Flexor digitorum superficialis m.
Median n.
Flexor pollicis brevis m.
Flexor pollicis longus m.
Abductor pollicis brevis m.
Opponens pollicis m.
Adductor pollicis m.
Metacarpals [I]
Extensor pollicis brevis m.
Dorsal interosseus m.
Extensor pollicis longus m.
Radial a.
Dorsal interosseus m.
Flexor digitorum profundus m.
Metacarpals [II]
Metacarpals [III]
Metacarpals [IV]
Extensor digitorum m.
Abductor digiti minimi m.
Flexor digiti minimi brevis m.
Opponens digiti minimi m.
Metacarpals [V]
Palmaris brevis m.
Ulnar a.
Palmar aponeurosis
Flexor pollicis brevis m.
Median n.
Abductor pollicis brevis m.
Flexor pollicis longus m.
Opponens pollicis m.
Flexor digitorum profundus m.
Metacarpals [I]
Extensor pollicis brevis m.
Princeps pollicis a.
Dorsal interosseus m.
Deep palmar branch
Dorsal interosseus m.
Adductor pollicis m.
Metacarpals [II]
Dorsal interosseus m.
Metacarpals [III]
Metacarpals [IV]
Extensor digitorum m.

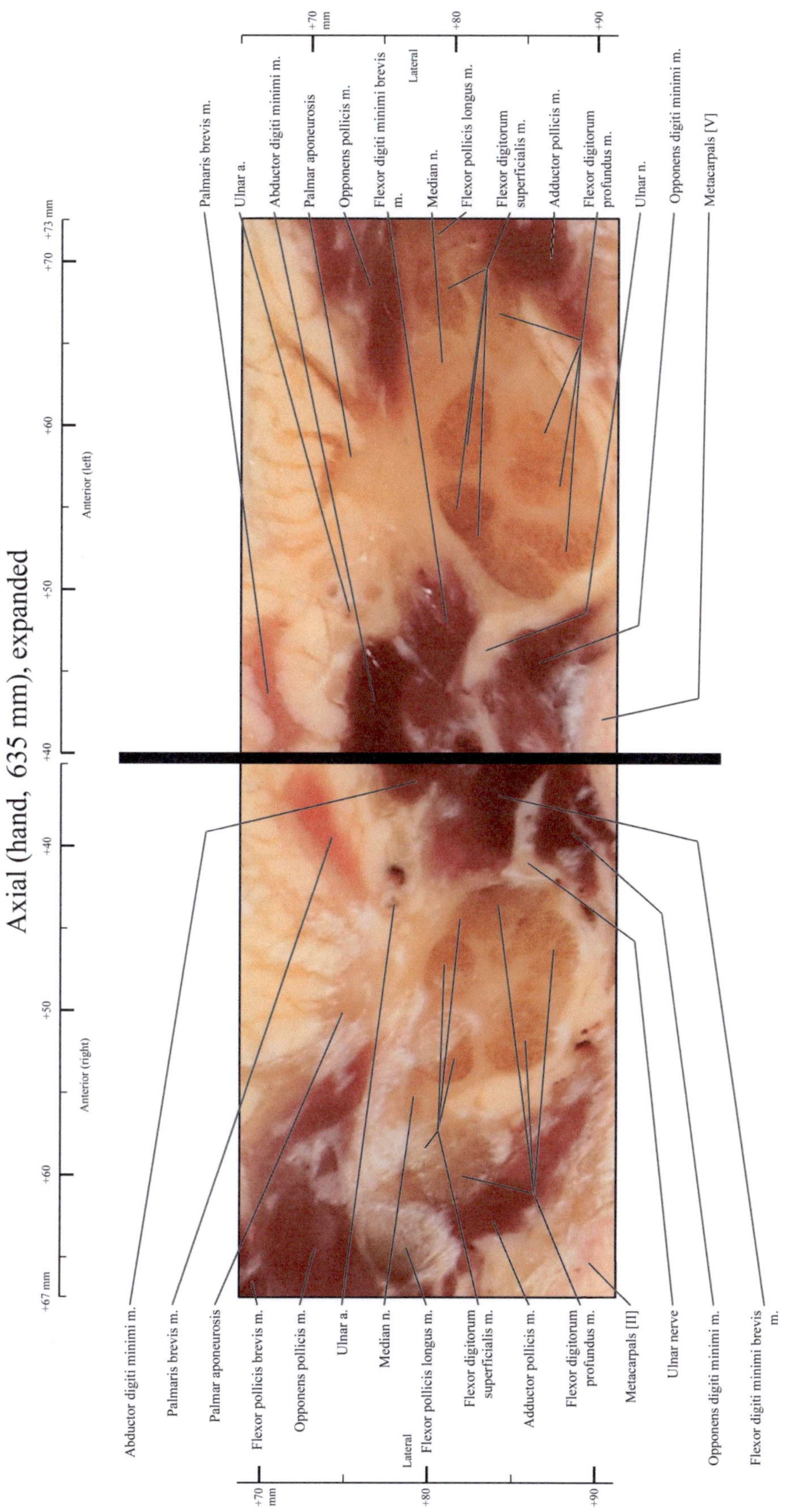
Axial (hand, 635 mm), expanded
Anterior (left)
Anterior (right)
+70 +73 mm
+60
+50
+40
+40
+50
+60
+67 mm
+70 mm
+80
+90
Lateral
Palmaris brevis m.
Ulnar a.
Abductor digiti minimi m.
Palmar aponeurosis
Opponens pollicis m.
Flexor digiti minimi brevis m.
Median n.
Flexor pollicis longus m.
Flexor digitorum superficialis m.
Adductor pollicis m.
Flexor digitorum profundus m.
Ulnar n.
Opponens digiti minimi m.
Metacarpals [V]
Abductor digiti minimi m.
Palmaris brevis m.
Palmar aponeurosis
Flexor pollicis brevis m.
Opponens pollicis m.
Ulnar a.
Median n.
Flexor pollicis longus m.
Flexor digitorum superficialis m.
Adductor pollicis m.
Flexor digitorum profundus m.
Metacarpals [II]
Ulnar nerve
Opponens digiti minimi m.
Flexor digiti minimi brevis m.
+70 mm
+80
+90
Lateral

Axial (hand, 640 mm)

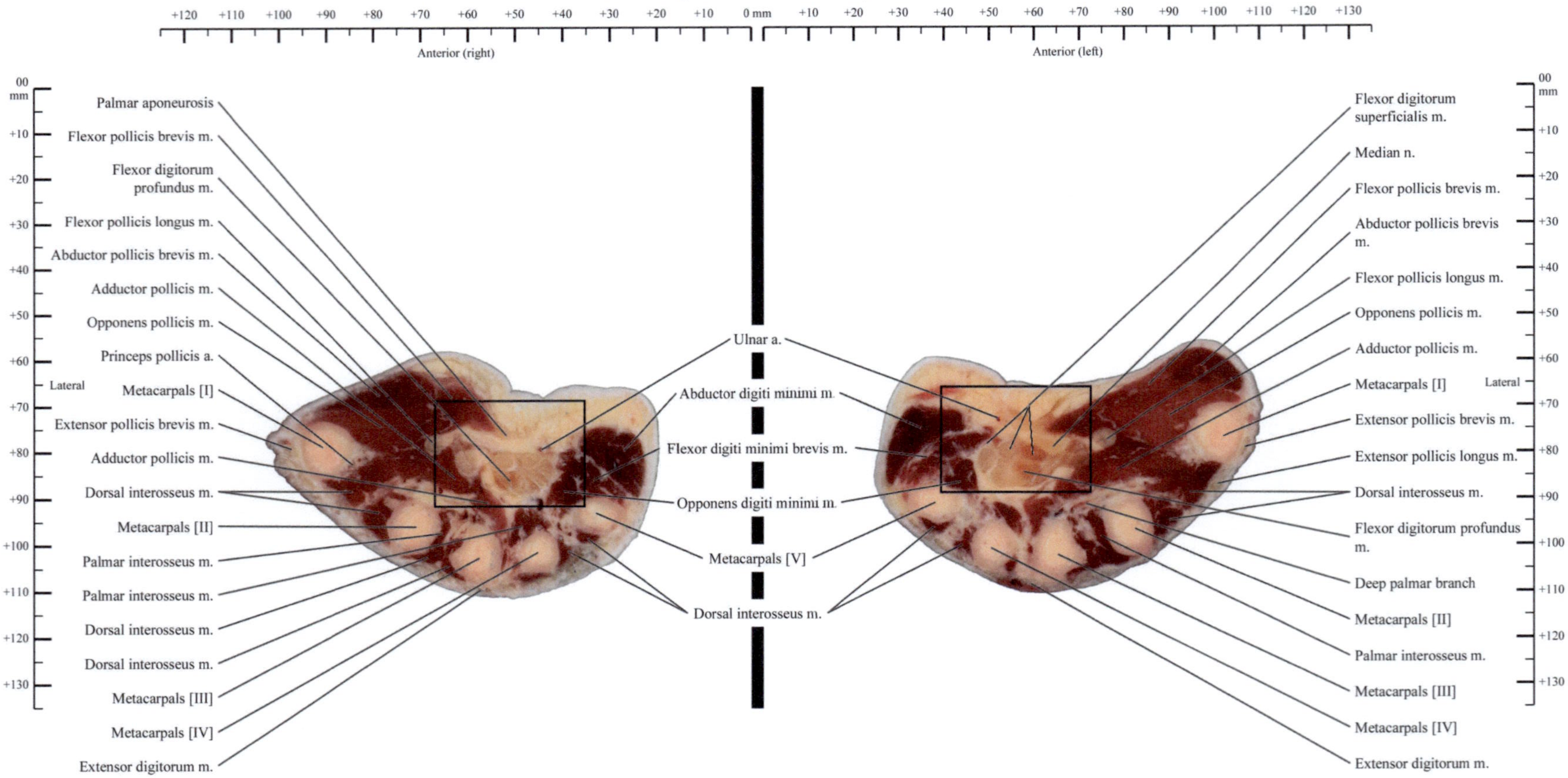

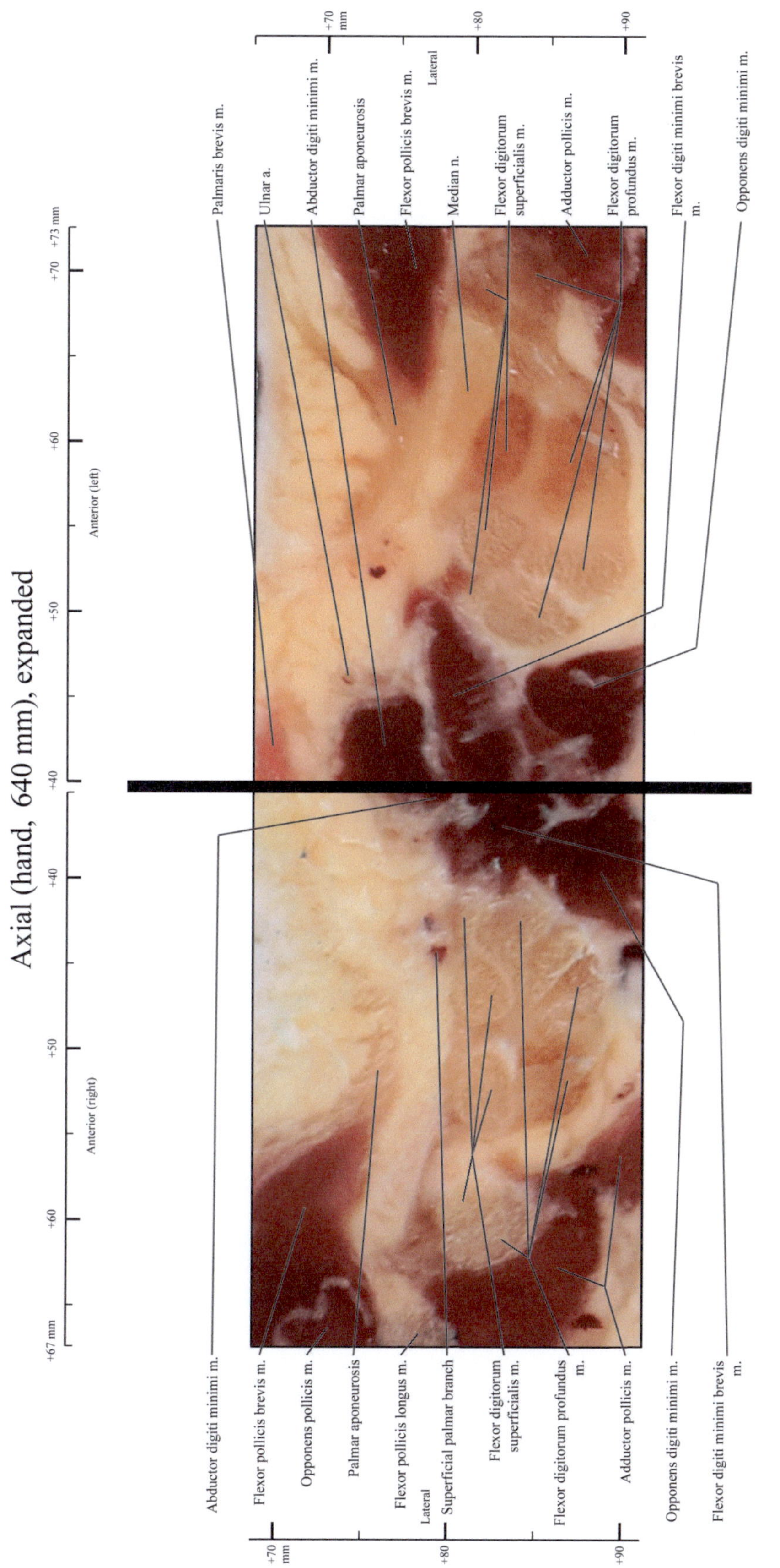

Axial (hand, 640 mm), expanded
Anterior (left)
+70 +73 mm
+60
+50
+40
Palmaris brevis m.
Ulnar a.
Abductor digiti minimi m.
Palmar aponeurosis
Flexor pollicis brevis m.
Lateral
Median n.
Flexor digitorum superficialis m.
Adductor pollicis m.
Flexor digitorum profundus m.
Flexor digiti minimi brevis m.
Opponens digiti minimi m.
+70 mm
+80
+90
Anterior (right)
+40
+50
+60
+67 mm
Abductor digiti minimi m.
Flexor pollicis brevis m.
Opponens pollicis m.
Palmar aponeurosis
Flexor pollicis longus m.
Lateral
Superficial palmar branch
Flexor digitorum superficialis m.
Flexor digitorum profundus m.
Adductor pollicis m.
Opponens digiti minimi m.
Flexor digiti minimi brevis m.
+70 mm
+80
+90

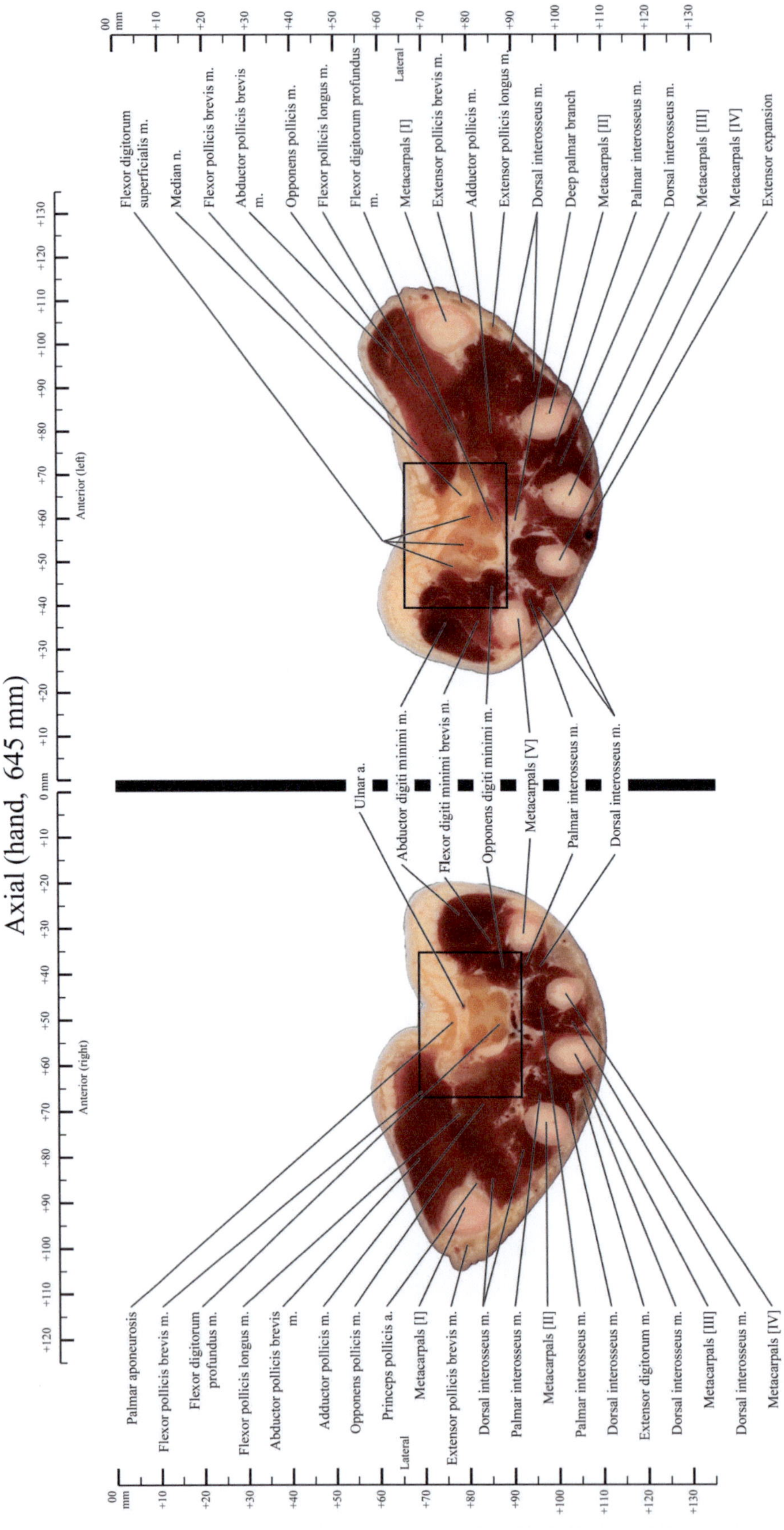

Axial (hand, 645 mm)
Anterior (left)
Anterior (right)
Lateral
Flexor digitorum superficialis m.
Median n.
Flexor pollicis brevis m.
Abductor pollicis brevis m.
Opponens pollicis m.
Flexor pollicis longus m.
Flexor digitorum profundus m.
Metacarpals [I]
Extensor pollicis brevis m.
Adductor pollicis m.
Extensor pollicis longus m.
Dorsal interosseus m.
Deep palmar branch
Metacarpals [II]
Palmar interosseus m.
Dorsal interosseus m.
Metacarpals [III]
Metacarpals [IV]
Extensor expansion
Ulnar a.
Abductor digiti minimi m.
Flexor digiti minimi brevis m.
Opponens digiti minimi m.
Metacarpals [V]
Palmar interosseus m.
Dorsal interosseus m.
Palmar aponeurosis
Flexor pollicis brevis m.
Flexor digitorum profundus m.
Flexor pollicis longus m.
Abductor pollicis brevis m.
Adductor pollicis m.
Opponens pollicis m.
Princeps pollicis a.
Metacarpals [I]
Extensor pollicis brevis m.
Dorsal interosseus m.
Palmar interosseus m.
Metacarpals [II]
Palmar interosseus m.
Dorsal interosseus m.
Extensor digitorum m.
Dorsal interosseus m.
Metacarpals [III]
Dorsal interosseus m.
Metacarpals [IV]

Axial (hand, 645 mm), expanded

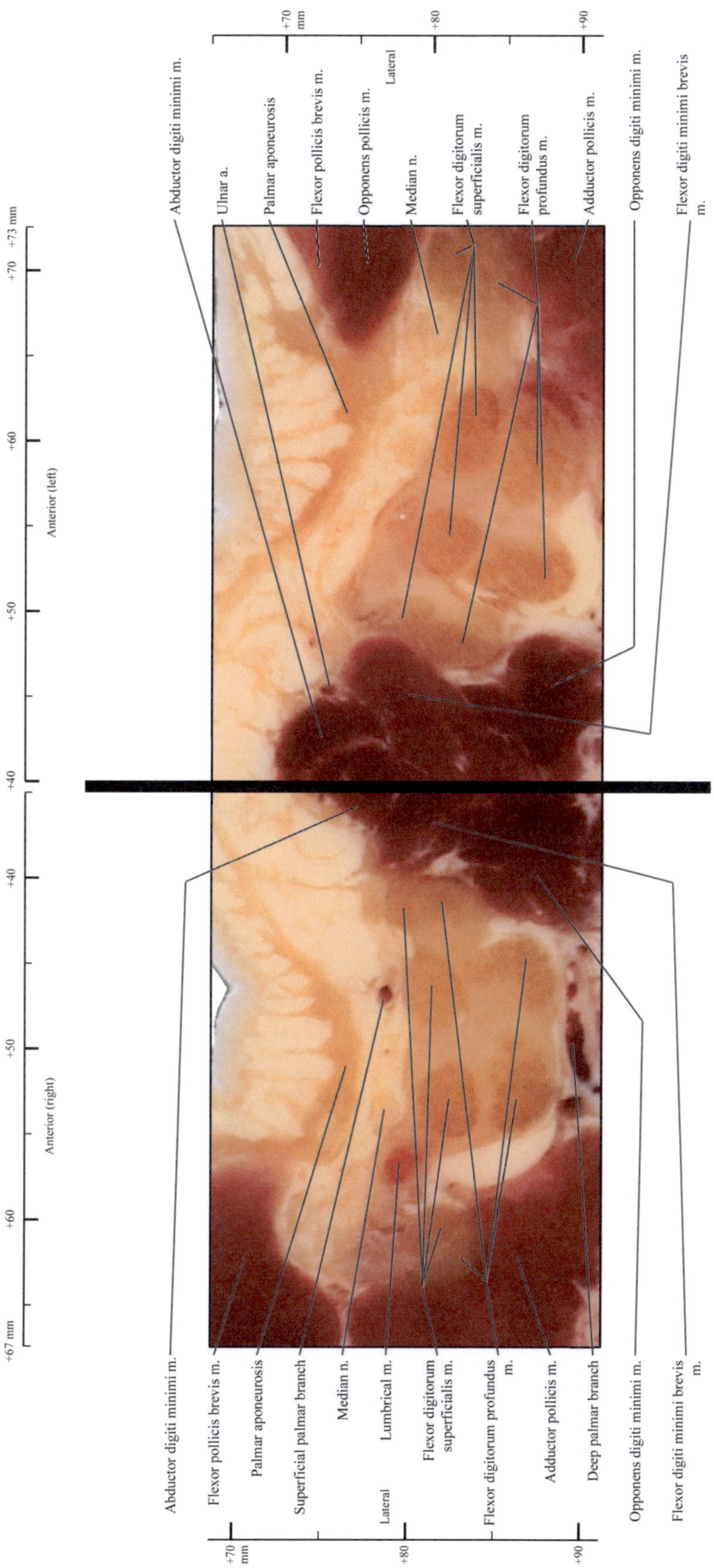

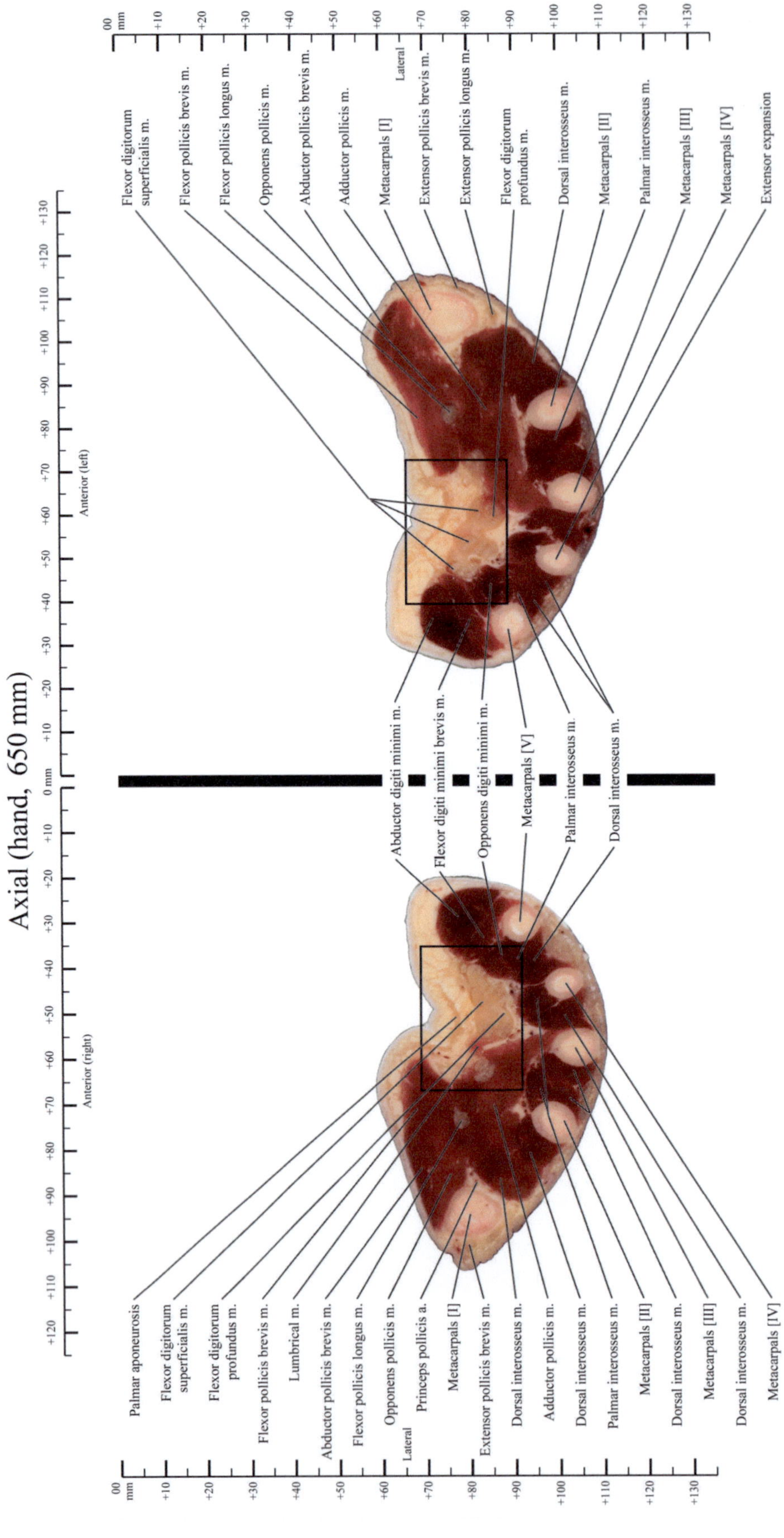
Axial (hand, 650 mm)
Anterior (left)
Anterior (right)
Lateral
Flexor digitorum superficialis m.
Flexor pollicis brevis m.
Flexor pollicis longus m.
Opponens pollicis m.
Abductor pollicis brevis m.
Adductor pollicis m.
Metacarpals [I]
Extensor pollicis brevis m.
Extensor pollicis longus m.
Flexor digitorum profundus m.
Dorsal interosseus m.
Metacarpals [II]
Palmar interosseus m.
Metacarpals [III]
Metacarpals [IV]
Extensor expansion
Abductor digiti minimi m.
Flexor digiti minimi brevis m.
Opponens digiti minimi m.
Metacarpals [V]
Palmar interosseus m.
Dorsal interosseus m.
Palmar aponeurosis
Flexor digitorum superficialis m.
Flexor digitorum profundus m.
Flexor pollicis brevis m.
Lumbrical m.
Abductor pollicis brevis m.
Flexor pollicis longus m.
Opponens pollicis m.
Princeps pollicis a.
Metacarpals [I]
Extensor pollicis brevis m.
Dorsal interosseus m.
Adductor pollicis m.
Dorsal interosseus m.
Palmar interosseus m.
Metacarpals [II]
Dorsal interosseus m.
Metacarpals [III]
Dorsal interosseus m.
Metacarpals [IV]

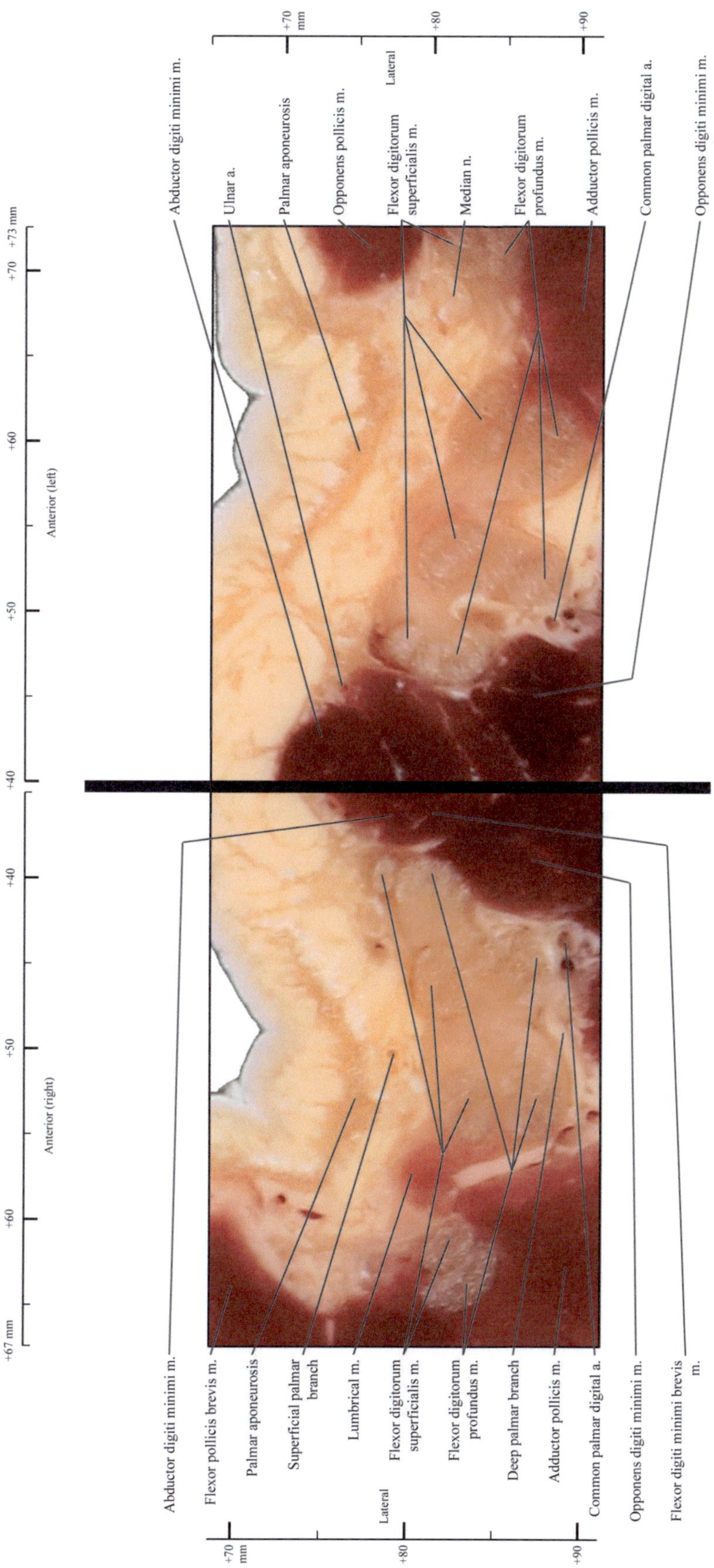
Abductor digiti minimi m.
Ulnar a.
Palmar aponeurosis
Opponens pollicis m.
Flexor digitorum superficialis m.
Median n.
Flexor digitorum profundus m.
Adductor pollicis m.
Common palmar digital a.
Opponens digiti minimi m.
Lateral
+70 mm
+80
+90
+70 +73 mm
+60
+50
+40
Anterior (left)
+40
+50
+60
+67 mm
Anterior (right)
Abductor digiti minimi m.
Flexor pollicis brevis m.
Palmar aponeurosis
Superficial palmar branch
Lumbrical m.
Flexor digitorum superficialis m.
Flexor digitorum profundus m.
Deep palmar branch
Adductor pollicis m.
Common palmar digital a.
Opponens digiti minimi m.
Flexor digiti minimi brevis m.
Lateral
+70 mm
+80
+90

Axial (hand, 650 mm), expanded

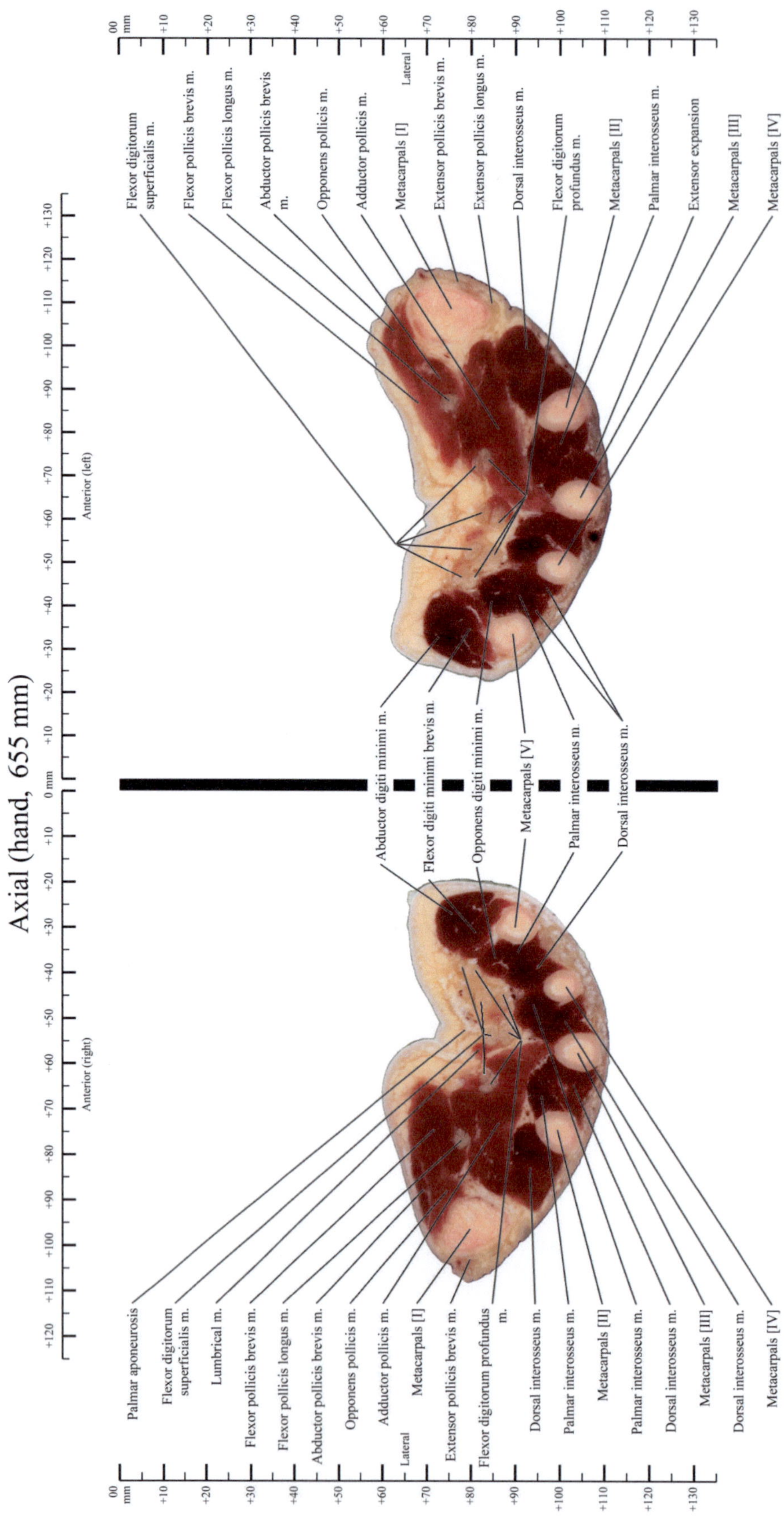

Axial (hand, 655 mm)
Anterior (left)
Anterior (right)
Lateral
Flexor digitorum superficialis m.
Flexor pollicis brevis m.
Flexor pollicis longus m.
Abductor pollicis brevis m.
Opponens pollicis m.
Adductor pollicis m.
Metacarpals [I]
Extensor pollicis brevis m.
Extensor pollicis longus m.
Dorsal interosseus m.
Flexor digitorum profundus m.
Metacarpals [II]
Palmar interosseus m.
Extensor expansion
Metacarpals [III]
Metacarpals [IV]
Abductor digiti minimi m.
Flexor digiti minimi brevis m.
Opponens digiti minimi m.
Metacarpals [V]
Palmar interosseus m.
Dorsal interosseus m.
Palmar aponeurosis
Flexor digitorum superficialis m.
Lumbrical m.
Flexor pollicis brevis m.
Flexor pollicis longus m.
Abductor pollicis brevis m.
Opponens pollicis m.
Adductor pollicis m.
Metacarpals [I]
Extensor pollicis brevis m.
Flexor digitorum profundus m.
Dorsal interosseus m.
Palmar interosseus m.
Metacarpals [II]
Palmar interosseus m.
Dorsal interosseus m.
Metacarpals [III]
Dorsal interosseus m.
Metacarpals [IV]

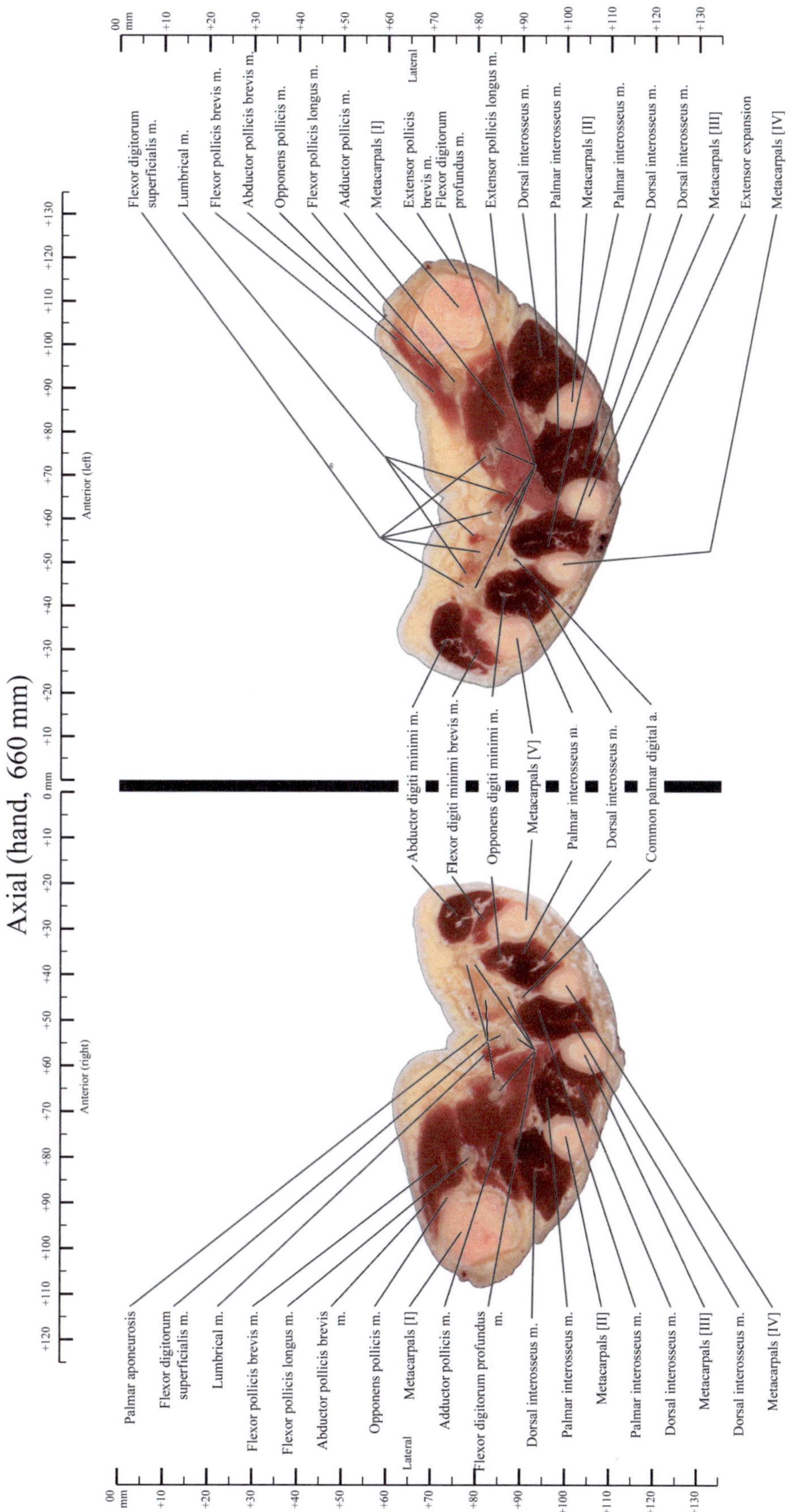

Axial (hand, 660 mm)
Anterior (left)
Anterior (right)
Lateral
Flexor digitorum superficialis m.
Lumbrical m.
Flexor pollicis brevis m.
Abductor pollicis brevis m.
Opponens pollicis m.
Flexor pollicis longus m.
Adductor pollicis m.
Metacarpals [I]
Extensor pollicis brevis m.
Flexor digitorum profundus m.
Extensor pollicis longus m.
Dorsal interosseus m.
Palmar interosseus m.
Metacarpals [II]
Palmar interosseus m.
Dorsal interosseus m.
Dorsal interosseus m.
Metacarpals [III]
Extensor expansion
Metacarpals [IV]
Abductor digiti minimi m.
Flexor digiti minimi brevis m.
Opponens digiti minimi m.
Metacarpals [V]
Palmar interosseus m.
Dorsal interosseus m.
Common palmar digital a.
Palmar aponeurosis
Flexor digitorum superficialis m.
Lumbrical m.
Flexor pollicis brevis m.
Flexor pollicis longus m.
Abductor pollicis brevis m.
Opponens pollicis m.
Metacarpals [I]
Adductor pollicis m.
Flexor digitorum profundus m.
Dorsal interosseus m.
Palmar interosseus m.
Metacarpals [II]
Palmar interosseus m.
Dorsal interosseus m.
Metacarpals [III]
Dorsal interosseus m.
Metacarpals [IV]

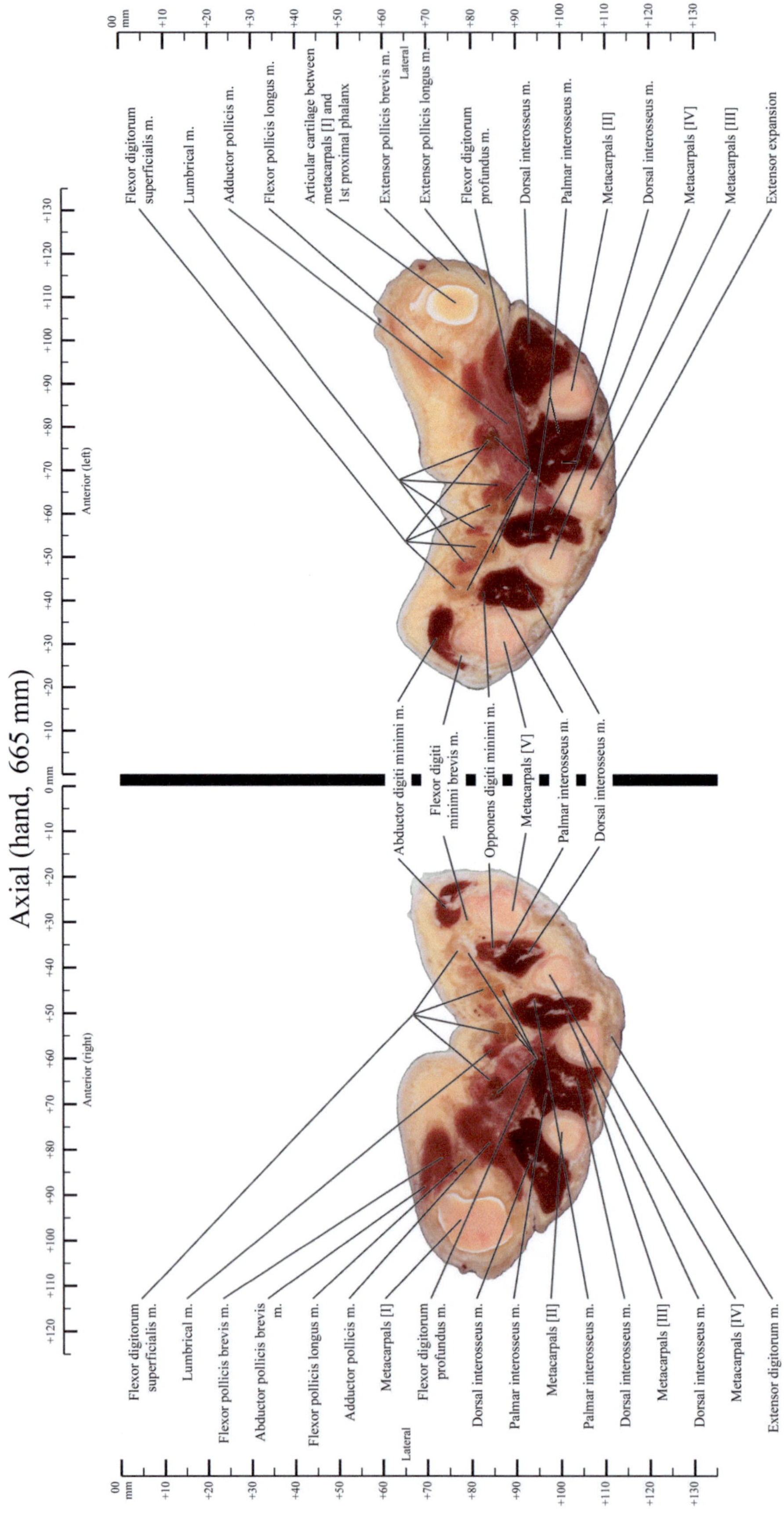
Axial (hand, 665 mm)
Anterior (left)
Anterior (right)
Lateral
Flexor digitorum superficialis m.
Lumbrical m.
Adductor pollicis m.
Flexor pollicis longus m.
Articular cartilage between metacarpals [I] and 1st proximal phalanx
Extensor pollicis brevis m.
Extensor pollicis longus m.
Flexor digitorum profundus m.
Dorsal interosseus m.
Palmar interosseus m.
Metacarpals [II]
Dorsal interosseus m.
Metacarpals [IV]
Metacarpals [III]
Extensor expansion
Abductor digiti minimi m.
Flexor digiti minimi brevis m.
Opponens digiti minimi m.
Metacarpals [V]
Palmar interosseus m.
Dorsal interosseus m.
Flexor digitorum superficialis m.
Lumbrical m.
Flexor pollicis brevis m.
Abductor pollicis brevis m.
Flexor pollicis longus m.
Adductor pollicis m.
Metacarpals [I]
Flexor digitorum profundus m.
Dorsal interosseus m.
Palmar interosseus m.
Metacarpals [II]
Palmar interosseus m.
Dorsal interosseus m.
Metacarpals [III]
Dorsal interosseus m.
Metacarpals [IV]
Extensor digitorum m.

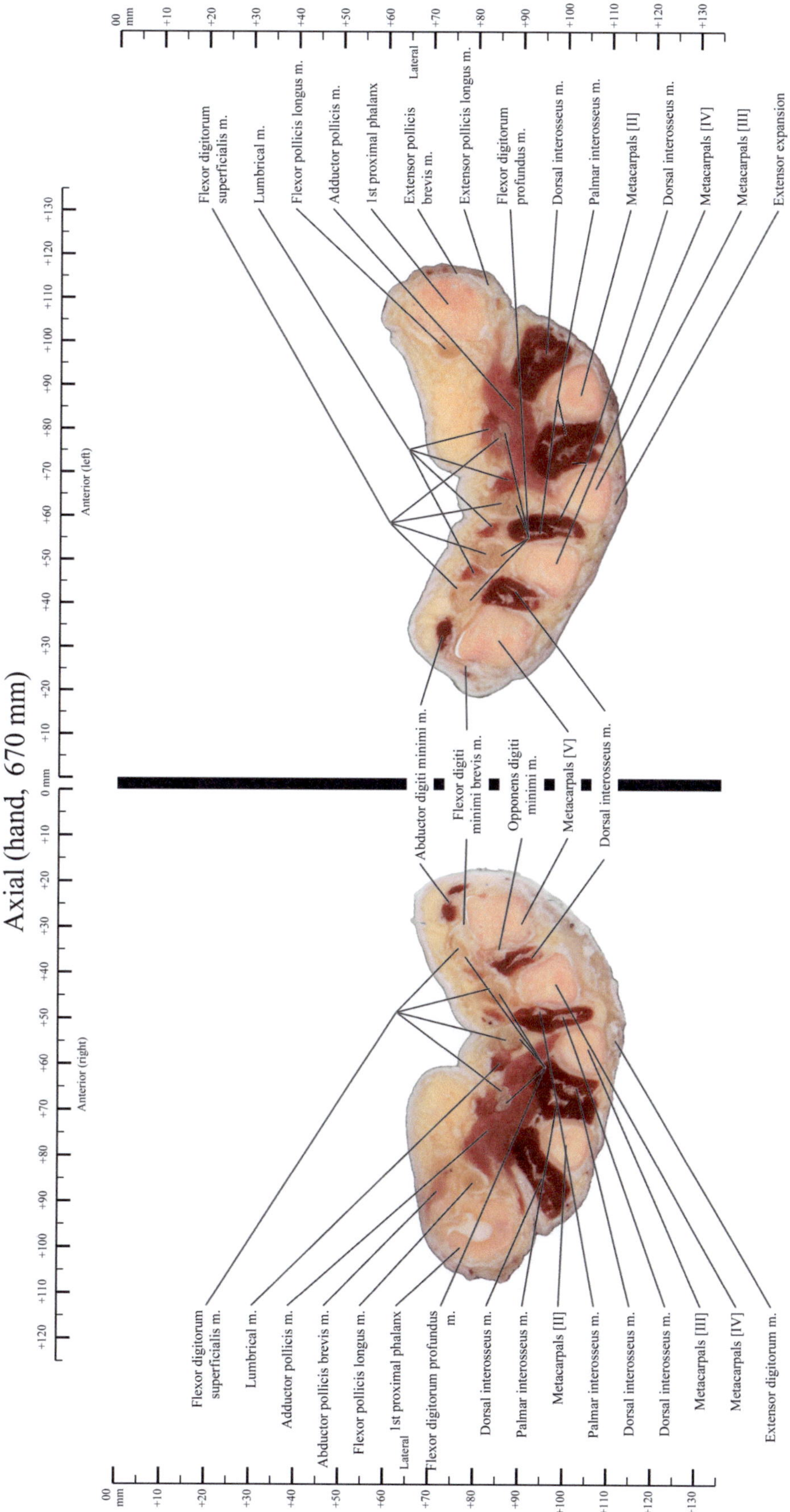
Axial (hand, 670 mm)
Anterior (left)
Anterior (right)
Lateral
Flexor digitorum superficialis m.
Lumbrical m.
Flexor pollicis longus m.
Adductor pollicis m.
1st proximal phalanx
Extensor pollicis brevis m.
Extensor pollicis longus m.
Flexor digitorum profundus m.
Dorsal interosseus m.
Palmar interosseus m.
Metacarpals [II]
Dorsal interosseus m.
Metacarpals [IV]
Metacarpals [III]
Extensor expansion
Abductor digiti minimi m.
Flexor digiti minimi brevis m.
Opponens digiti minimi m.
Metacarpals [V]
Dorsal interosseus m.
Flexor digitorum superficialis m.
Lumbrical m.
Adductor pollicis m.
Abductor pollicis brevis m.
Flexor pollicis longus m.
1st proximal phalanx
Flexor digitorum profundus m.
Dorsal interosseus m.
Palmar interosseus m.
Metacarpals [II]
Palmar interosseus m.
Dorsal interosseus m.
Dorsal interosseus m.
Metacarpals [III]
Metacarpals [IV]
Extensor digitorum m.

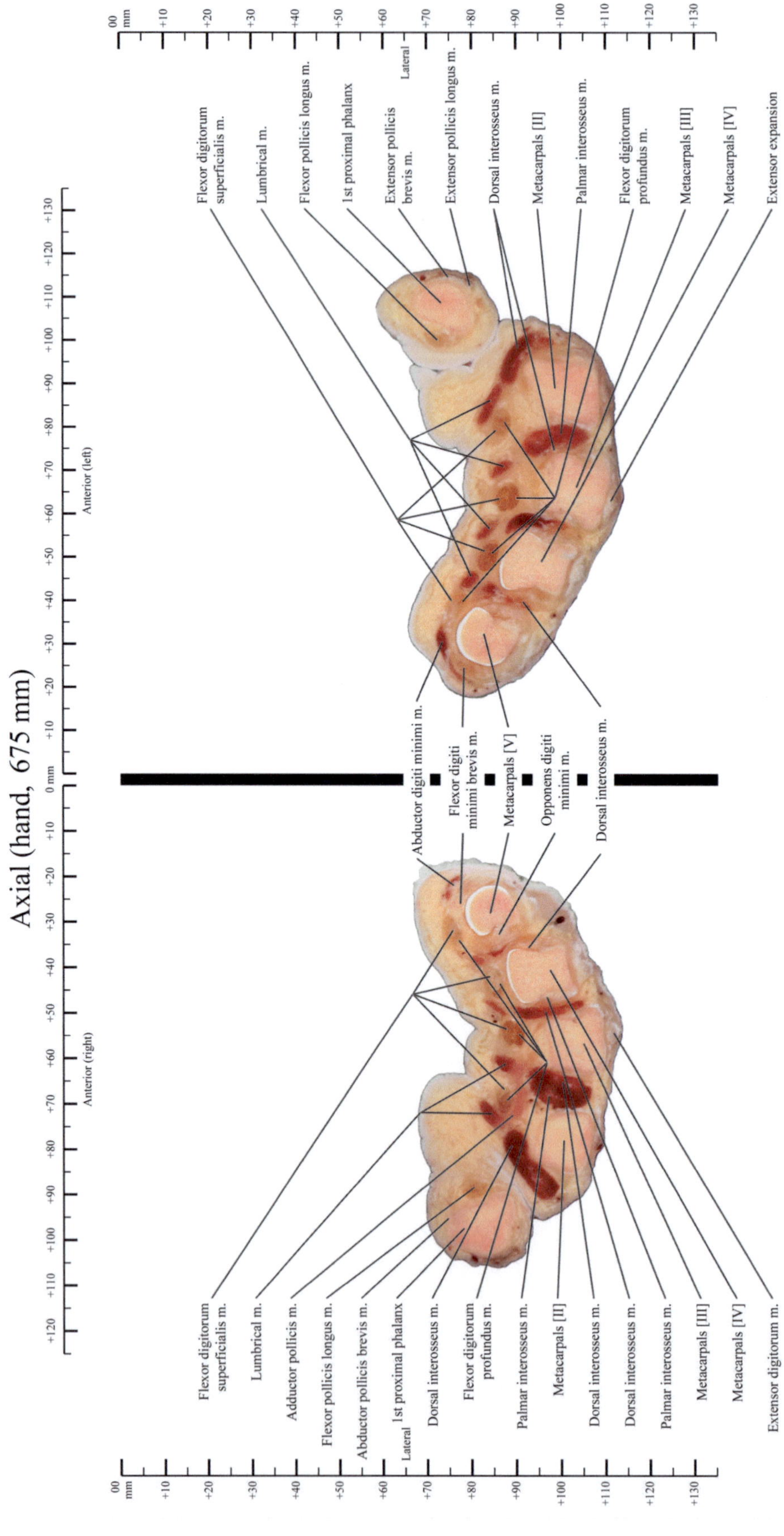
Axial (hand, 675 mm)
Anterior (left)
Anterior (right)
Lateral
Flexor digitorum superficialis m.
Lumbrical m.
Flexor pollicis longus m.
1st proximal phalanx
Extensor pollicis brevis m.
Extensor pollicis longus m.
Dorsal interosseus m.
Metacarpals [II]
Palmar interosseus m.
Flexor digitorum profundus m.
Metacarpals [III]
Metacarpals [IV]
Extensor expansion
Abductor digiti minimi m.
Flexor digiti minimi brevis m.
Metacarpals [V]
Opponens digiti minimi m.
Dorsal interosseus m.
Flexor digitorum superficialis m.
Lumbrical m.
Adductor pollicis m.
Flexor pollicis longus m.
Abductor pollicis brevis m.
1st proximal phalanx
Dorsal interosseus m.
Flexor digitorum profundus m.
Palmar interosseus m.
Metacarpals [II]
Dorsal interosseus m.
Dorsal interosseus m.
Palmar interosseus m.
Metacarpals [III]
Metacarpals [IV]
Extensor digitorum m.

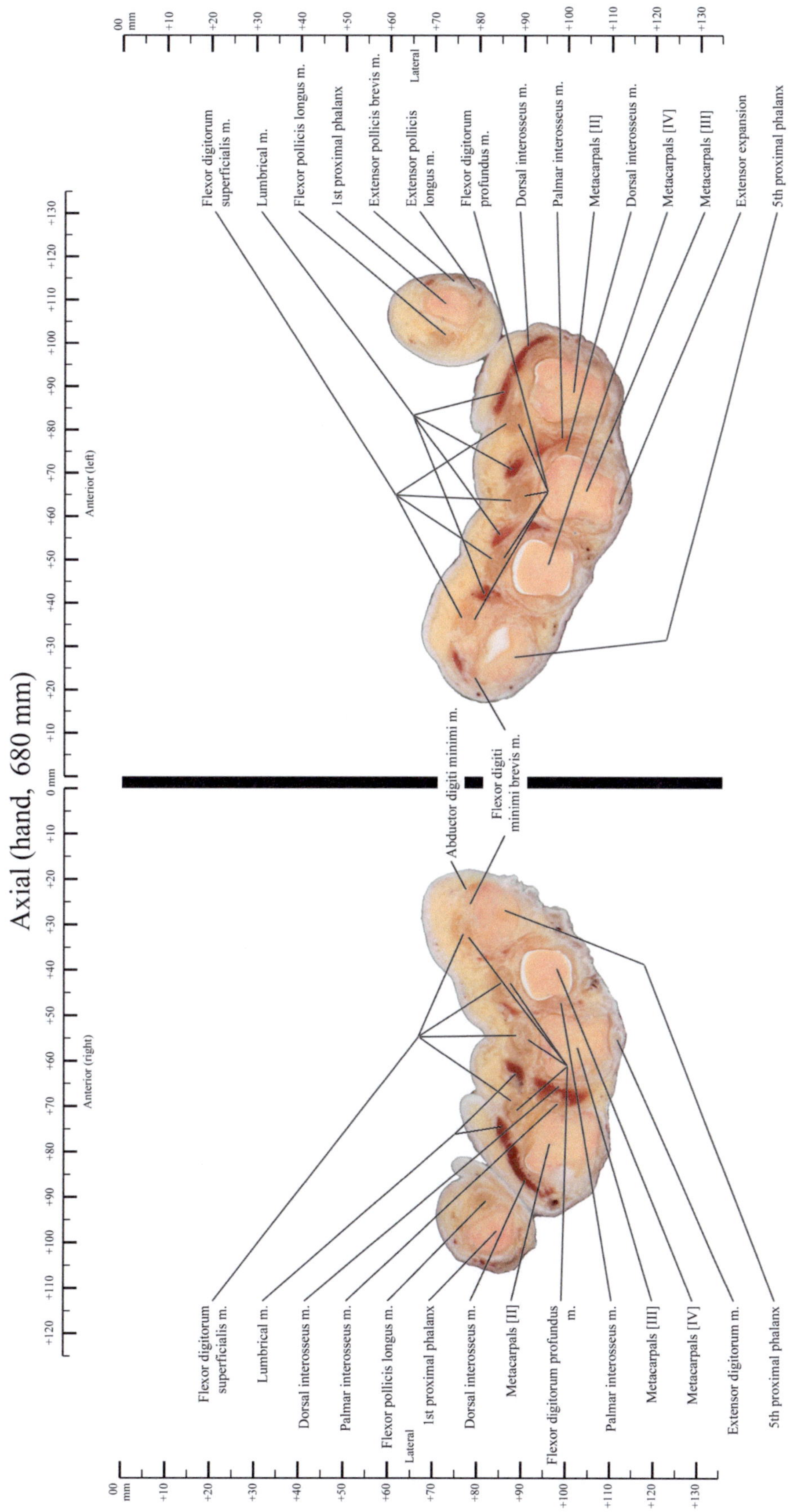
Axial (hand, 680 mm)
Anterior (left)
Lateral
Flexor digitorum superficialis m.
Lumbrical m.
Flexor pollicis longus m.
1st proximal phalanx
Extensor pollicis brevis m.
Extensor pollicis longus m.
Flexor digitorum profundus m.
Dorsal interosseus m.
Palmar interosseus m.
Metacarpals [II]
Dorsal interosseus m.
Metacarpals [IV]
Metacarpals [III]
Extensor expansion
5th proximal phalanx
Abductor digiti minimi m.
Flexor digiti minimi brevis m.
Anterior (right)
Lateral
Flexor digitorum superficialis m.
Lumbrical m.
Dorsal interosseus m.
Palmar interosseus m.
Flexor pollicis longus m.
1st proximal phalanx
Dorsal interosseus m.
Metacarpals [II]
Flexor digitorum profundus m.
Palmar interosseus m.
Metacarpals [III]
Metacarpals [IV]
Extensor digitorum m.
5th proximal phalanx

Axial (hand, 685 mm)

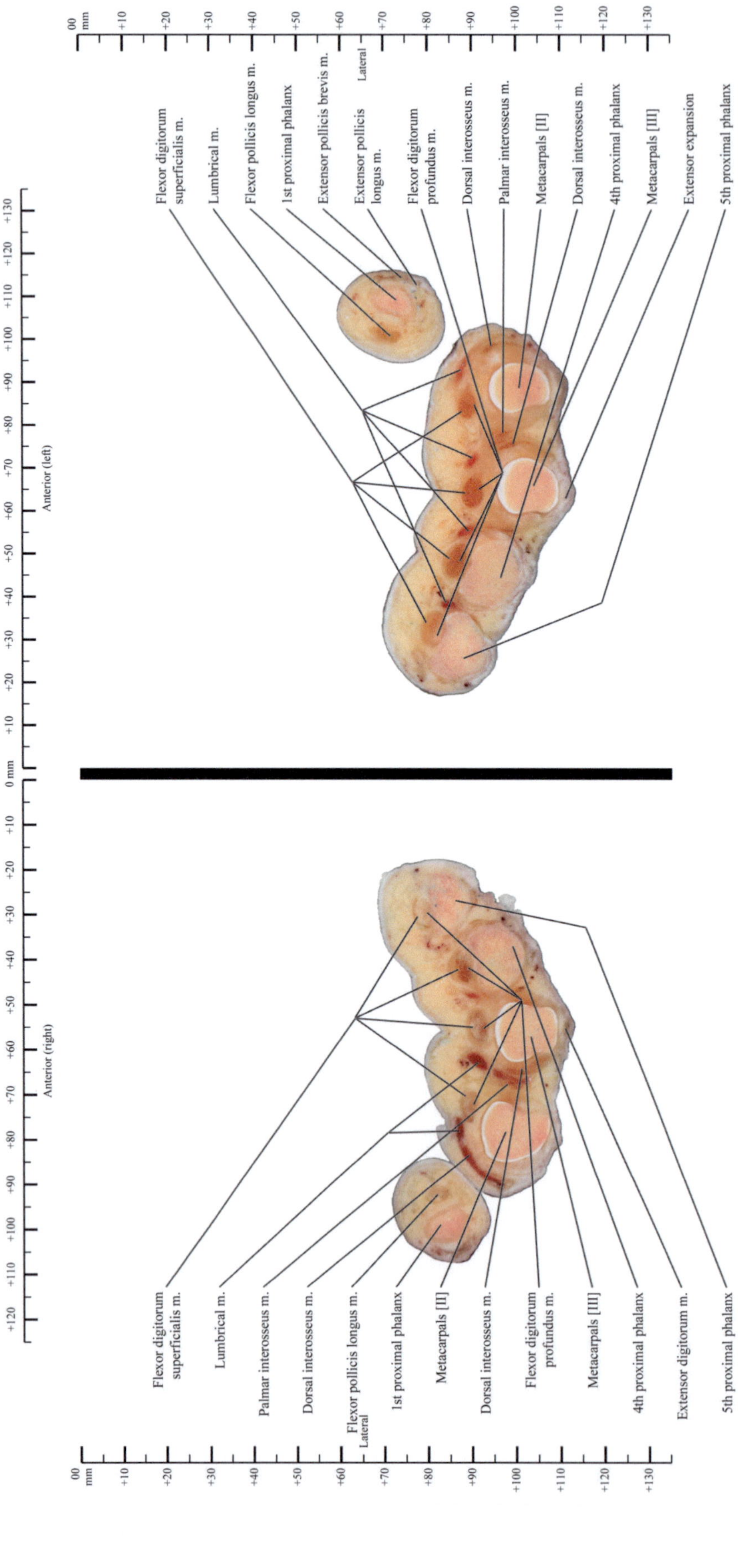

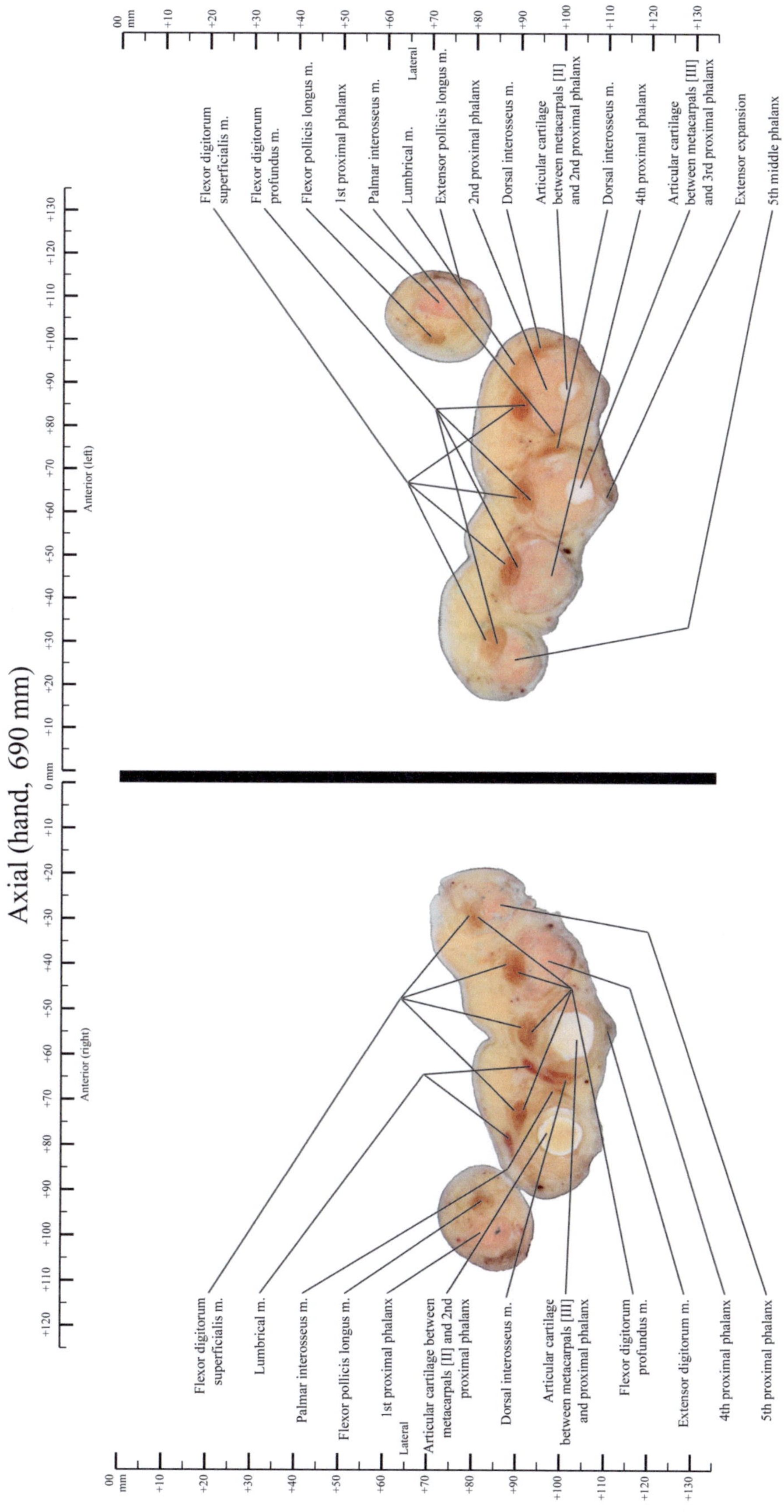
Axial (hand, 690 mm)
Anterior (left)
Lateral
Flexor digitorum superficialis m.
Flexor digitorum profundus m.
Flexor pollicis longus m.
1st proximal phalanx
Palmar interosseus m.
Lumbrical m.
Extensor pollicis longus m.
2nd proximal phalanx
Dorsal interosseus m.
Articular cartilage between metacarpals [II] and 2nd proximal phalanx
Dorsal interosseus m.
4th proximal phalanx
Articular cartilage between metacarpals [III] and 3rd proximal phalanx
Extensor expansion
5th middle phalanx
Anterior (right)
Lateral
Flexor digitorum superficialis m.
Lumbrical m.
Palmar interosseus m.
Flexor pollicis longus m.
1st proximal phalanx
Articular cartilage between metacarpals [II] and 2nd proximal phalanx
Dorsal interosseus m.
Articular cartilage between metacarpals [III] and proximal phalanx
Flexor digitorum profundus m.
Extensor digitorum m.
4th proximal phalanx
5th proximal phalanx

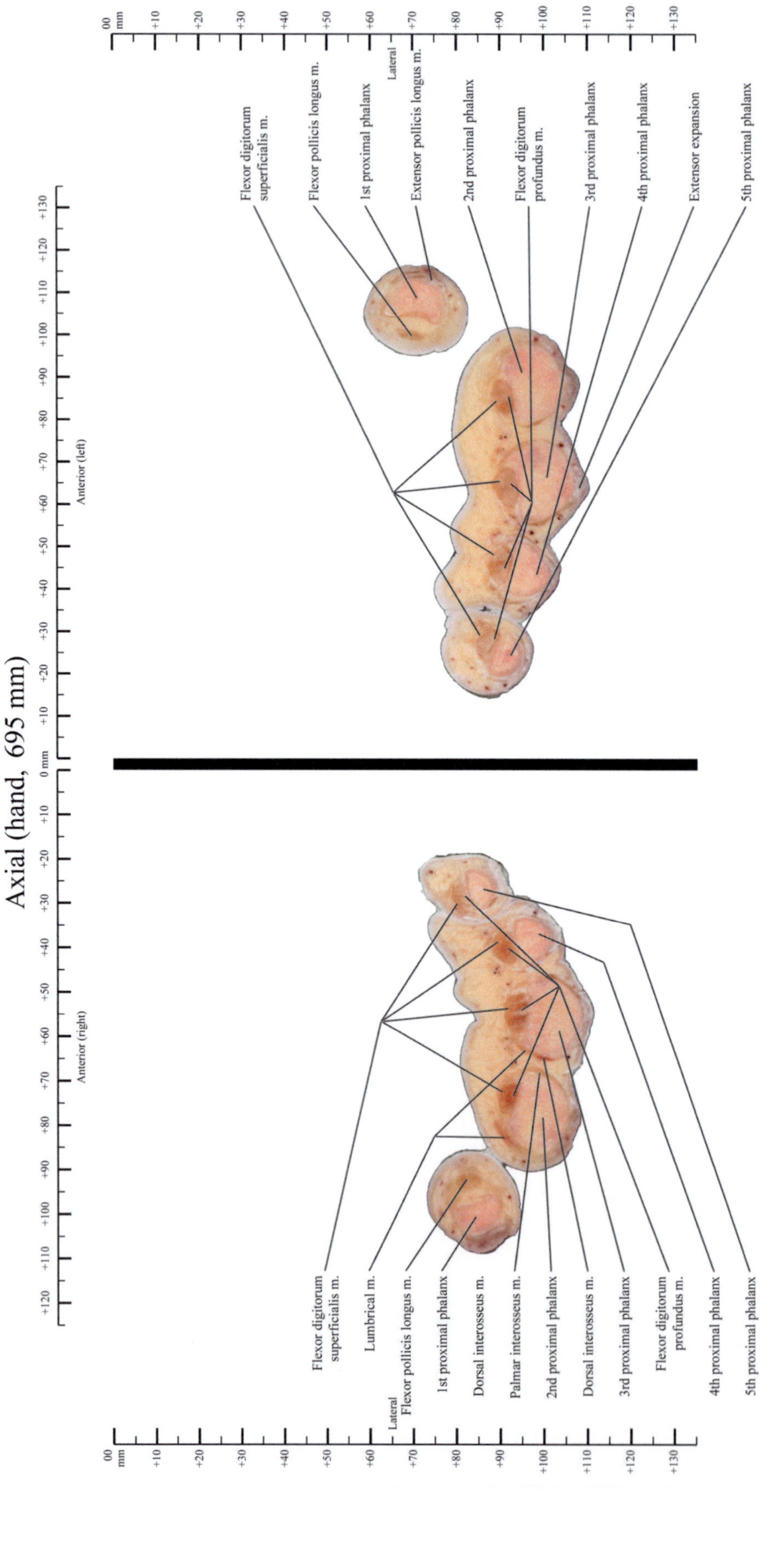
Axial (hand, 695 mm)
Anterior (left)
Anterior (right)
Lateral
Flexor digitorum superficialis m.
Flexor pollicis longus m.
1st proximal phalanx
Extensor pollicis longus m.
2nd proximal phalanx
Flexor digitorum profundus m.
3rd proximal phalanx
4th proximal phalanx
Extensor expansion
5th proximal phalanx
Lumbrical m.
Dorsal interosseus m.
Palmar interosseus m.
mm

Axial (hand, 700 mm)

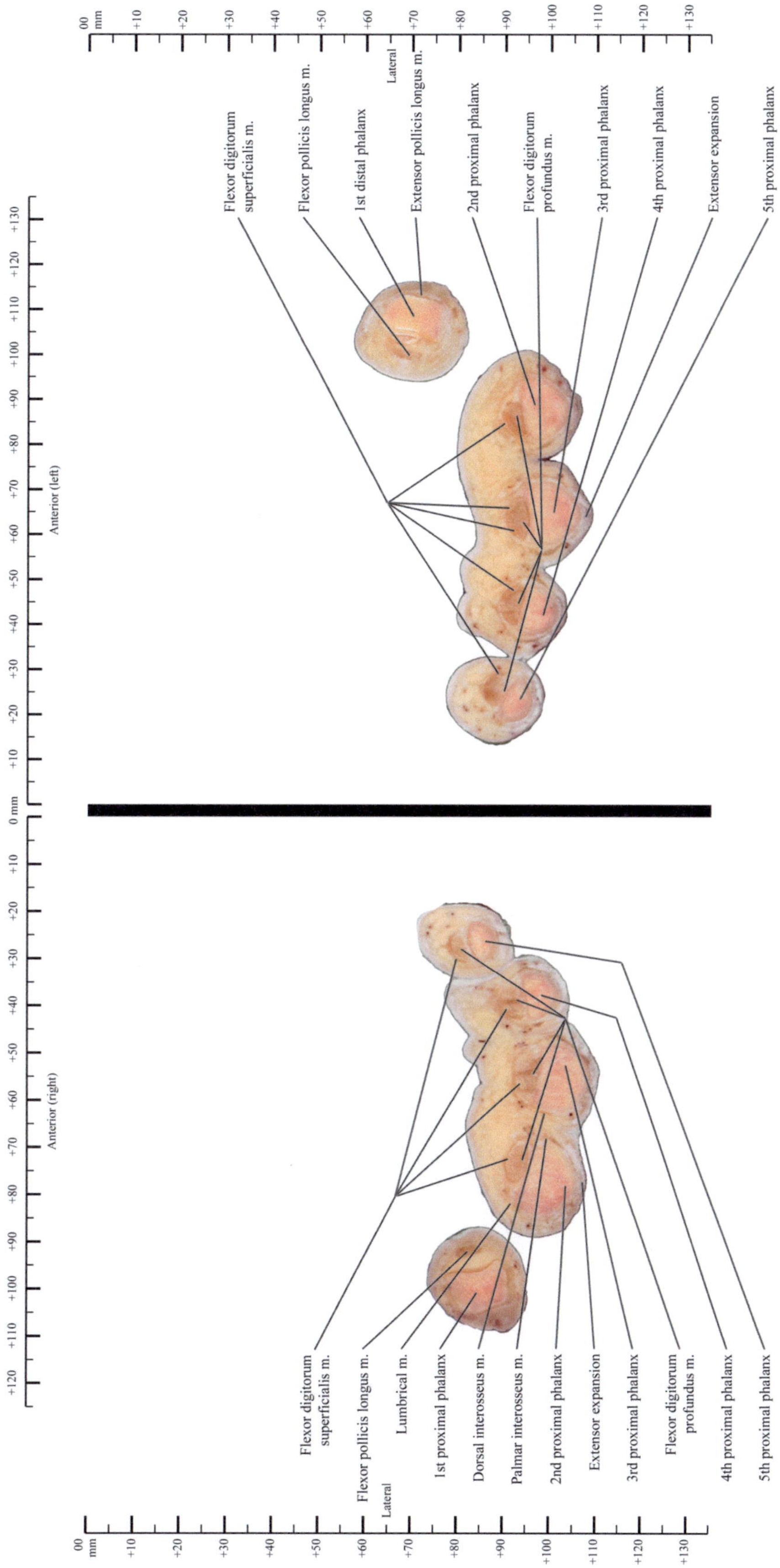

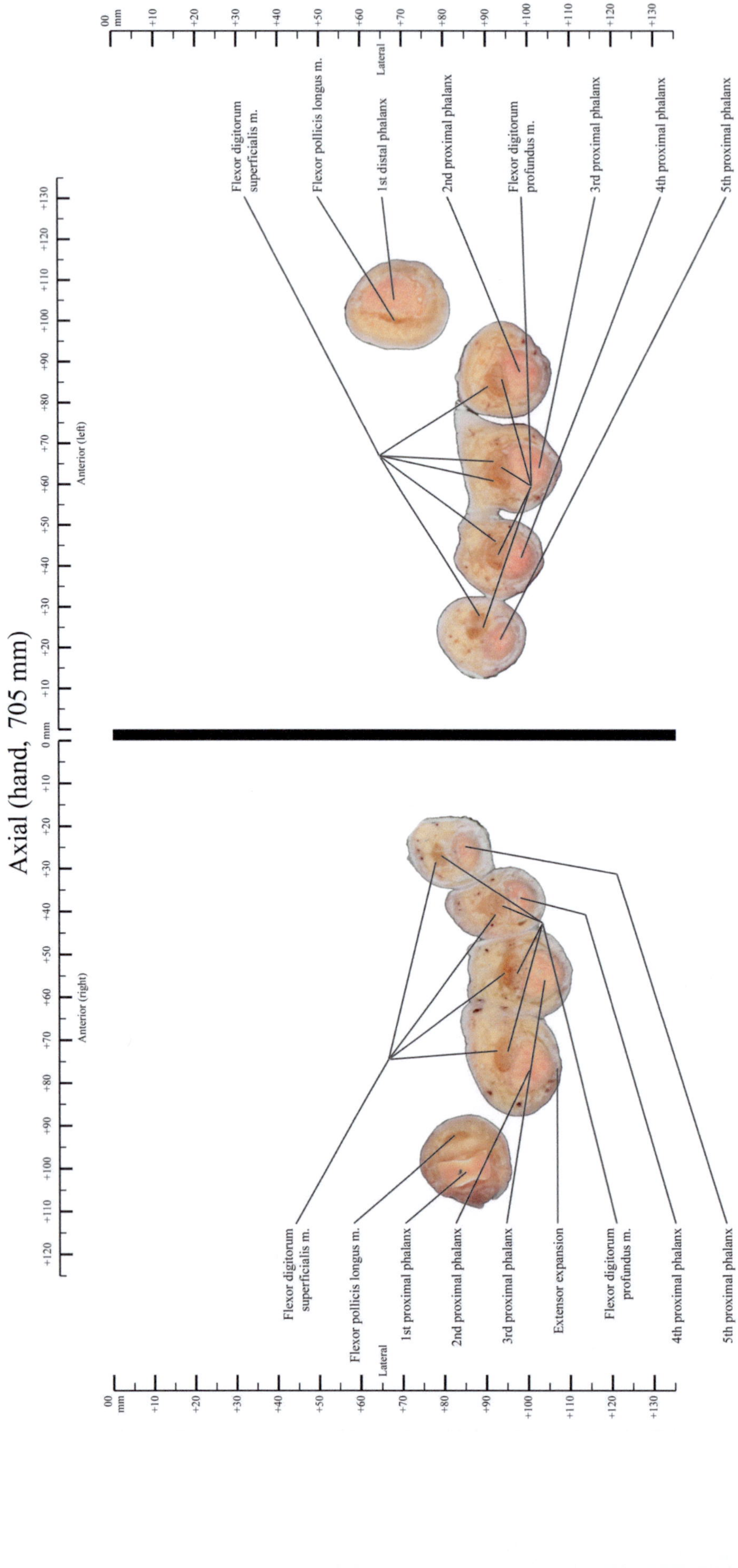
Axial (hand, 705 mm)
Anterior (left)
Lateral
Flexor digitorum superficialis m.
Flexor pollicis longus m.
1st distal phalanx
2nd proximal phalanx
Flexor digitorum profundus m.
3rd proximal phalanx
4th proximal phalanx
5th proximal phalanx
Anterior (right)
Lateral
Flexor digitorum superficialis m.
Flexor pollicis longus m.
1st proximal phalanx
2nd proximal phalanx
3rd proximal phalanx
Extensor expansion
Flexor digitorum profundus m.
4th proximal phalanx
5th proximal phalanx

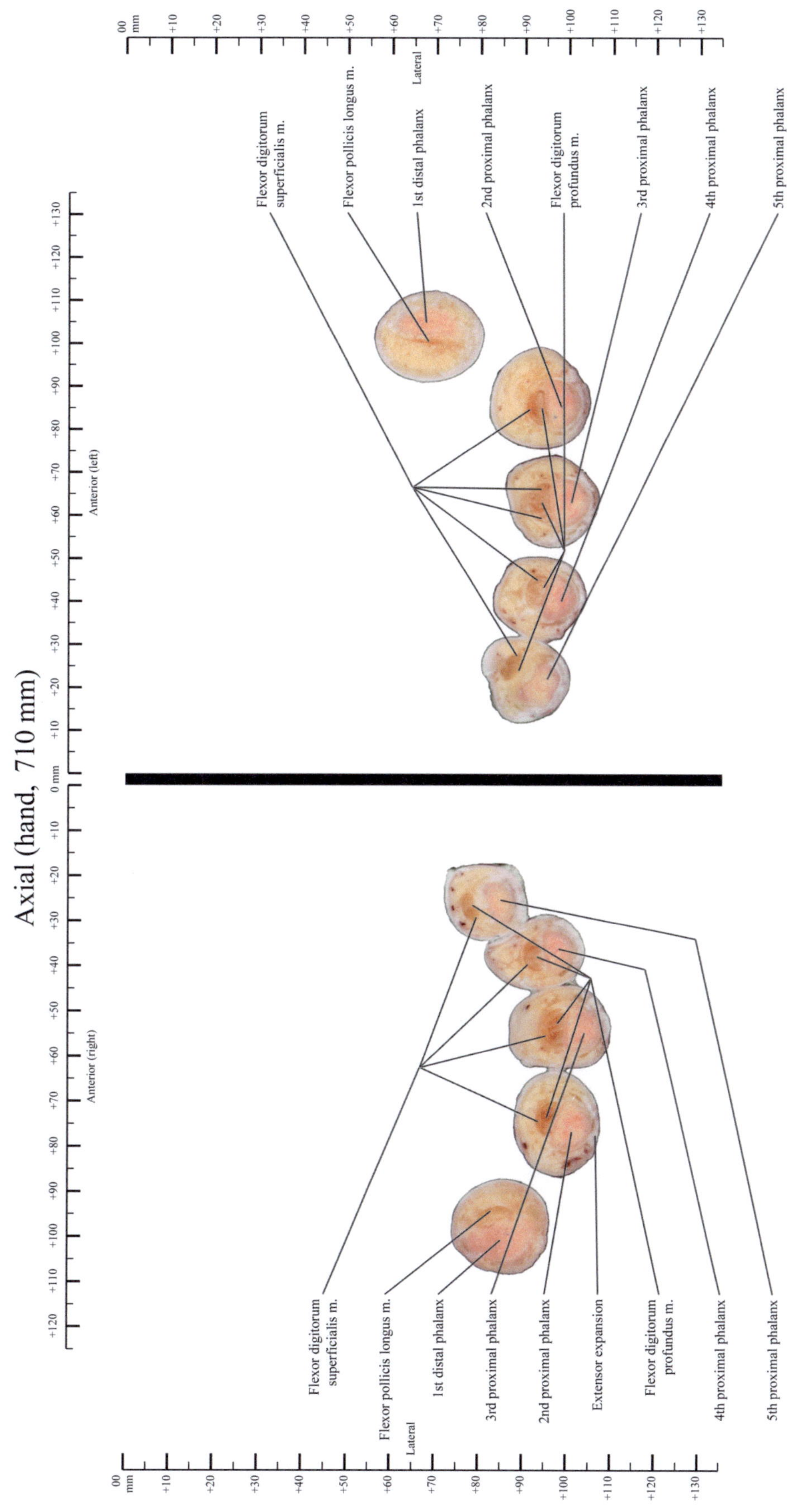
Axial (hand, 710 mm)
Anterior (left)
Anterior (right)
Lateral
Flexor digitorum superficialis m.
Flexor pollicis longus m.
1st distal phalanx
2nd proximal phalanx
Flexor digitorum profundus m.
3rd proximal phalanx
4th proximal phalanx
5th proximal phalanx
Extensor expansion

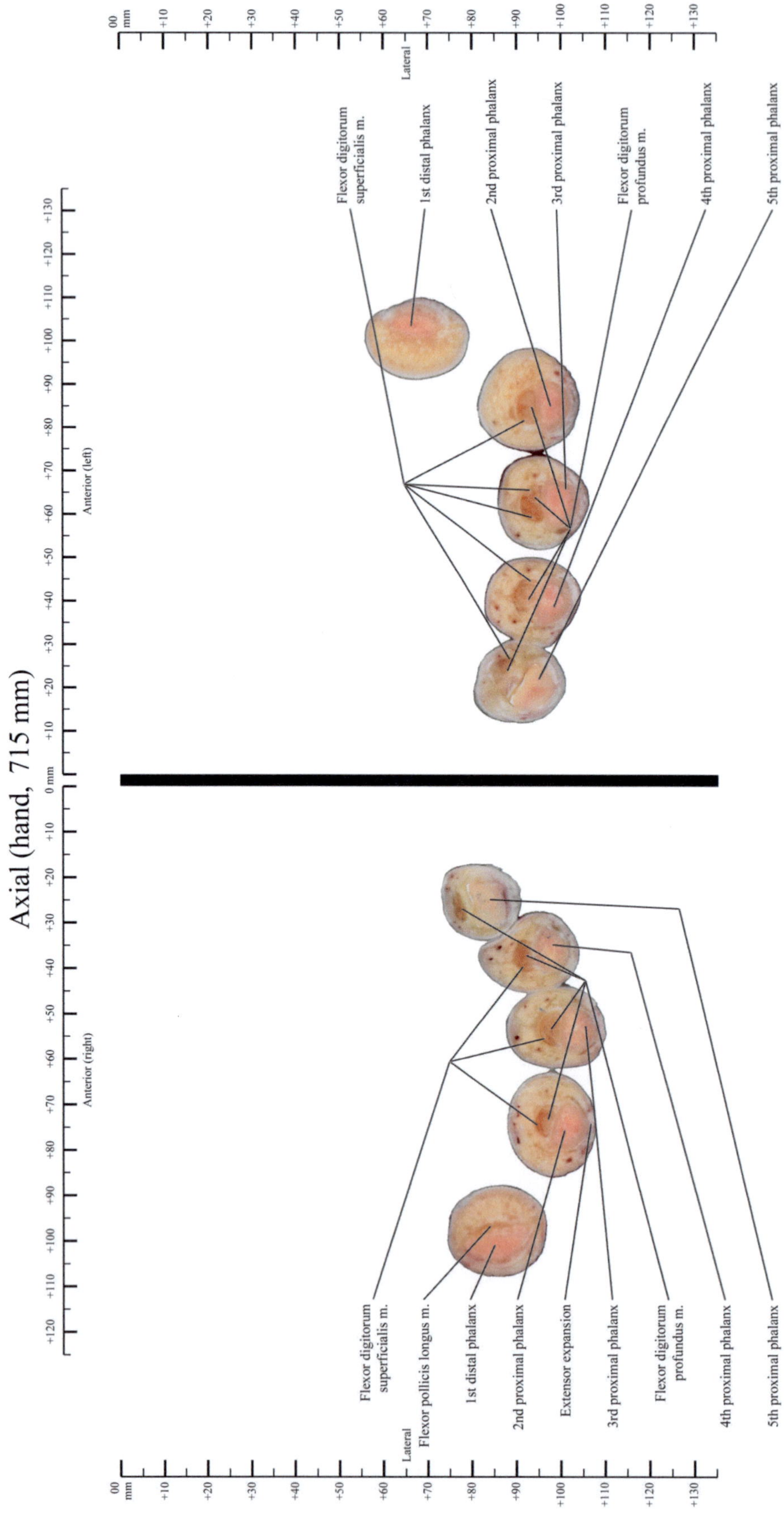
Axial (hand, 715 mm)
Anterior (left)
Anterior (right)
Lateral
Flexor digitorum superficialis m.
1st distal phalanx
2nd proximal phalanx
3rd proximal phalanx
Flexor digitorum profundus m.
4th proximal phalanx
5th proximal phalanx
Flexor pollicis longus m.
Extensor expansion
0 mm
00 mm
+10
+20
+30
+40
+50
+60
+70
+80
+90
+100
+110
+120
+130

Axial (hand, 720 mm)

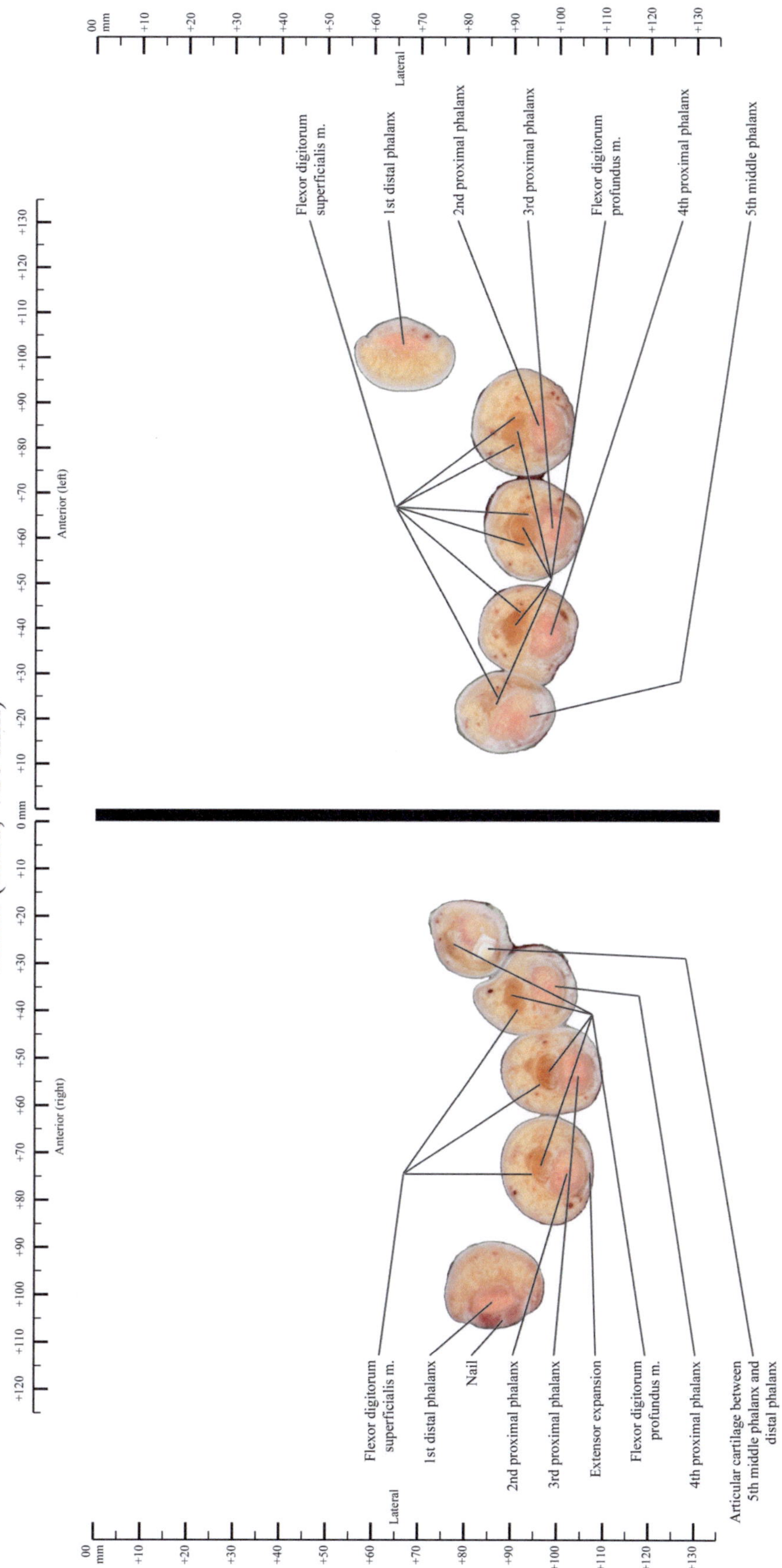

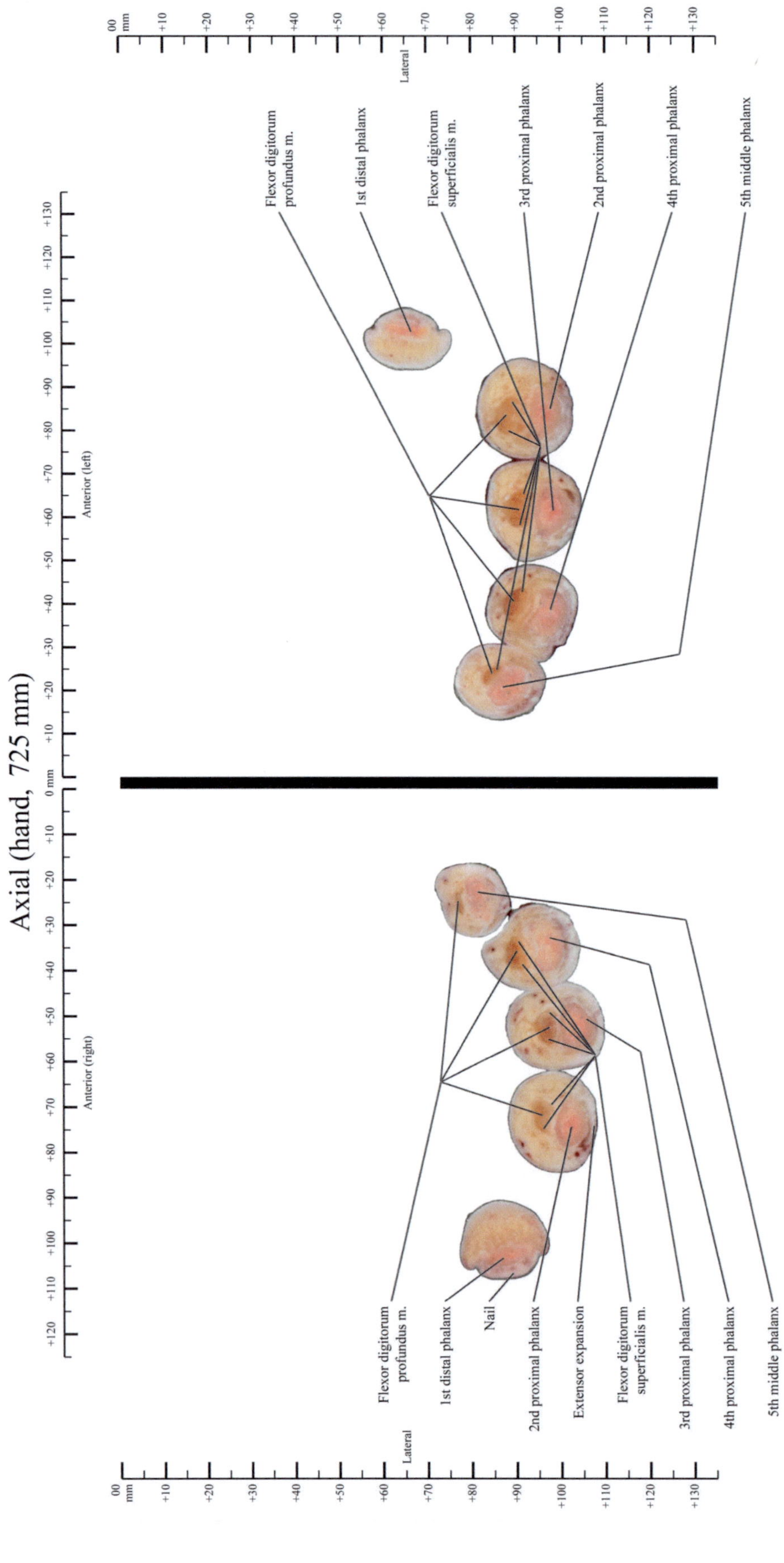

Axial (hand, 725 mm)
Anterior (left)
Anterior (right)
Lateral
Lateral
Flexor digitorum profundus m.
1st distal phalanx
Flexor digitorum superficialis m.
3rd proximal phalanx
2nd proximal phalanx
4th proximal phalanx
5th middle phalanx
Flexor digitorum profundus m.
1st distal phalanx
Nail
2nd proximal phalanx
Extensor expansion
Flexor digitorum superficialis m.
3rd proximal phalanx
4th proximal phalanx
5th middle phalanx

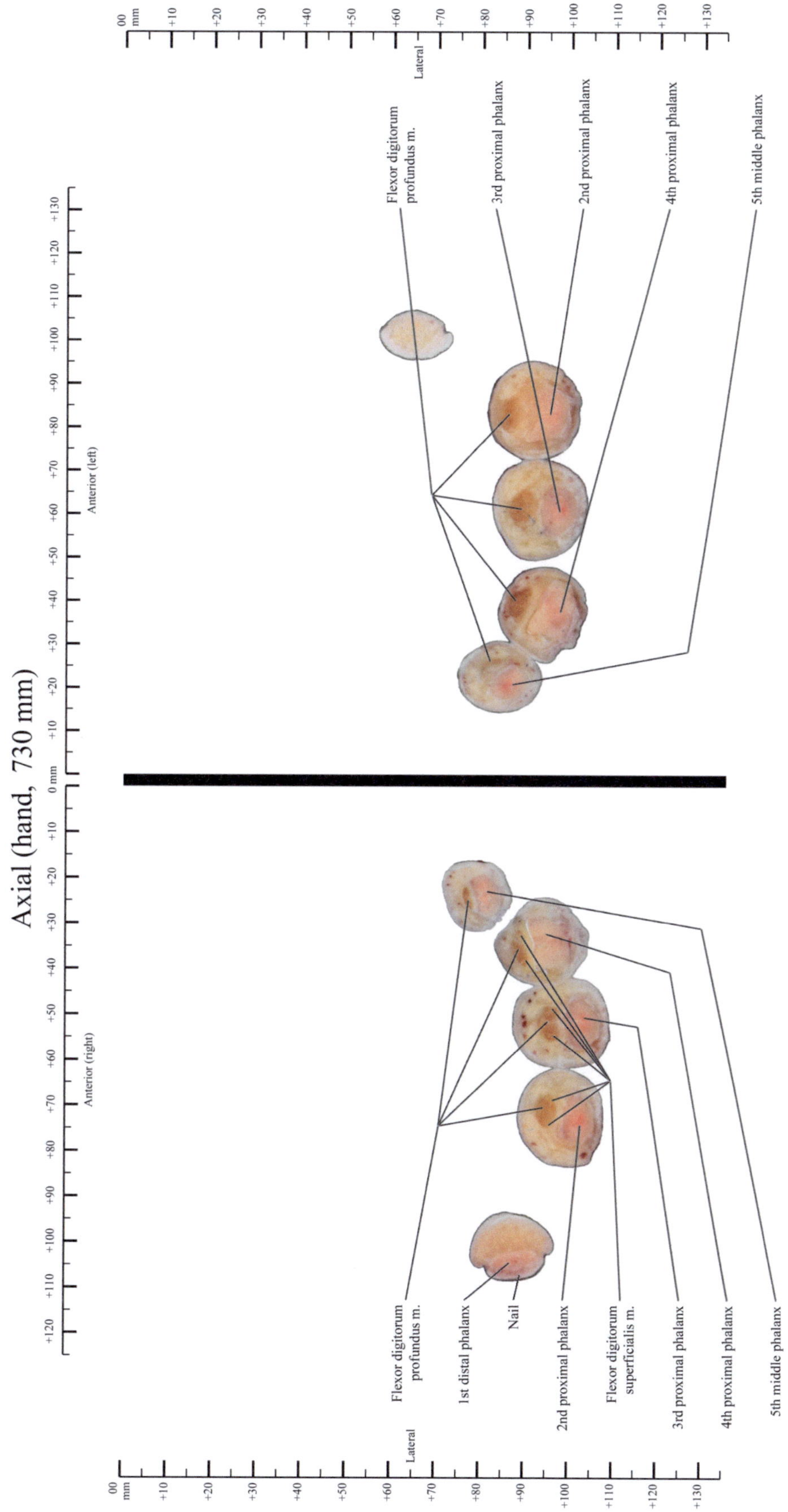
Axial (hand, 730 mm)
Anterior (left)
Anterior (right)
Lateral
Lateral
Flexor digitorum profundus m.
3rd proximal phalanx
2nd proximal phalanx
4th proximal phalanx
5th middle phalanx
Flexor digitorum profundus m.
1st distal phalanx
Nail
2nd proximal phalanx
Flexor digitorum superficialis m.
3rd proximal phalanx
4th proximal phalanx
5th middle phalanx

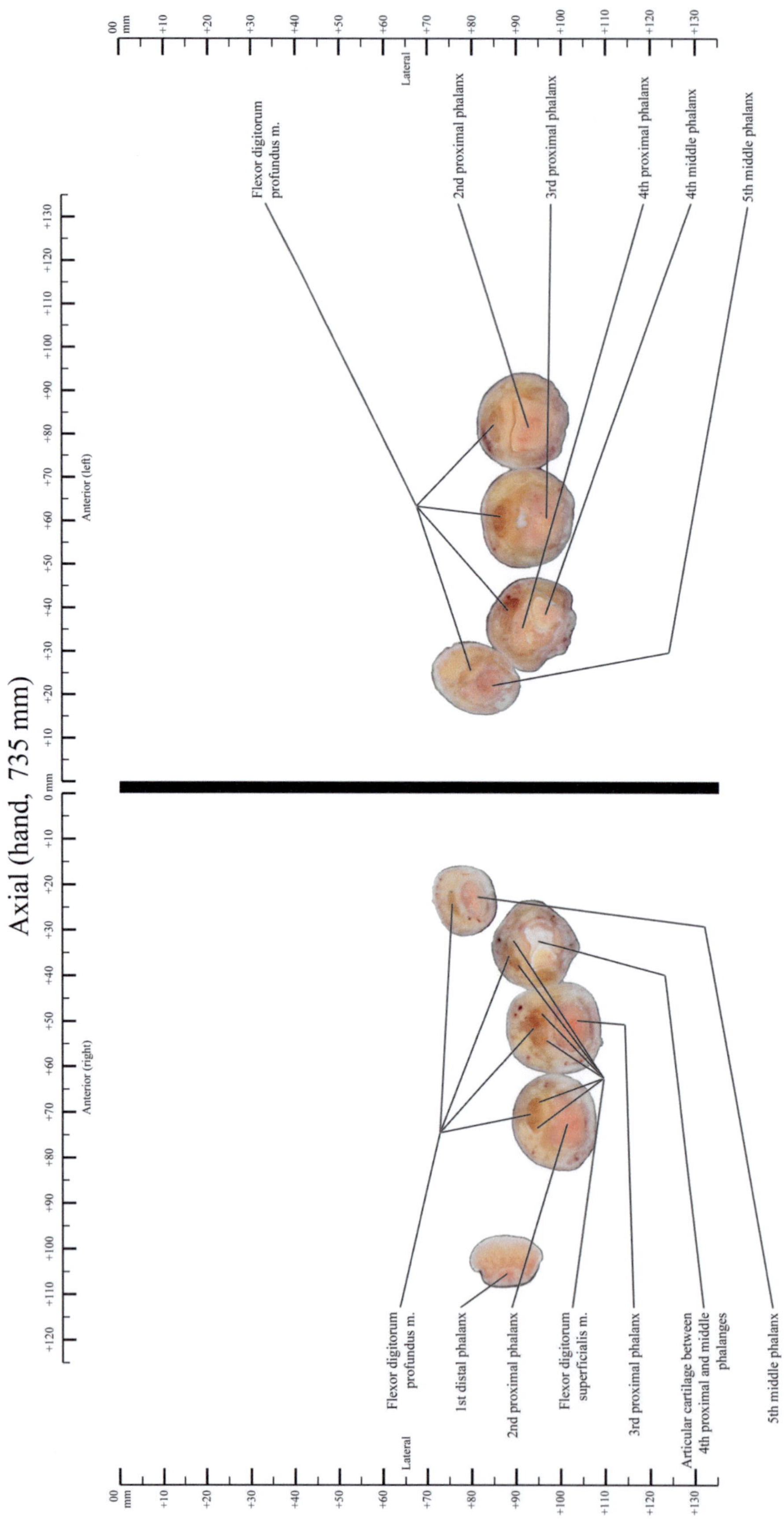
Axial (hand, 735 mm)
Anterior (left)
Anterior (right)
Lateral
Lateral
Flexor digitorum profundus m.
2nd proximal phalanx
3rd proximal phalanx
4th proximal phalanx
4th middle phalanx
5th middle phalanx
Flexor digitorum profundus m.
1st distal phalanx
2nd proximal phalanx
Flexor digitorum superficialis m.
3rd proximal phalanx
Articular cartilage between 4th proximal and middle phalanges
5th middle phalanx

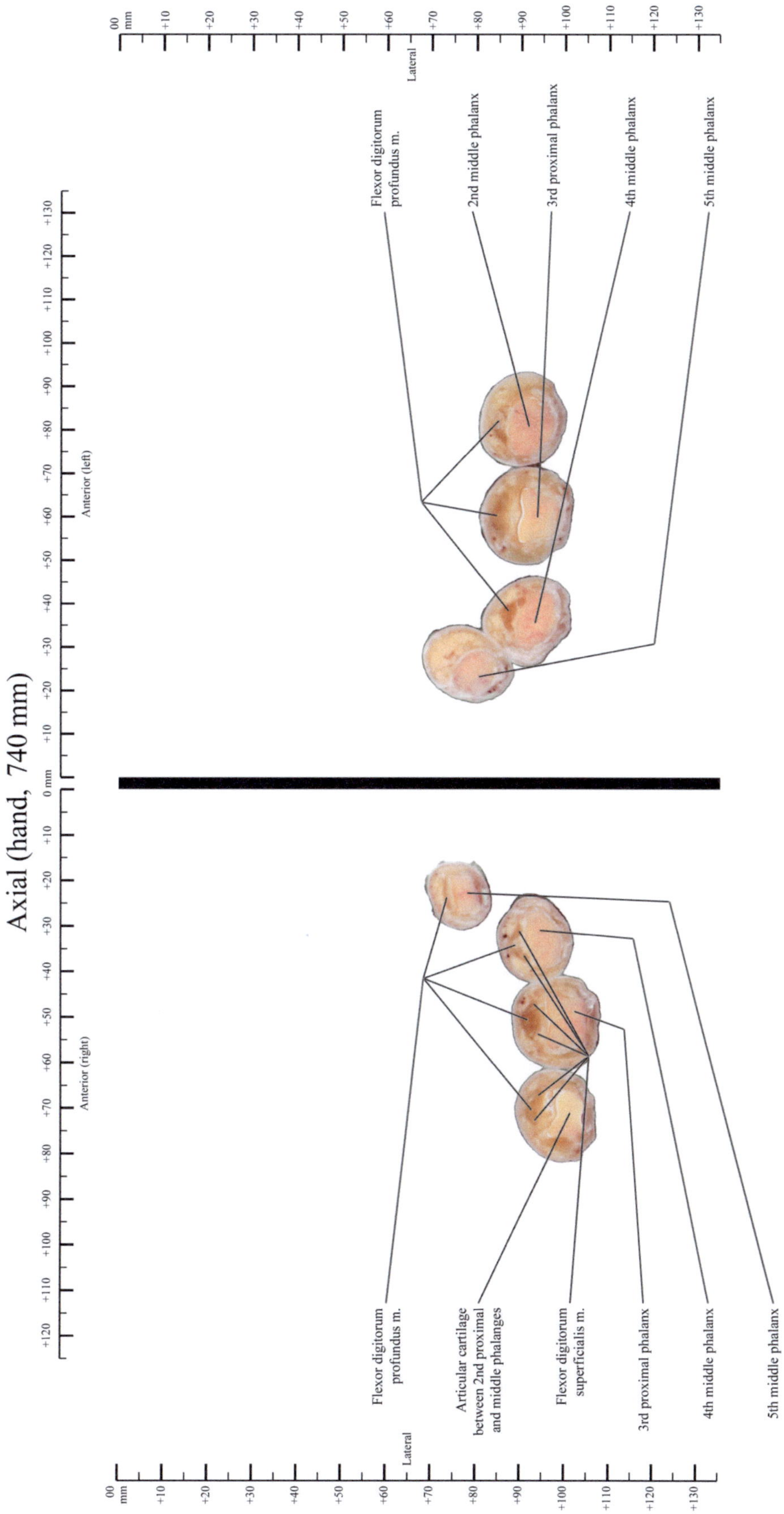
Axial (hand, 740 mm)
Anterior (left)
Anterior (right)
Lateral
Flexor digitorum profundus m.
2nd middle phalanx
3rd proximal phalanx
4th middle phalanx
5th middle phalanx
Flexor digitorum profundus m.
Articular cartilage between 2nd proximal and middle phalanges
Flexor digitorum superficialis m.
3rd proximal phalanx
4th middle phalanx
5th middle phalanx

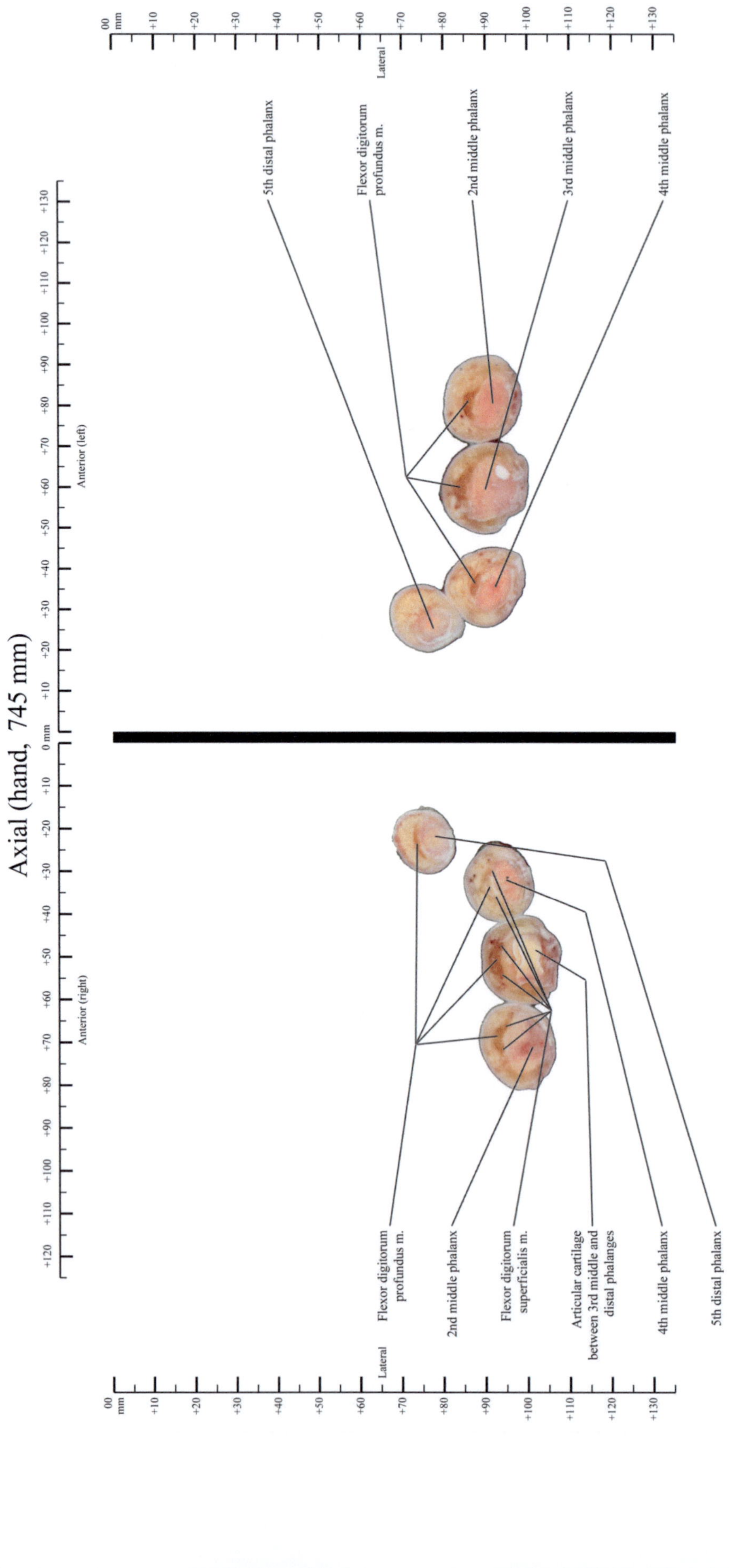
Axial (hand, 745 mm)
Anterior (left)
Anterior (right)
Lateral
5th distal phalanx
Flexor digitorum profundus m.
2nd middle phalanx
3rd middle phalanx
4th middle phalanx
Flexor digitorum superficialis m.
Articular cartilage between 3rd middle and distal phalanges

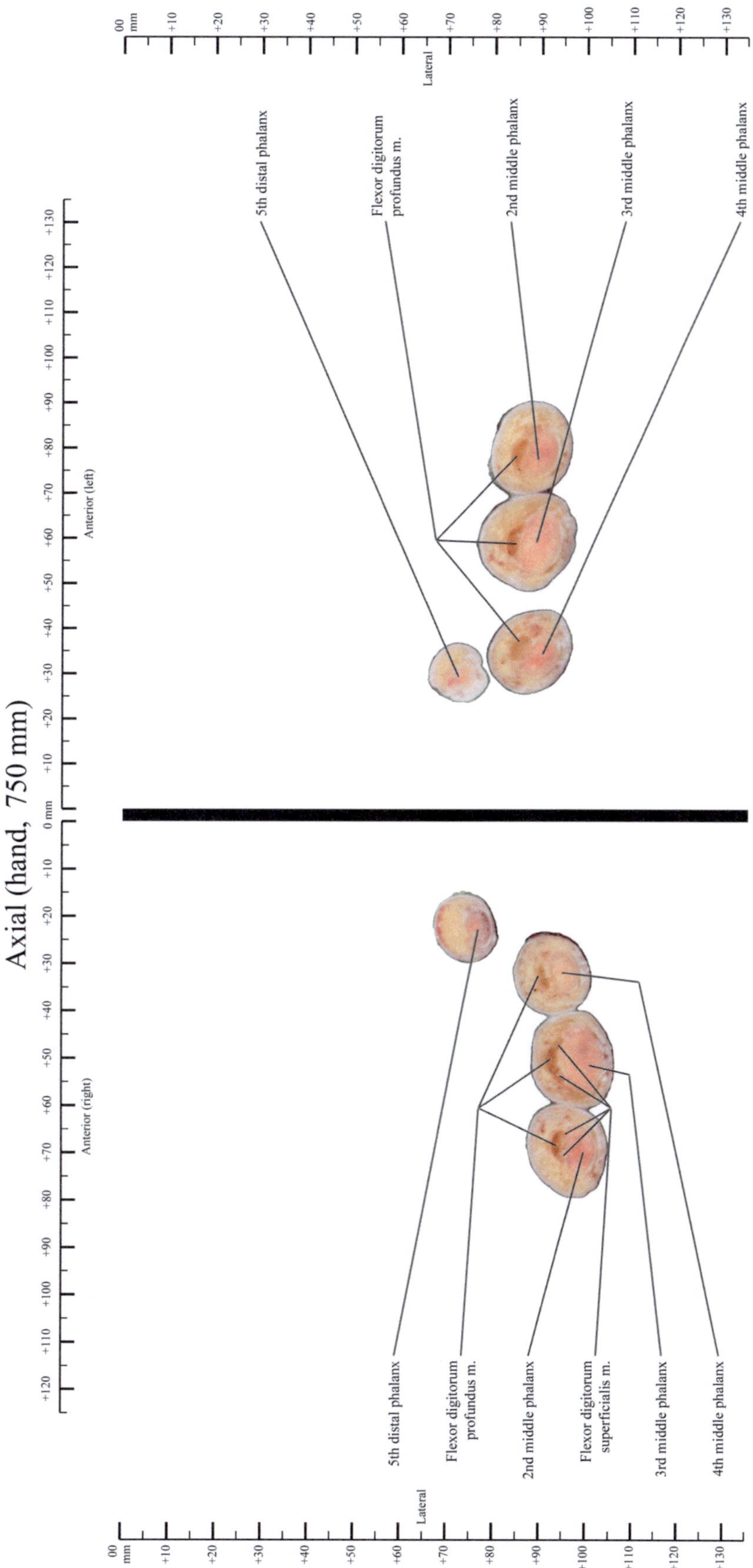
Axial (hand, 750 mm)
Anterior (left)
Anterior (right)
Lateral
5th distal phalanx
Flexor digitorum profundus m.
2nd middle phalanx
3rd middle phalanx
4th middle phalanx
Flexor digitorum superficialis m.

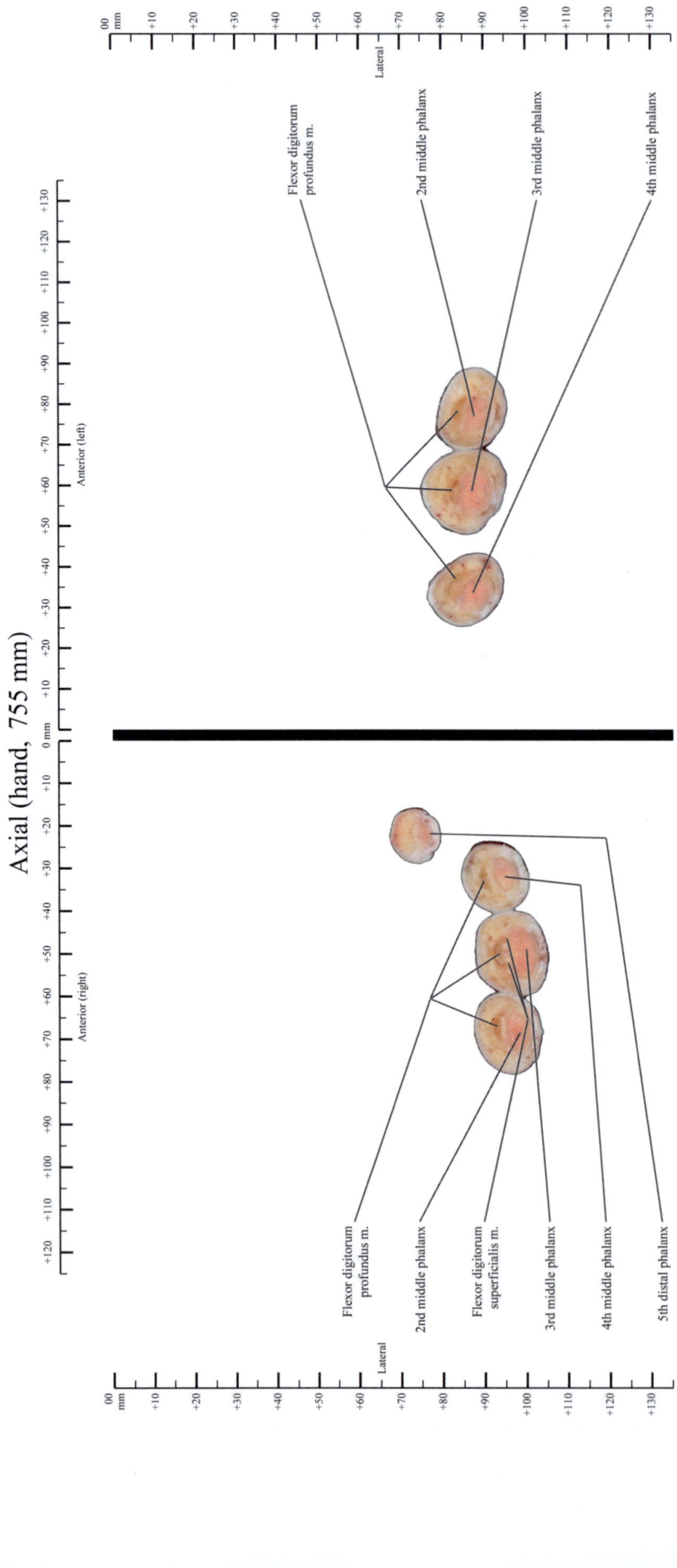
Axial (hand, 755 mm)
Anterior (left)
Anterior (right)
Lateral
Lateral
Flexor digitorum profundus m.
2nd middle phalanx
3rd middle phalanx
4th middle phalanx
Flexor digitorum profundus m.
2nd middle phalanx
Flexor digitorum superficialis m.
3rd middle phalanx
4th middle phalanx
5th distal phalanx

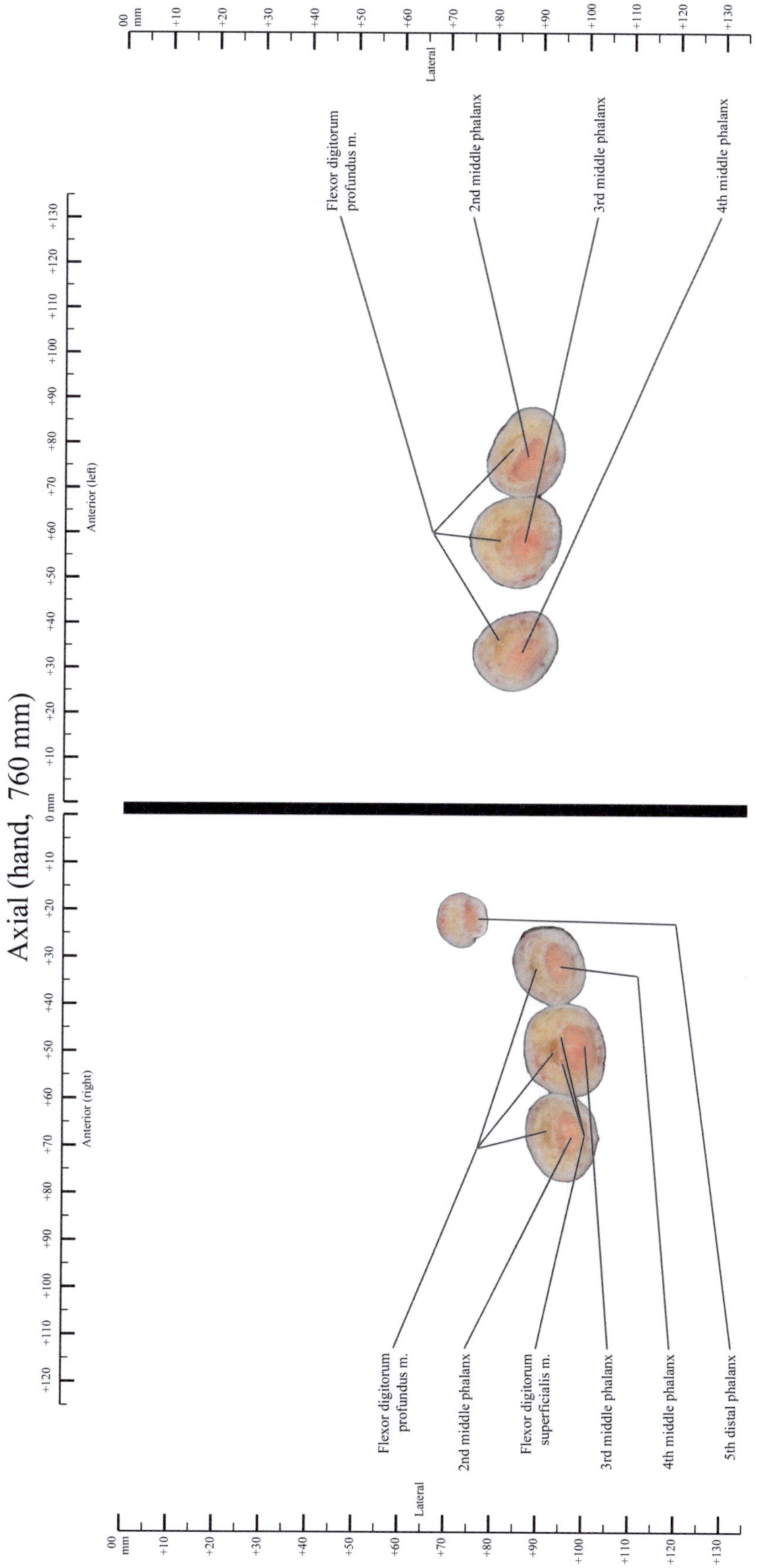
Axial (hand, 760 mm)
Anterior (left)
Anterior (right)
Lateral
Flexor digitorum profundus m.
2nd middle phalanx
3rd middle phalanx
4th middle phalanx
Flexor digitorum profundus m.
2nd middle phalanx
Flexor digitorum superficialis m.
3rd middle phalanx
4th middle phalanx
5th distal phalanx

Axial (hand, 765 mm)

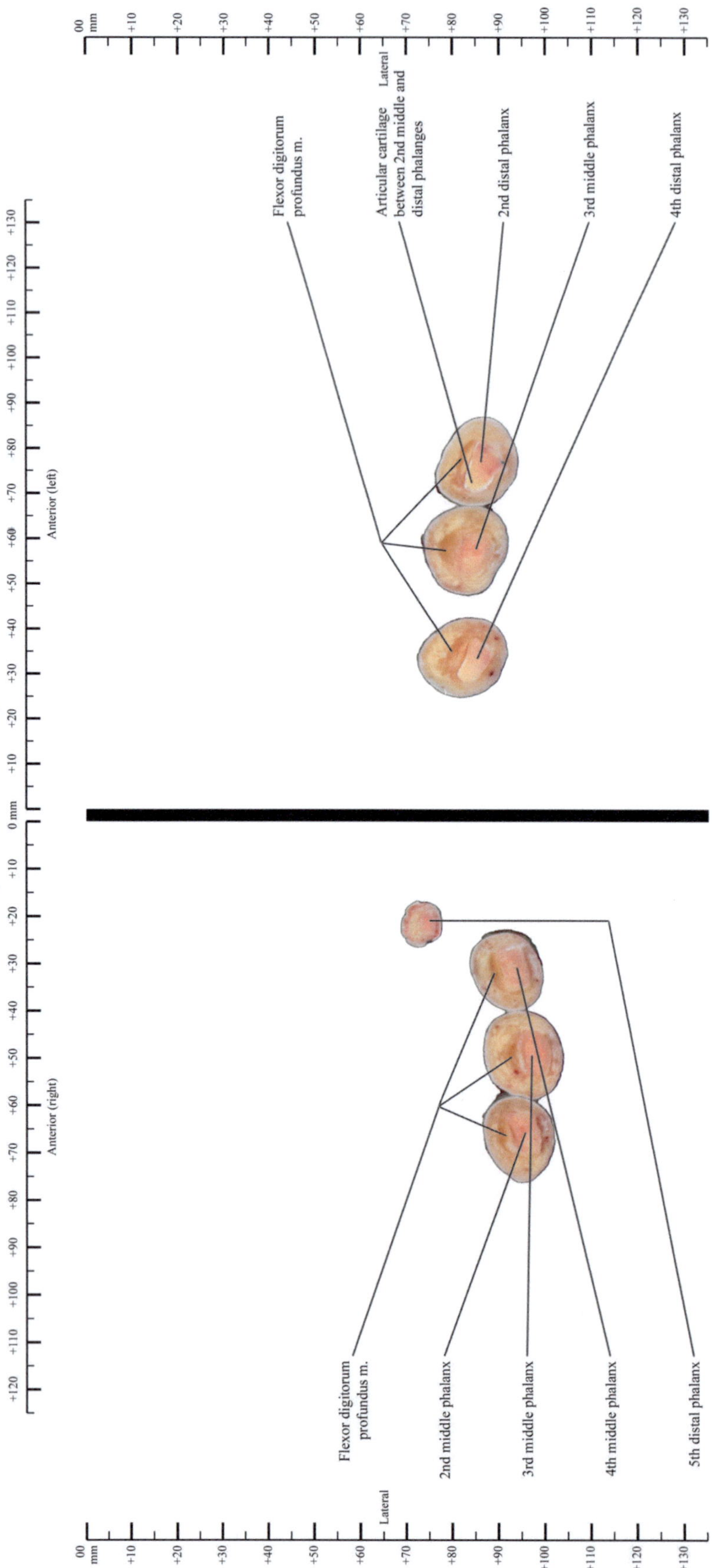

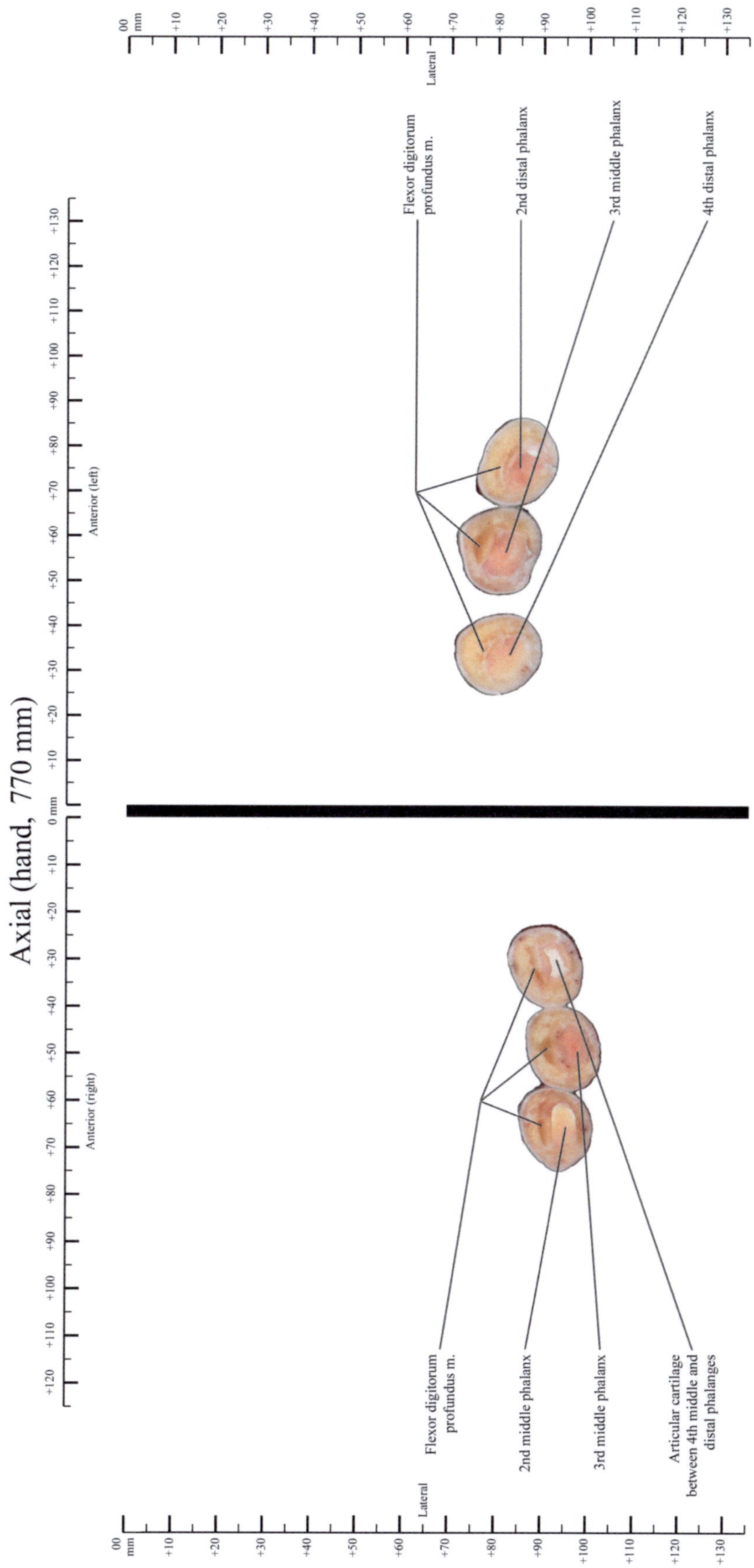
Axial (hand, 770 mm)
Anterior (left)
Anterior (right)
Lateral
Flexor digitorum profundus m.
2nd distal phalanx
3rd middle phalanx
4th distal phalanx
Lateral
Flexor digitorum profundus m.
2nd middle phalanx
3rd middle phalanx
Articular cartilage between 4th middle and distal phalanges

Axial (hand, 775 mm)

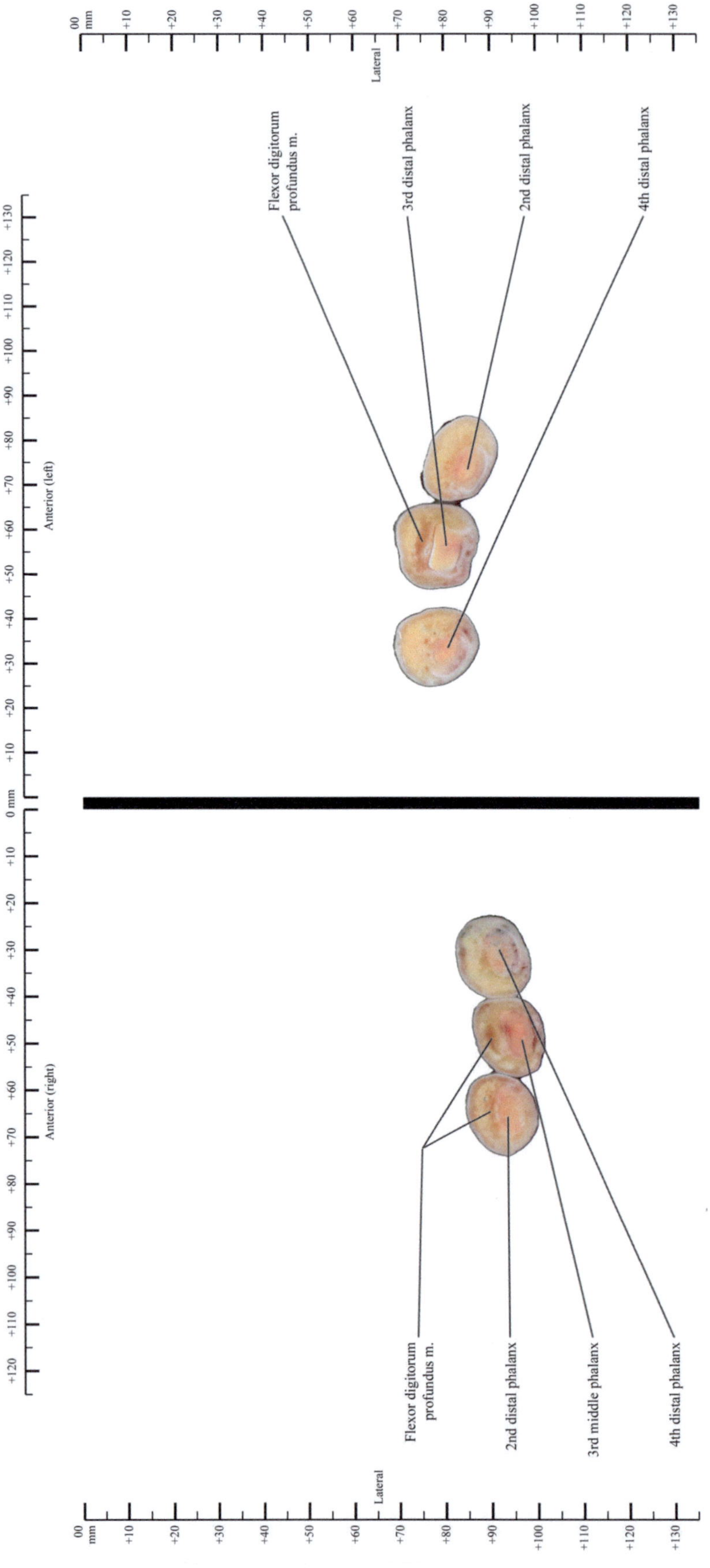

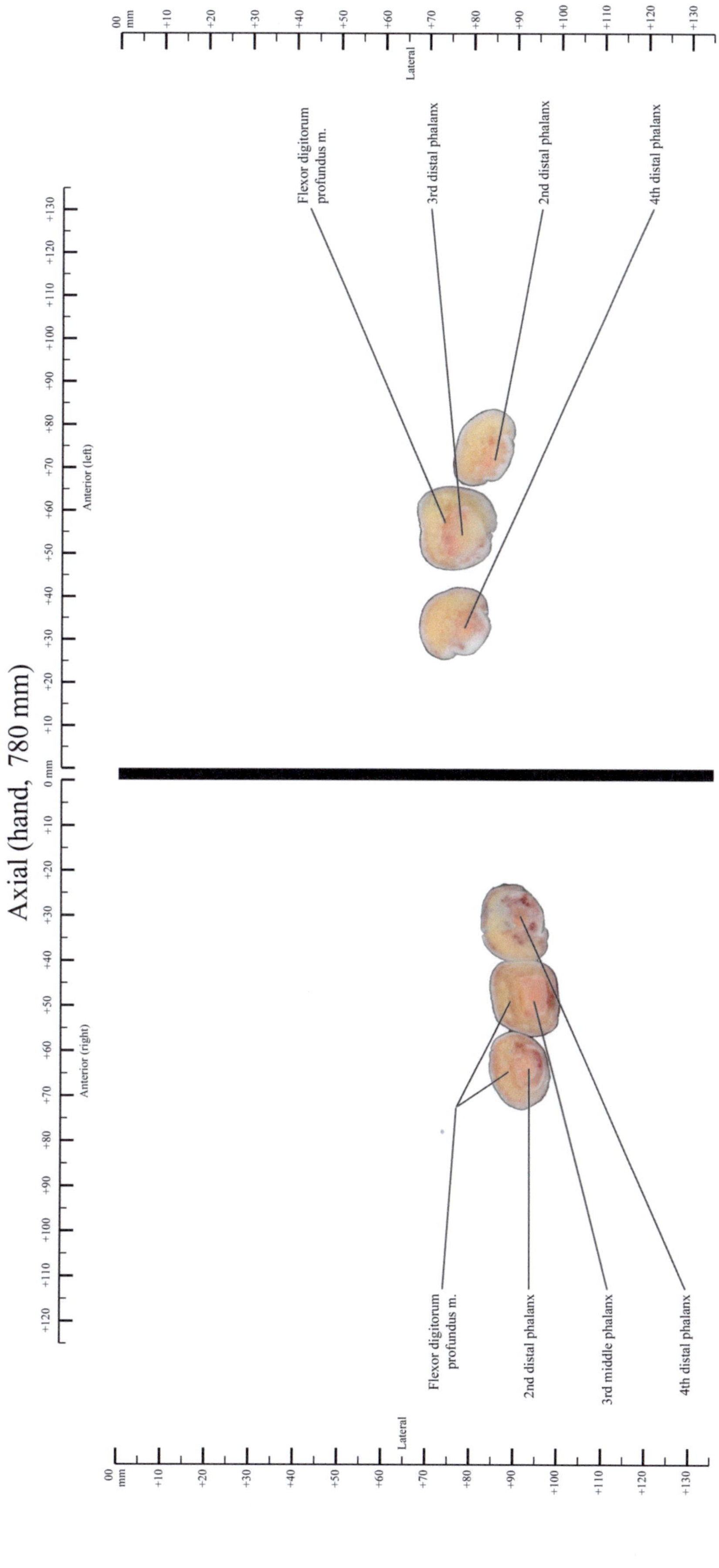
Axial (hand, 780 mm)
Anterior (left)
Anterior (right)
Lateral
Flexor digitorum profundus m.
3rd distal phalanx
2nd distal phalanx
4th distal phalanx
Flexor digitorum profundus m.
2nd distal phalanx
3rd middle phalanx
4th distal phalanx

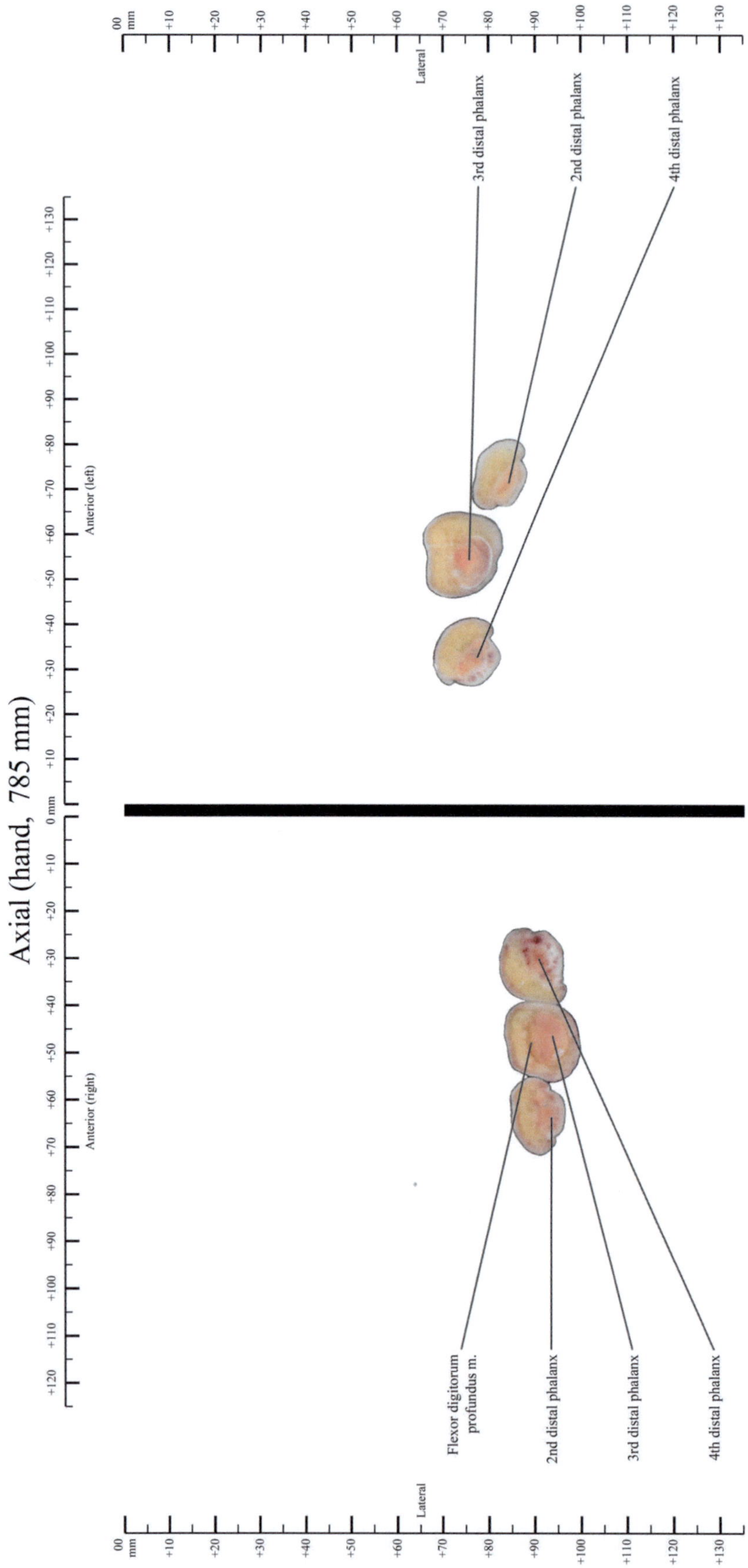

Axial (hand, 785 mm)
Anterior (left)
Anterior (right)
Lateral
Lateral
3rd distal phalanx
2nd distal phalanx
4th distal phalanx
Flexor digitorum profundus m.
2nd distal phalanx
3rd distal phalanx
4th distal phalanx

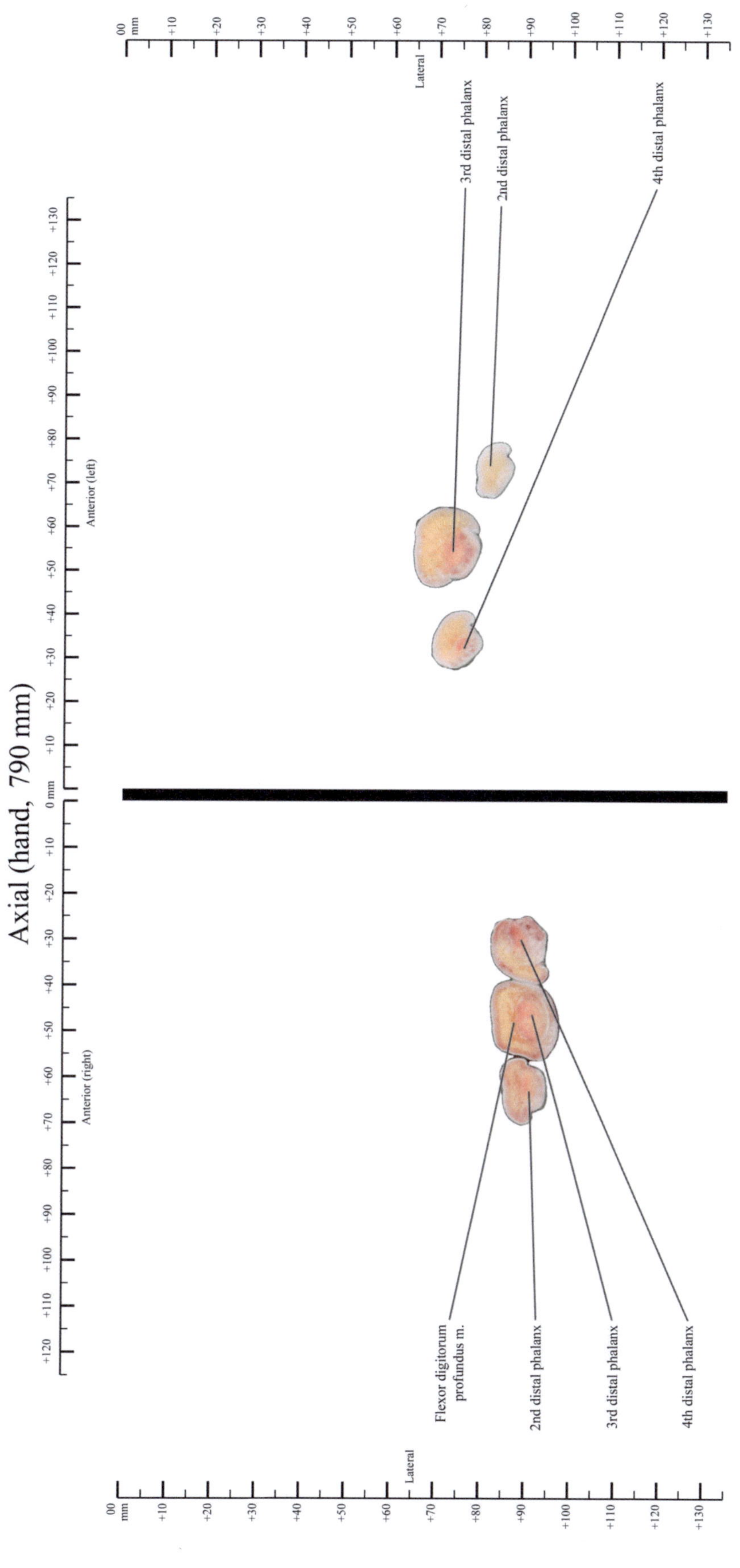
Axial (hand, 790 mm)
Anterior (left)
Anterior (right)
Lateral
3rd distal phalanx
2nd distal phalanx
4th distal phalanx
Flexor digitorum profundus m.
2nd distal phalanx
3rd distal phalanx
4th distal phalanx
Lateral

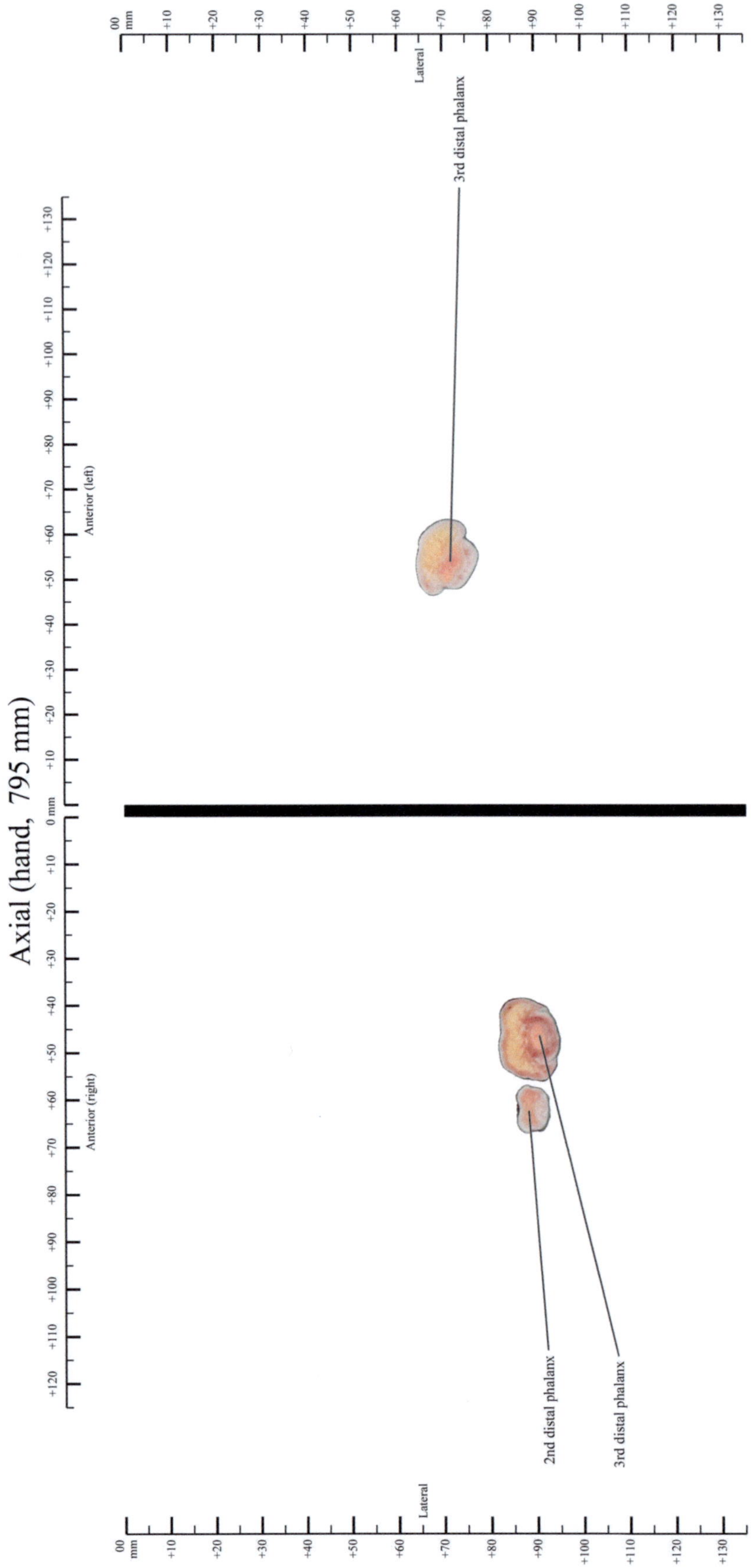
Axial (hand, 795 mm)
Anterior (left)
Anterior (right)
Lateral
Lateral
3rd distal phalanx
2nd distal phalanx
3rd distal phalanx

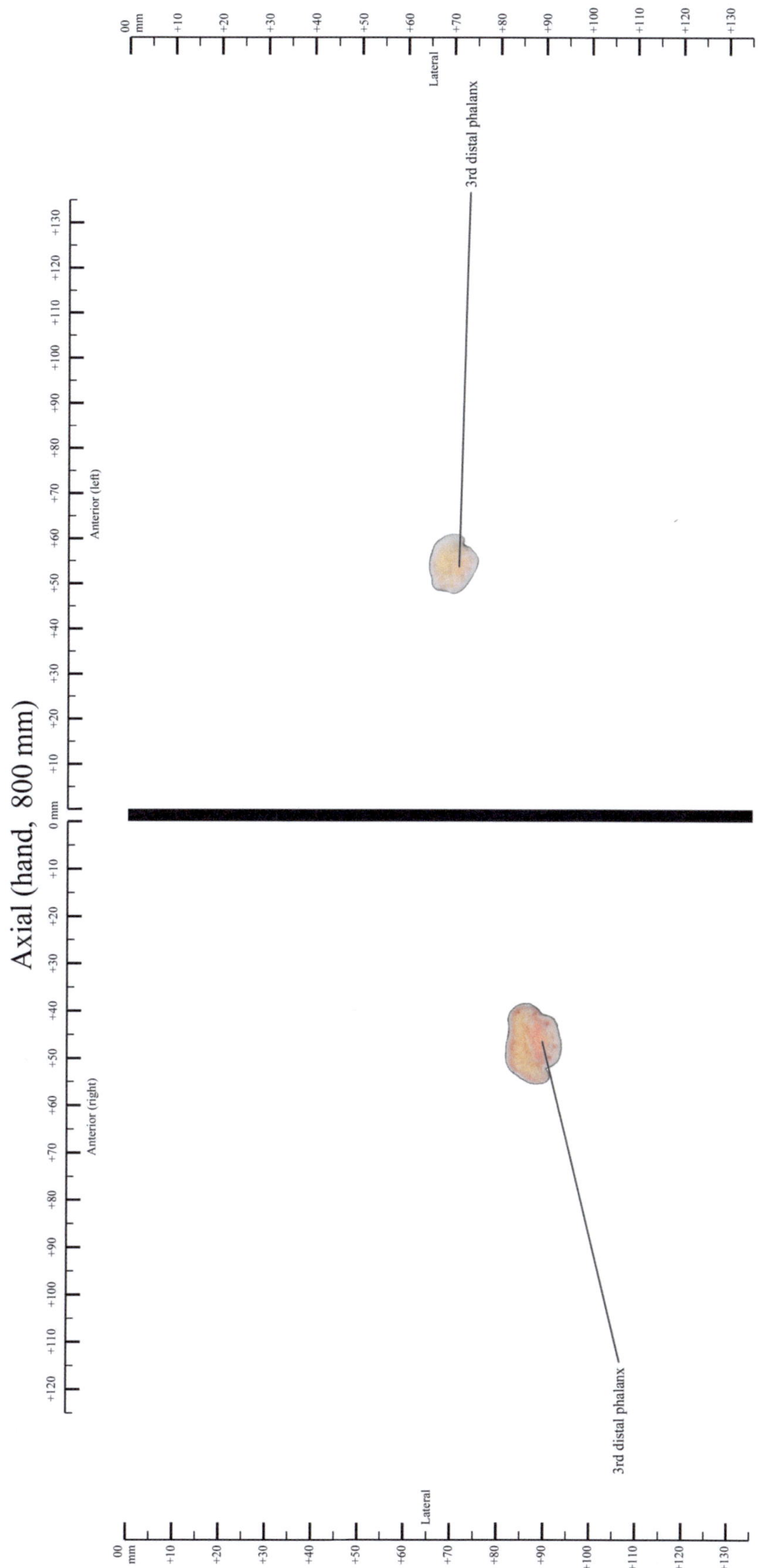
Axial (hand, 800 mm)
Anterior (left)
Anterior (right)
Lateral
Lateral
3rd distal phalanx
3rd distal phalanx

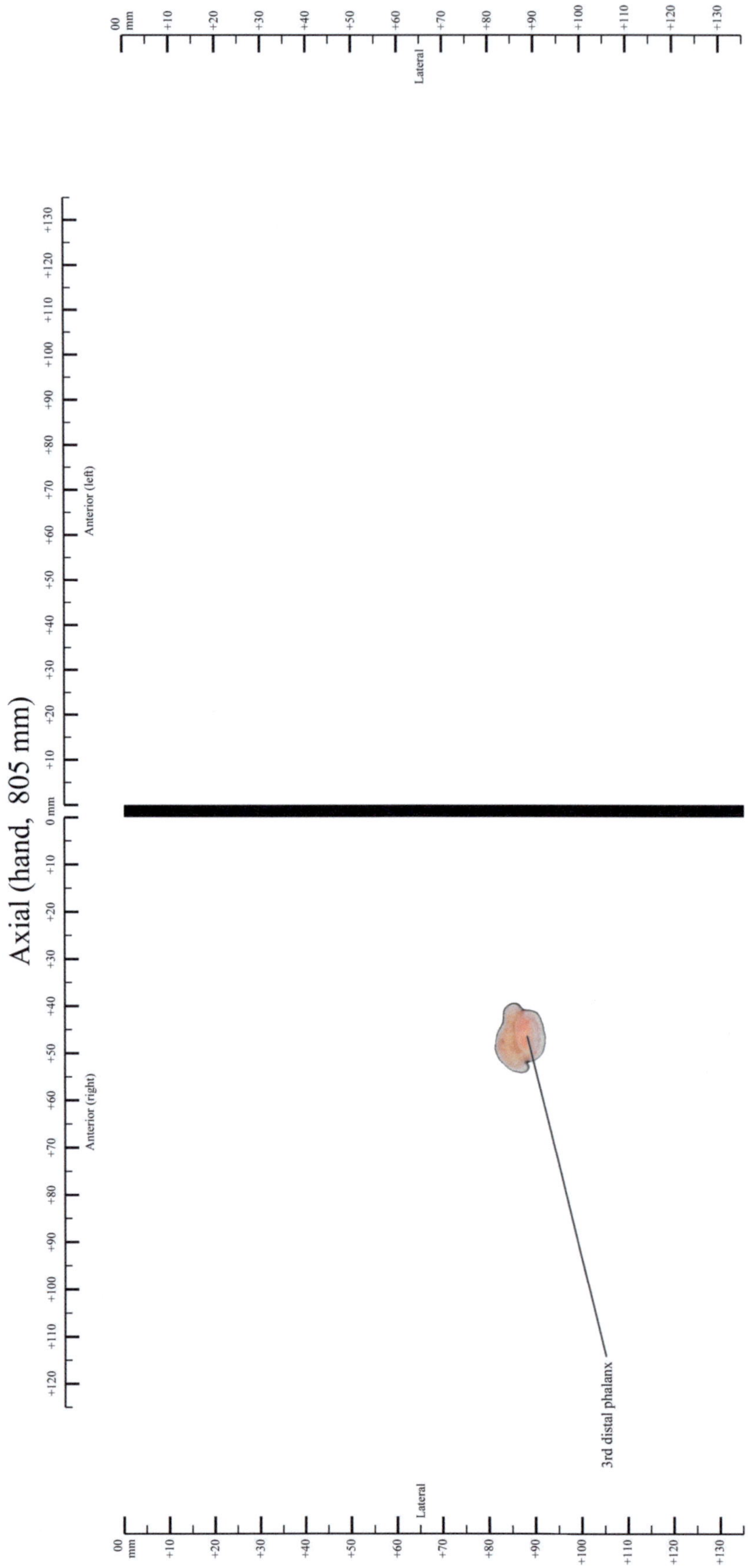
Axial (hand, 805 mm)
Anterior (left)
Anterior (right)
Lateral
Lateral
3rd distal phalanx

3 Coronal (Arm) and Sagittal (Forearm and Hand) Planes of Upper Limbs

All images, 1.5 mm intervals
All images, 0.06 mm × 0.06 mm sized-pixel

Abbreviation: ant., anterior; a., artery; inf., inferior; int., internal; lat., lateral; m., m.; med., medial; n., nerve; nu., nucleus; post., posterior; sup., superior; tr., tract

J. S. Park, *Cross-Sectional Atlas of Human Upper Limbs*, Cross-Sectional Atlas, https://doi.org/10.1007/978-981-95-1900-2_3

Coronal (arm) and sagittal (forearm and hand) (35.0 mm)

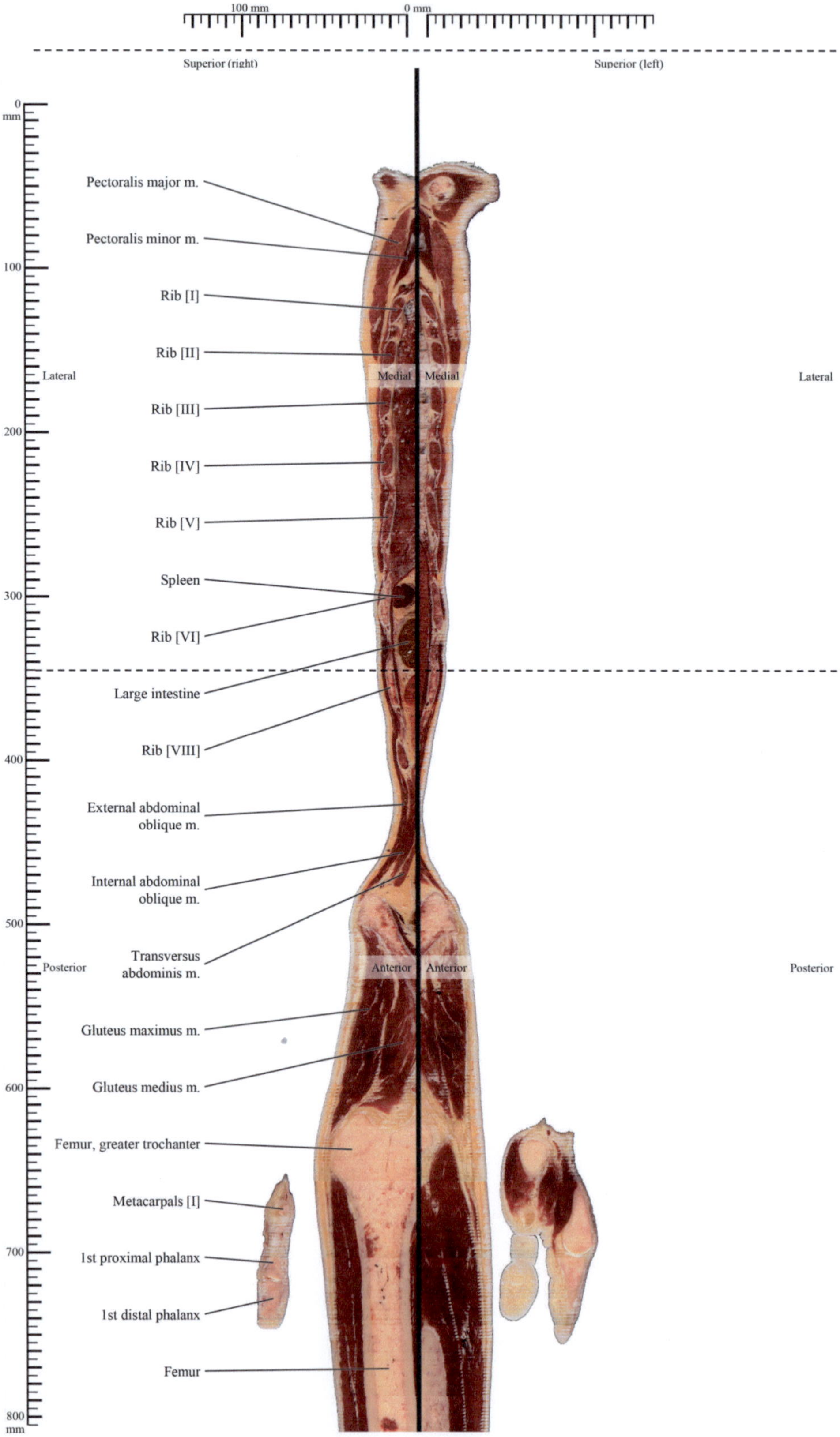

Coronal (arm) and sagittal (forearm and hand) (36.5 mm)

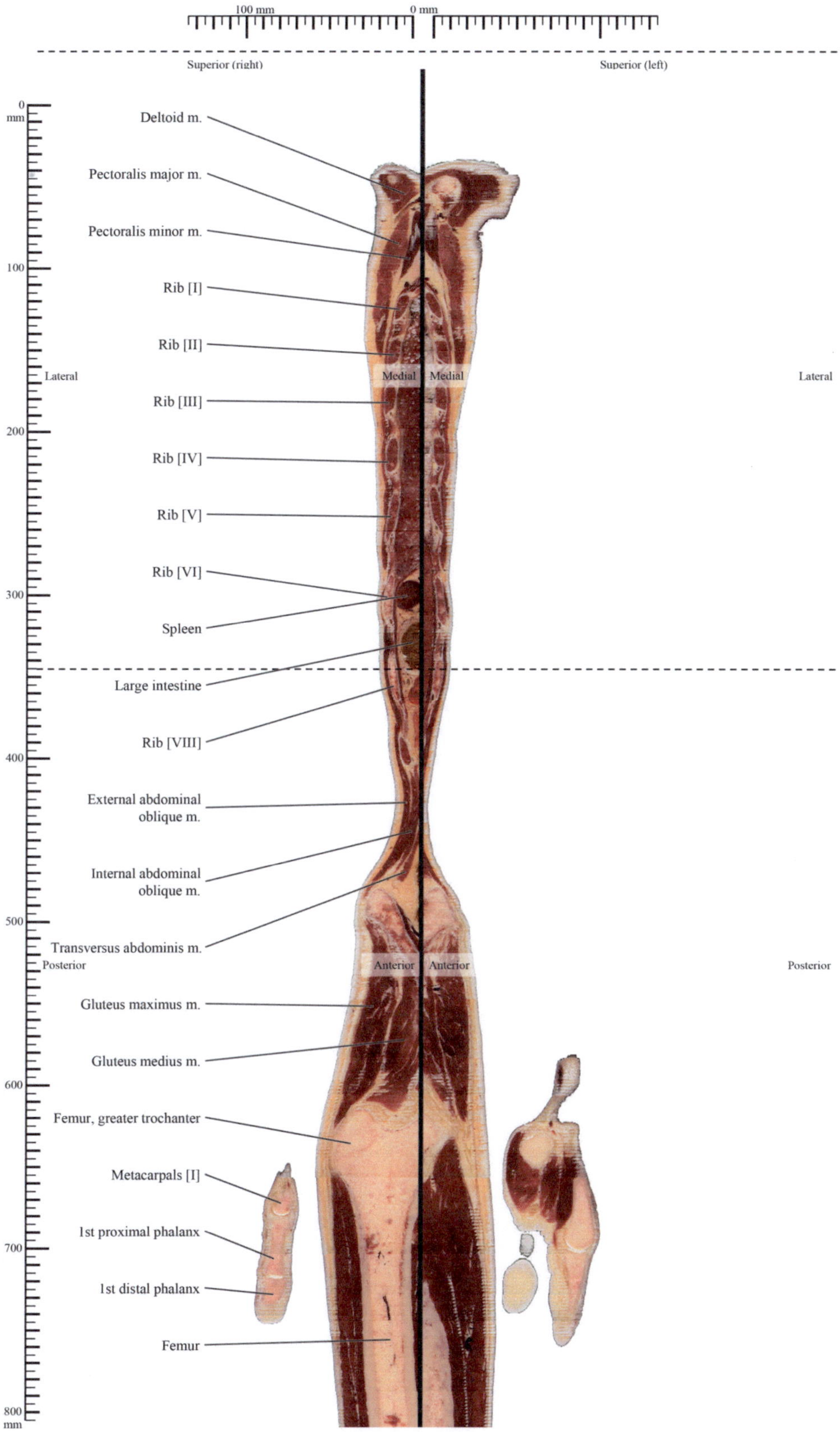

Coronal (arm) and sagittal (forearm and hand)
(38.0 mm)

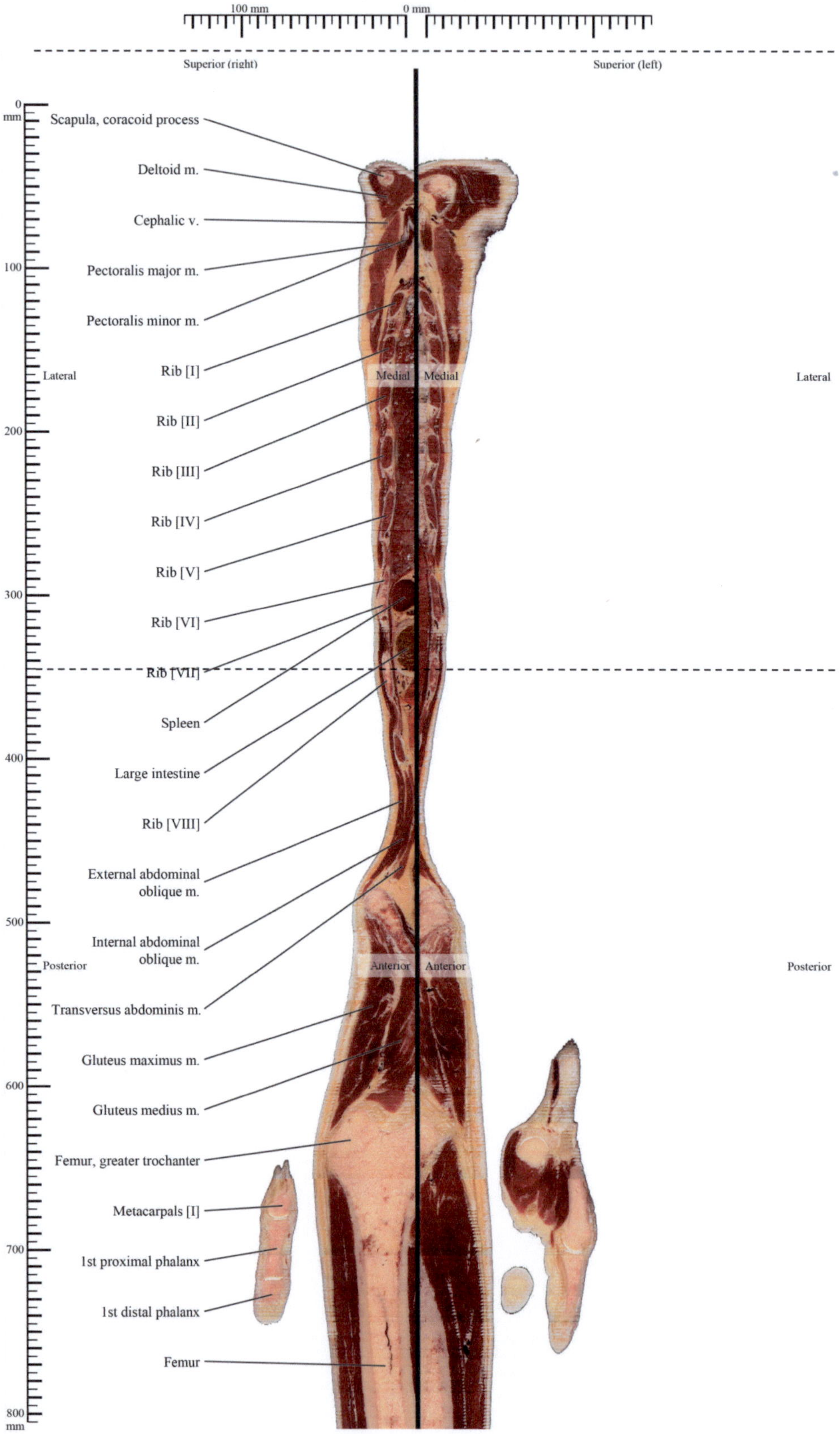

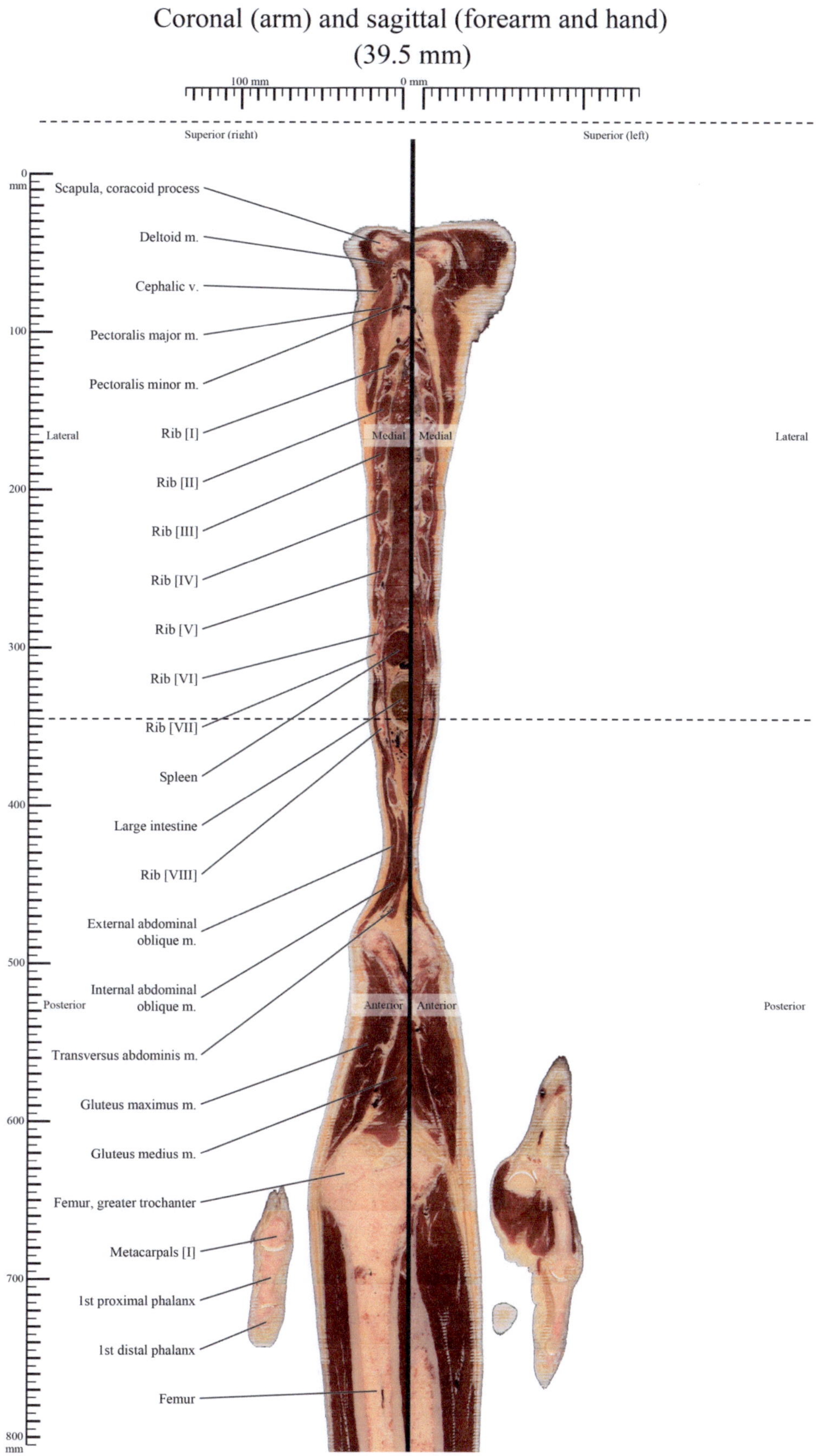
Coronal (arm) and sagittal (forearm and hand)
(39.5 mm)
100 mm
0 mm
Superior (right)
Superior (left)
0 mm
100
200
300
400
500
600
700
800 mm
Scapula, coracoid process
Deltoid m.
Cephalic v.
Pectoralis major m.
Pectoralis minor m.
Rib [I]
Rib [II]
Rib [III]
Rib [IV]
Rib [V]
Rib [VI]
Rib [VII]
Spleen
Large intestine
Rib [VIII]
External abdominal oblique m.
Internal abdominal oblique m.
Transversus abdominis m.
Gluteus maximus m.
Gluteus medius m.
Femur, greater trochanter
Metacarpals [I]
1st proximal phalanx
1st distal phalanx
Femur
Lateral
Medial
Medial
Lateral
Posterior
Anterior
Anterior
Posterior

Coronal (arm) and sagittal (forearm and hand) (41.0 mm)

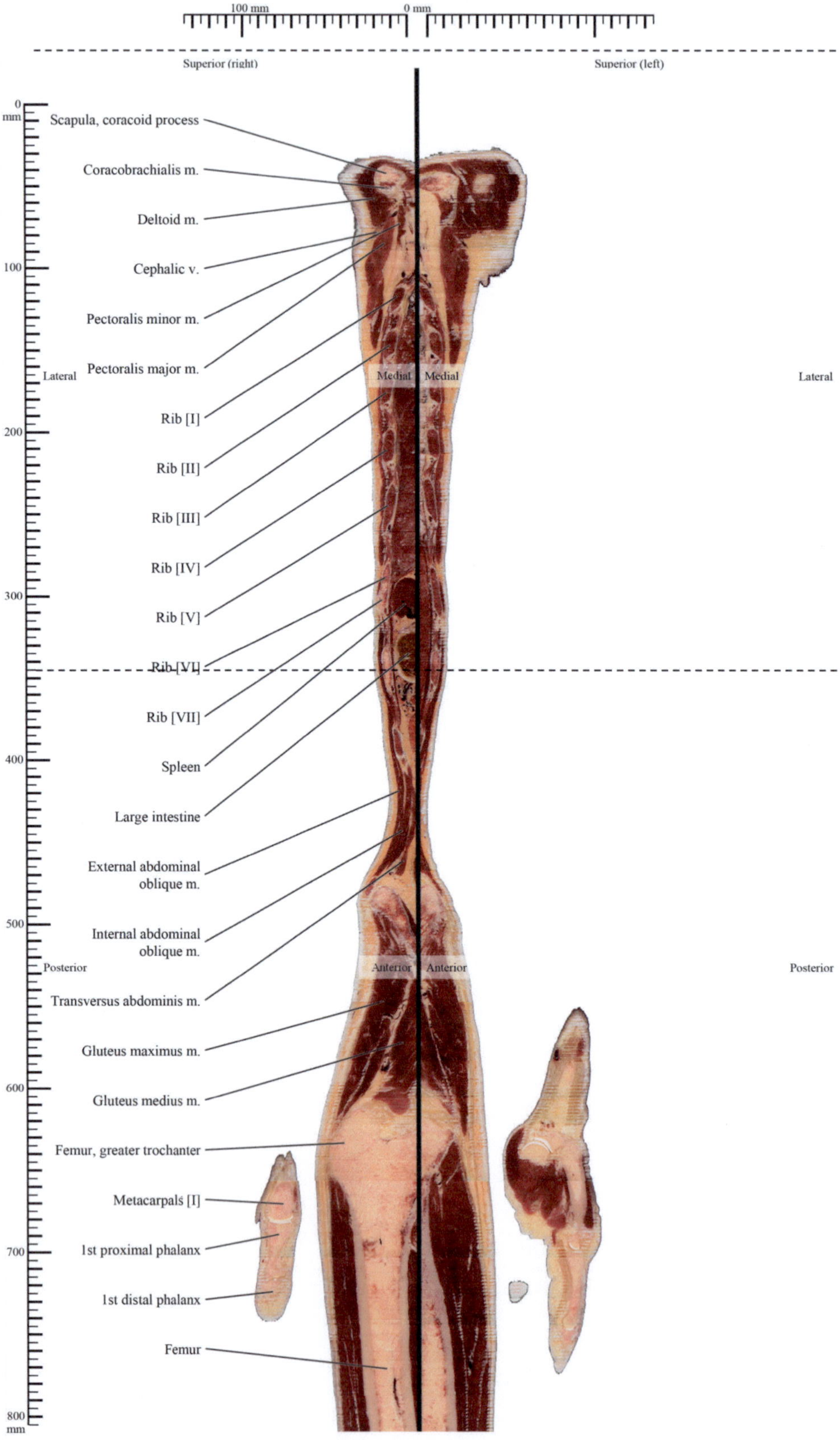

Coronal (arm) and sagittal (forearm and hand)
(42.5 mm)

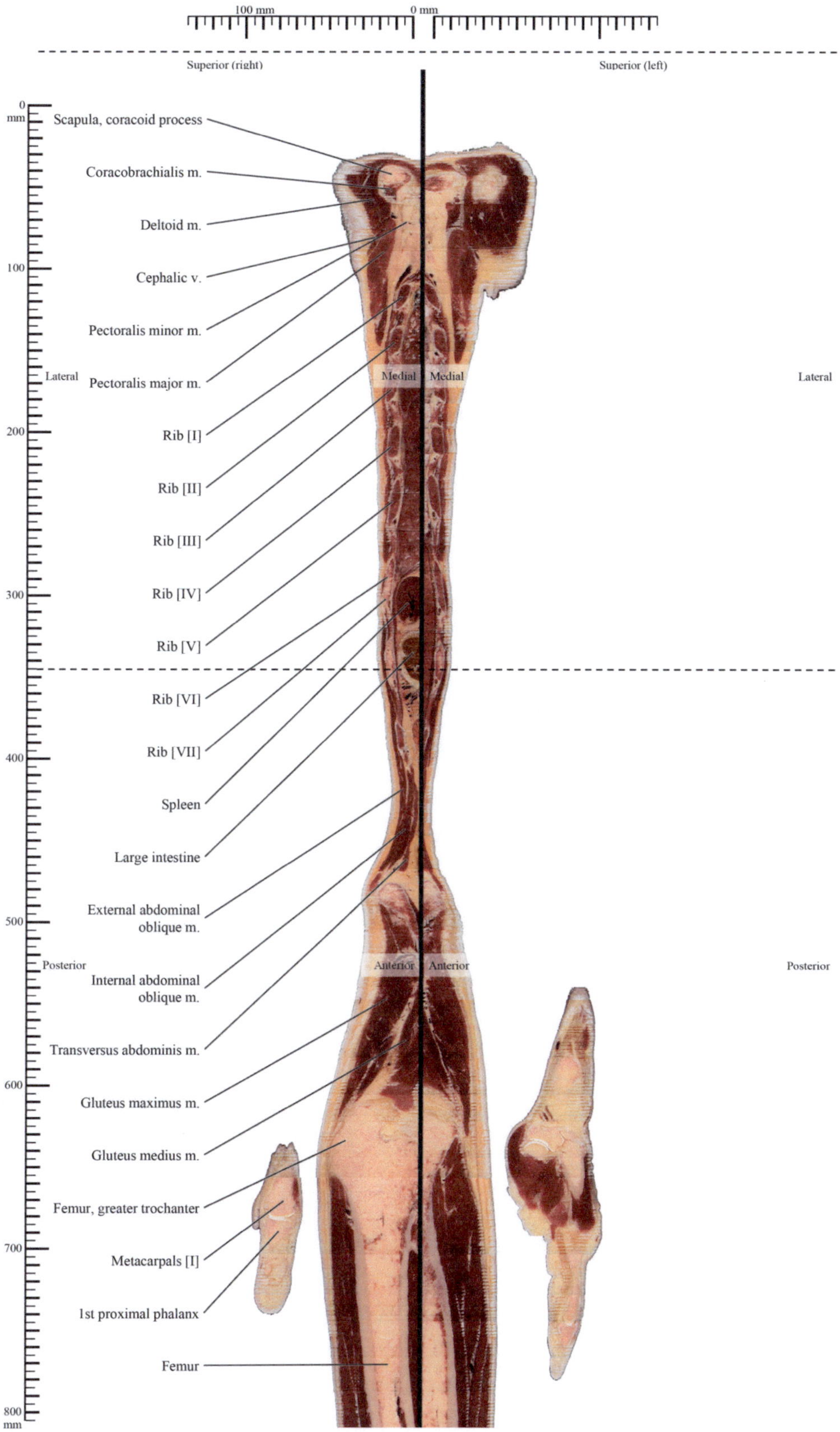

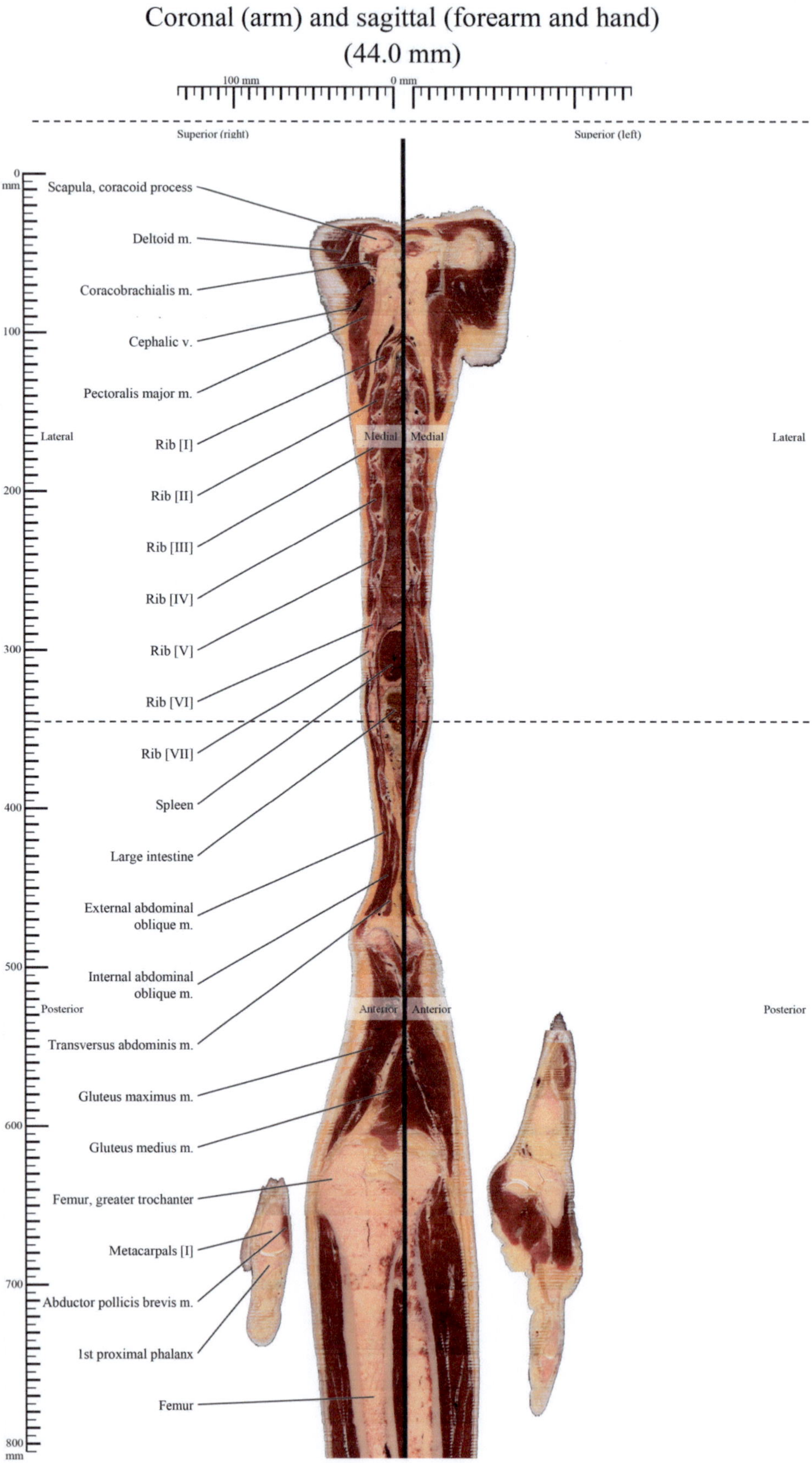
Coronal (arm) and sagittal (forearm and hand)
(44.0 mm)
100 mm
0 mm
Superior (right)
Superior (left)
0 mm
100
200
300
400
500
600
700
800 mm
Lateral
Medial
Medial
Lateral
Posterior
Anterior
Anterior
Posterior
Scapula, coracoid process
Deltoid m.
Coracobrachialis m.
Cephalic v.
Pectoralis major m.
Rib [I]
Rib [II]
Rib [III]
Rib [IV]
Rib [V]
Rib [VI]
Rib [VII]
Spleen
Large intestine
External abdominal oblique m.
Internal abdominal oblique m.
Transversus abdominis m.
Gluteus maximus m.
Gluteus medius m.
Femur, greater trochanter
Metacarpals [I]
Abductor pollicis brevis m.
1st proximal phalanx
Femur

Coronal (arm) and sagittal (forearm and hand) (45.5 mm)

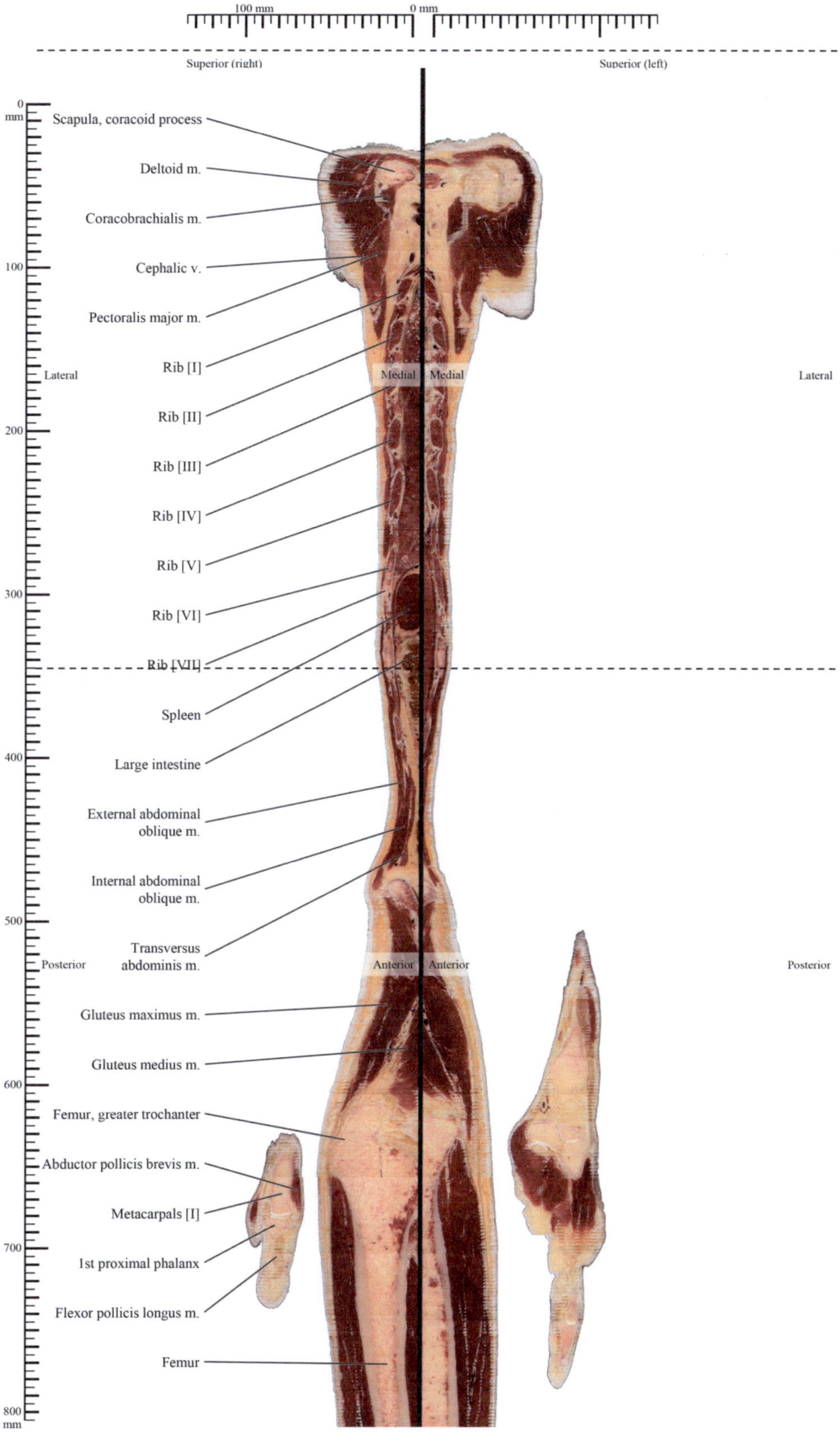

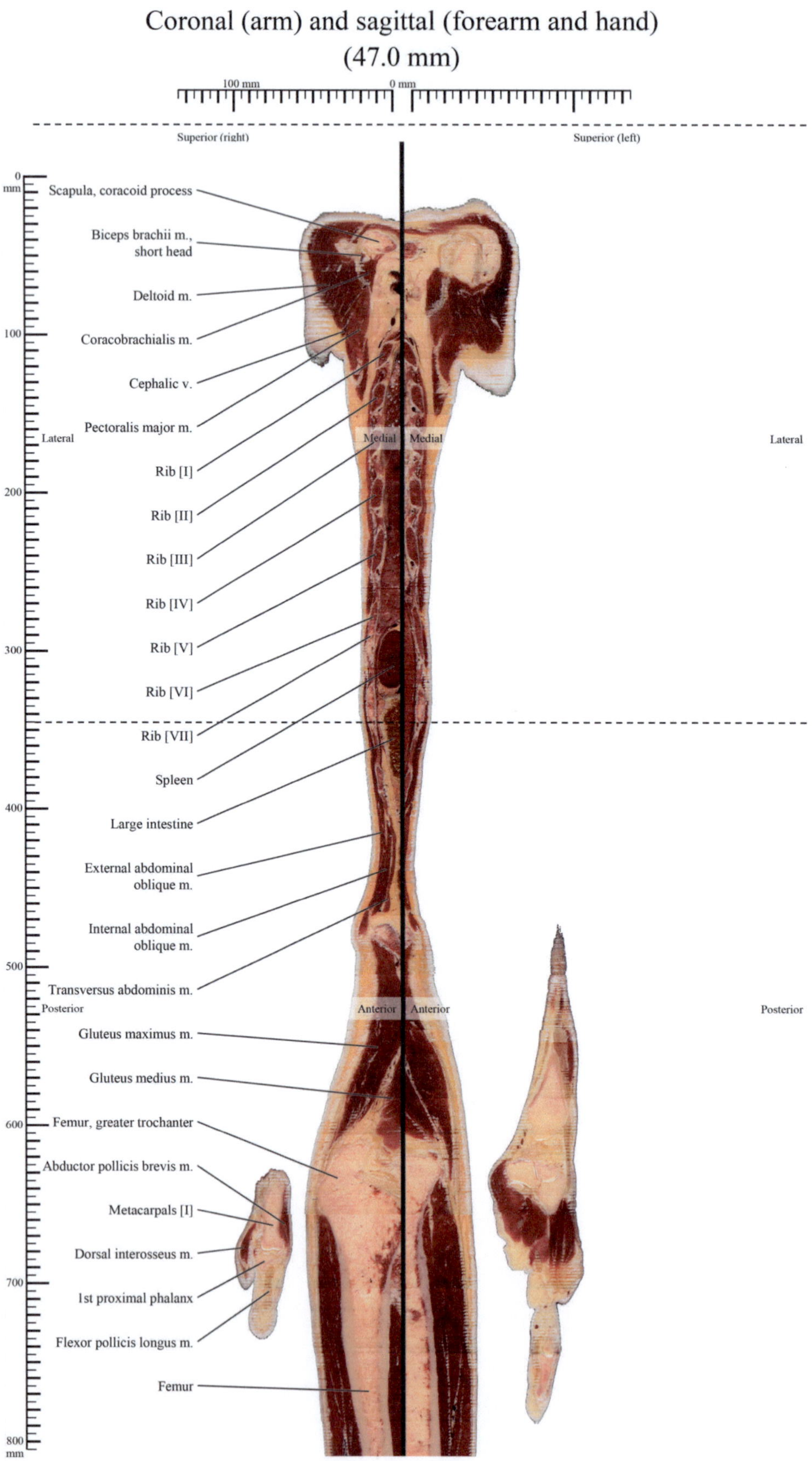
Coronal (arm) and sagittal (forearm and hand)
(47.0 mm)
100 mm
0 mm
Superior (right)
Superior (left)
0 mm
100
200
300
400
500
600
700
800 mm
Scapula, coracoid process
Biceps brachii m., short head
Deltoid m.
Coracobrachialis m.
Cephalic v.
Pectoralis major m.
Lateral
Medial
Medial
Lateral
Rib [I]
Rib [II]
Rib [III]
Rib [IV]
Rib [V]
Rib [VI]
Rib [VII]
Spleen
Large intestine
External abdominal oblique m.
Internal abdominal oblique m.
Transversus abdominis m.
Posterior
Anterior
Anterior
Posterior
Gluteus maximus m.
Gluteus medius m.
Femur, greater trochanter
Abductor pollicis brevis m.
Metacarpals [I]
Dorsal interosseus m.
1st proximal phalanx
Flexor pollicis longus m.
Femur

Coronal (arm) and sagittal (forearm and hand) (48.5 mm)

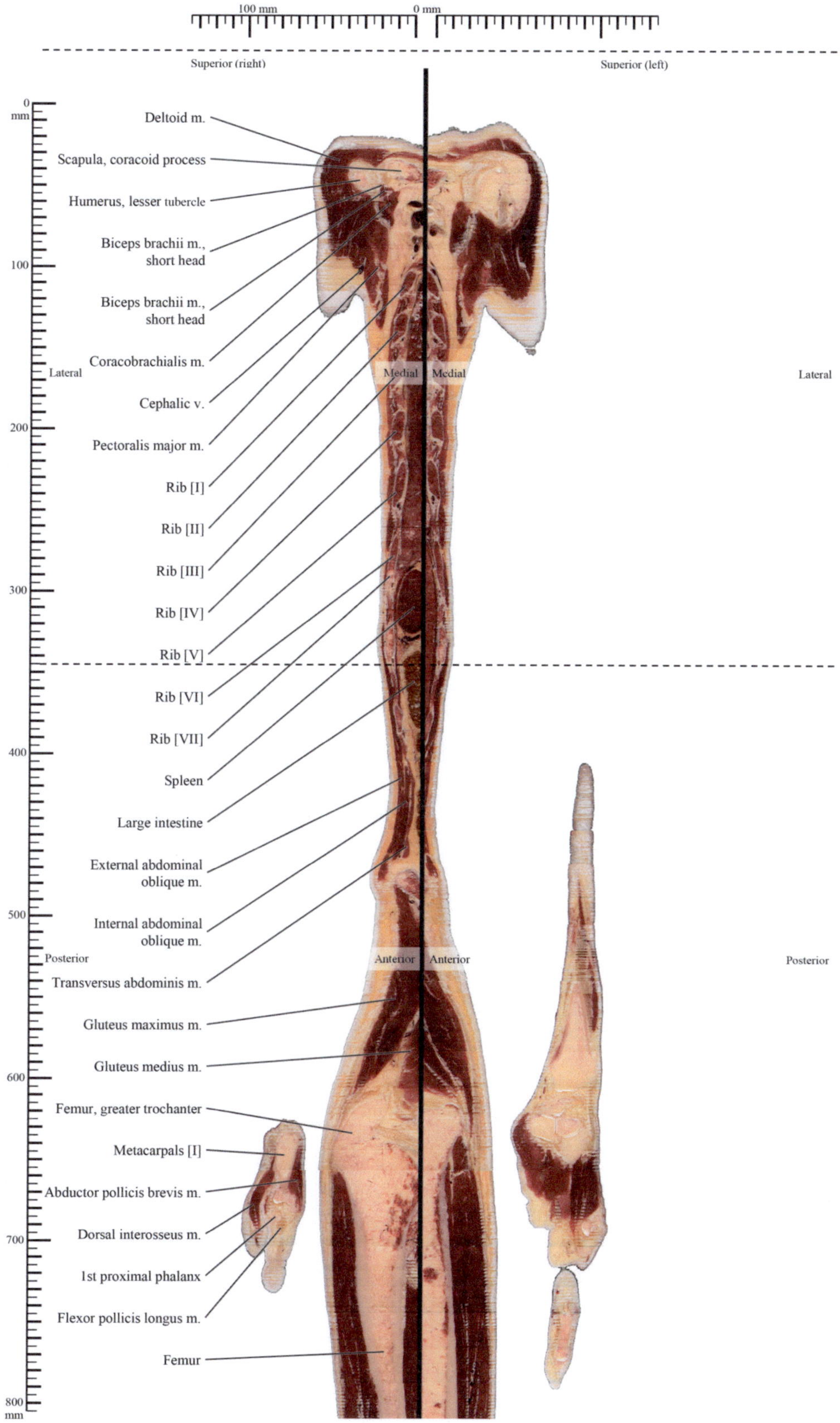

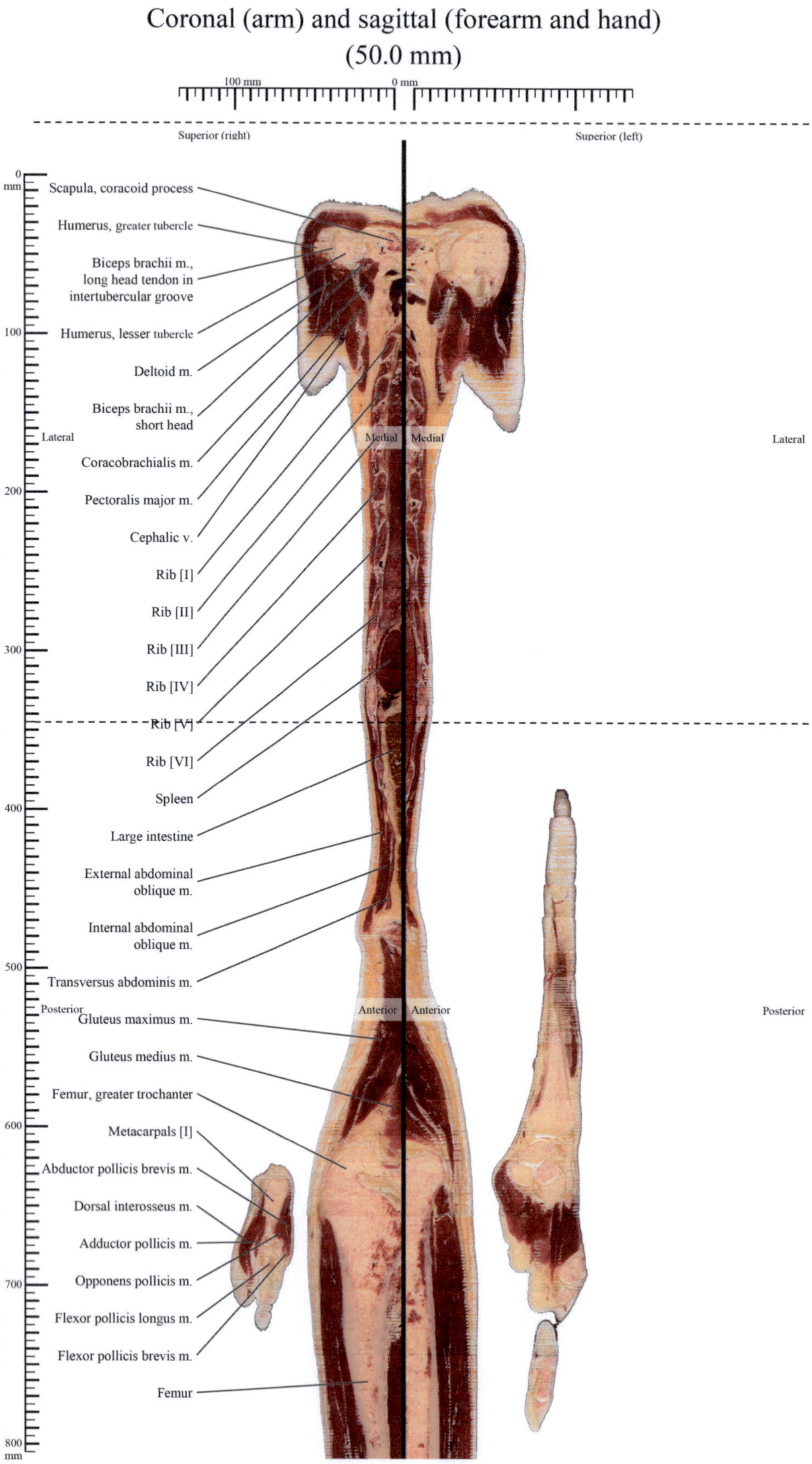
Coronal (arm) and sagittal (forearm and hand)
(50.0 mm)
100 mm
0 mm
Superior (right)
Superior (left)
0 mm
100
200
300
400
500
600
700
800 mm
Lateral
Medial
Medial
Lateral
Posterior
Anterior
Anterior
Posterior
Scapula, coracoid process
Humerus, greater tubercle
Biceps brachii m., long head tendon in intertubercular groove
Humerus, lesser tubercle
Deltoid m.
Biceps brachii m., short head
Coracobrachialis m.
Pectoralis major m.
Cephalic v.
Rib [I]
Rib [II]
Rib [III]
Rib [IV]
Rib [V]
Rib [VI]
Spleen
Large intestine
External abdominal oblique m.
Internal abdominal oblique m.
Transversus abdominis m.
Gluteus maximus m.
Gluteus medius m.
Femur, greater trochanter
Metacarpals [I]
Abductor pollicis brevis m.
Dorsal interosseus m.
Adductor pollicis m.
Opponens pollicis m.
Flexor pollicis longus m.
Flexor pollicis brevis m.
Femur

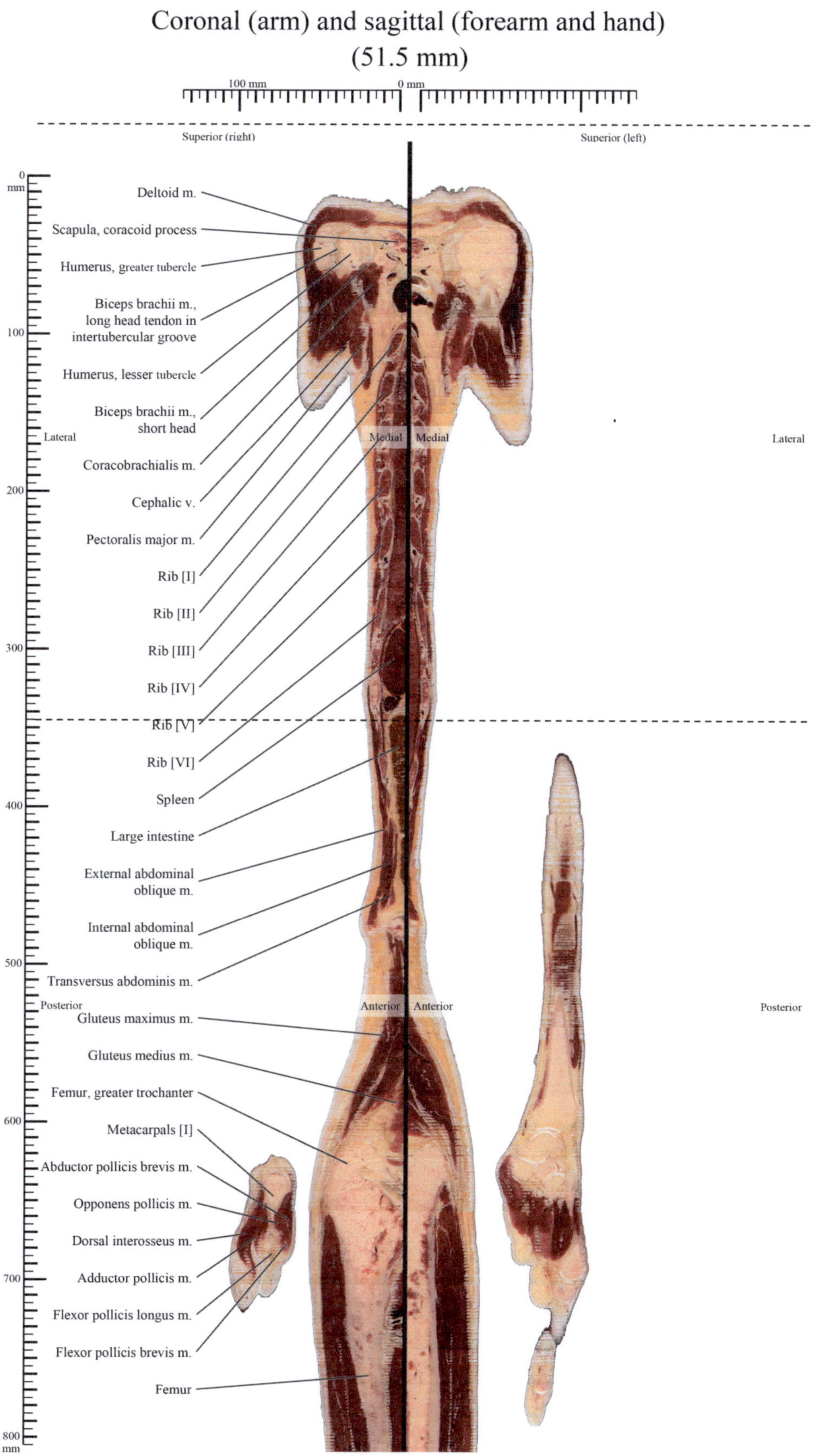
Coronal (arm) and sagittal (forearm and hand)
(51.5 mm)
100 mm
0 mm
Superior (right)
Superior (left)
0 mm
100
200
300
400
500
600
700
800 mm
Deltoid m.
Scapula, coracoid process
Humerus, greater tubercle
Biceps brachii m., long head tendon in intertubercular groove
Humerus, lesser tubercle
Biceps brachii m., short head
Lateral
Medial
Medial
Lateral
Coracobrachialis m.
Cephalic v.
Pectoralis major m.
Rib [I]
Rib [II]
Rib [III]
Rib [IV]
Rib [V]
Rib [VI]
Spleen
Large intestine
External abdominal oblique m.
Internal abdominal oblique m.
Transversus abdominis m.
Posterior
Anterior
Anterior
Posterior
Gluteus maximus m.
Gluteus medius m.
Femur, greater trochanter
Metacarpals [I]
Abductor pollicis brevis m.
Opponens pollicis m.
Dorsal interosseus m.
Adductor pollicis m.
Flexor pollicis longus m.
Flexor pollicis brevis m.
Femur

Coronal (arm) and sagittal (forearm and hand) (53.0 mm)

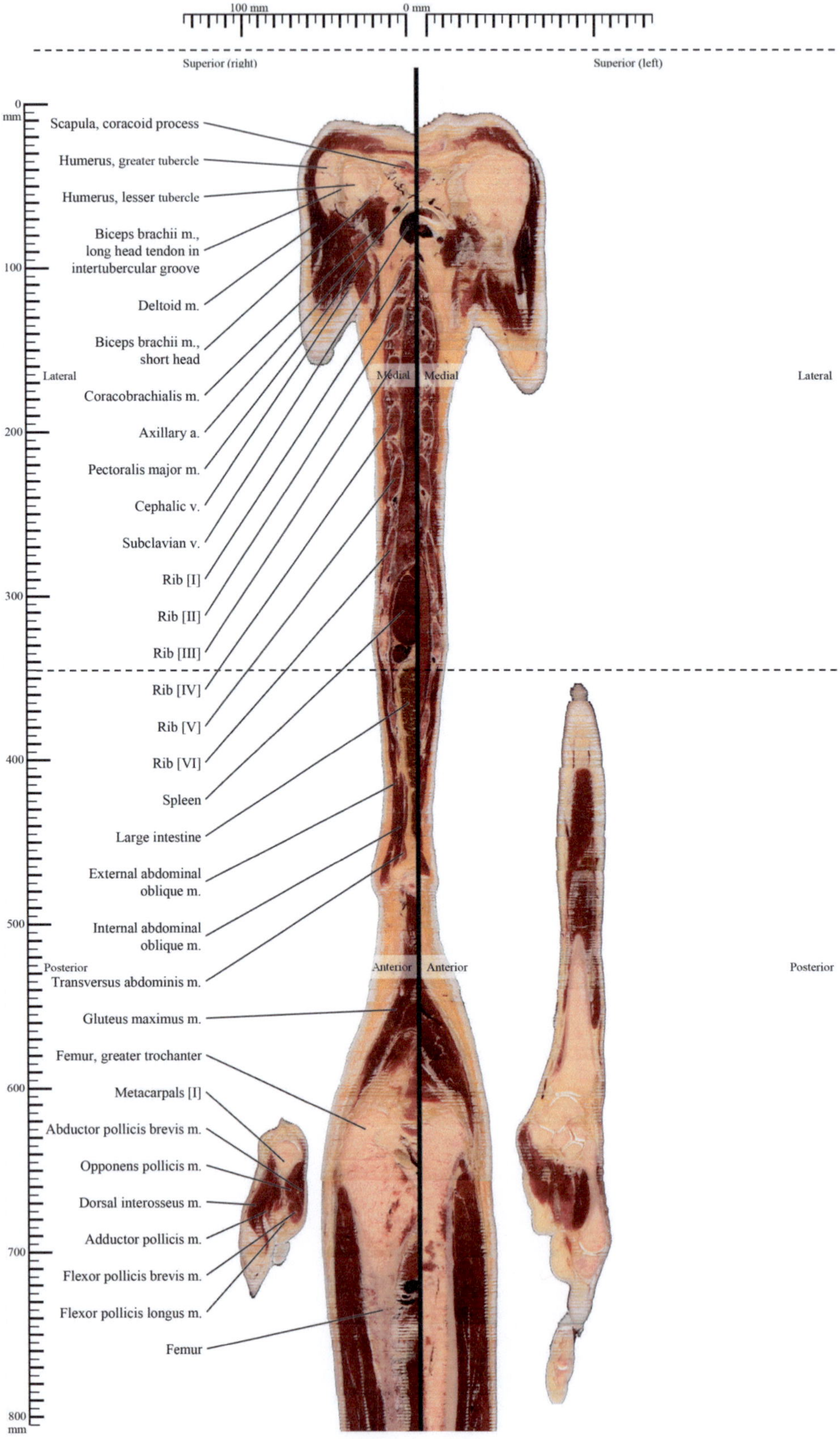

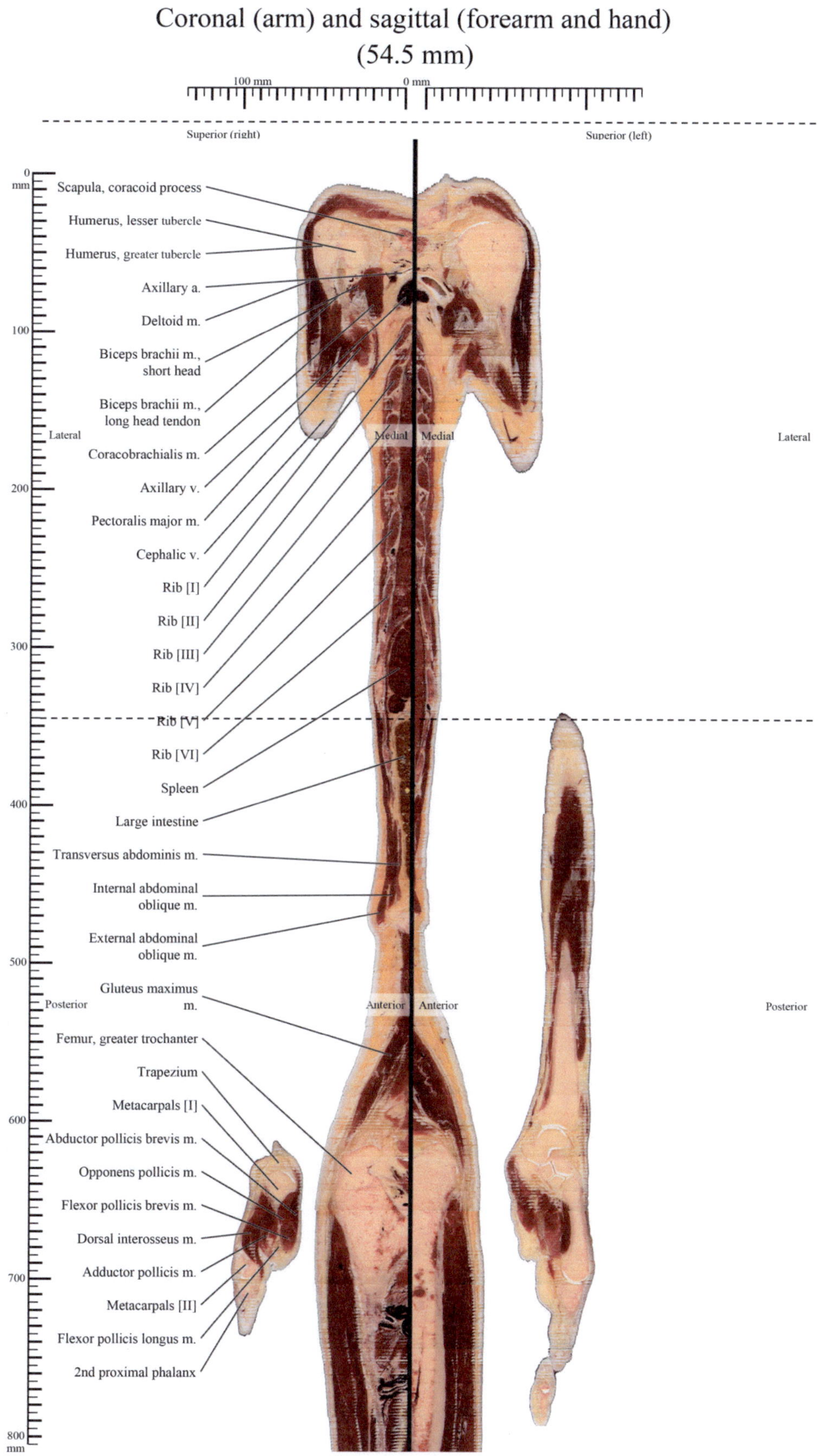
Coronal (arm) and sagittal (forearm and hand)
(54.5 mm)
100 mm
0 mm
Superior (right)
Superior (left)
0 mm
100
200
300
400
500
600
700
800 mm
Lateral
Medial
Medial
Lateral
Posterior
Anterior
Anterior
Posterior
Scapula, coracoid process
Humerus, lesser tubercle
Humerus, greater tubercle
Axillary a.
Deltoid m.
Biceps brachii m., short head
Biceps brachii m., long head tendon
Coracobrachialis m.
Axillary v.
Pectoralis major m.
Cephalic v.
Rib [I]
Rib [II]
Rib [III]
Rib [IV]
Rib [V]
Rib [VI]
Spleen
Large intestine
Transversus abdominis m.
Internal abdominal oblique m.
External abdominal oblique m.
Gluteus maximus m.
Femur, greater trochanter
Trapezium
Metacarpals [I]
Abductor pollicis brevis m.
Opponens pollicis m.
Flexor pollicis brevis m.
Dorsal interosseus m.
Adductor pollicis m.
Metacarpals [II]
Flexor pollicis longus m.
2nd proximal phalanx

Coronal (arm) and sagittal (forearm and hand) (56.0 mm)

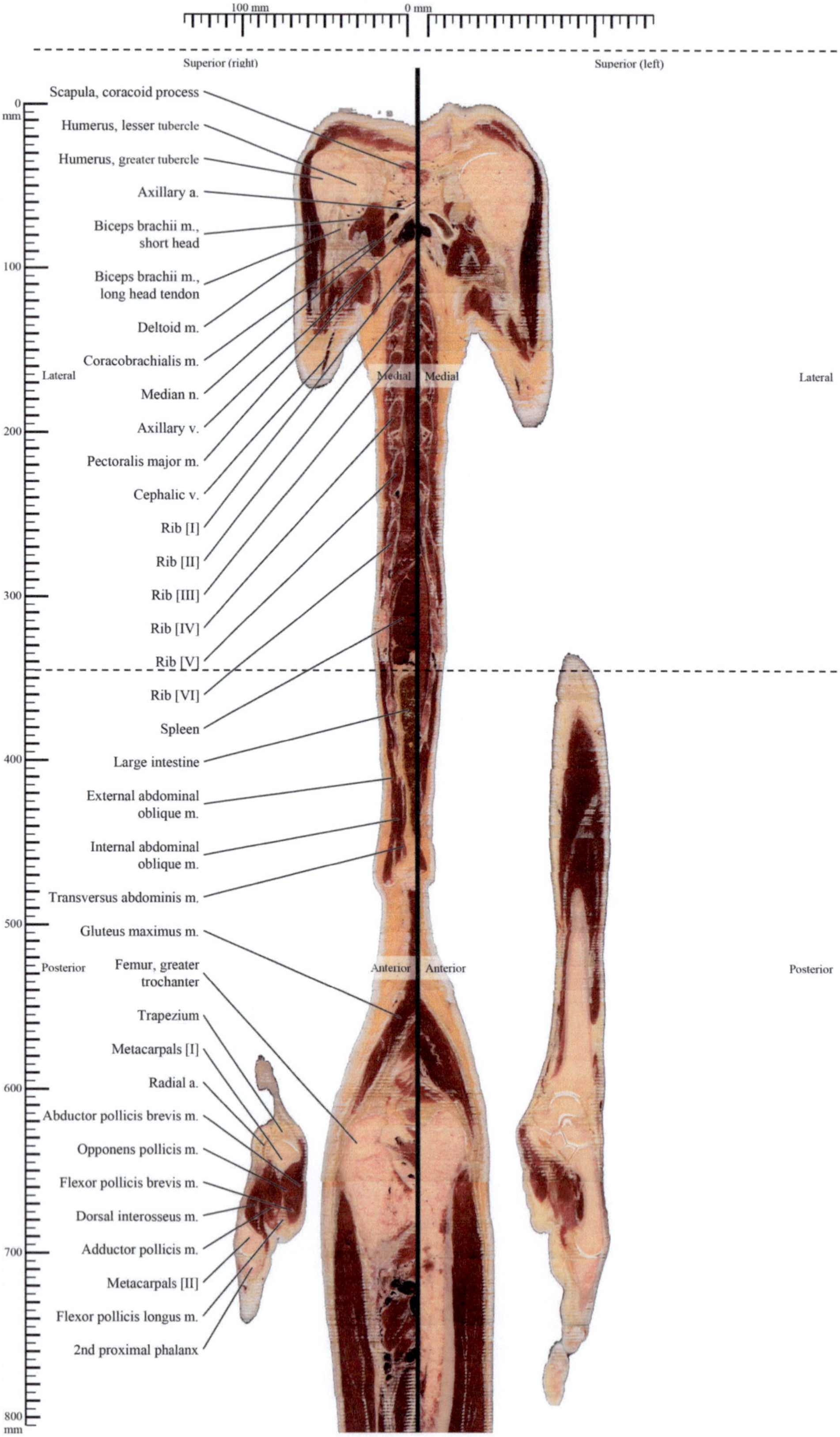

Coronal (arm) and sagittal (forearm and hand)
(57.5 mm)

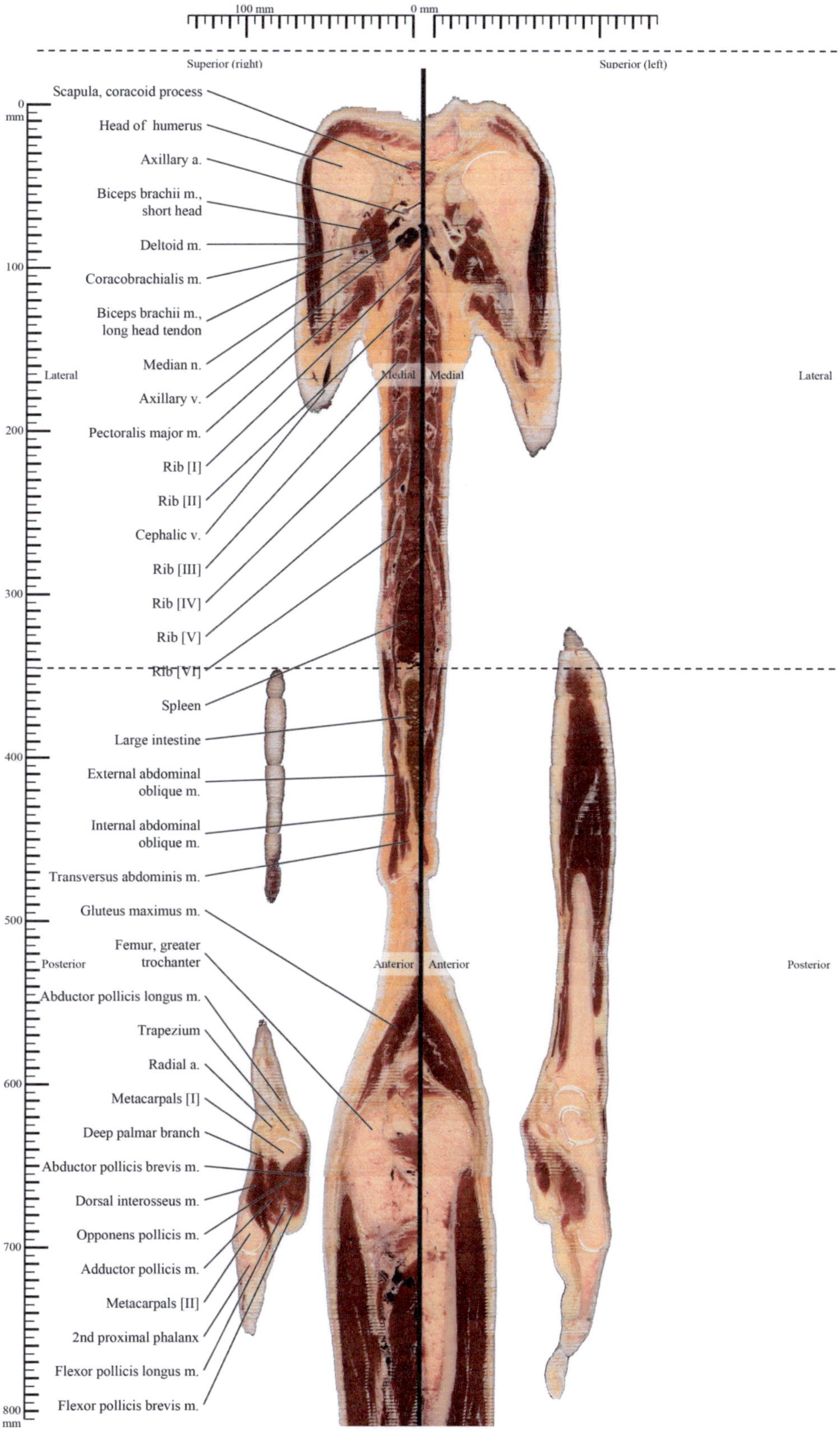

Coronal (arm) and sagittal (forearm and hand) (59.0 mm)

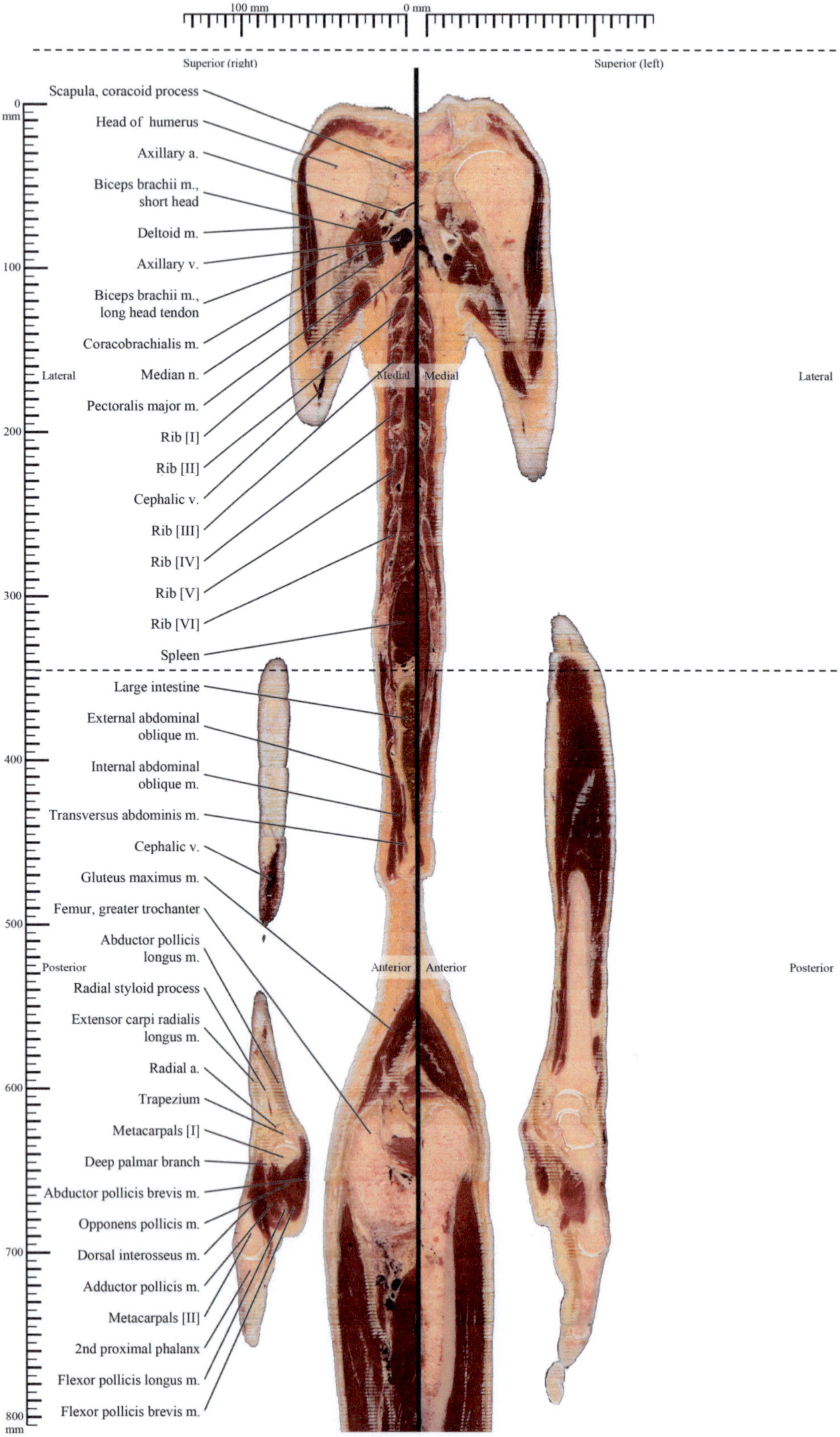

Coronal (arm) and sagittal (forearm and hand) (60.5 mm)

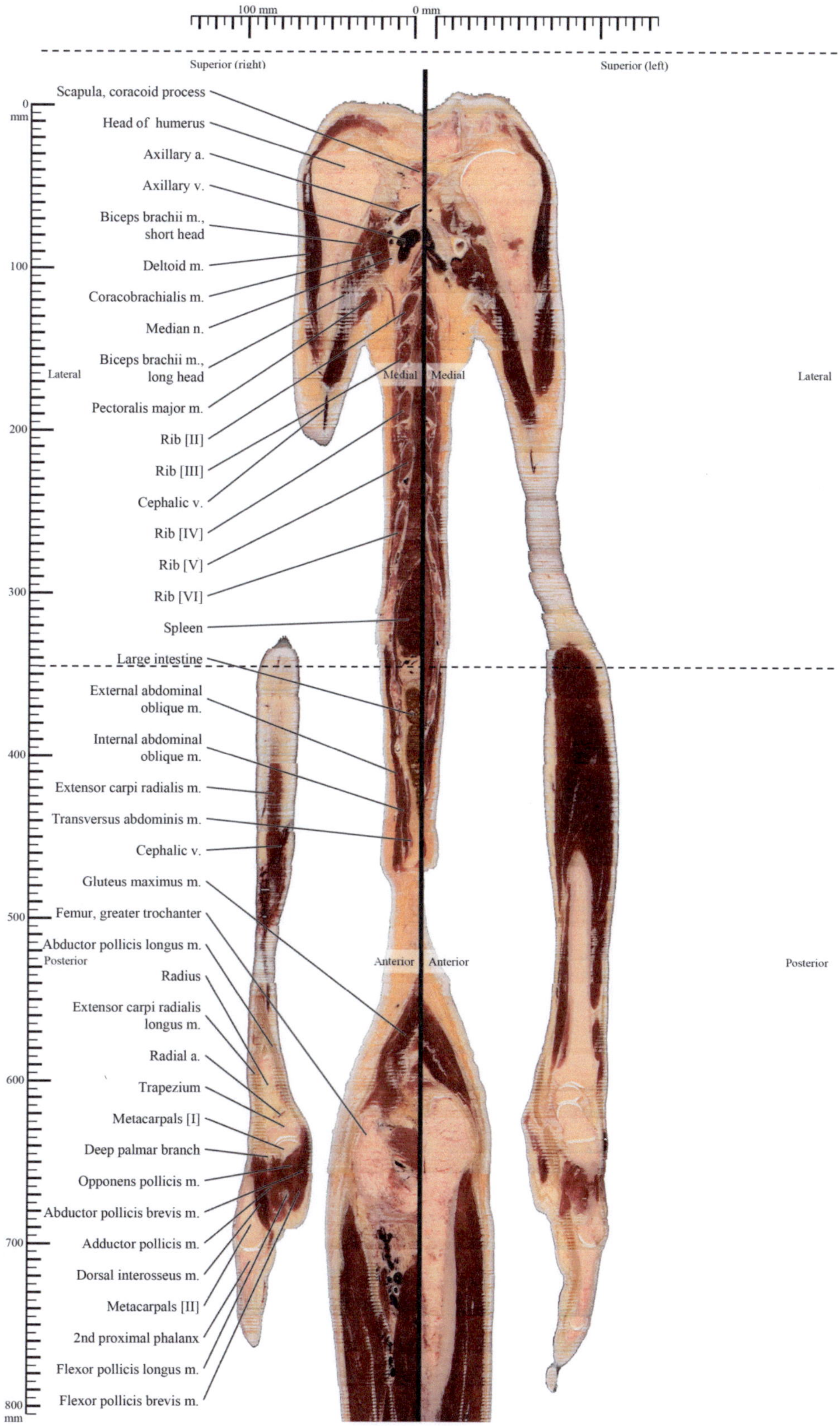
100 mm
0 mm
Superior (right)
Superior (left)
0 mm
100
200
300
400
500
600
700
800 mm
Lateral
Medial
Medial
Lateral
Posterior
Anterior
Anterior
Posterior
Scapula, coracoid process
Head of humerus
Axillary a.
Axillary v.
Biceps brachii m., short head
Deltoid m.
Coracobrachialis m.
Median n.
Biceps brachii m., long head
Pectoralis major m.
Rib [II]
Rib [III]
Cephalic v.
Rib [IV]
Rib [V]
Rib [VI]
Spleen
Large intestine
External abdominal oblique m.
Internal abdominal oblique m.
Extensor carpi radialis m.
Transversus abdominis m.
Cephalic v.
Gluteus maximus m.
Femur, greater trochanter
Abductor pollicis longus m.
Radius
Extensor carpi radialis longus m.
Radial a.
Trapezium
Metacarpals [I]
Deep palmar branch
Opponens pollicis m.
Abductor pollicis brevis m.
Adductor pollicis m.
Dorsal interosseus m.
Metacarpals [II]
2nd proximal phalanx
Flexor pollicis longus m.
Flexor pollicis brevis m.

Coronal (arm) and sagittal (forearm and hand) (62.0 mm)

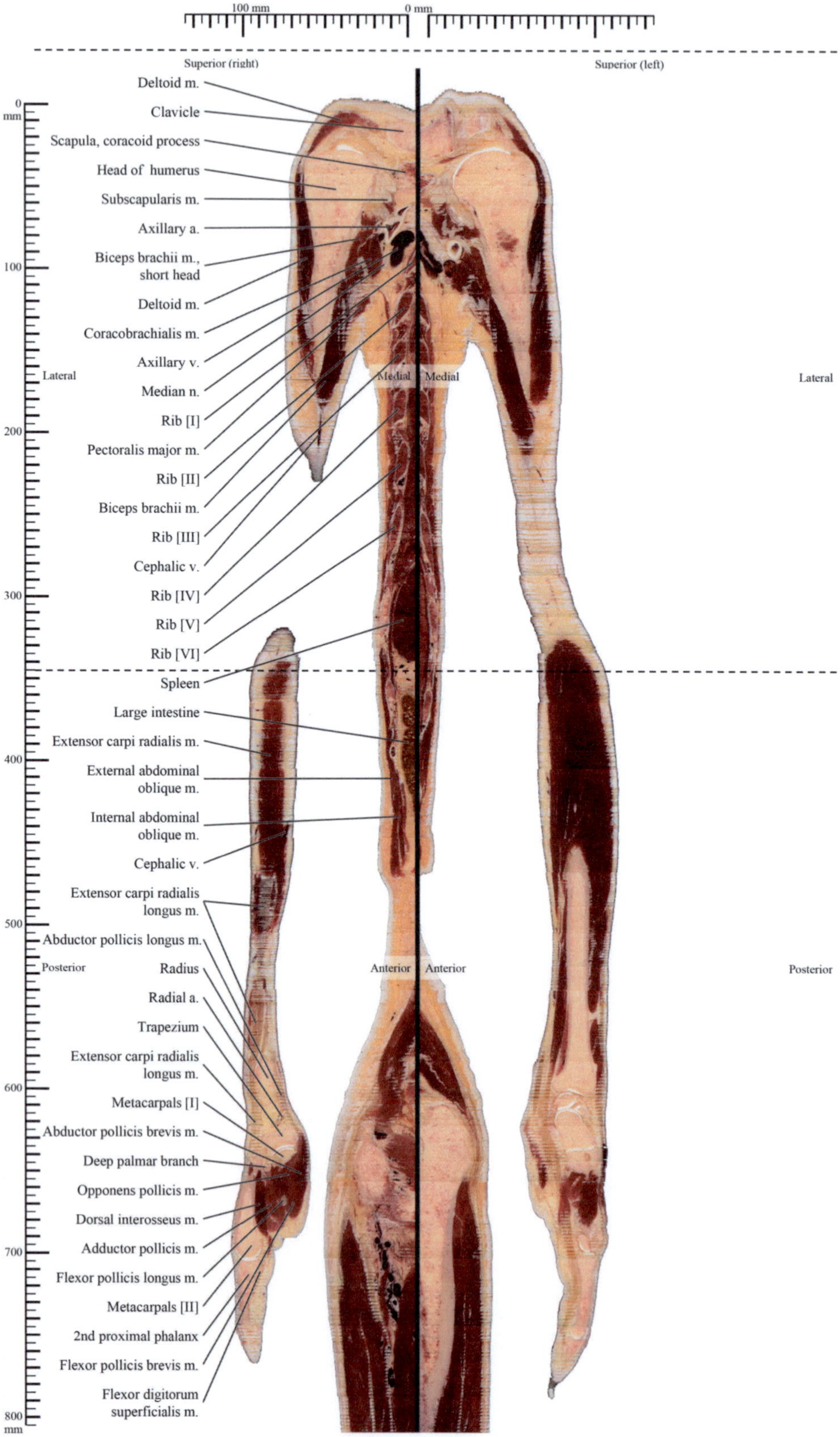
100 mm
0 mm
Superior (right)
Superior (left)
0 mm
100
200
300
400
500
600
700
800 mm
Lateral
Medial
Medial
Lateral
Posterior
Anterior
Anterior
Posterior
Deltoid m.
Clavicle
Scapula, coracoid process
Head of humerus
Subscapularis m.
Axillary a.
Biceps brachii m., short head
Deltoid m.
Coracobrachialis m.
Axillary v.
Median n.
Rib [I]
Pectoralis major m.
Rib [II]
Biceps brachii m.
Rib [III]
Cephalic v.
Rib [IV]
Rib [V]
Rib [VI]
Spleen
Large intestine
Extensor carpi radialis m.
External abdominal oblique m.
Internal abdominal oblique m.
Cephalic v.
Extensor carpi radialis longus m.
Abductor pollicis longus m.
Radius
Radial a.
Trapezium
Extensor carpi radialis longus m.
Metacarpals [I]
Abductor pollicis brevis m.
Deep palmar branch
Opponens pollicis m.
Dorsal interosseus m.
Adductor pollicis m.
Flexor pollicis longus m.
Metacarpals [II]
2nd proximal phalanx
Flexor pollicis brevis m.
Flexor digitorum superficialis m.

Coronal (arm) and sagittal (forearm and hand)
(63.5 mm)

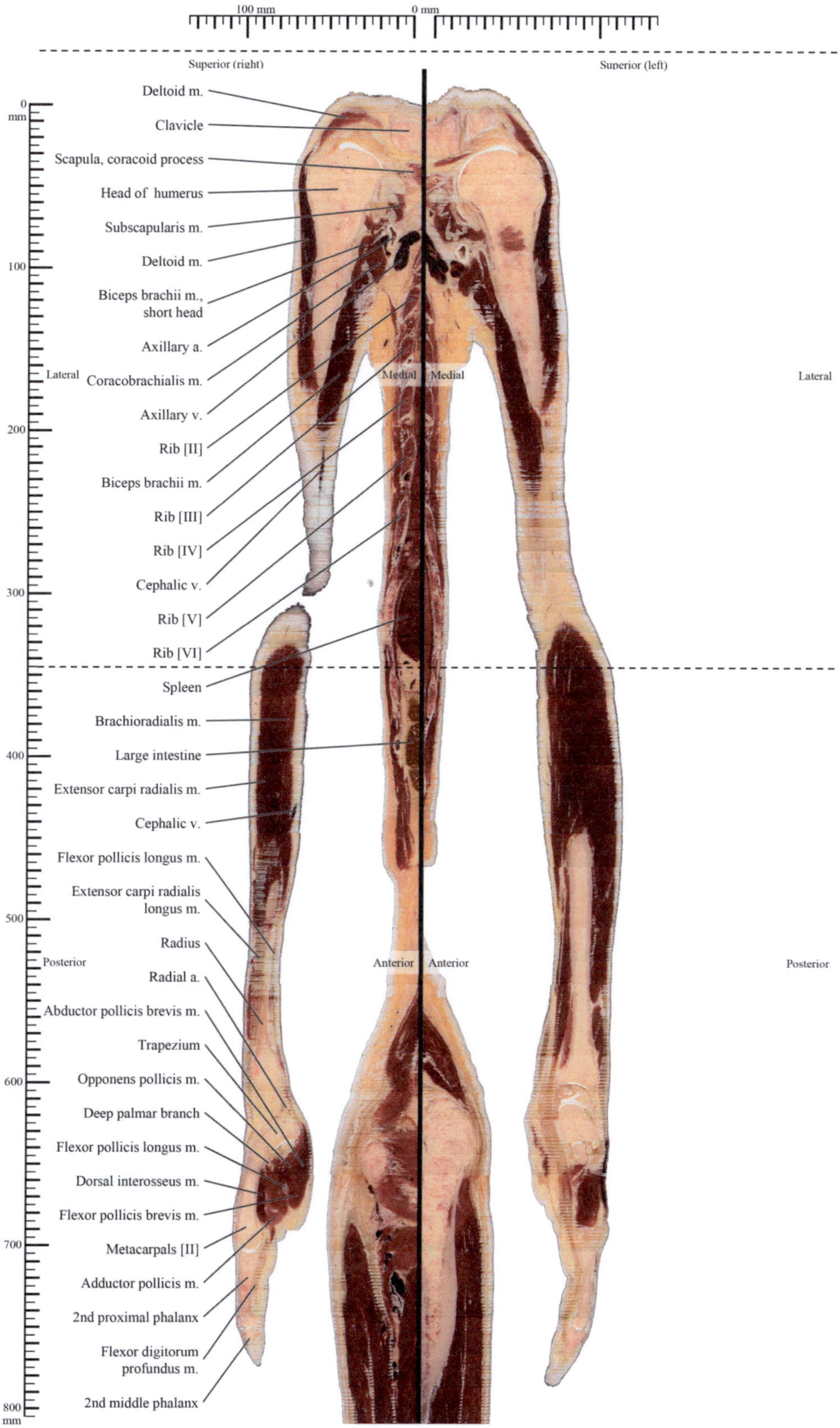

Coronal (arm) and sagittal (forearm and hand) (65.0 mm)

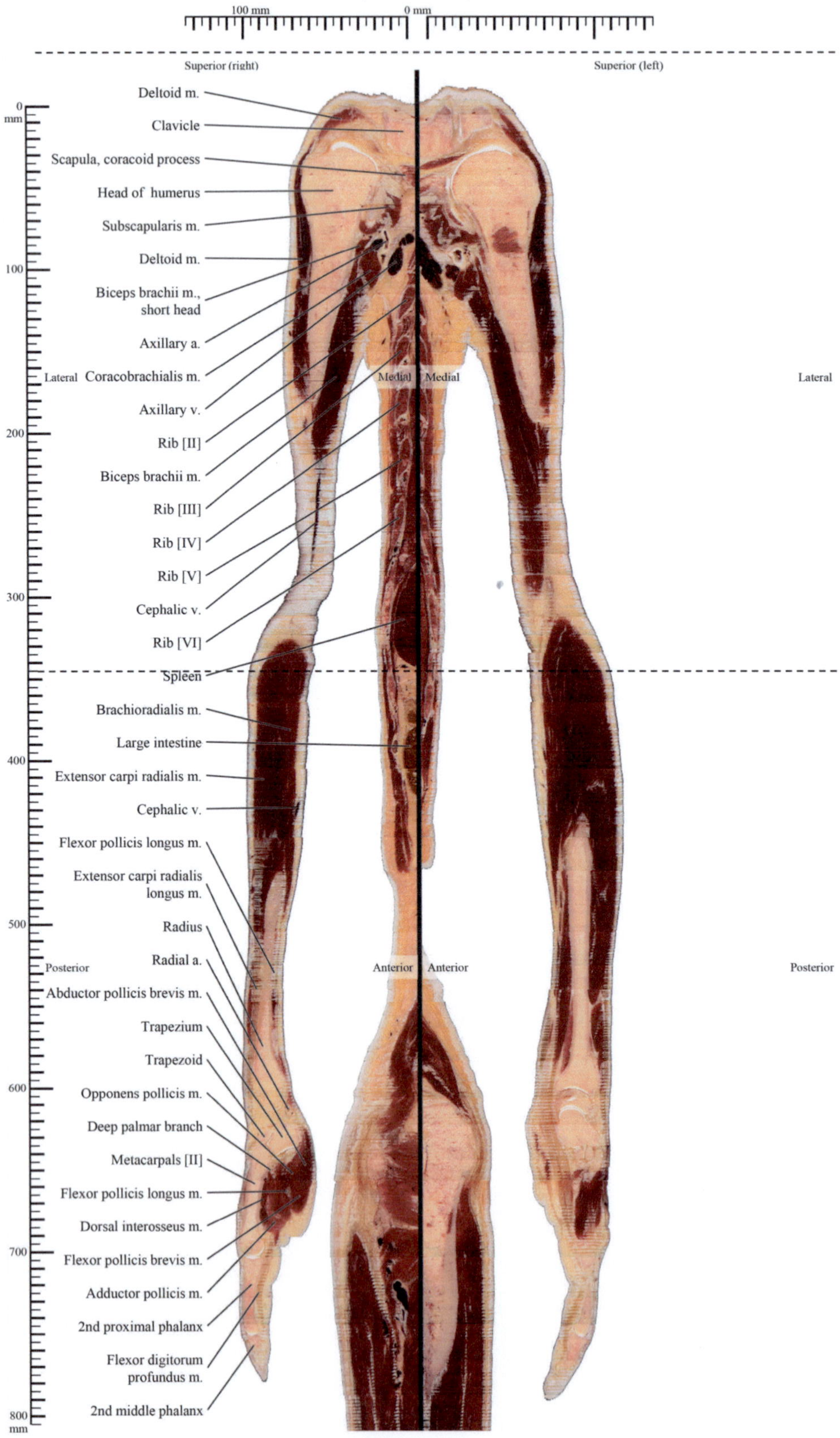

Coronal (arm) and sagittal (forearm and hand) (66.5 mm)

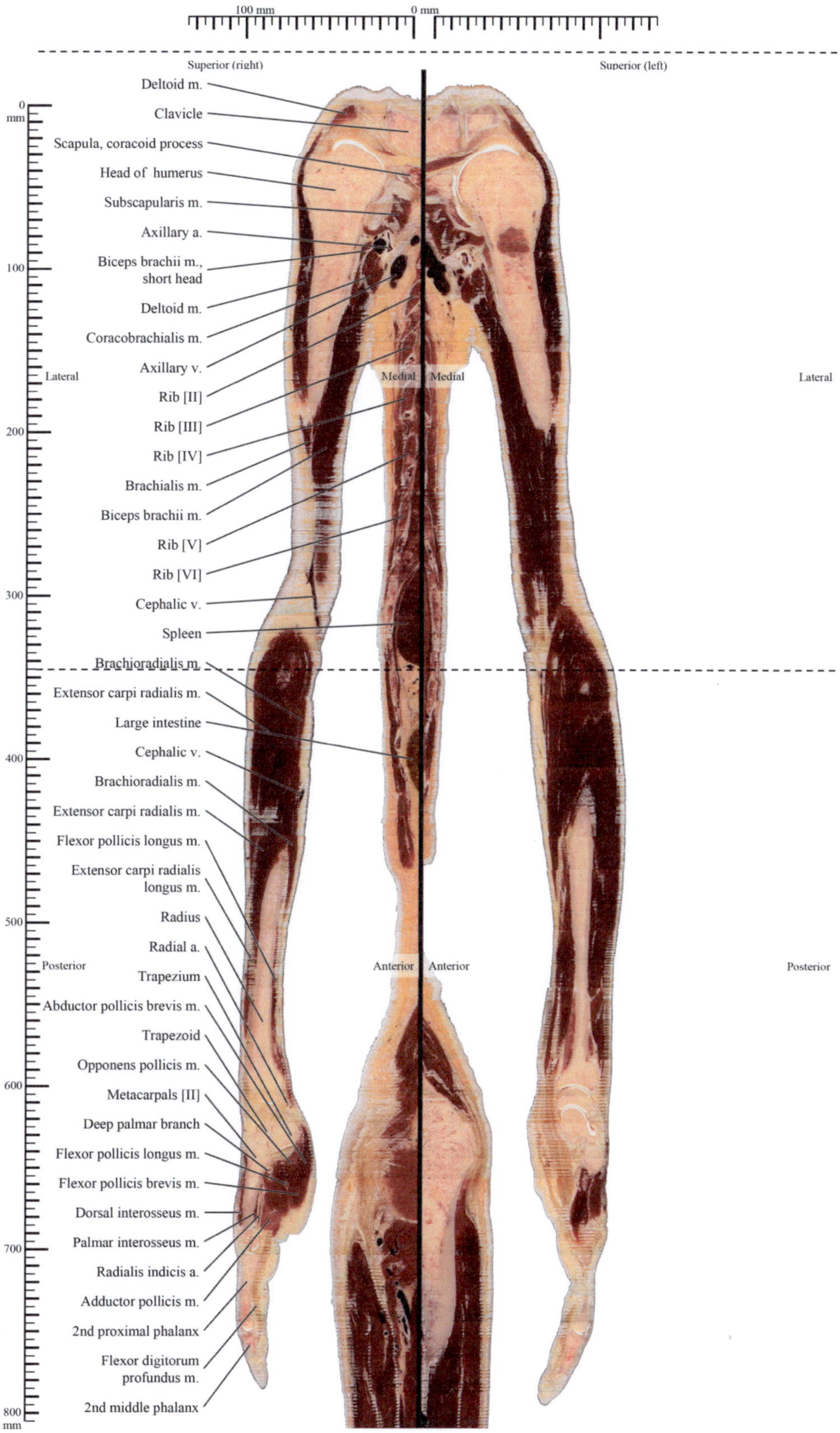

Coronal (arm) and sagittal (forearm and hand) (68.0 mm)

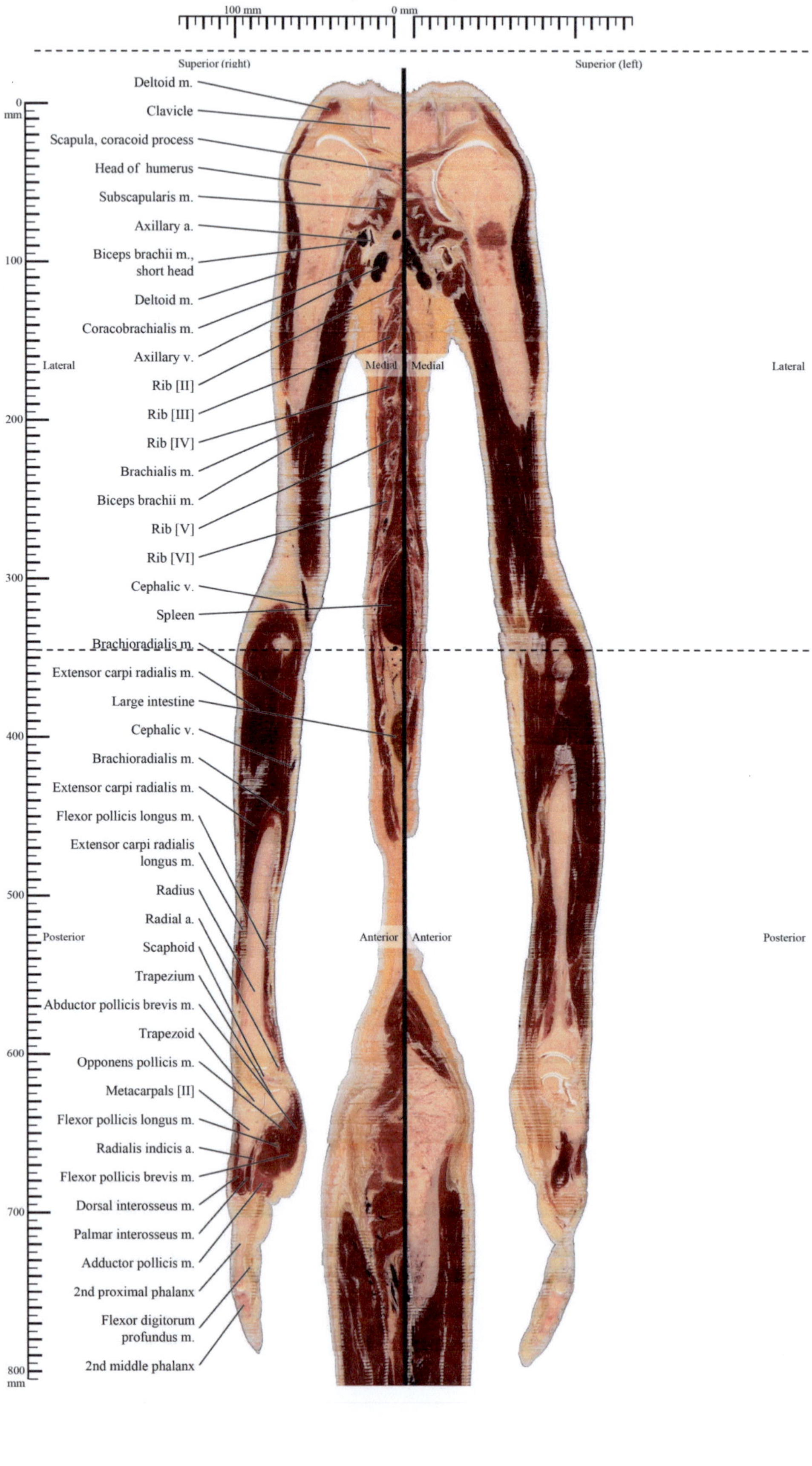

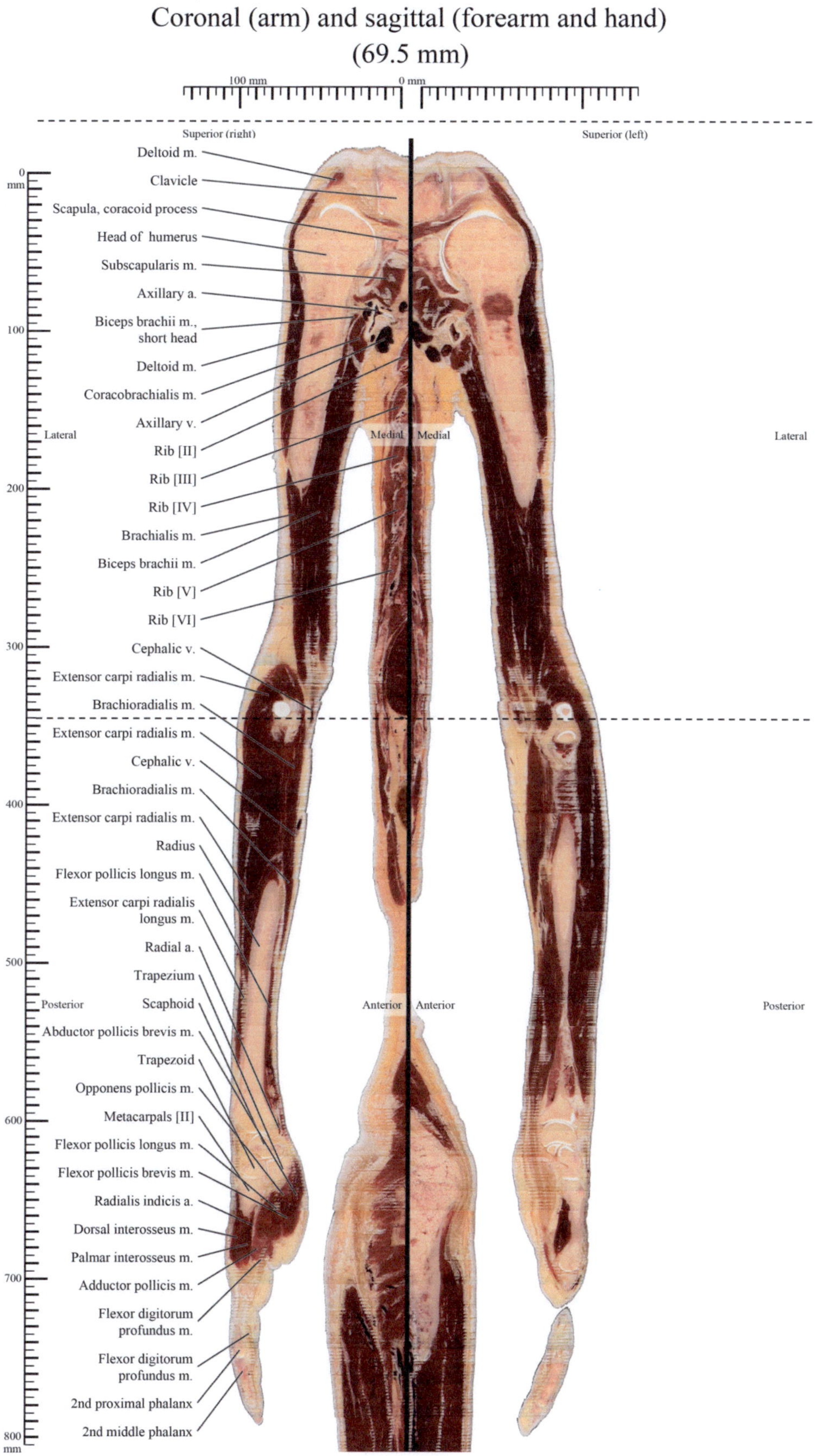
Coronal (arm) and sagittal (forearm and hand)
(69.5 mm)
100 mm
0 mm
Superior (right)
Superior (left)
Lateral
Medial
Medial
Lateral
Posterior
Anterior
Anterior
Posterior
Deltoid m.
Clavicle
Scapula, coracoid process
Head of humerus
Subscapularis m.
Axillary a.
Biceps brachii m., short head
Deltoid m.
Coracobrachialis m.
Axillary v.
Rib [II]
Rib [III]
Rib [IV]
Brachialis m.
Biceps brachii m.
Rib [V]
Rib [VI]
Cephalic v.
Extensor carpi radialis m.
Brachioradialis m.
Extensor carpi radialis m.
Cephalic v.
Brachioradialis m.
Extensor carpi radialis m.
Radius
Flexor pollicis longus m.
Extensor carpi radialis longus m.
Radial a.
Trapezium
Scaphoid
Abductor pollicis brevis m.
Trapezoid
Opponens pollicis m.
Metacarpals [II]
Flexor pollicis longus m.
Flexor pollicis brevis m.
Radialis indicis a.
Dorsal interosseus m.
Palmar interosseus m.
Adductor pollicis m.
Flexor digitorum profundus m.
Flexor digitorum profundus m.
2nd proximal phalanx
2nd middle phalanx
0 mm
100
200
300
400
500
600
700
800 mm

Coronal (arm) and sagittal (forearm and hand) (71.0 mm)

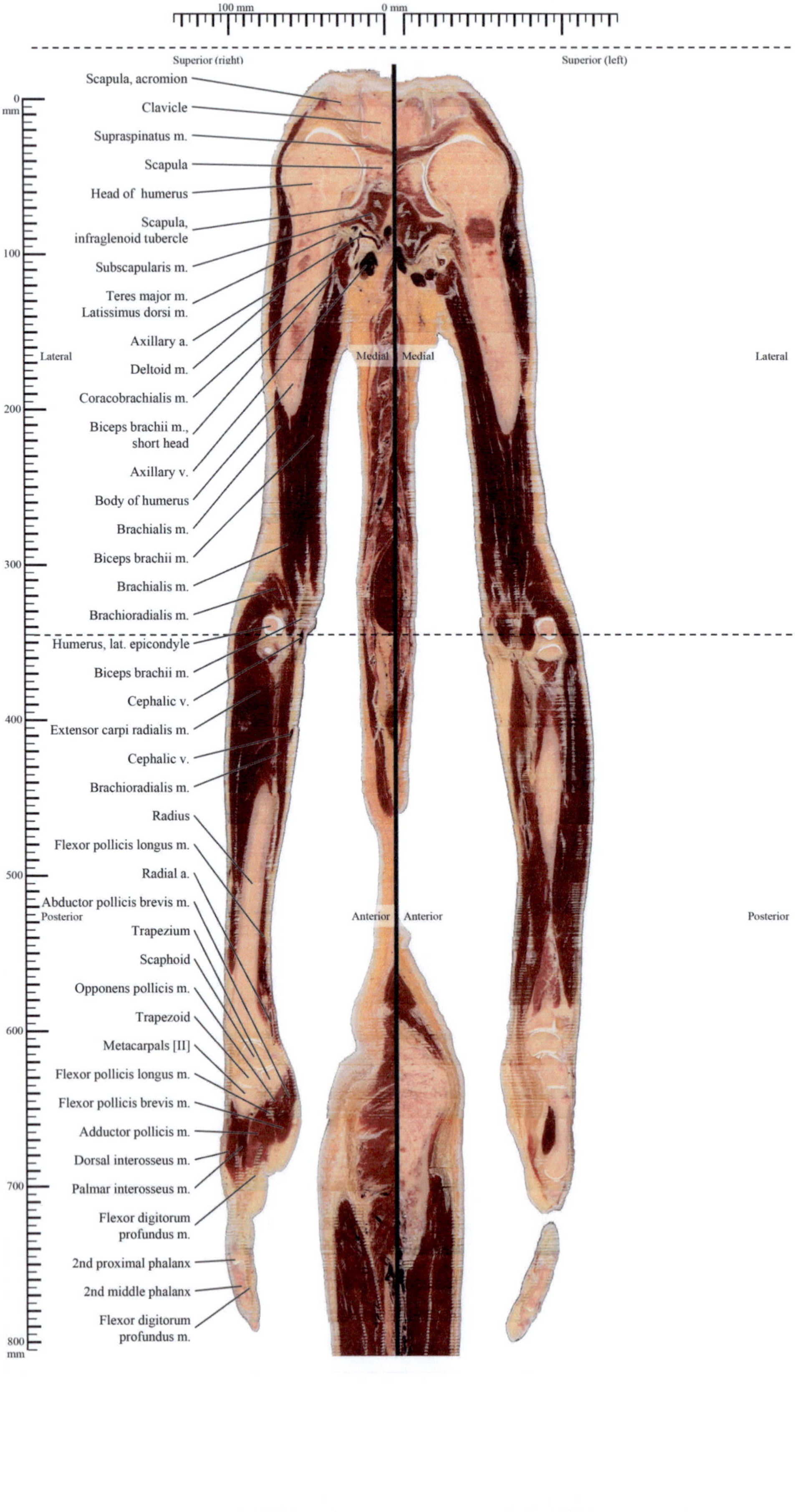

Coronal (arm) and sagittal (forearm and hand) (72.5 mm)

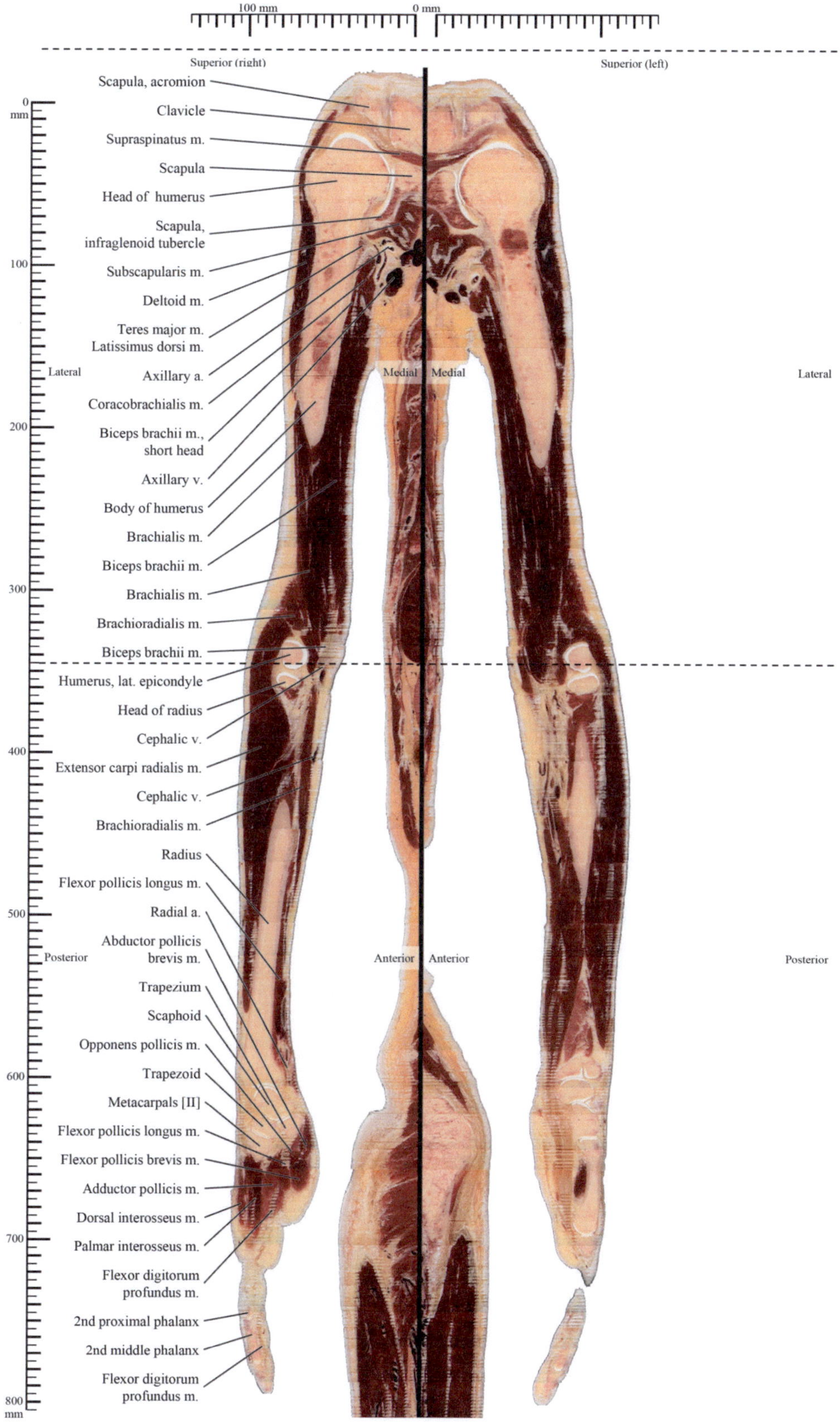

Coronal (arm) and sagittal (forearm and hand) (74.0 mm)

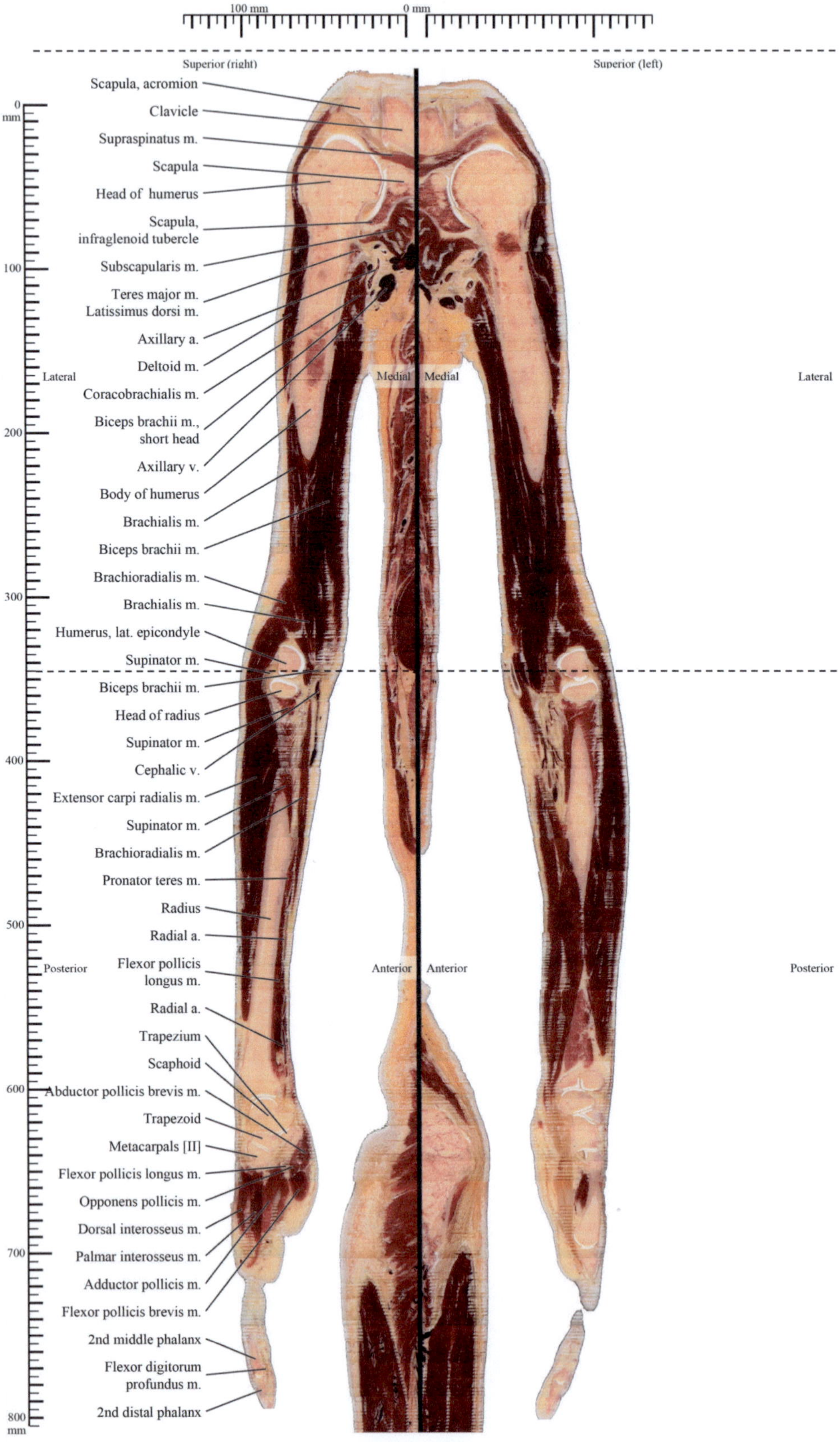

Coronal (arm) and sagittal (forearm and hand) (75.5 mm)

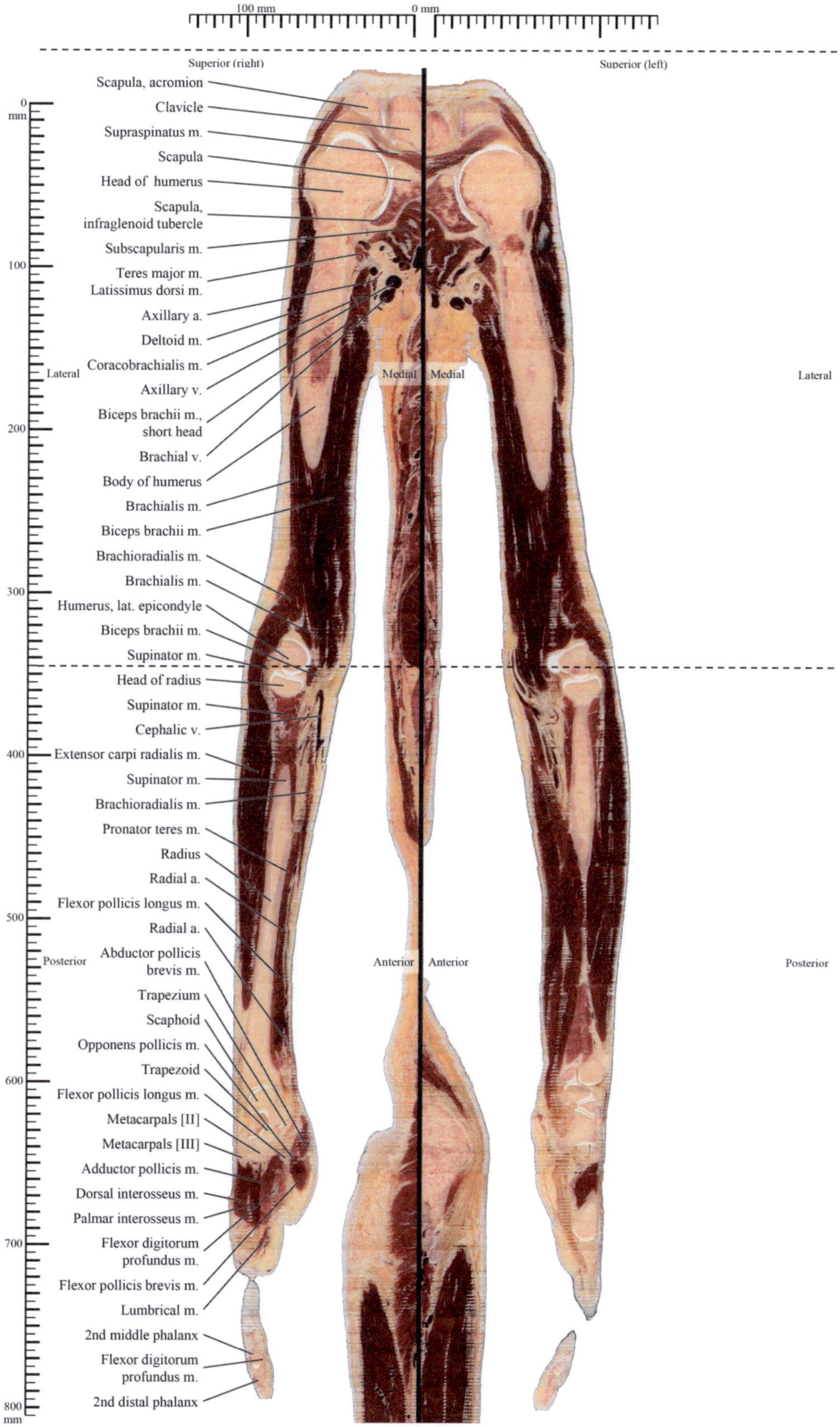

Coronal (arm) and sagittal (forearm and hand) (77.0 mm)

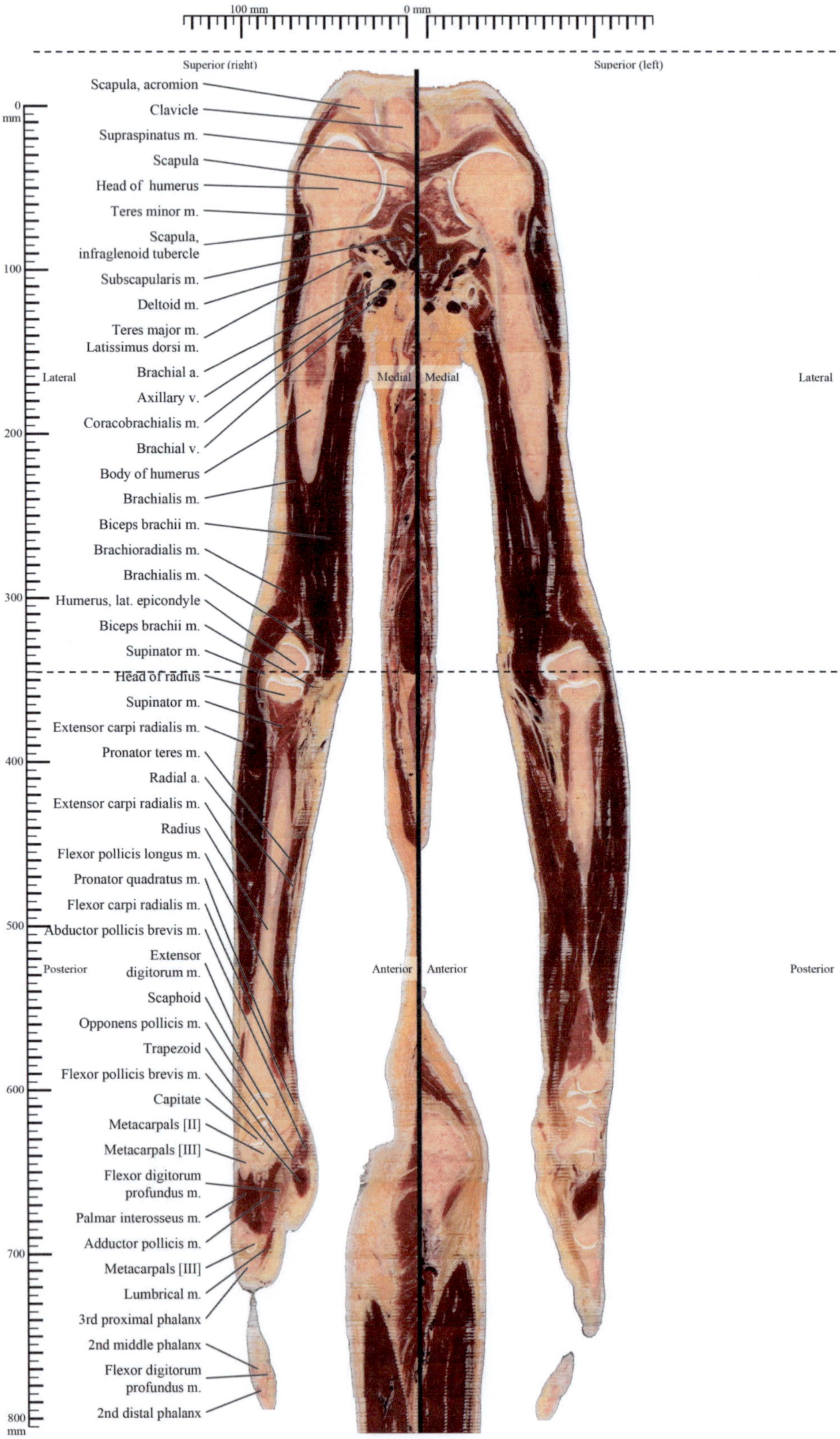
100 mm
0 mm
Superior (right)
Superior (left)
Scapula, acromion
Clavicle
Supraspinatus m.
Scapula
Head of humerus
Teres minor m.
Scapula, infraglenoid tubercle
Subscapularis m.
Deltoid m.
Teres major m.
Latissimus dorsi m.
Brachial a.
Axillary v.
Coracobrachialis m.
Brachial v.
Body of humerus
Brachialis m.
Biceps brachii m.
Brachioradialis m.
Brachialis m.
Humerus, lat. epicondyle
Biceps brachii m.
Supinator m.
Head of radius
Supinator m.
Extensor carpi radialis m.
Pronator teres m.
Radial a.
Extensor carpi radialis m.
Radius
Flexor pollicis longus m.
Pronator quadratus m.
Flexor carpi radialis m.
Abductor pollicis brevis m.
Extensor digitorum m.
Scaphoid
Opponens pollicis m.
Trapezoid
Flexor pollicis brevis m.
Capitate
Metacarpals [II]
Metacarpals [III]
Flexor digitorum profundus m.
Palmar interosseus m.
Adductor pollicis m.
Metacarpals [III]
Lumbrical m.
3rd proximal phalanx
2nd middle phalanx
Flexor digitorum profundus m.
2nd distal phalanx
Lateral
Medial
Medial
Lateral
Posterior
Anterior
Anterior
Posterior
0 mm
100
200
300
400
500
600
700
800 mm

Coronal (arm) and sagittal (forearm and hand) (78.5 mm)

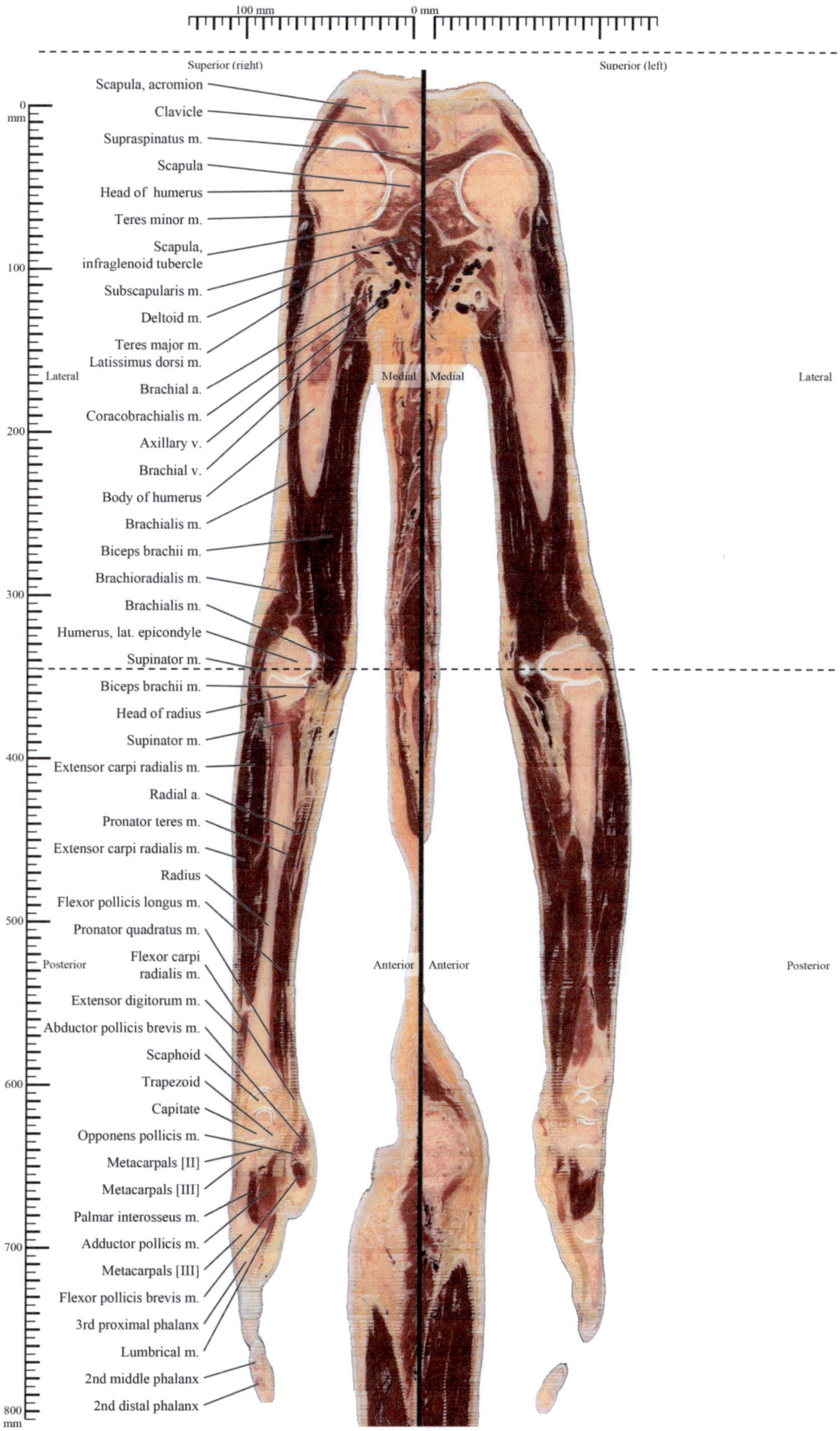

Coronal (arm) and sagittal (forearm and hand) (80.0 mm)

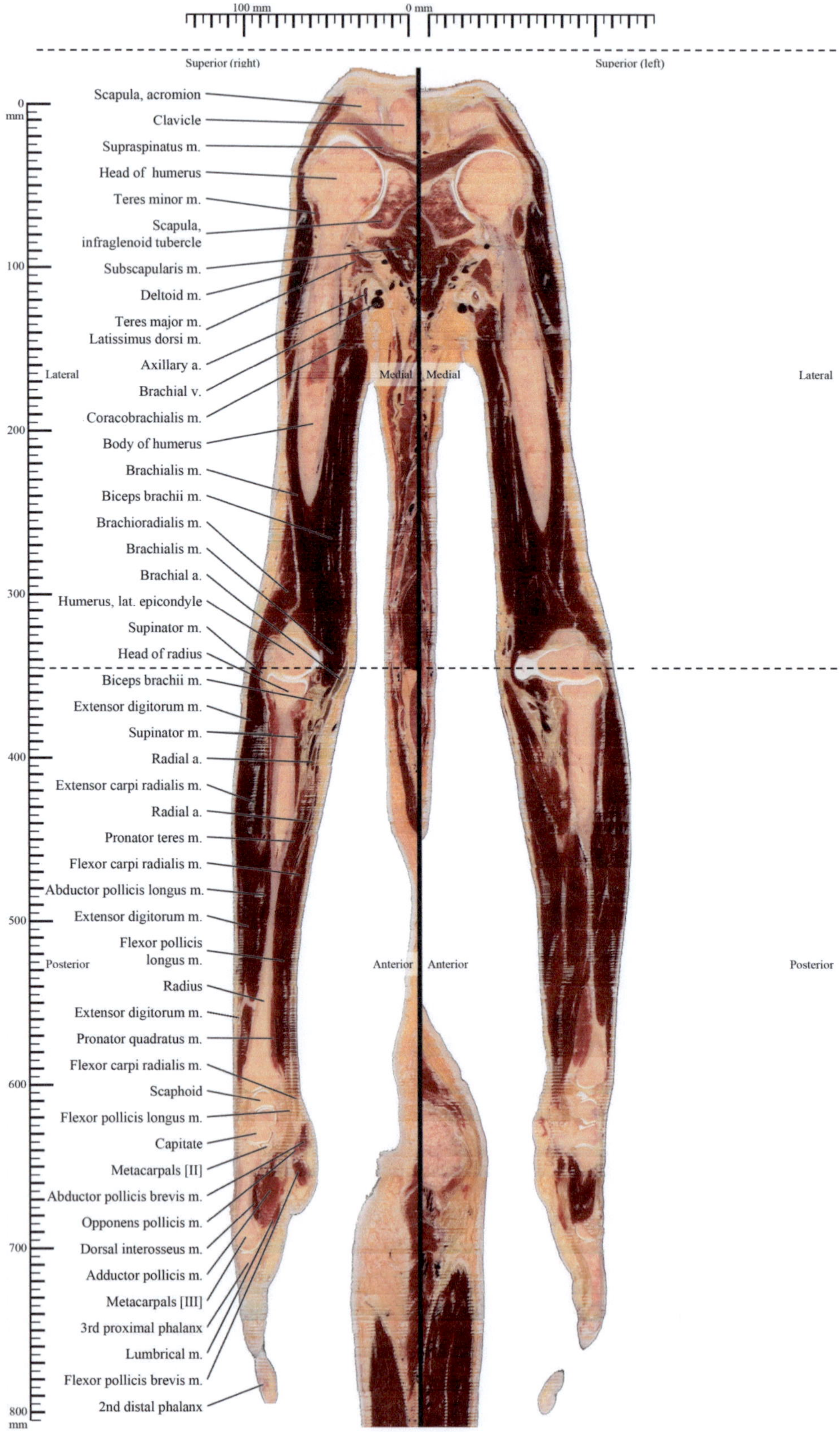

Coronal (arm) and sagittal (forearm and hand) (81.5 mm)

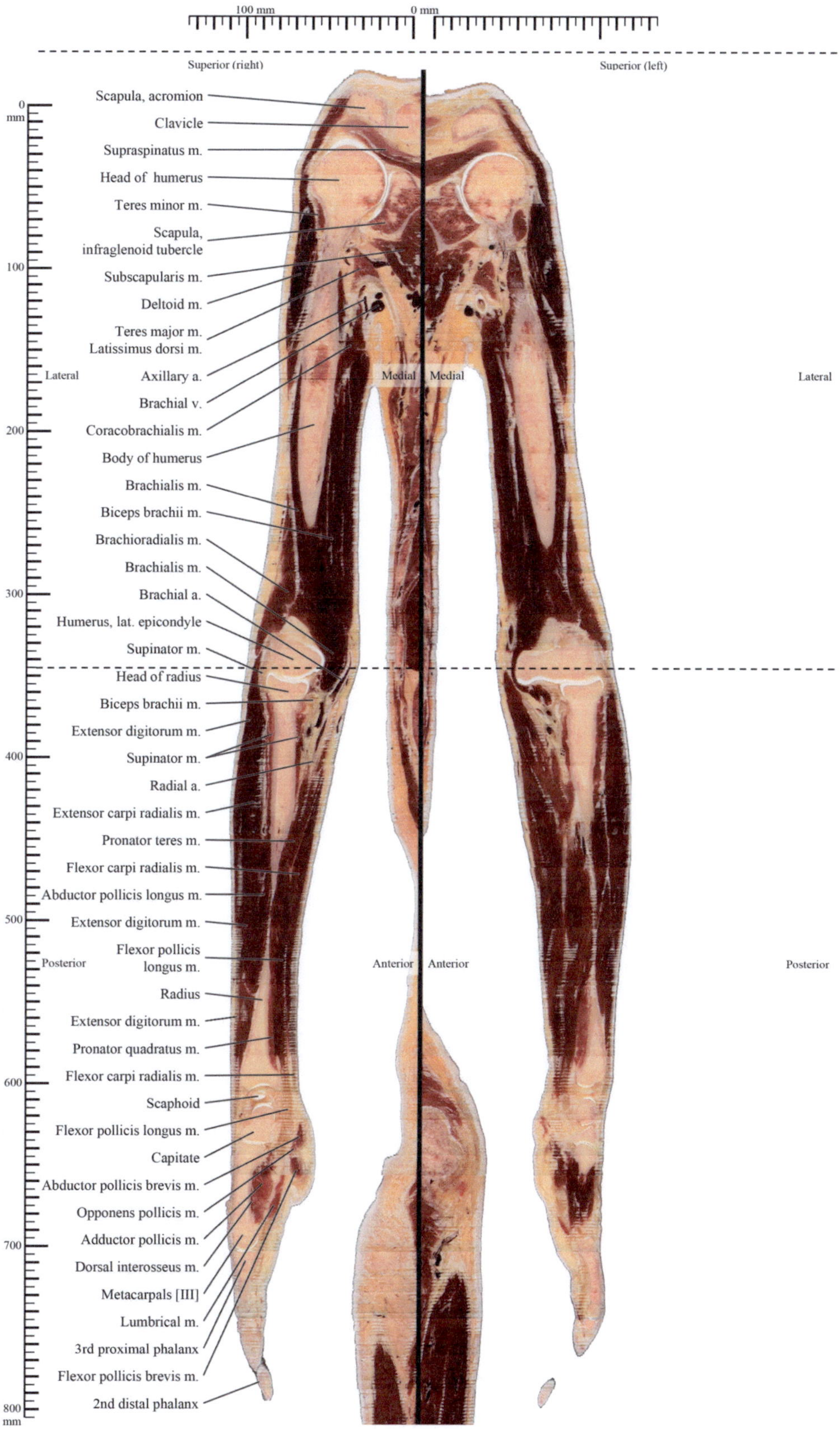

Coronal (arm) and sagittal (forearm and hand) (83.0 mm)

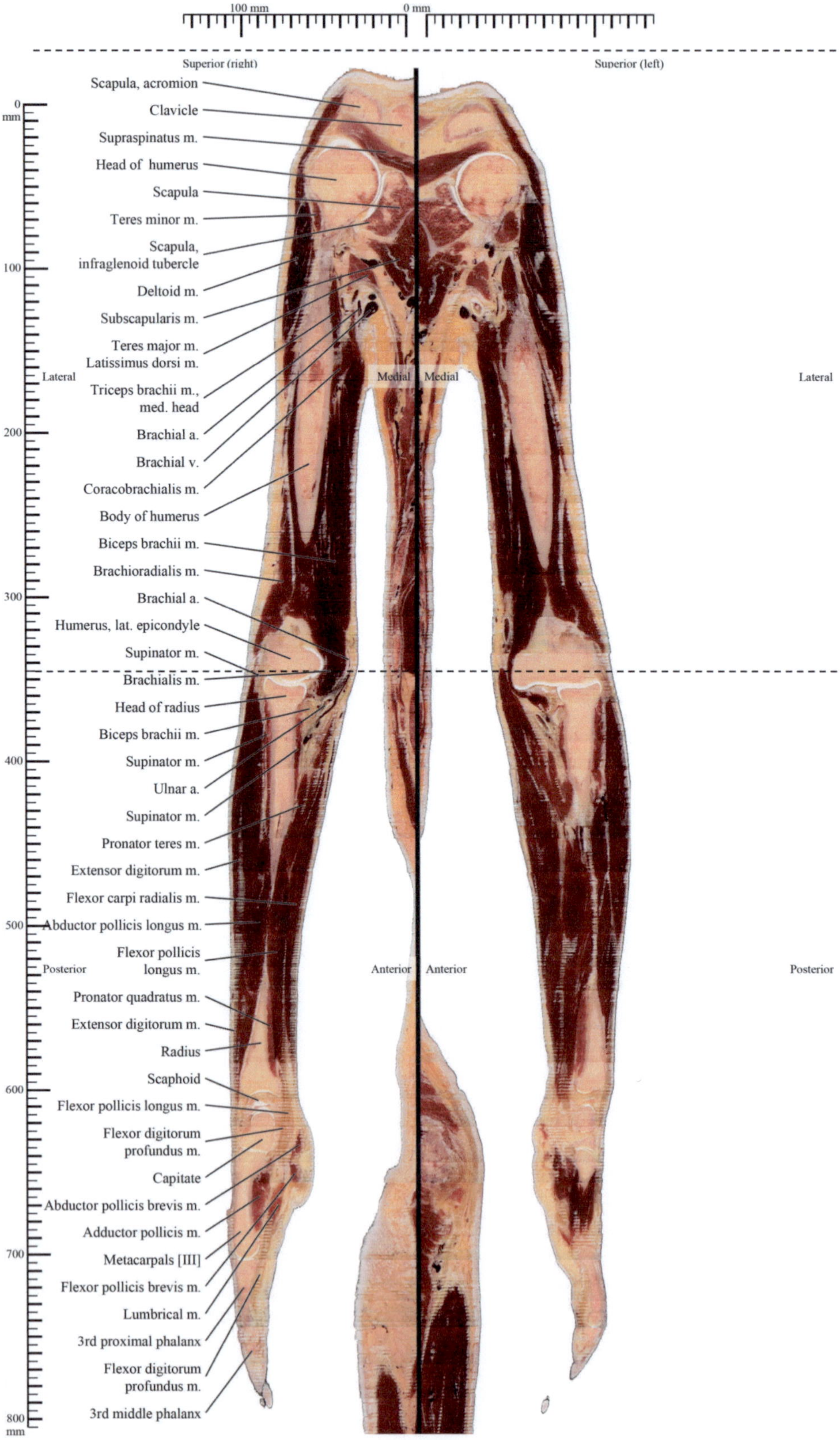

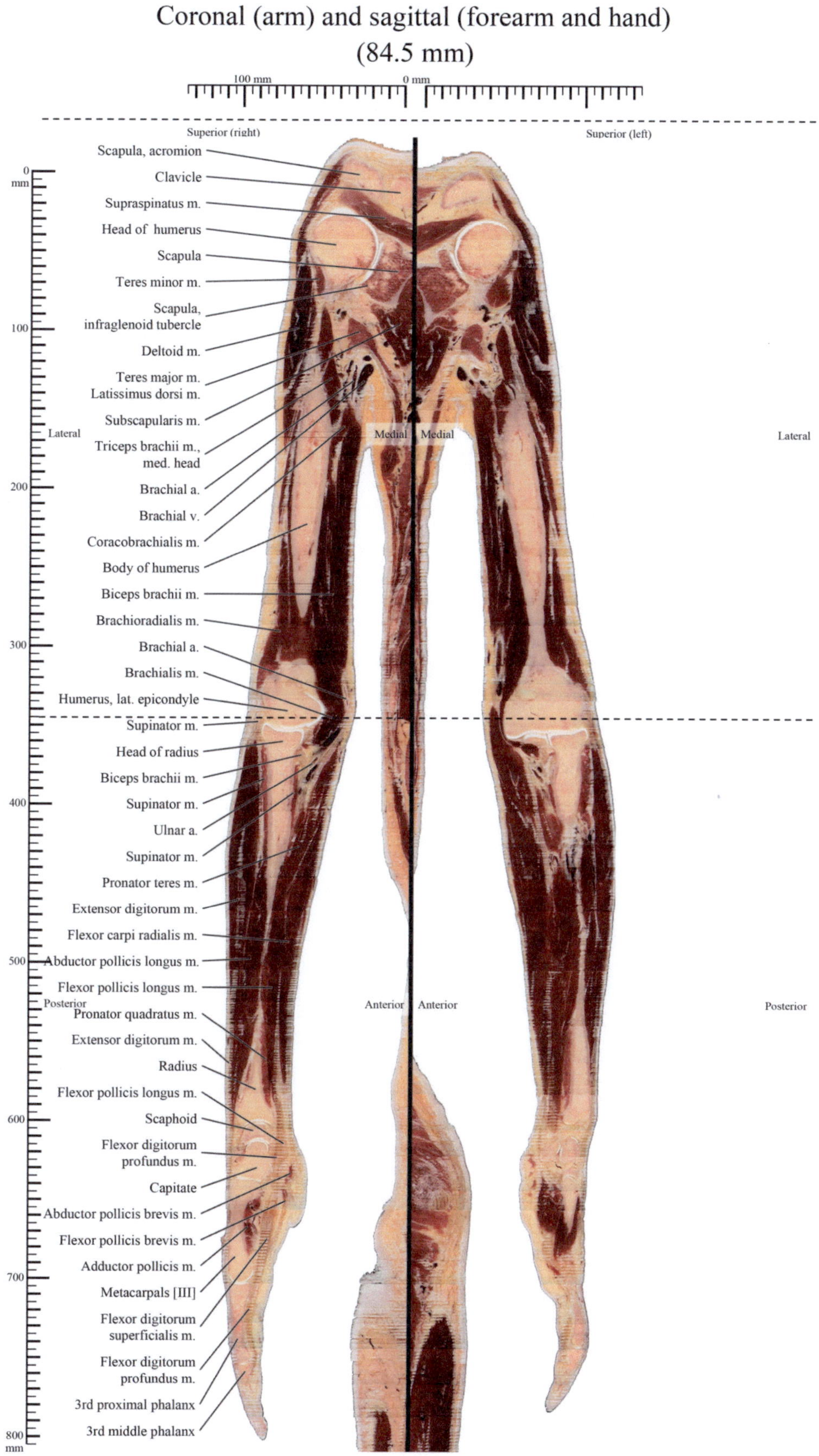
Coronal (arm) and sagittal (forearm and hand)
(84.5 mm)
100 mm
0 mm
Superior (right)
Superior (left)
Scapula, acromion
Clavicle
Supraspinatus m.
Head of humerus
Scapula
Teres minor m.
Scapula, infraglenoid tubercle
Deltoid m.
Teres major m.
Latissimus dorsi m.
Subscapularis m.
Triceps brachii m., med. head
Brachial a.
Brachial v.
Coracobrachialis m.
Body of humerus
Biceps brachii m.
Brachioradialis m.
Brachial a.
Brachialis m.
Humerus, lat. epicondyle
Supinator m.
Head of radius
Biceps brachii m.
Supinator m.
Ulnar a.
Supinator m.
Pronator teres m.
Extensor digitorum m.
Flexor carpi radialis m.
Abductor pollicis longus m.
Flexor pollicis longus m.
Pronator quadratus m.
Extensor digitorum m.
Radius
Flexor pollicis longus m.
Scaphoid
Flexor digitorum profundus m.
Capitate
Abductor pollicis brevis m.
Flexor pollicis brevis m.
Adductor pollicis m.
Metacarpals [III]
Flexor digitorum superficialis m.
Flexor digitorum profundus m.
3rd proximal phalanx
3rd middle phalanx
Lateral
Medial
Medial
Lateral
Posterior
Anterior
Anterior
Posterior
0 mm
100
200
300
400
500
600
700
800 mm

Coronal (arm) and sagittal (forearm and hand) (86.0 mm)

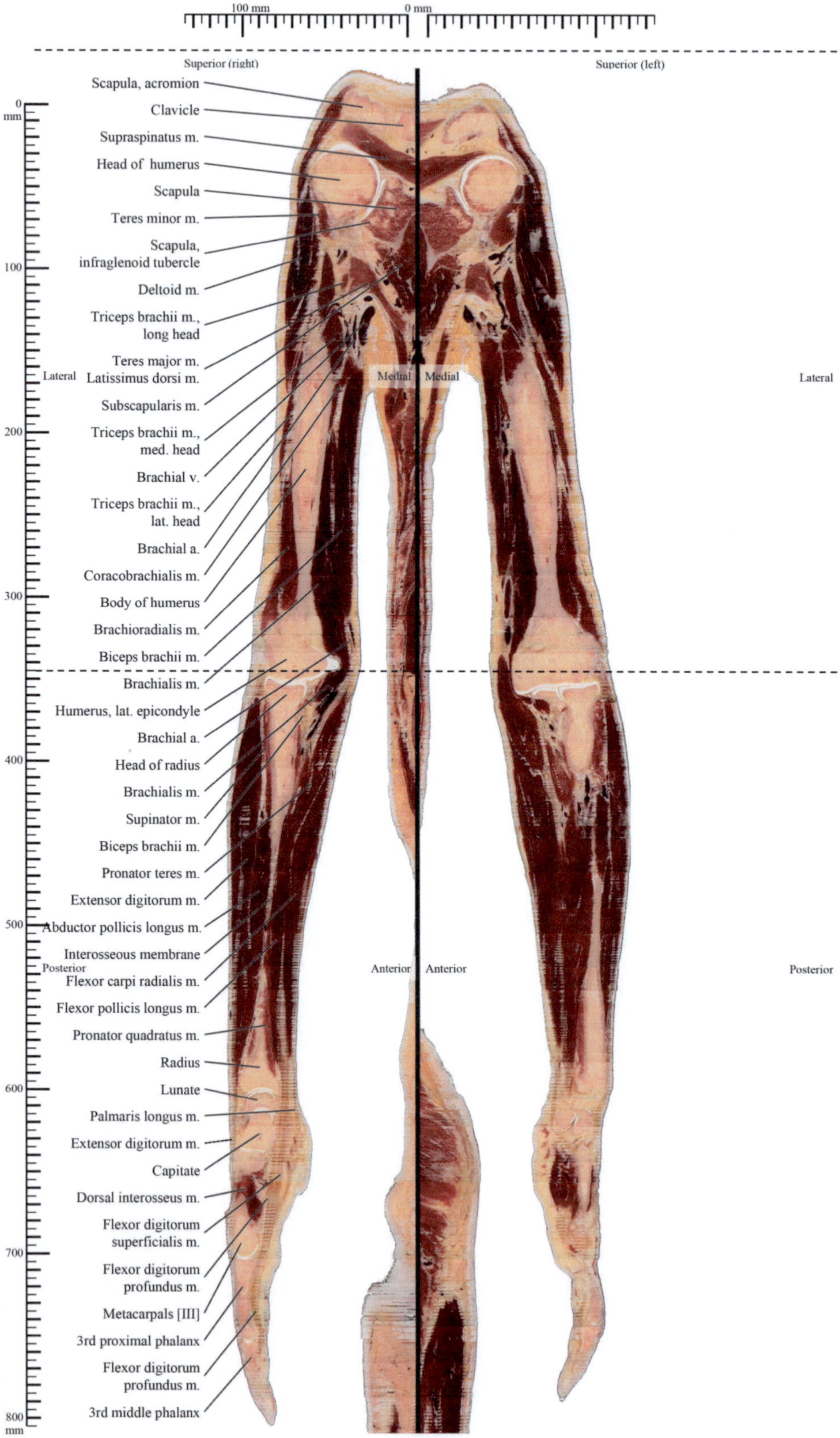

Coronal (arm) and sagittal (forearm and hand) (87.5 mm)

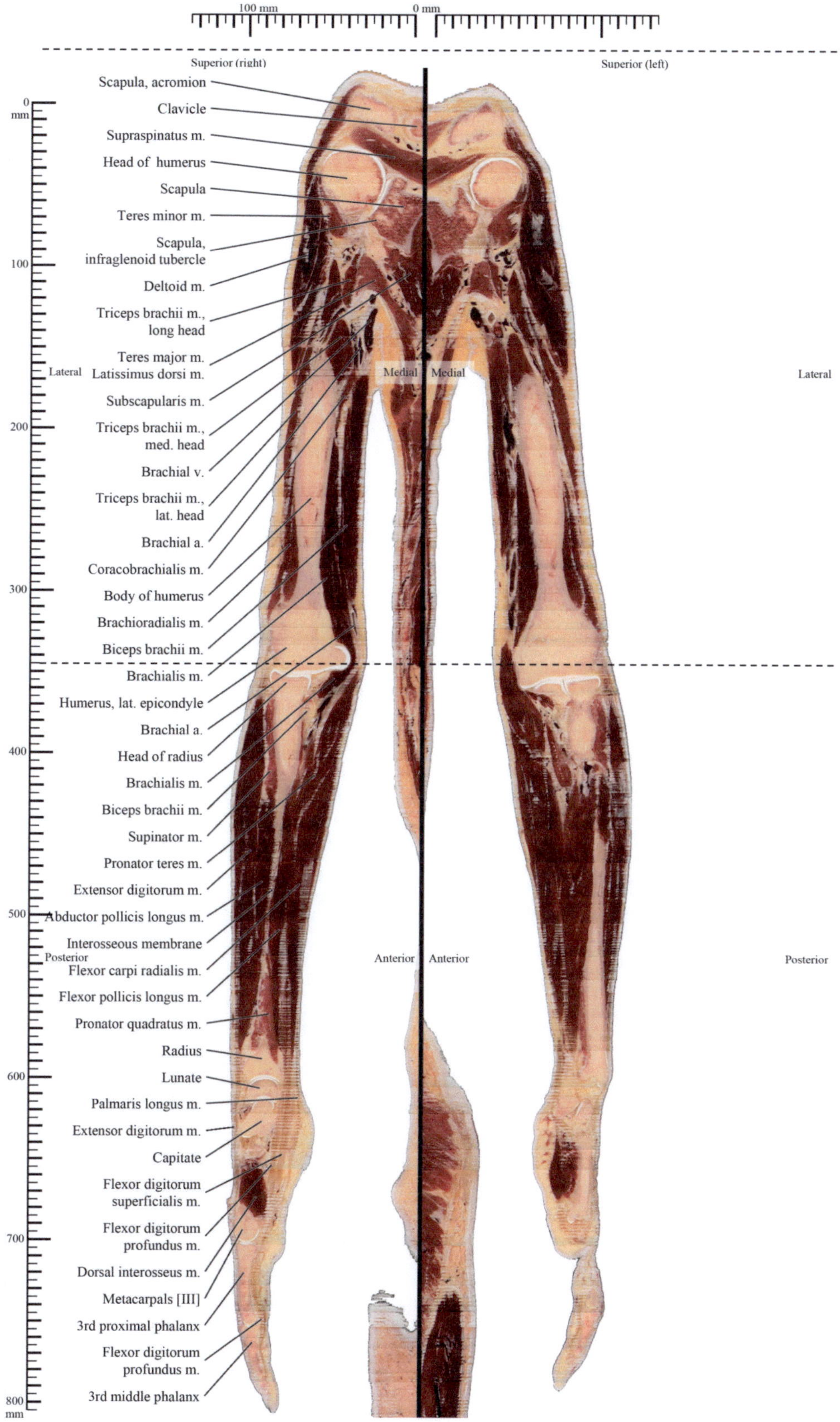

Coronal (arm) and sagittal (forearm and hand) (89.0 mm)

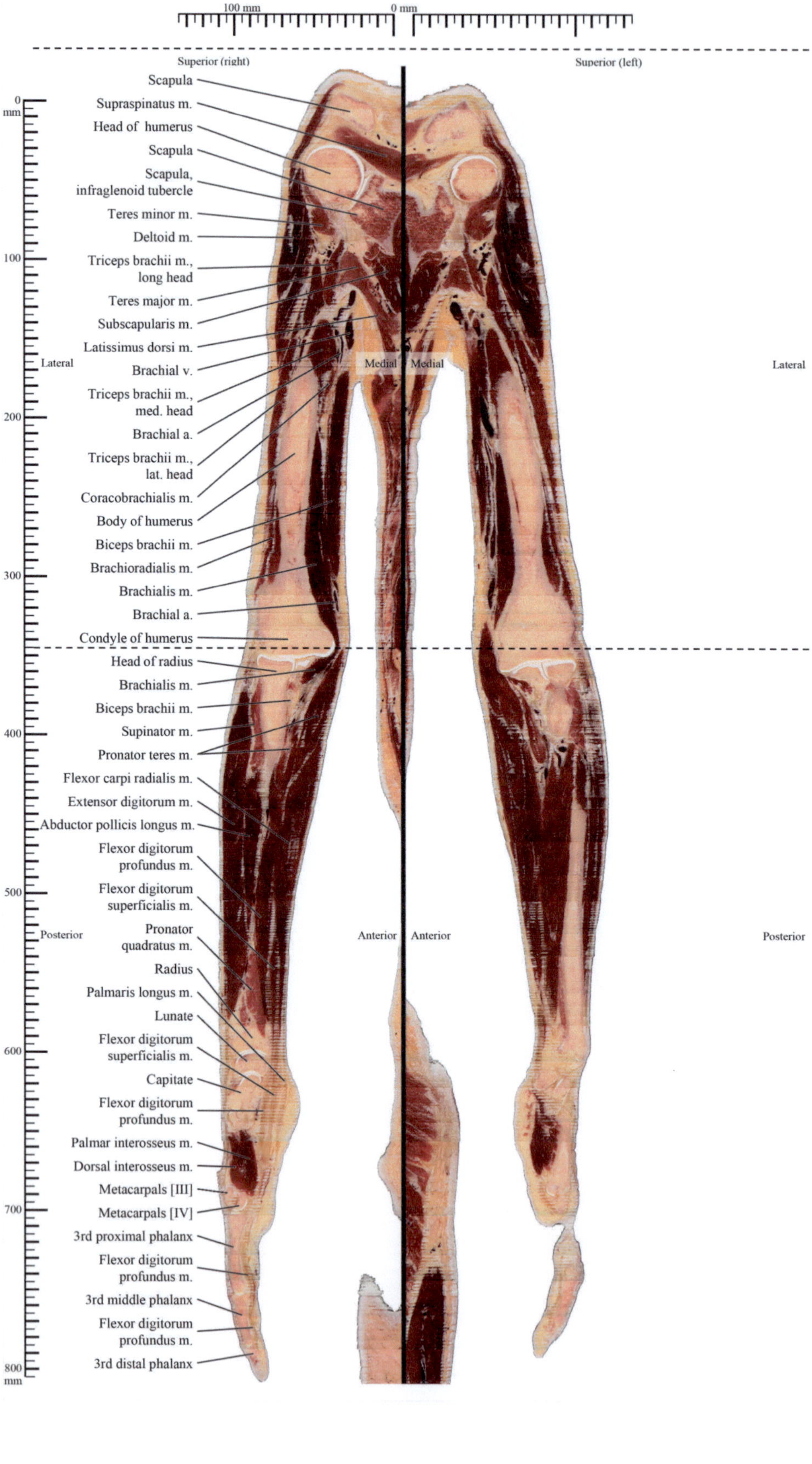

Coronal (arm) and sagittal (forearm and hand) (90.5 mm)

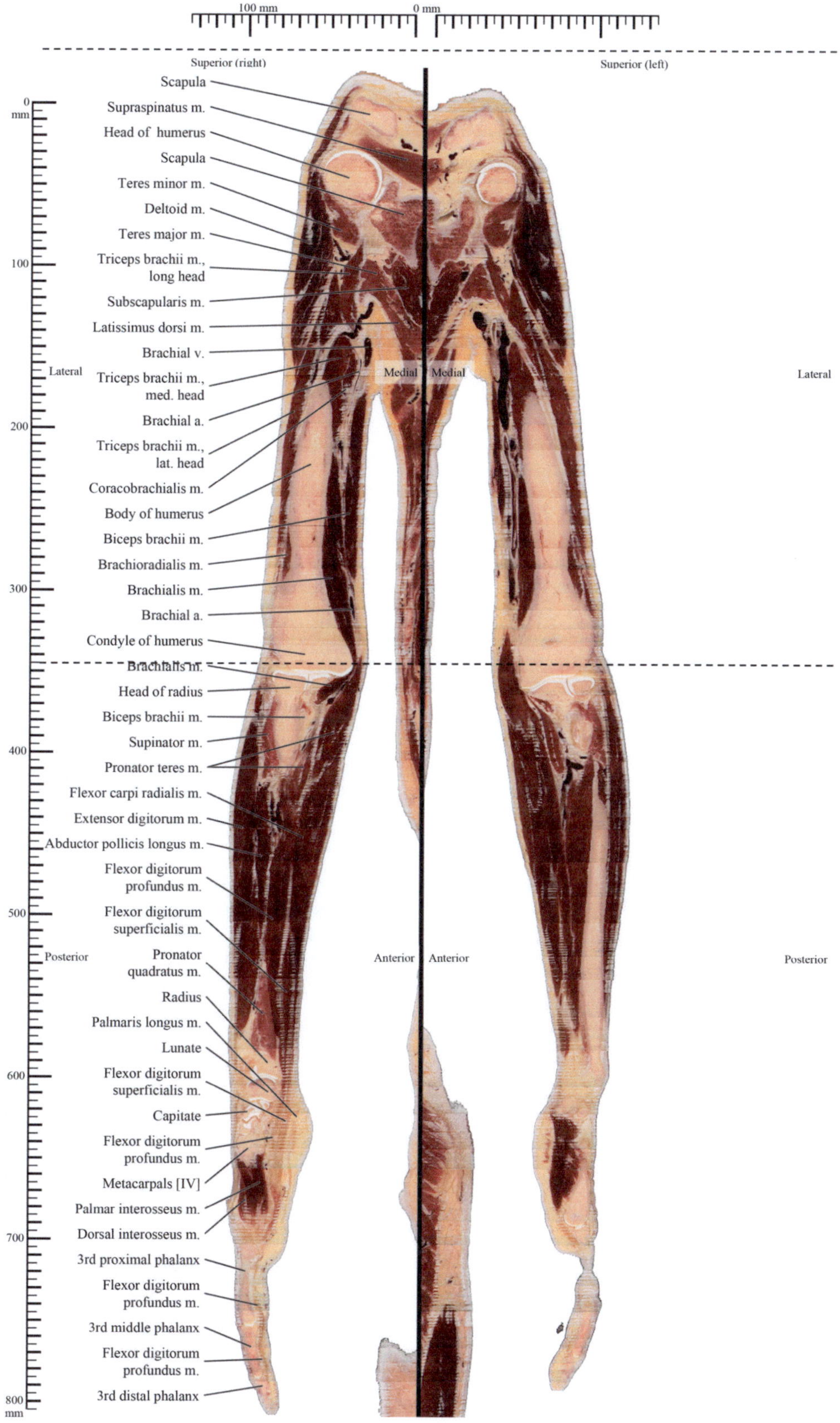

Coronal (arm) and sagittal (forearm and hand)
(92.0 mm)

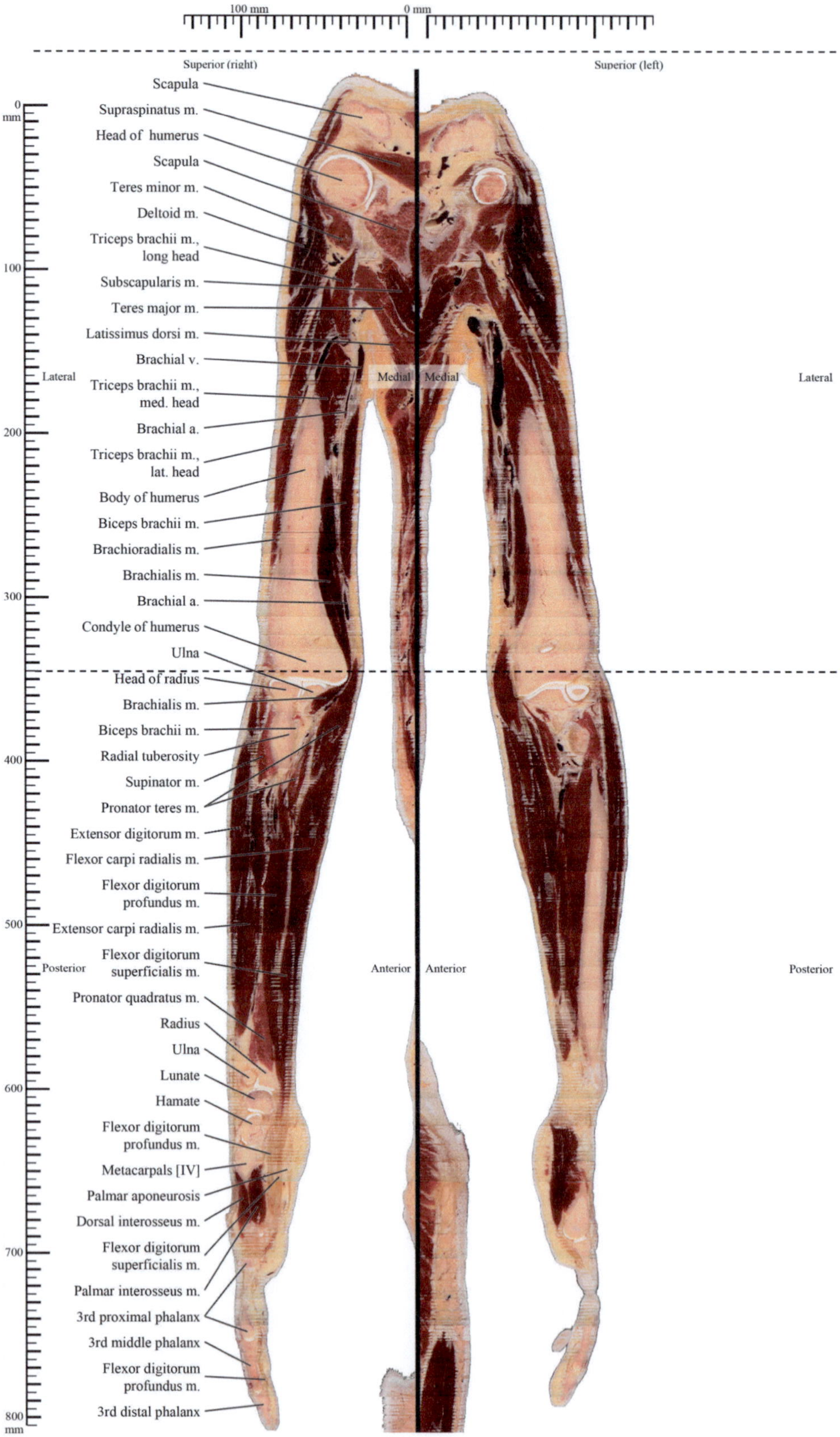

Coronal (arm) and sagittal (forearm and hand) (93.5 mm)

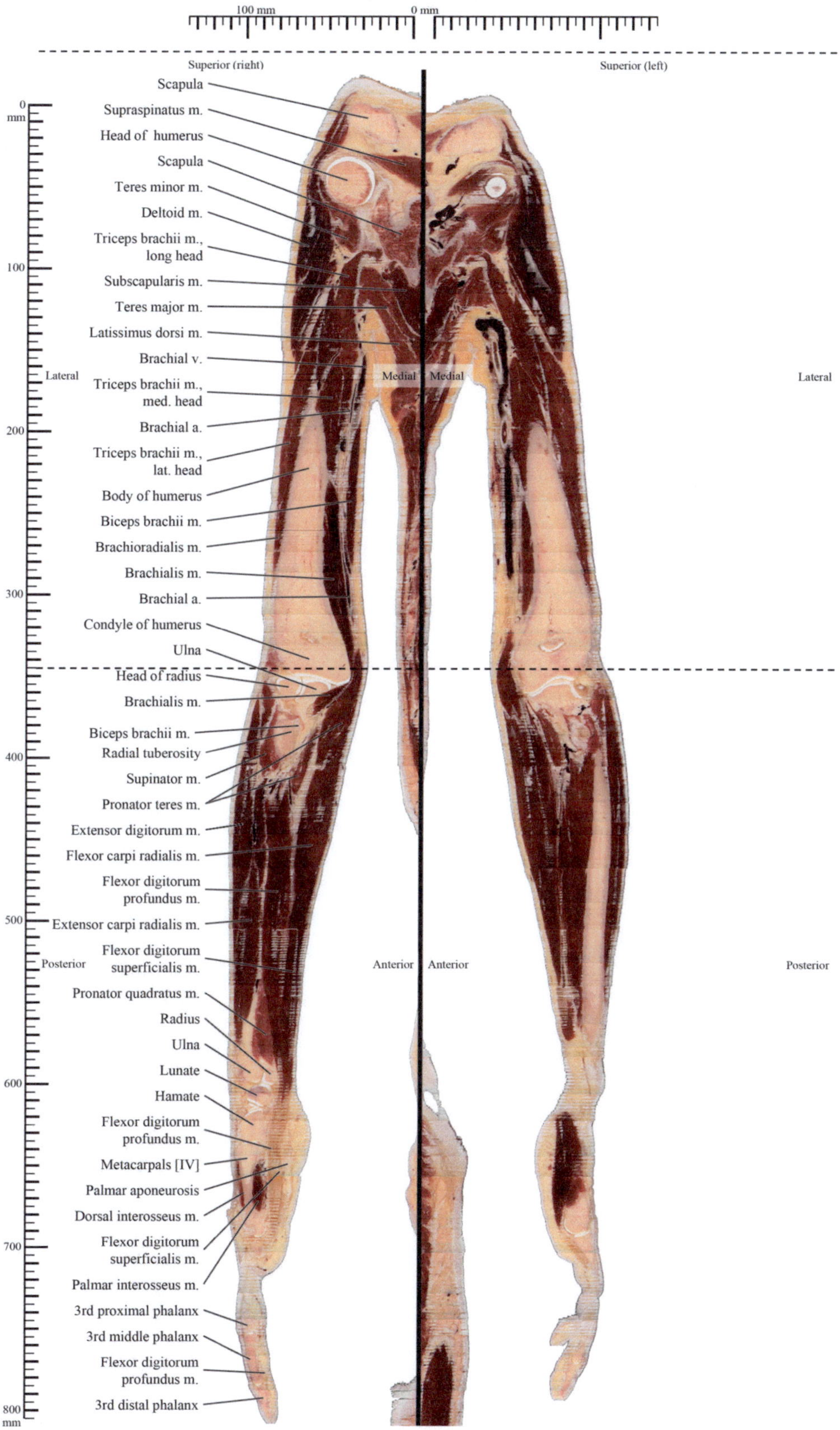

Coronal (arm) and sagittal (forearm and hand) (95.0 mm)

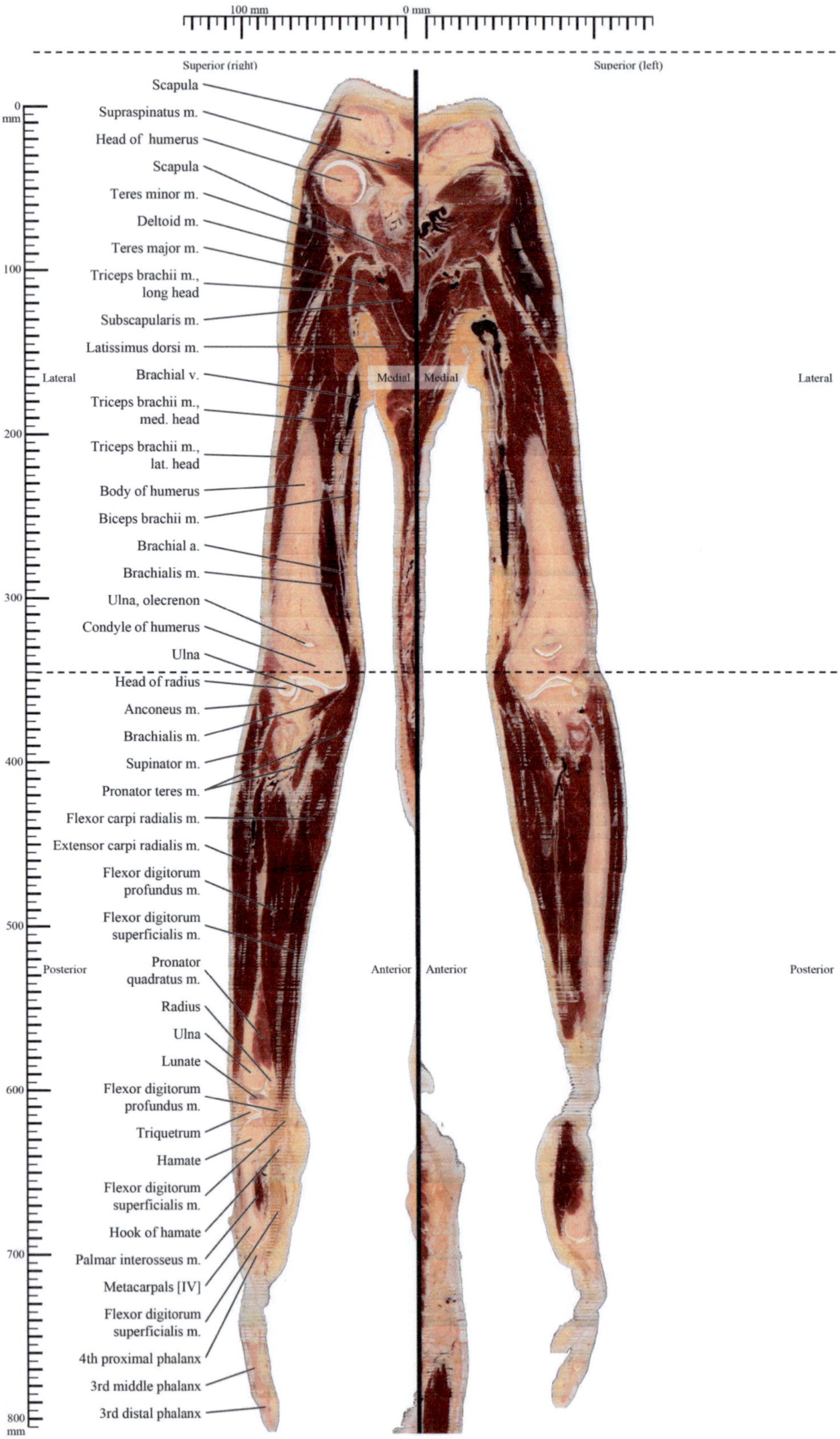

Coronal (arm) and sagittal (forearm and hand) (96.5 mm)

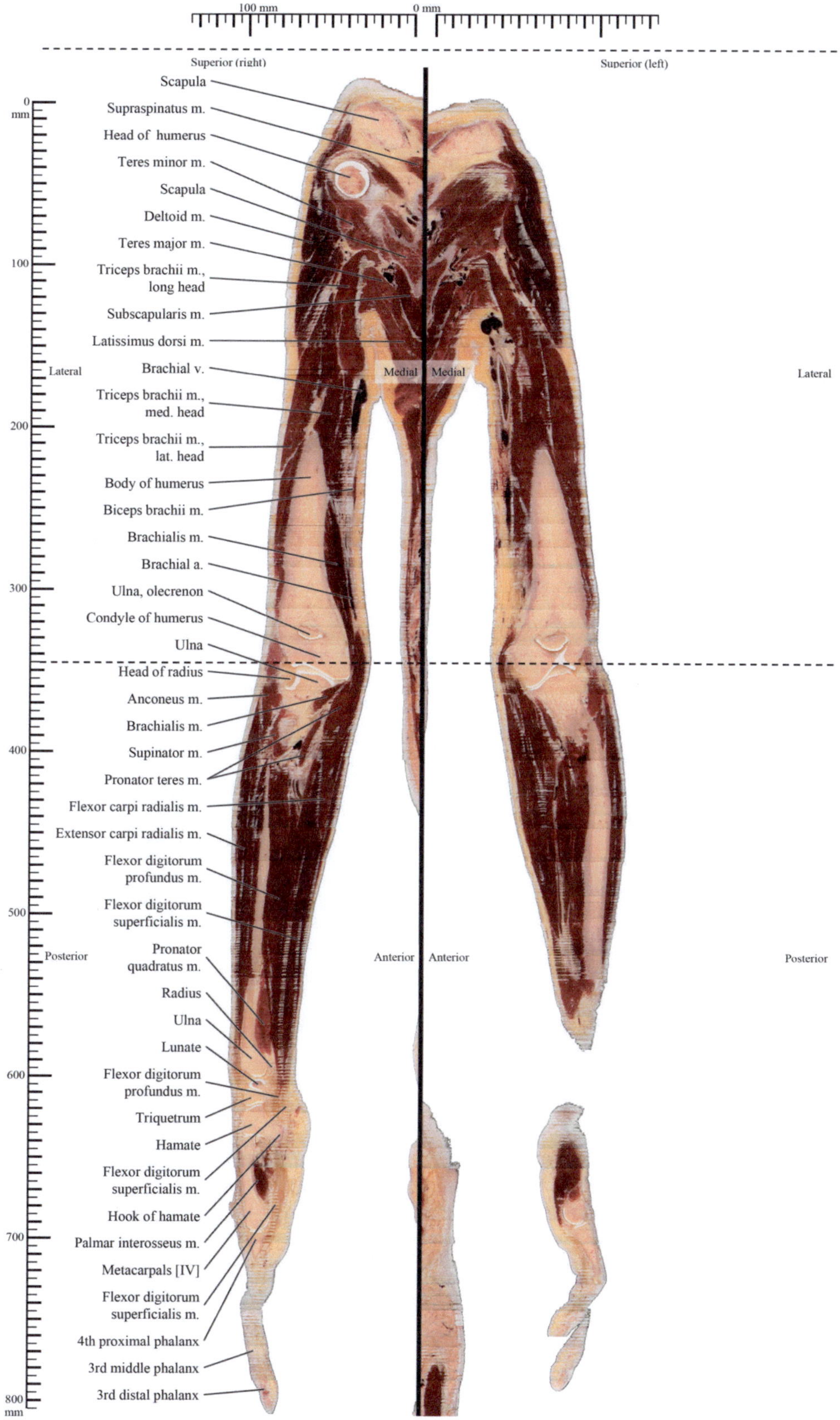

Coronal (arm) and sagittal (forearm and hand) (98.0 mm)

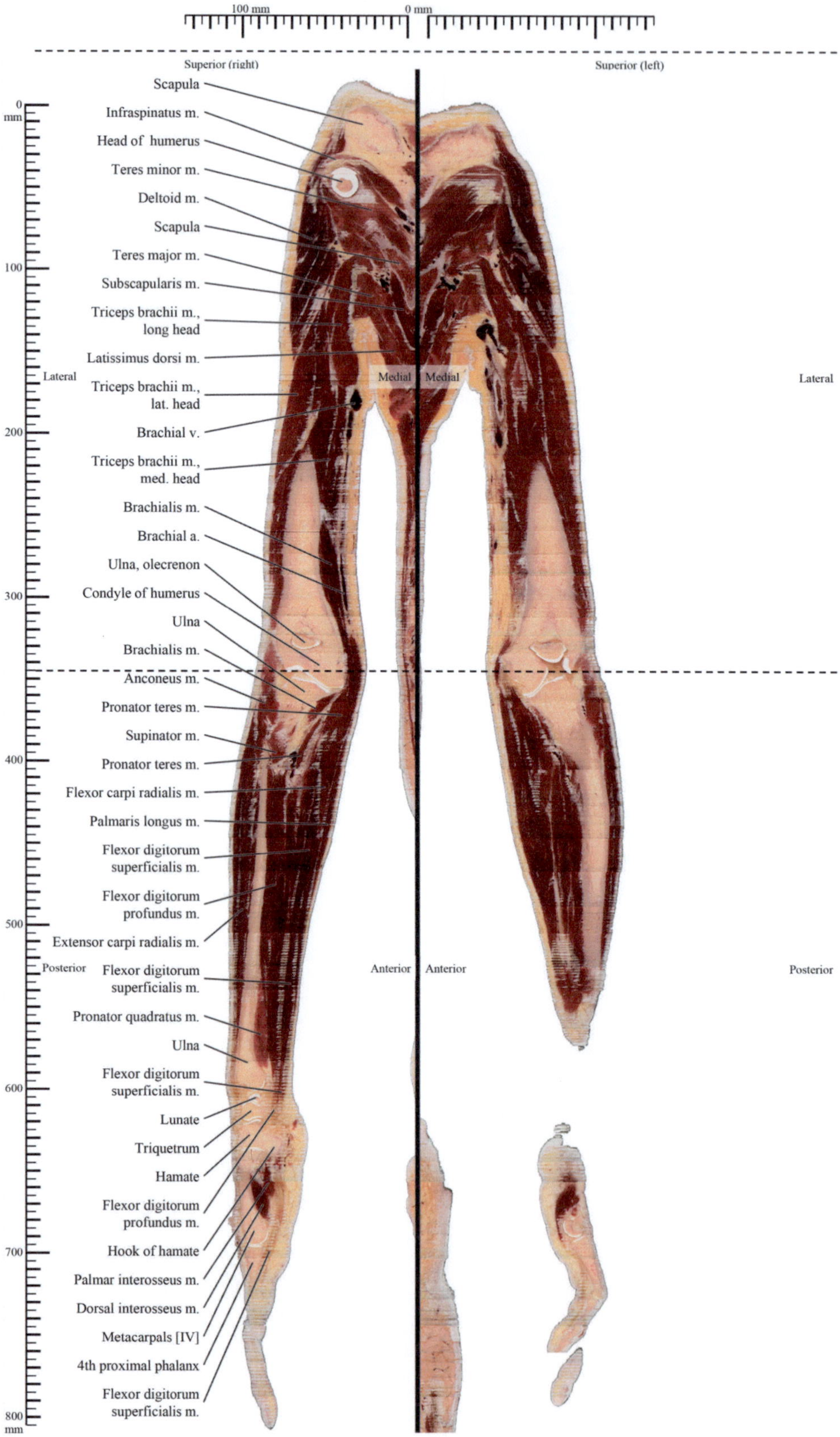

Coronal (arm) and sagittal (forearm and hand) (99.5 mm)

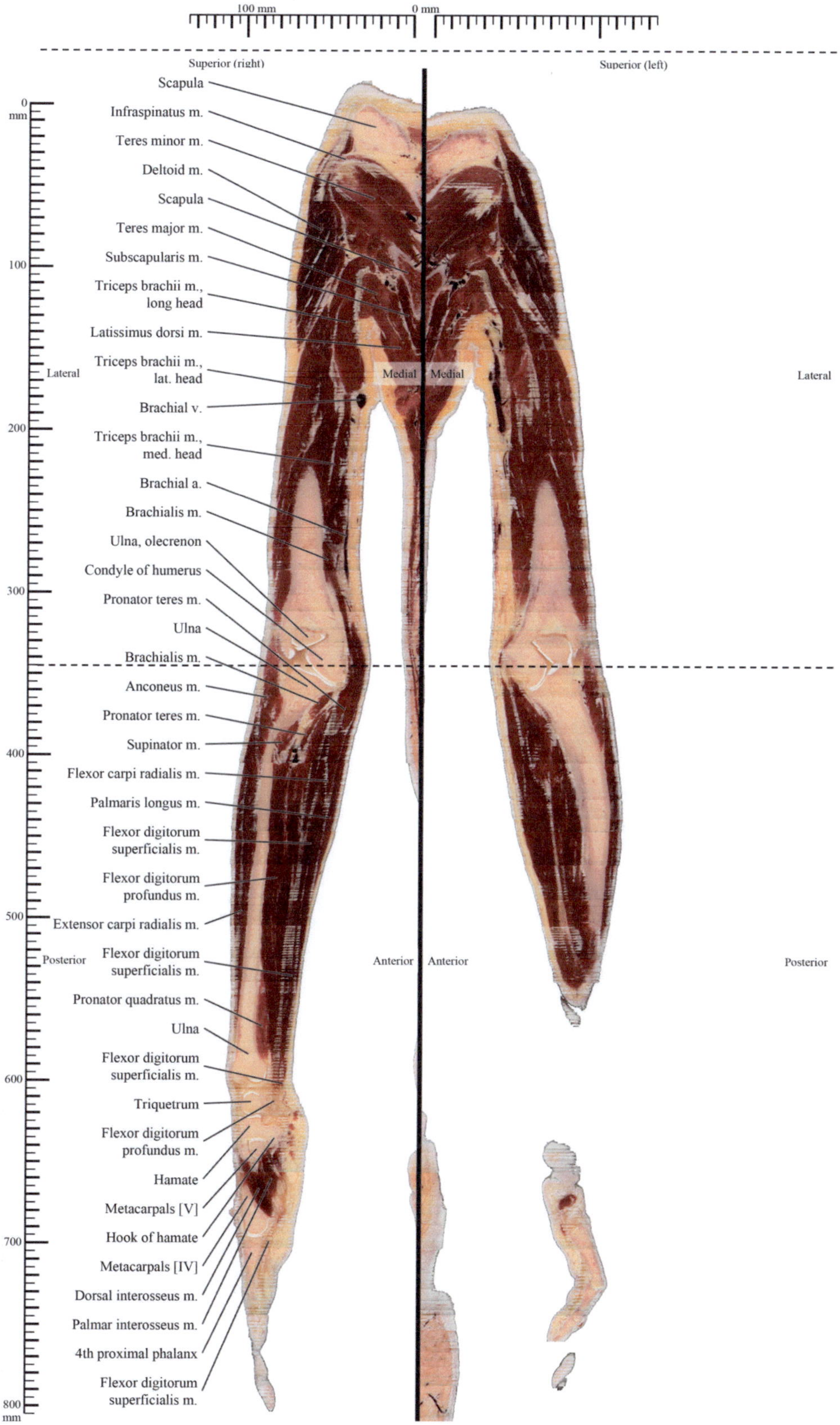

Coronal (arm) and sagittal (forearm and hand) (101.0 mm)

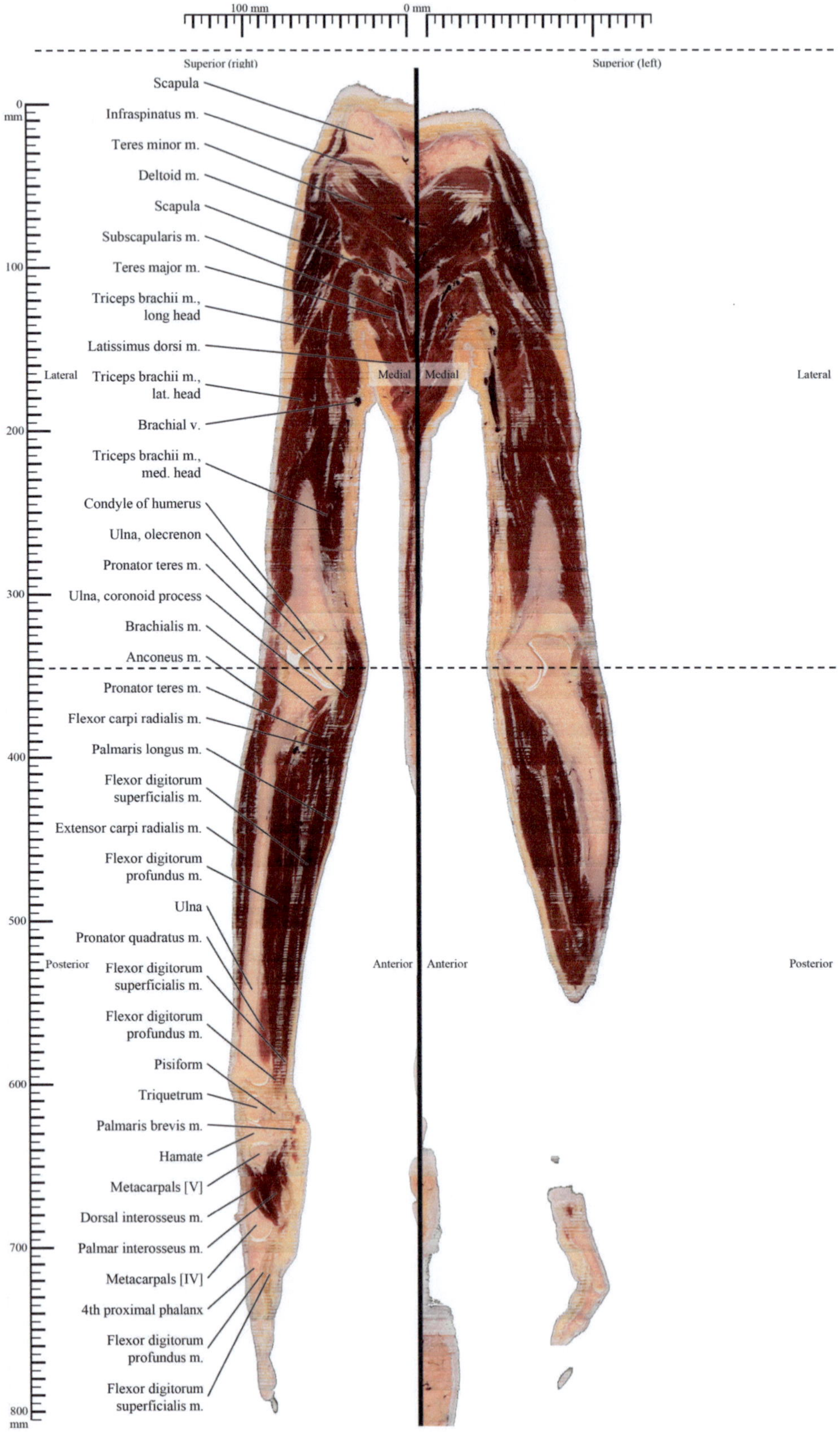

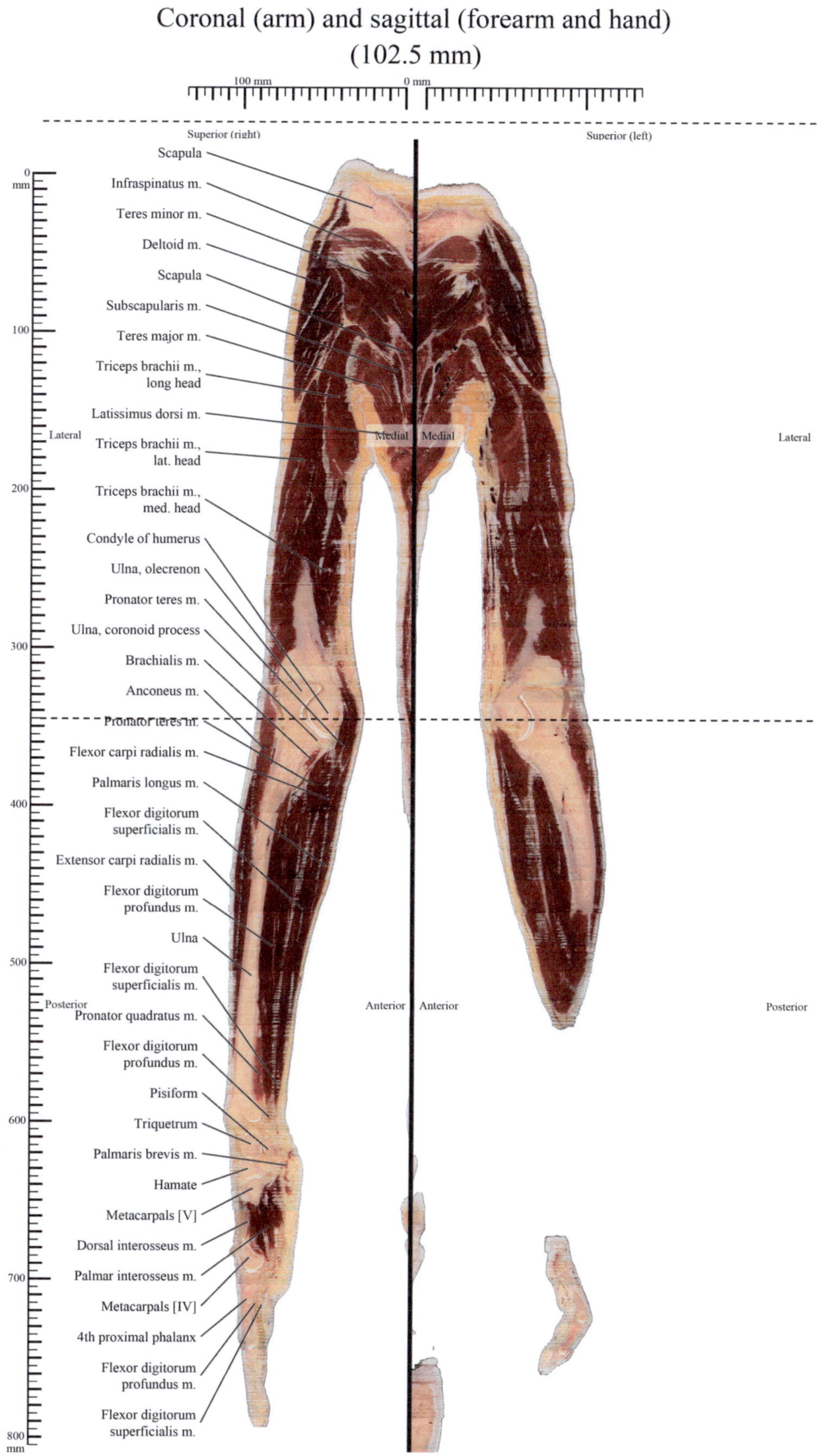
Coronal (arm) and sagittal (forearm and hand)
(102.5 mm)
100 mm
0 mm
Superior (right)
Superior (left)
Scapula
Infraspinatus m.
Teres minor m.
Deltoid m.
Scapula
Subscapularis m.
Teres major m.
Triceps brachii m., long head
Latissimus dorsi m.
Triceps brachii m., lat. head
Triceps brachii m., med. head
Condyle of humerus
Ulna, olecrenon
Pronator teres m.
Ulna, coronoid process
Brachialis m.
Anconeus m.
Pronator teres m.
Flexor carpi radialis m.
Palmaris longus m.
Flexor digitorum superficialis m.
Extensor carpi radialis m.
Flexor digitorum profundus m.
Ulna
Flexor digitorum superficialis m.
Pronator quadratus m.
Flexor digitorum profundus m.
Pisiform
Triquetrum
Palmaris brevis m.
Hamate
Metacarpals [V]
Dorsal interosseus m.
Palmar interosseus m.
Metacarpals [IV]
4th proximal phalanx
Flexor digitorum profundus m.
Flexor digitorum superficialis m.
Lateral
Medial
Medial
Lateral
Posterior
Anterior
Anterior
Posterior
0 mm
100
200
300
400
500
600
700
800 mm

Coronal (arm) and sagittal (forearm and hand) (104.0 mm)

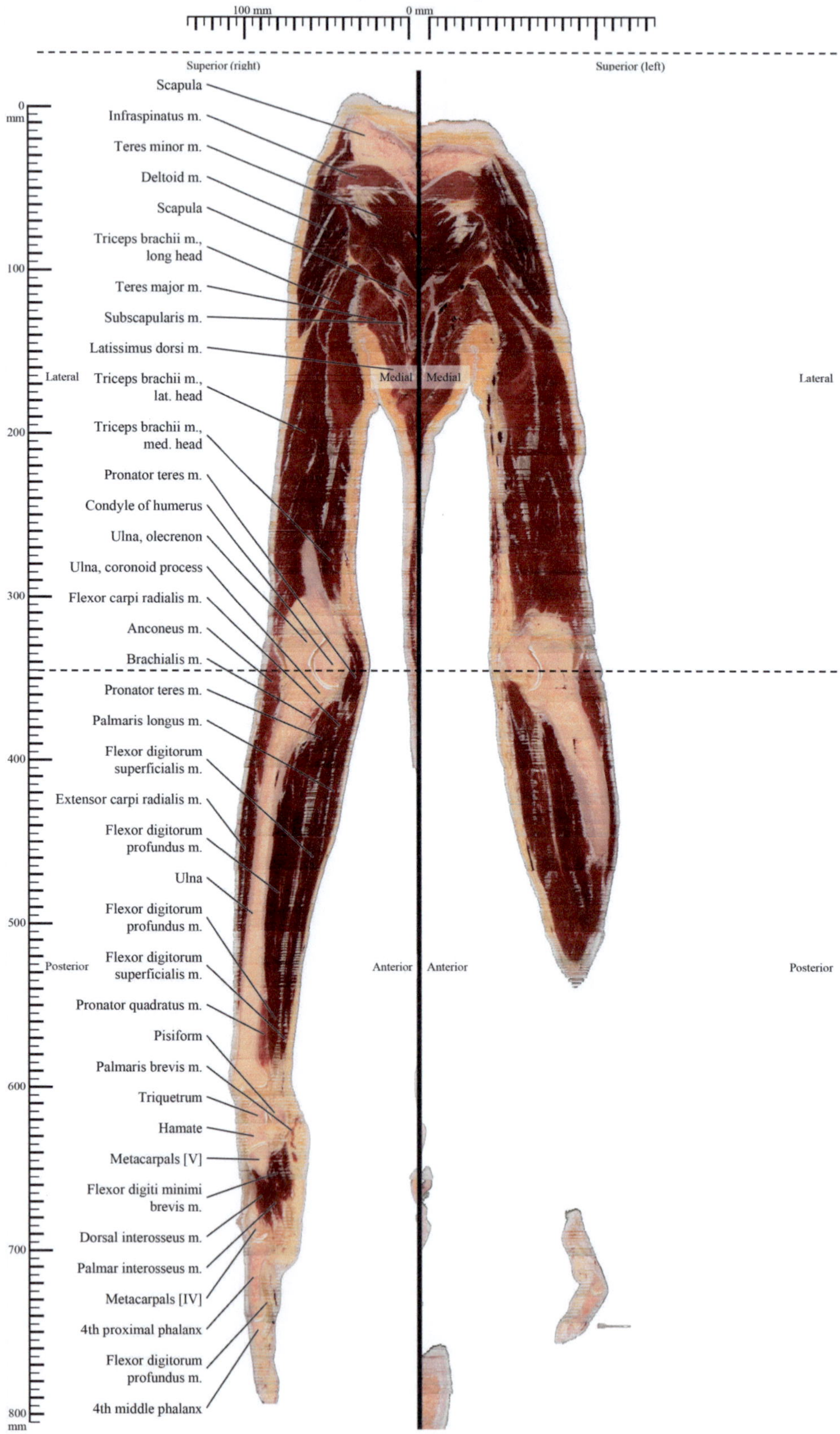

Coronal (arm) and sagittal (forearm and hand) (105.5 mm)

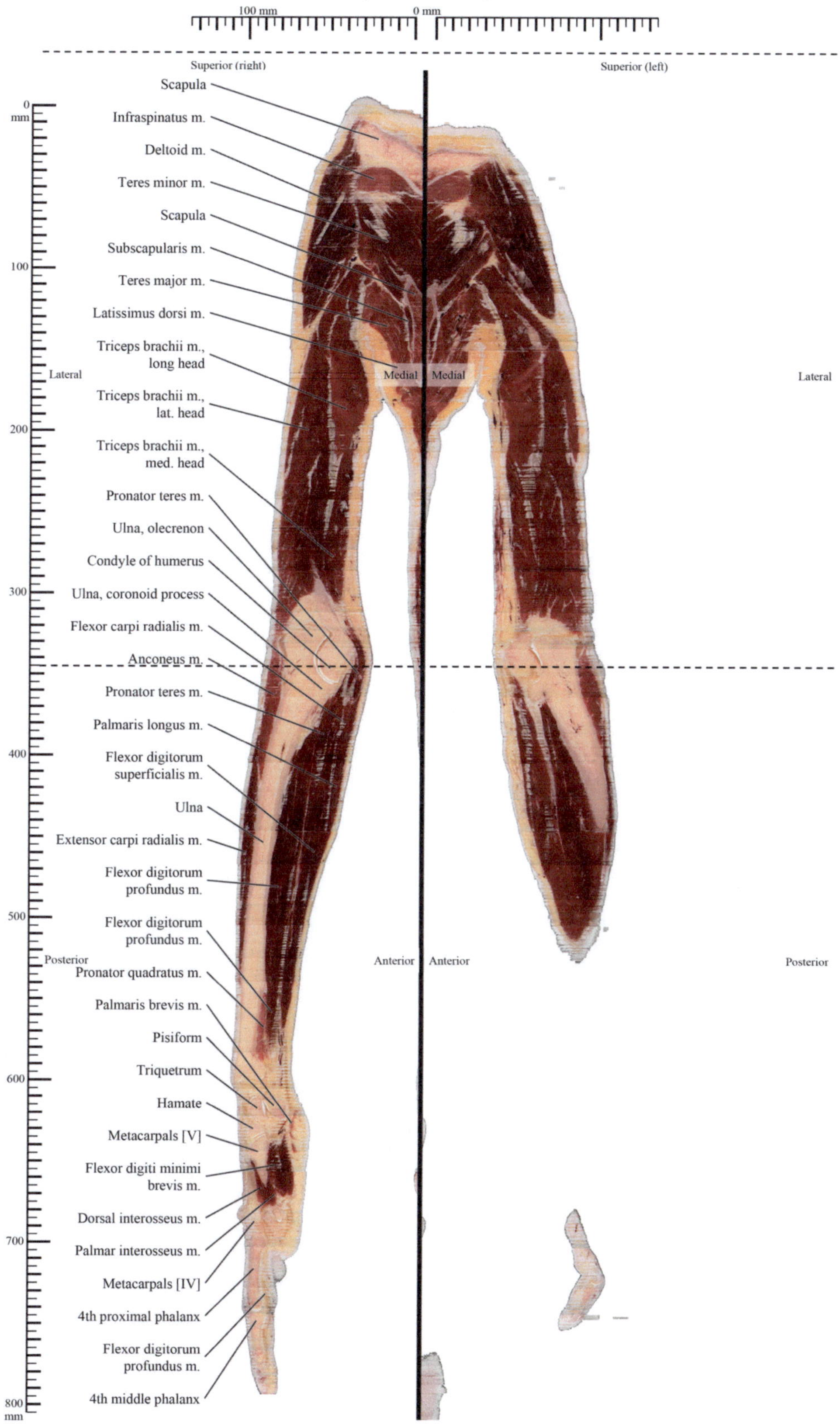

Coronal (arm) and sagittal (forearm and hand) (107.0 mm)

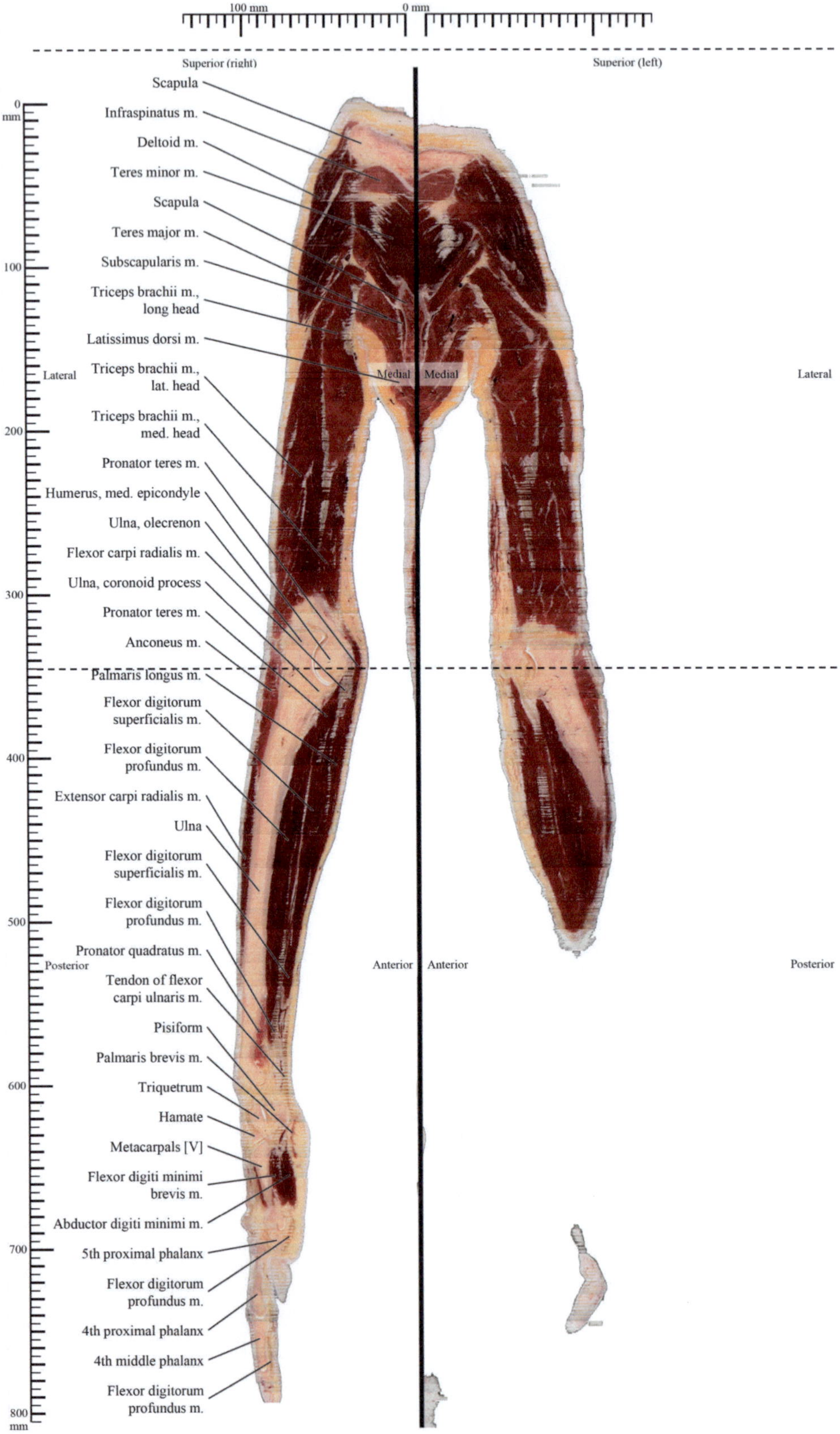

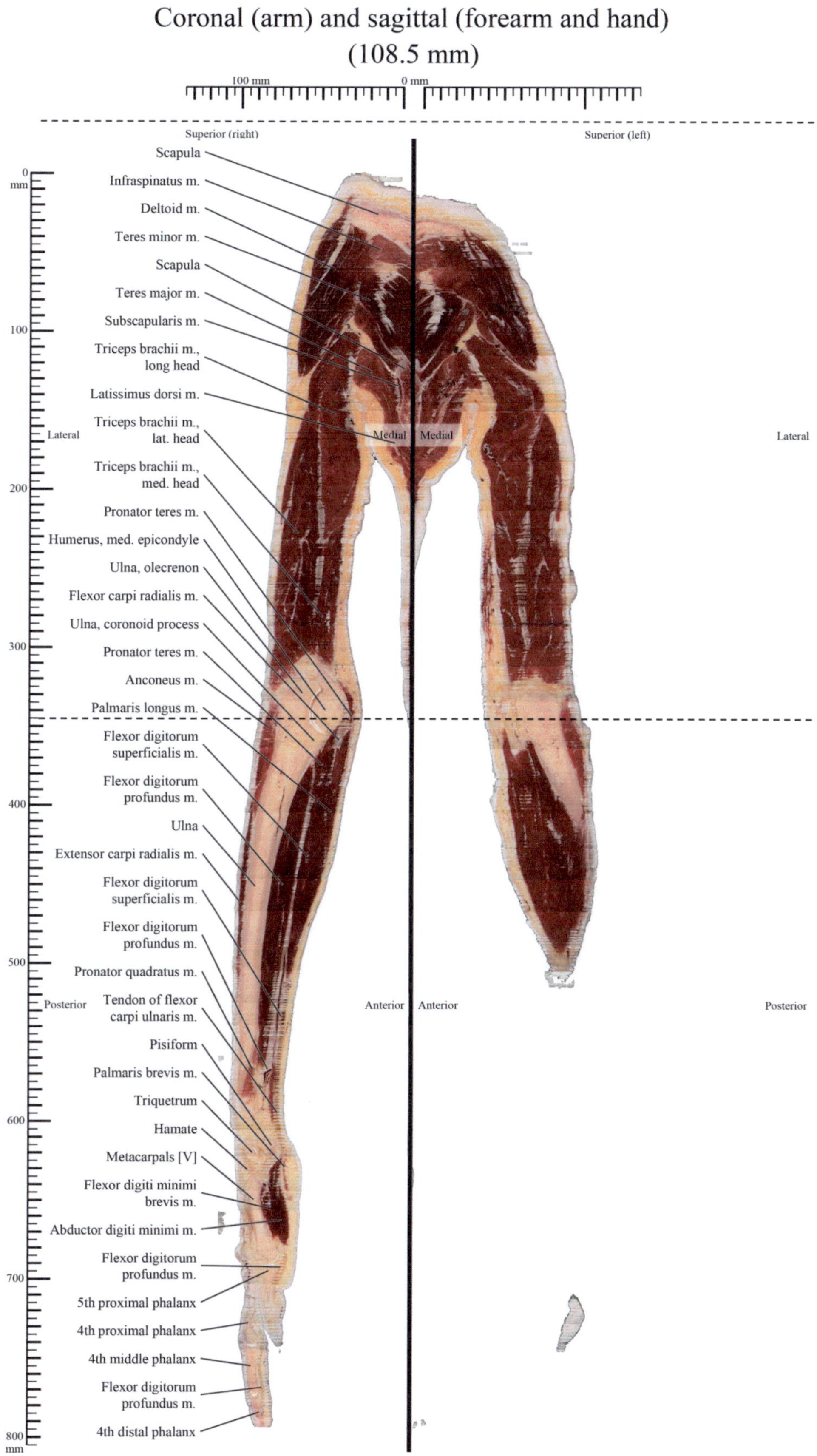
Coronal (arm) and sagittal (forearm and hand)
(108.5 mm)
100 mm
0 mm
Superior (right)
Superior (left)
Scapula
Infraspinatus m.
Deltoid m.
Teres minor m.
Scapula
Teres major m.
Subscapularis m.
Triceps brachii m., long head
Latissimus dorsi m.
Triceps brachii m., lat. head
Triceps brachii m., med. head
Pronator teres m.
Humerus, med. epicondyle
Ulna, olecrenon
Flexor carpi radialis m.
Ulna, coronoid process
Pronator teres m.
Anconeus m.
Palmaris longus m.
Flexor digitorum superficialis m.
Flexor digitorum profundus m.
Ulna
Extensor carpi radialis m.
Flexor digitorum superficialis m.
Flexor digitorum profundus m.
Pronator quadratus m.
Tendon of flexor carpi ulnaris m.
Pisiform
Palmaris brevis m.
Triquetrum
Hamate
Metacarpals [V]
Flexor digiti minimi brevis m.
Abductor digiti minimi m.
Flexor digitorum profundus m.
5th proximal phalanx
4th proximal phalanx
4th middle phalanx
Flexor digitorum profundus m.
4th distal phalanx
Lateral
Medial
Medial
Lateral
Posterior
Anterior
Anterior
Posterior
0 mm
100
200
300
400
500
600
700
800 mm

Coronal (arm) and sagittal (forearm and hand) (110.0 mm)

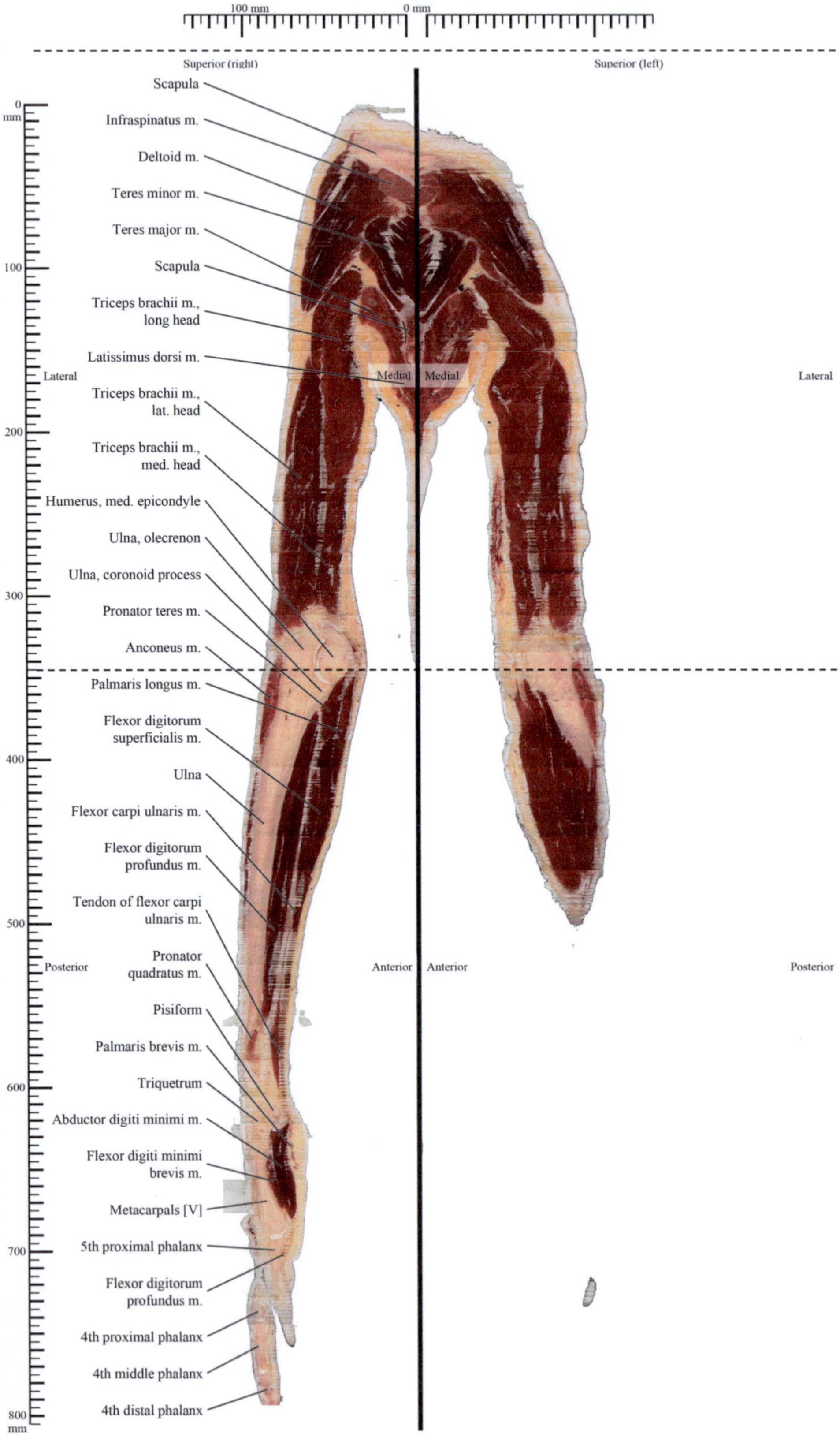

Coronal (arm) and sagittal (forearm and hand) (111.5 mm)

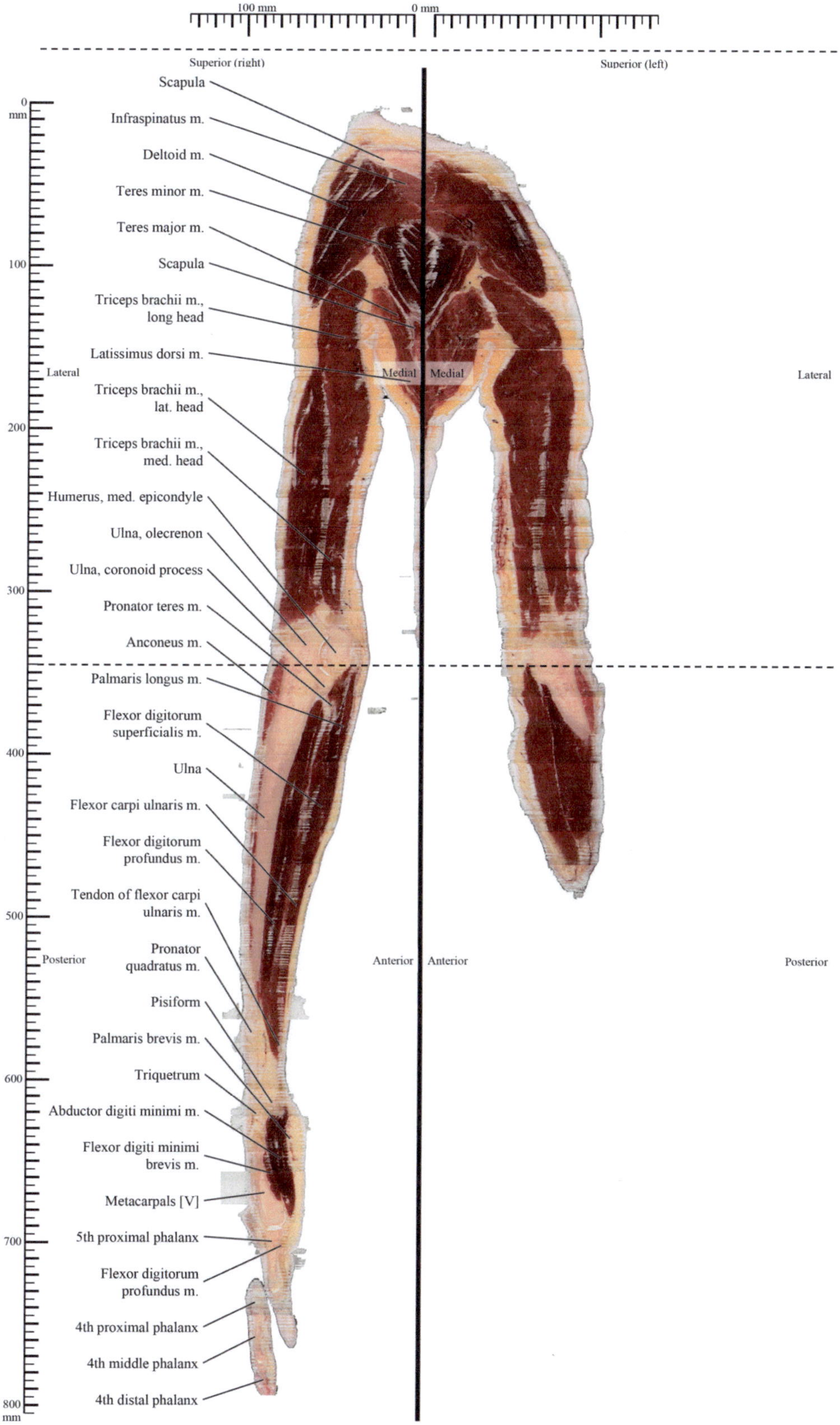

Coronal (arm) and sagittal (forearm and hand) (113.0 mm)

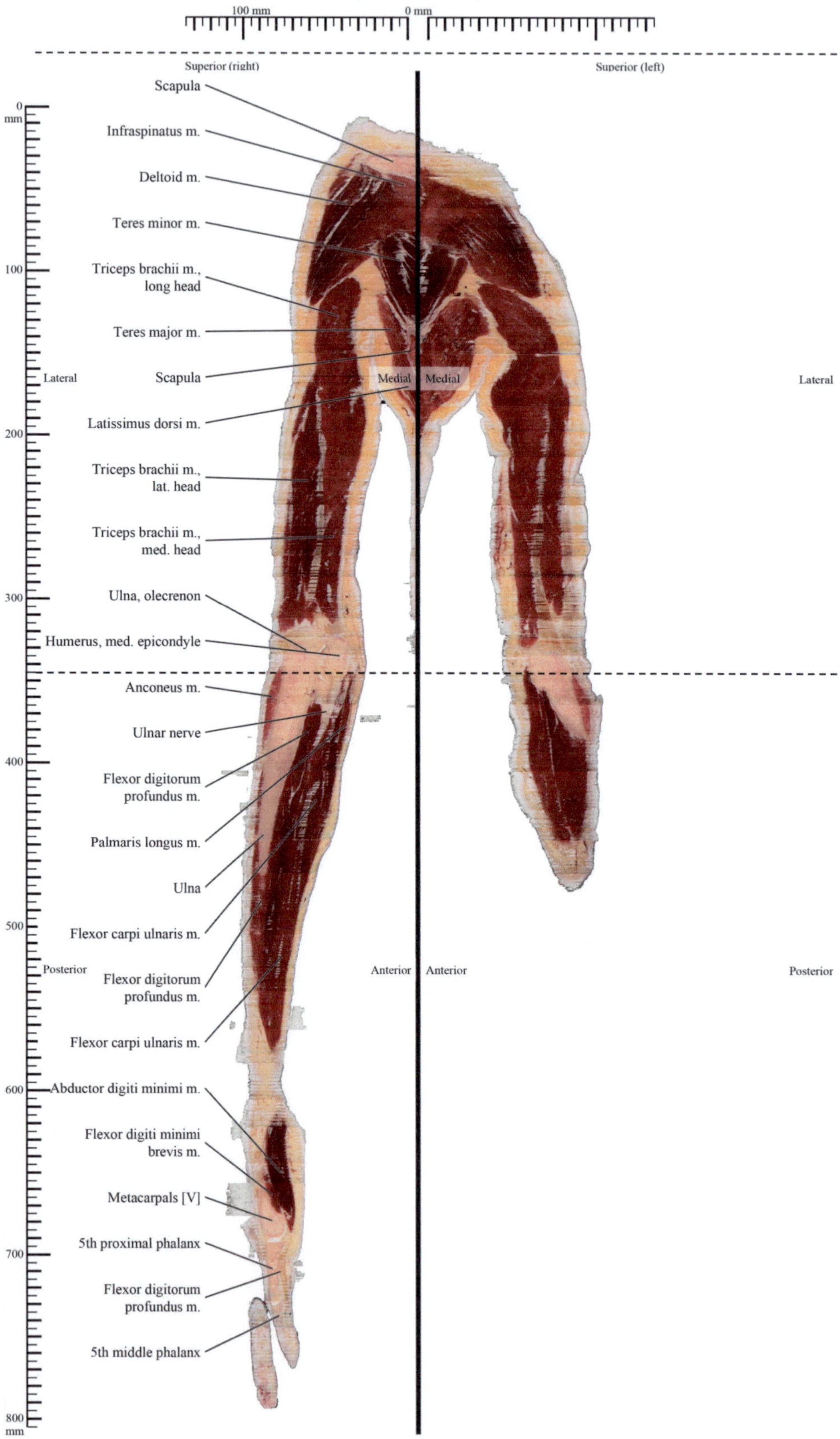

Coronal (arm) and sagittal (forearm and hand) (114.5 mm)

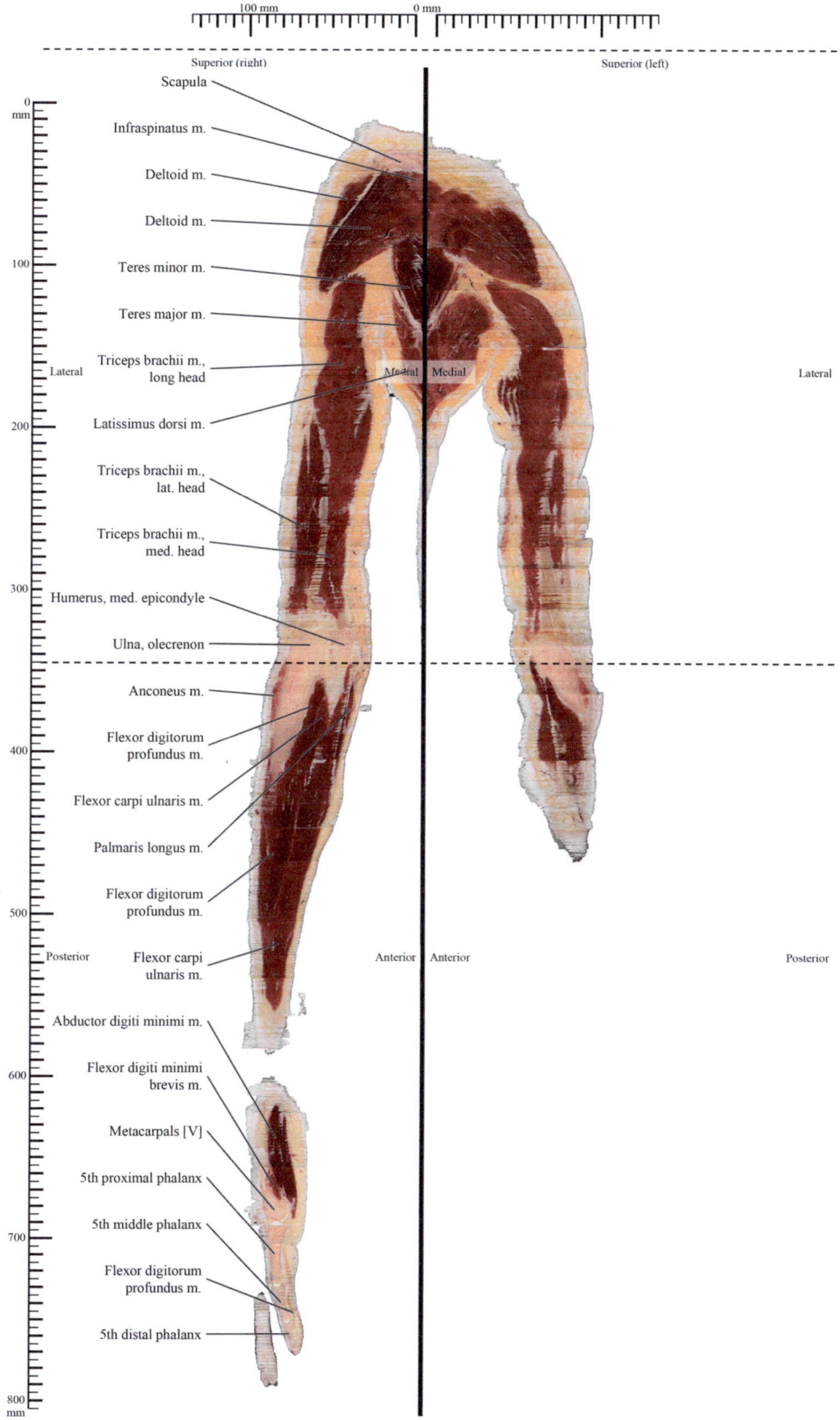

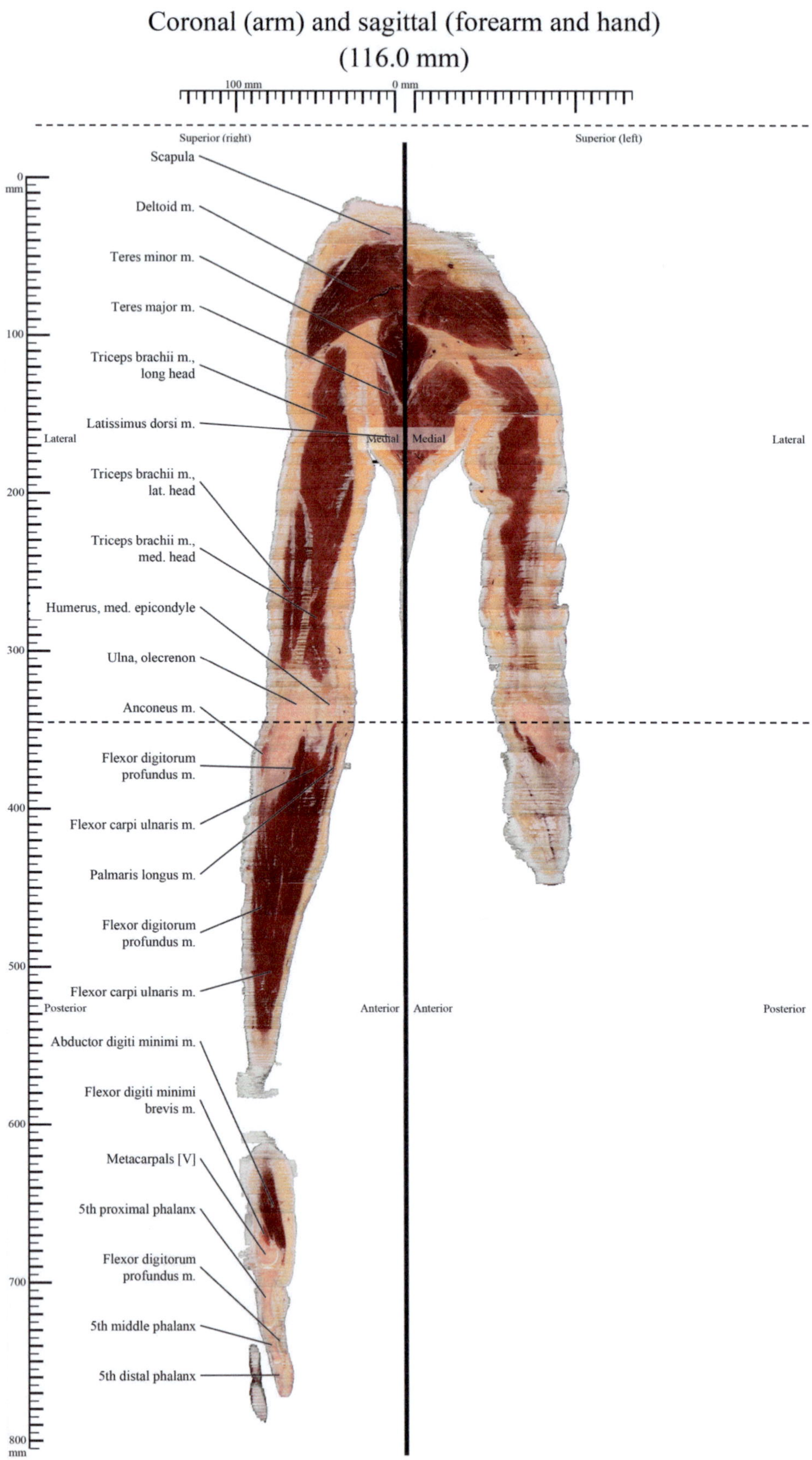
Coronal (arm) and sagittal (forearm and hand)
(116.0 mm)
100 mm
0 mm
Superior (right)
Superior (left)
Scapula
Deltoid m.
Teres minor m.
Teres major m.
Triceps brachii m., long head
Latissimus dorsi m.
Lateral
Medial
Medial
Lateral
Triceps brachii m., lat. head
Triceps brachii m., med. head
Humerus, med. epicondyle
Ulna, olecrenon
Anconeus m.
Flexor digitorum profundus m.
Flexor carpi ulnaris m.
Palmaris longus m.
Flexor digitorum profundus m.
Flexor carpi ulnaris m.
Posterior
Anterior
Anterior
Posterior
Abductor digiti minimi m.
Flexor digiti minimi brevis m.
Metacarpals [V]
5th proximal phalanx
Flexor digitorum profundus m.
5th middle phalanx
5th distal phalanx
0 mm
100
200
300
400
500
600
700
800 mm

Coronal (arm) and sagittal (forearm and hand)
(117.5 mm)

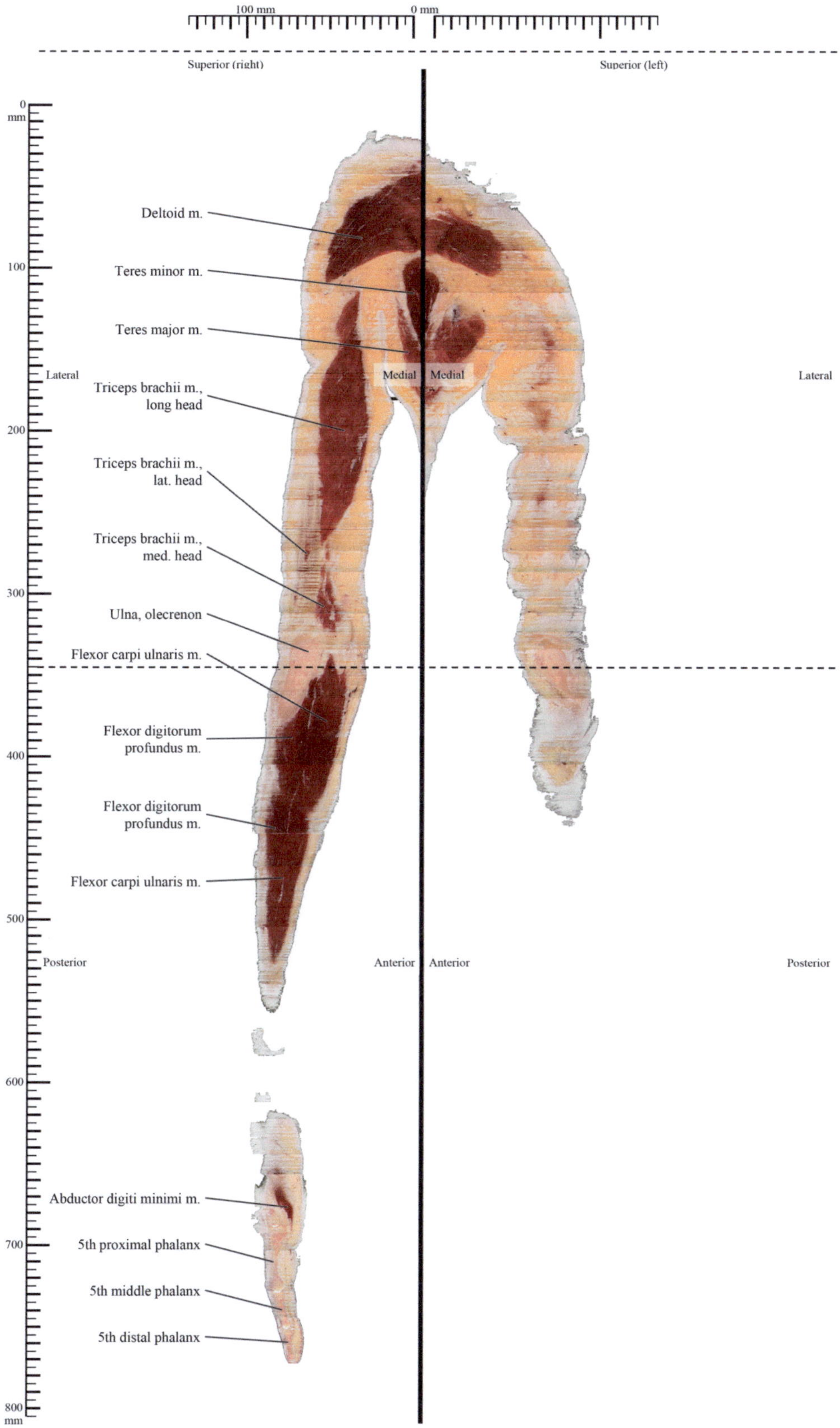

Coronal (arm) and sagittal (forearm and hand) (119 mm)

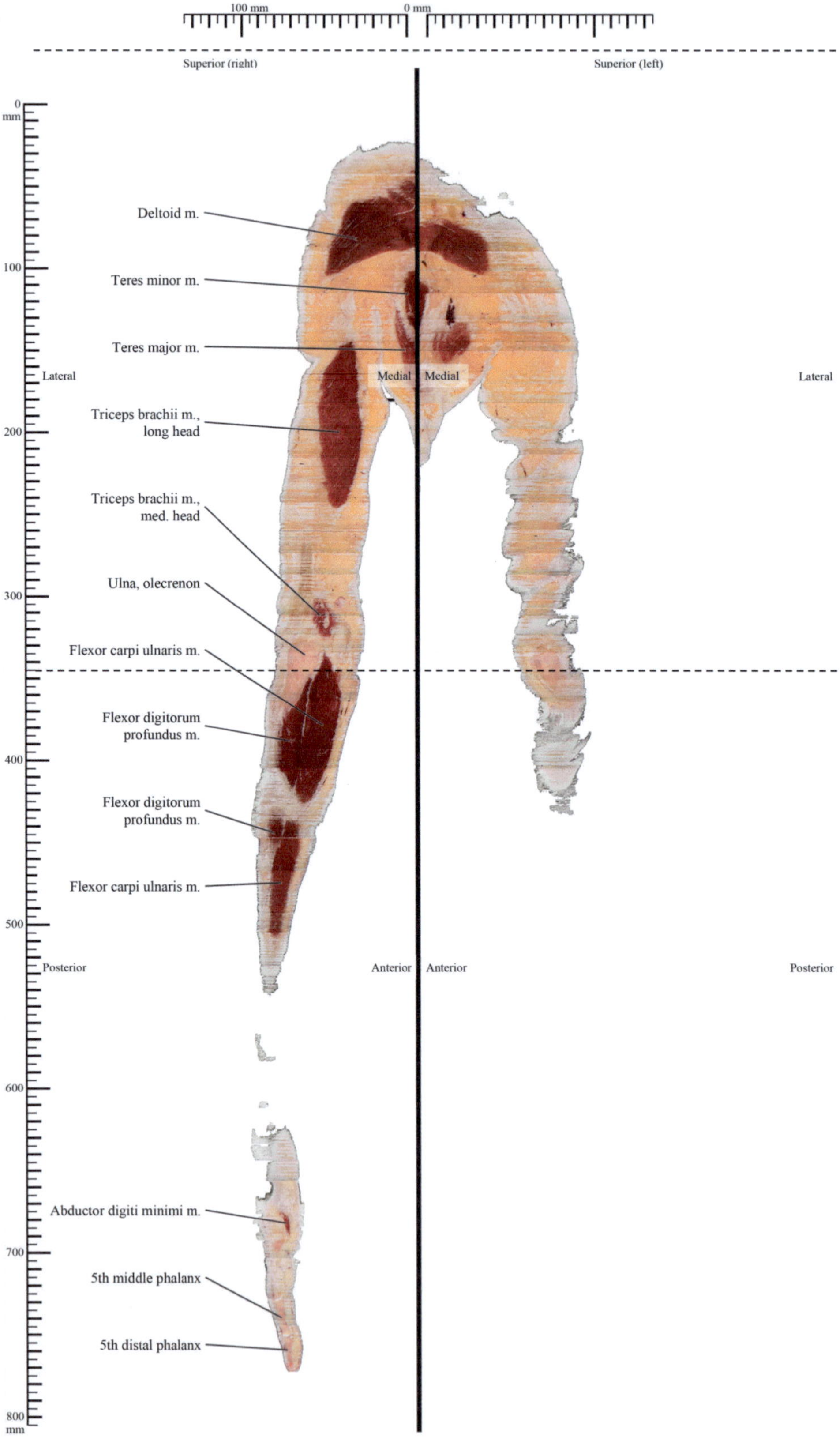

Sagittal (Arm) and Coronal (Forearm and Hand) Planes of Left Upper Limb

4

All images, 1.5 mm intervals
All images, 0.06 mm × 0.06 mm sized-pixel

Abbreviation: ant., anterior; a., artery; inf., inferior; int., internal; lat., lateral; m., m.; med., medial; n., nerve; nu., nucleus; post., posterior; sup., superior; tr., tract

J. S. Park, *Cross-Sectional Atlas of Human Upper Limbs*, Cross-Sectional Atlas, https://doi.org/10.1007/978-981-95-1900-2_4

Sagittal (right arm) and coronal (right forearm and hand)
(105.0 mm)

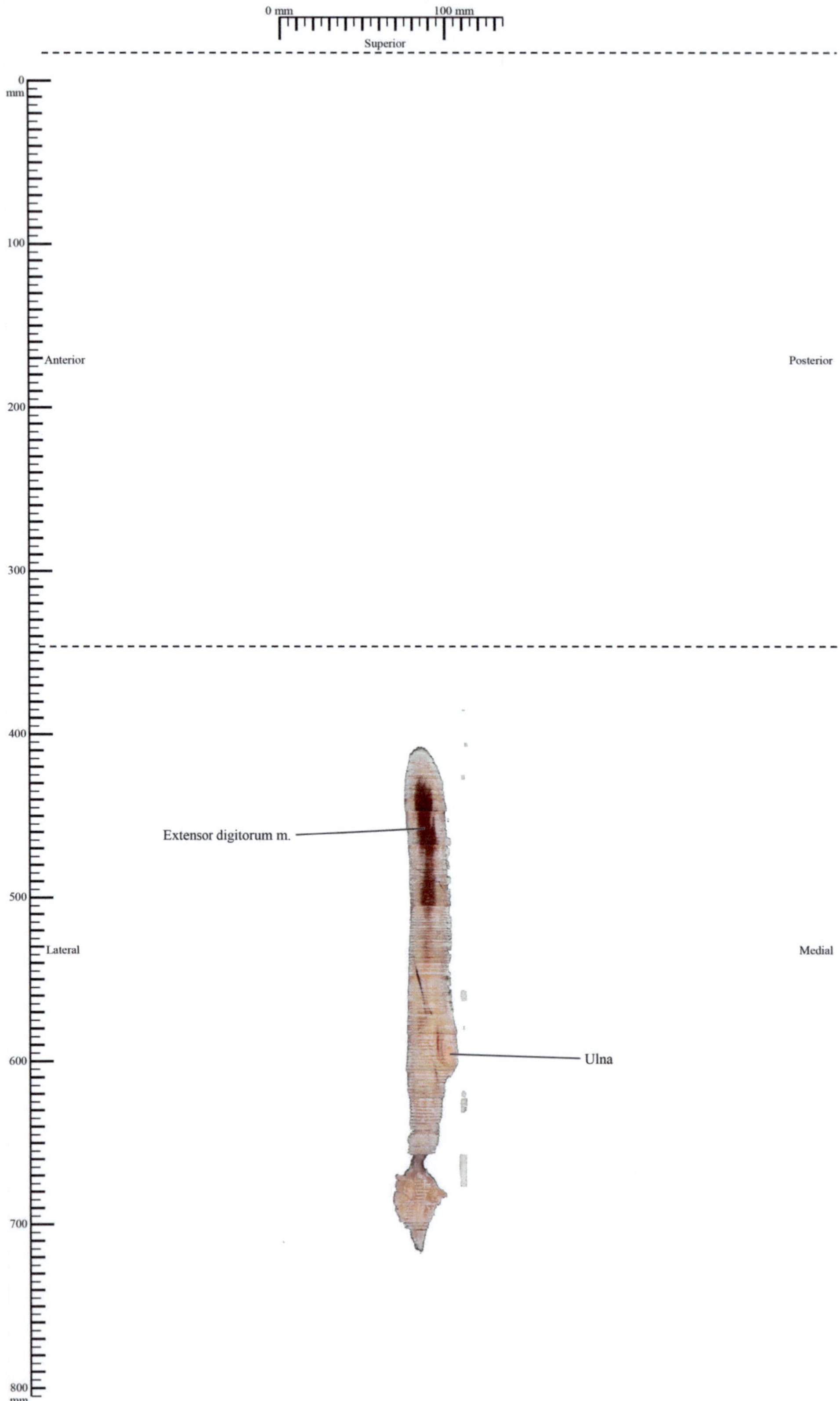

Sagittal (right arm) and coronal (right forearm and hand) (103.5 mm)

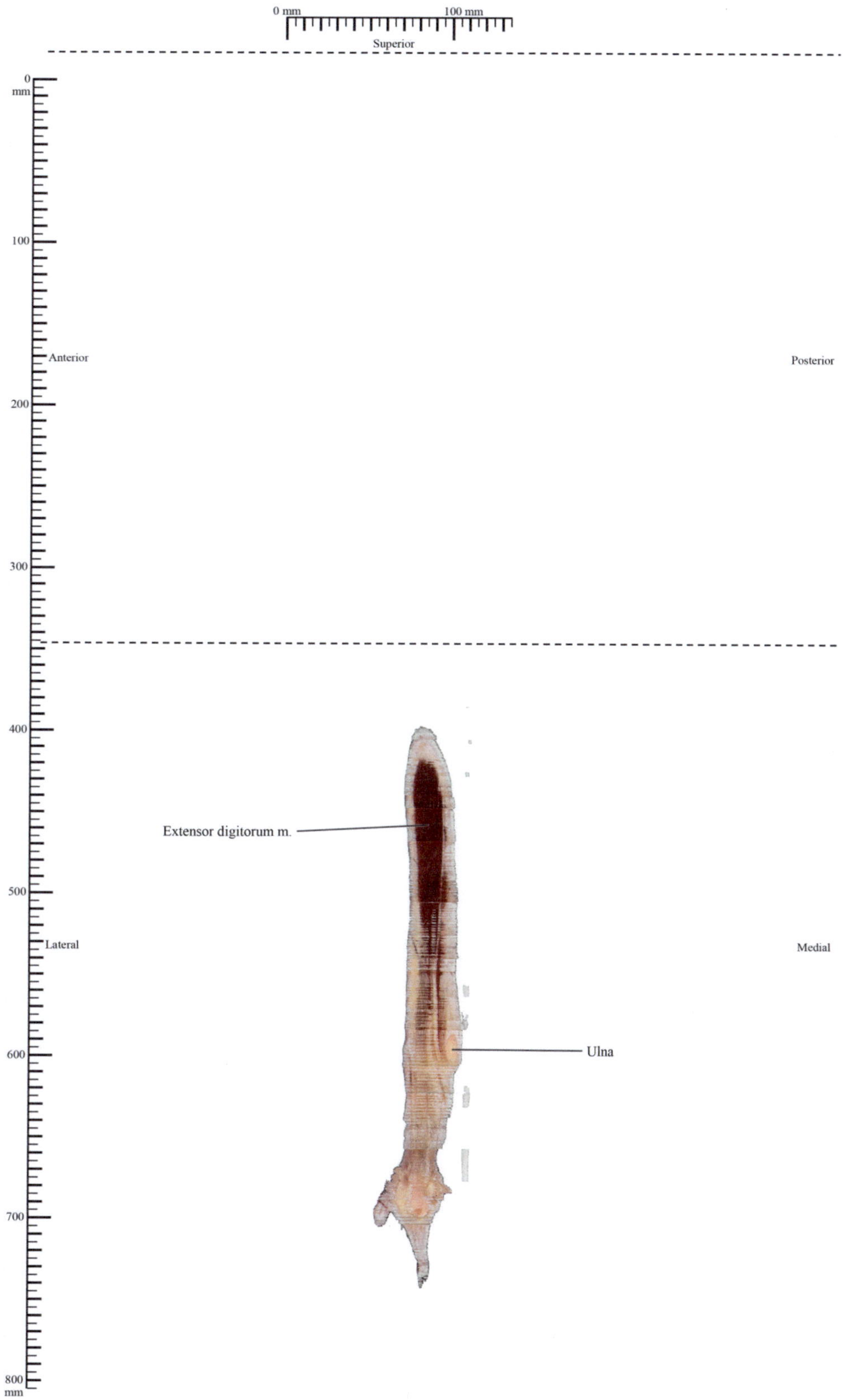

Sagittal (right arm) and coronal (right forearm and hand) (102.0 mm)

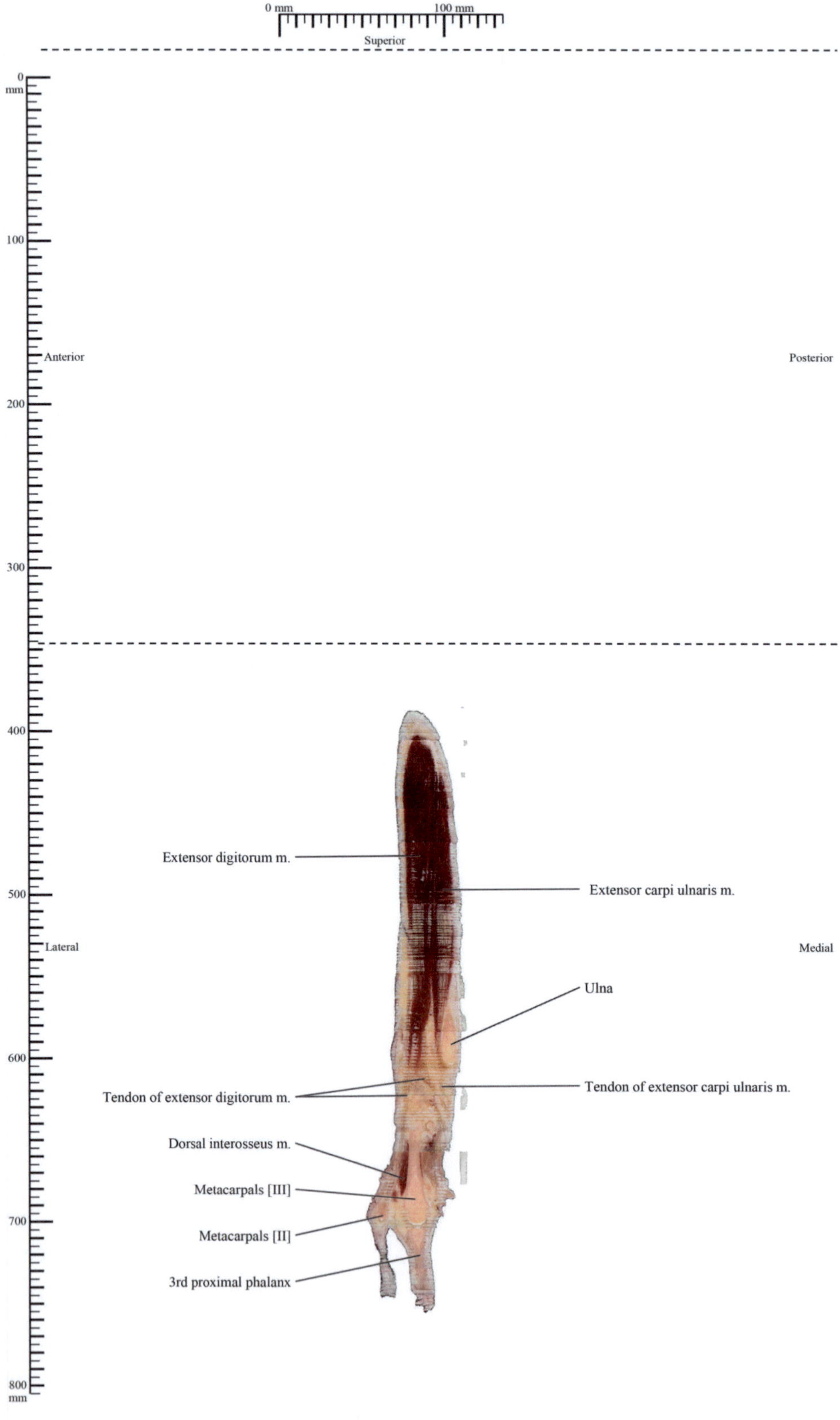

Sagittal (right arm) and coronal (right forearm and hand) (100.5 mm)

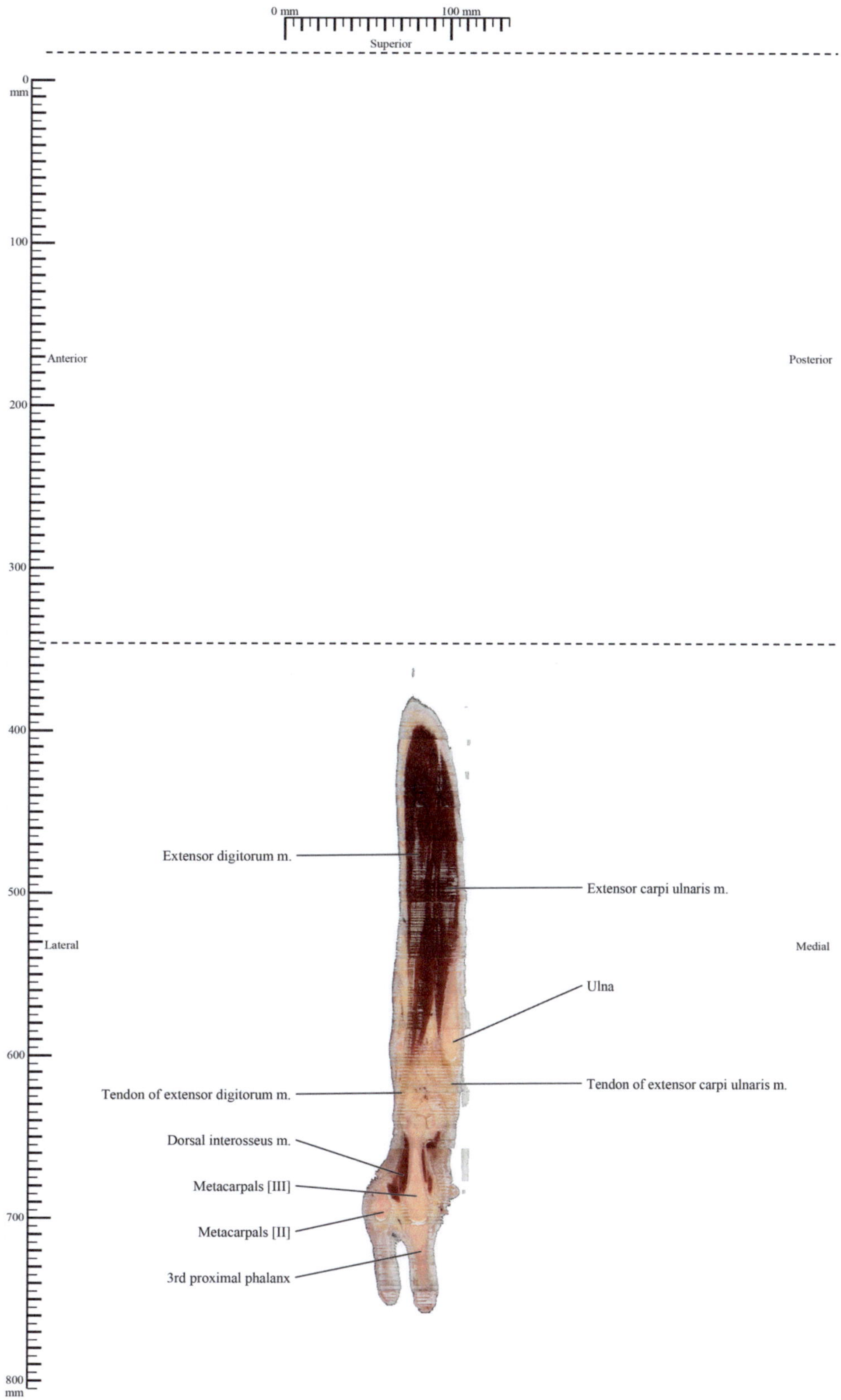

Sagittal (right arm) and coronal (right forearm and hand) (99.0 mm)

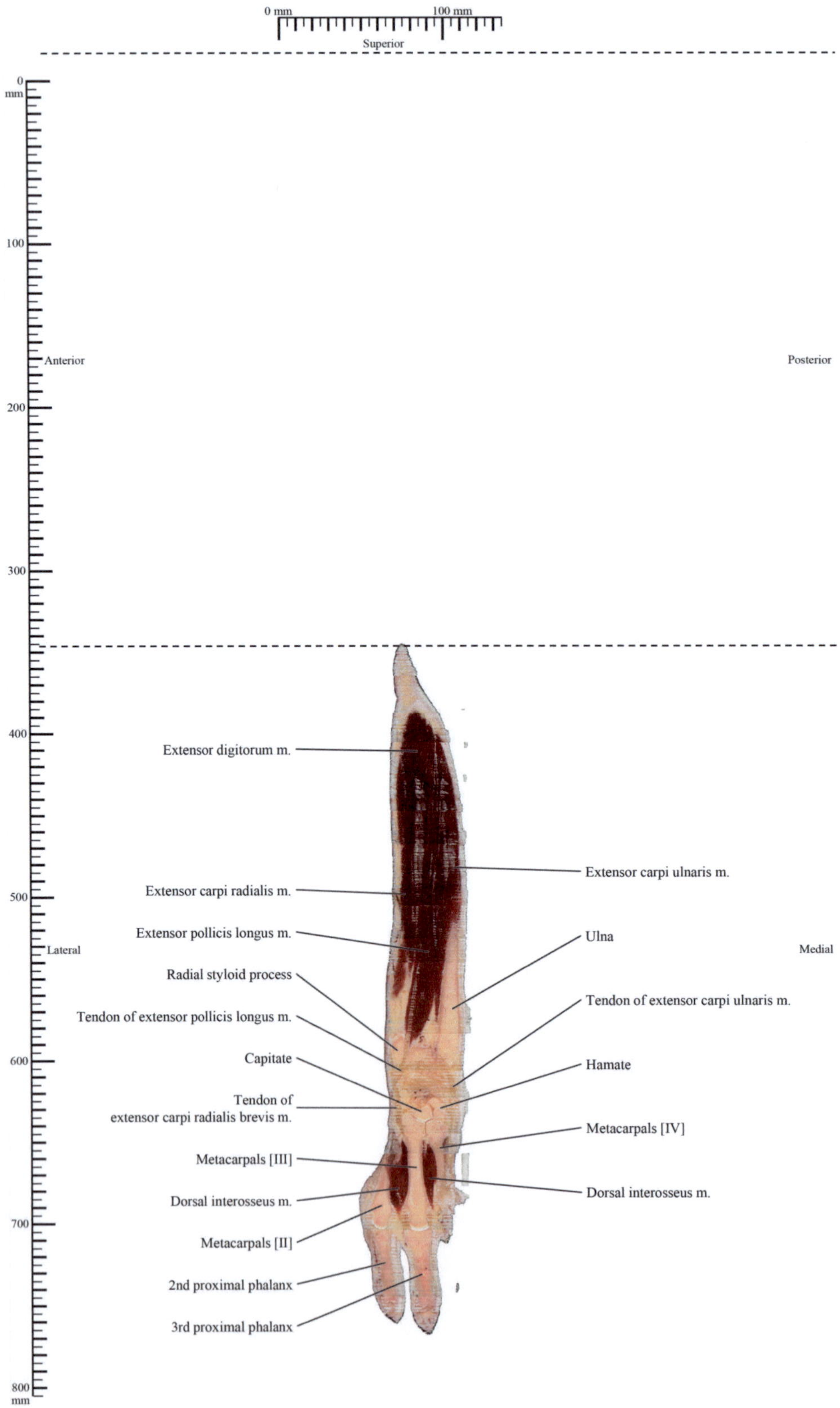

Sagittal (right arm) and coronal (right forearm and hand) (97.5 mm)

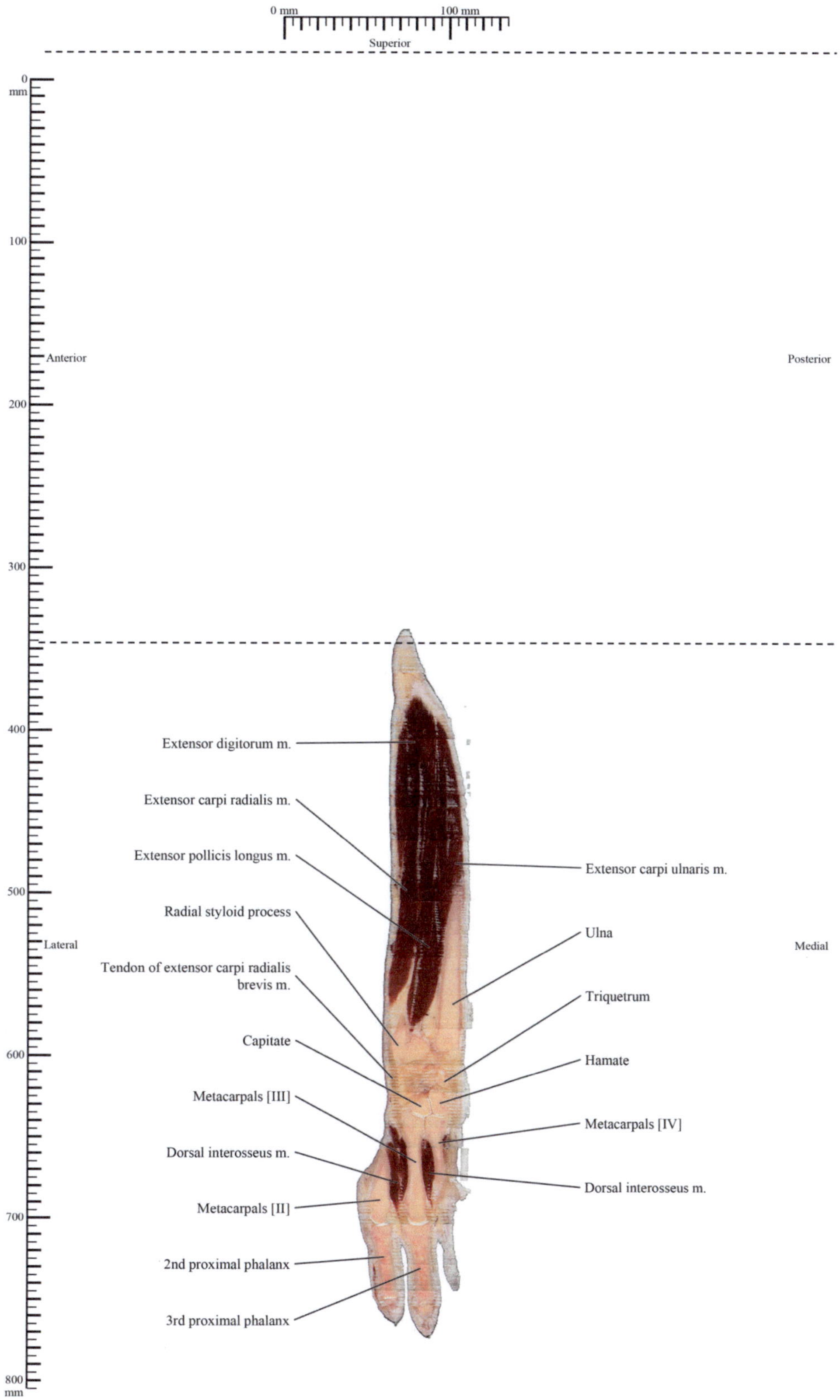

Sagittal (right arm) and coronal (right forearm and hand) (95.0 mm)

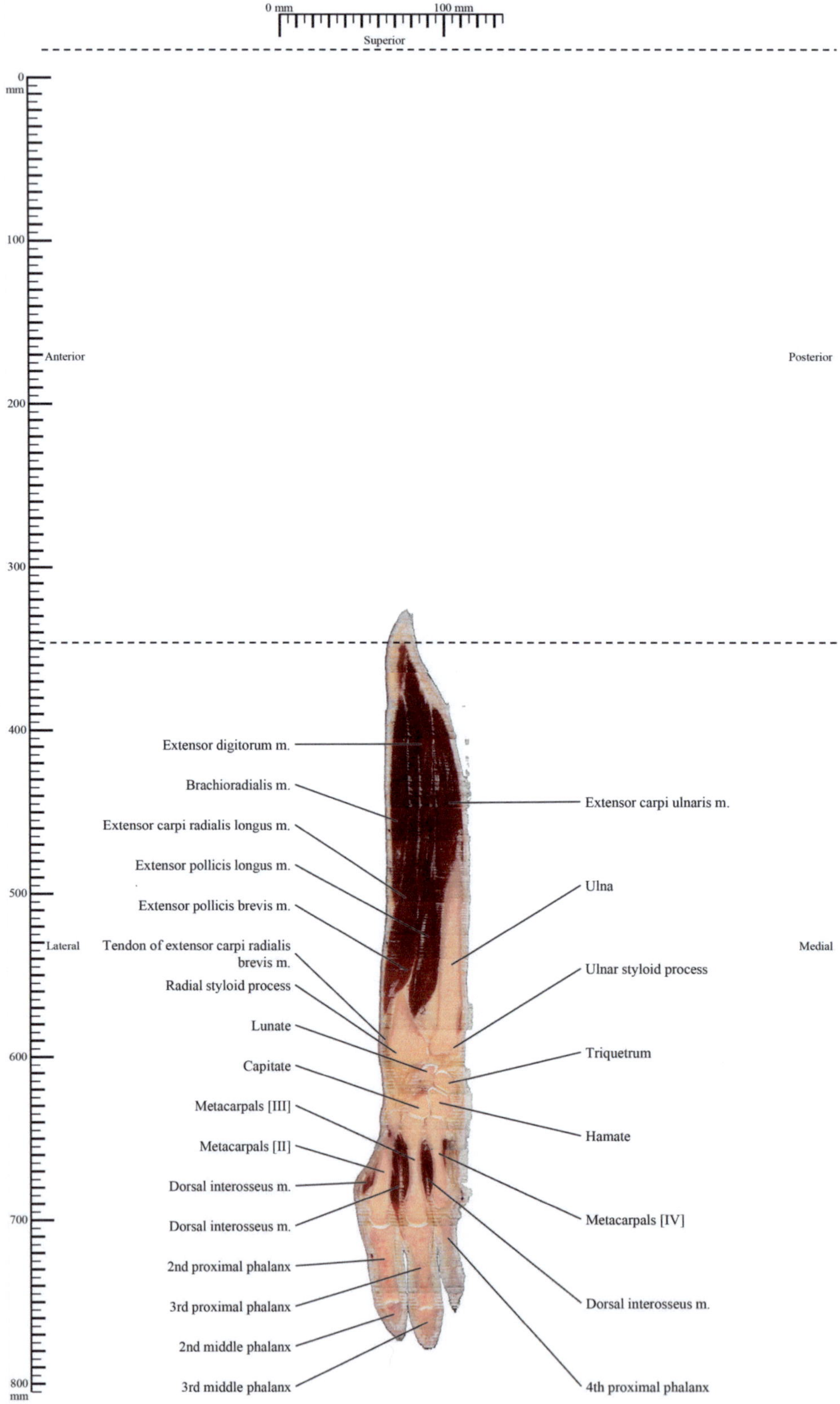

Sagittal (right arm) and coronal (right forearm and hand) (94.5 mm)

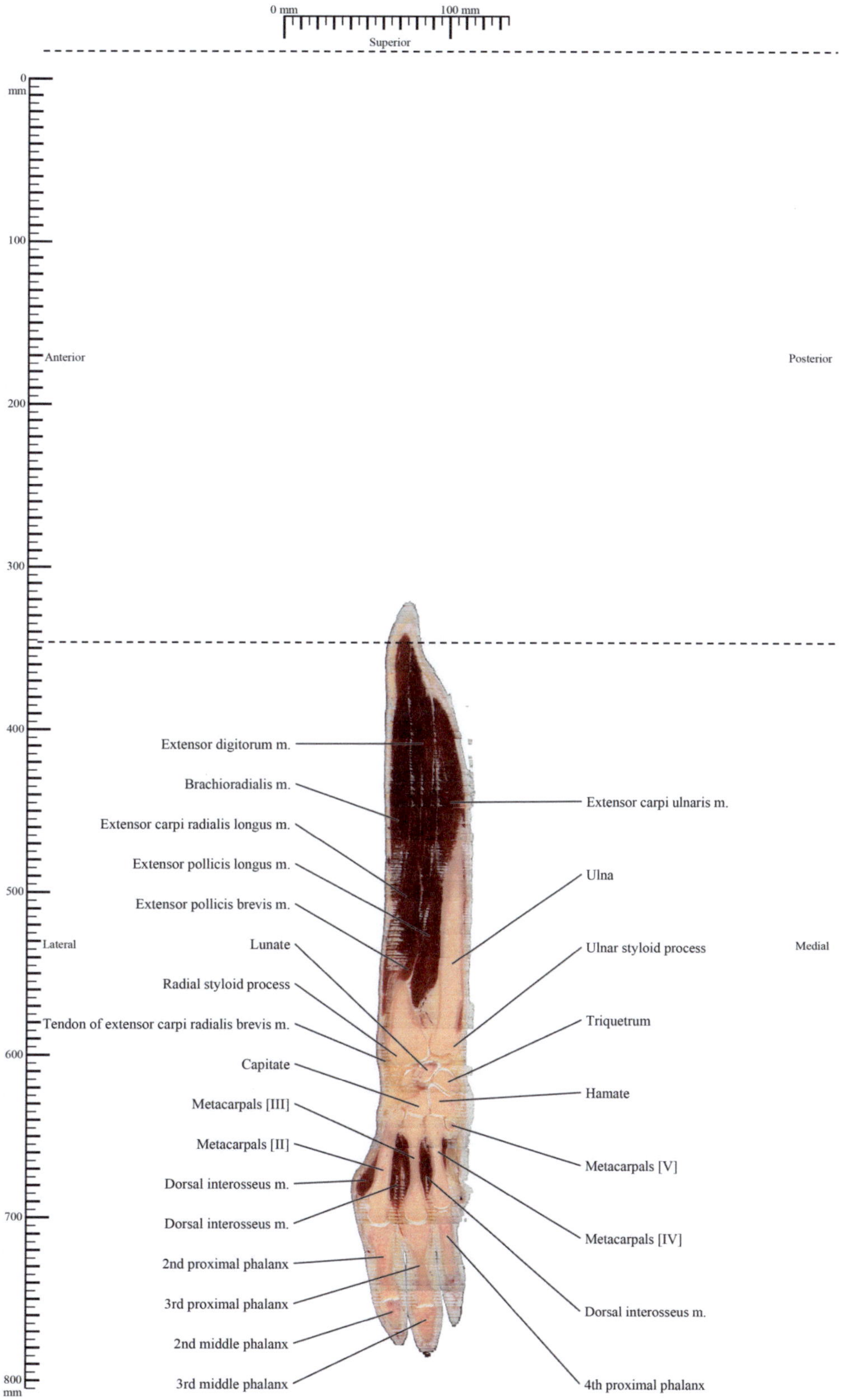

Sagittal (right arm) and coronal (right forearm and hand) (93.0 mm)

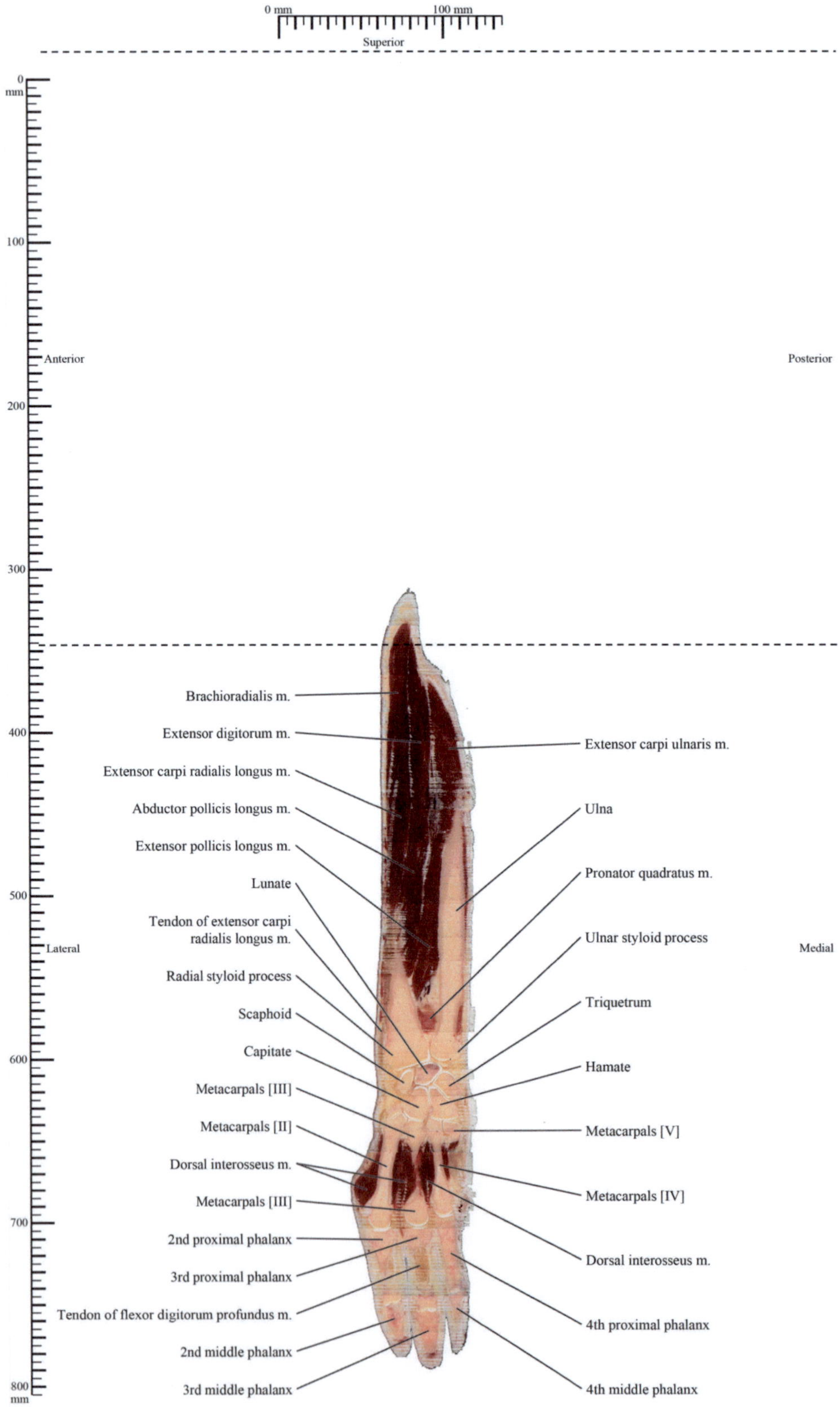

Sagittal (right arm) and coronal (right forearm and hand) (91.5 mm)

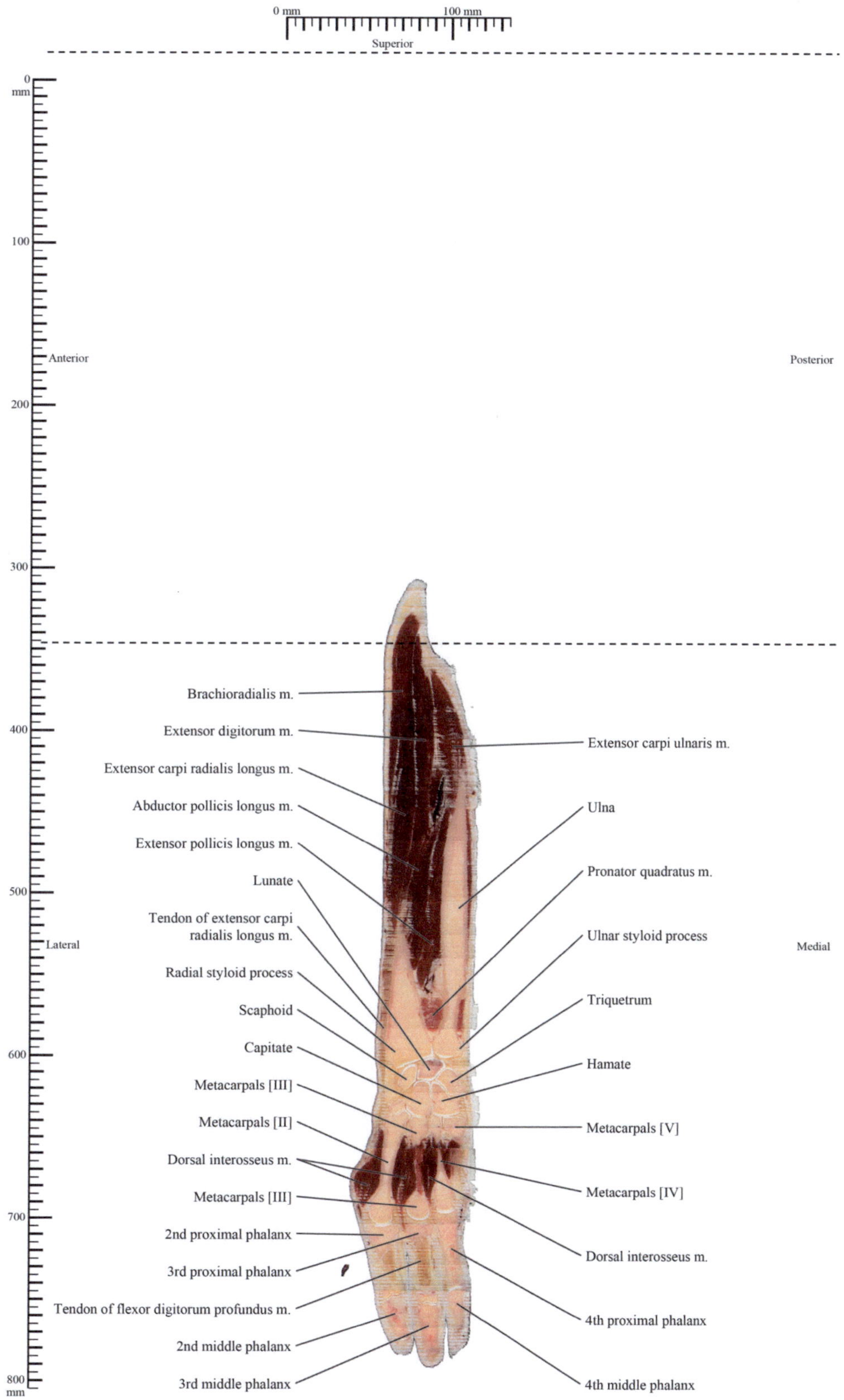

Sagittal (right arm) and coronal (right forearm and hand) (90.0 mm)

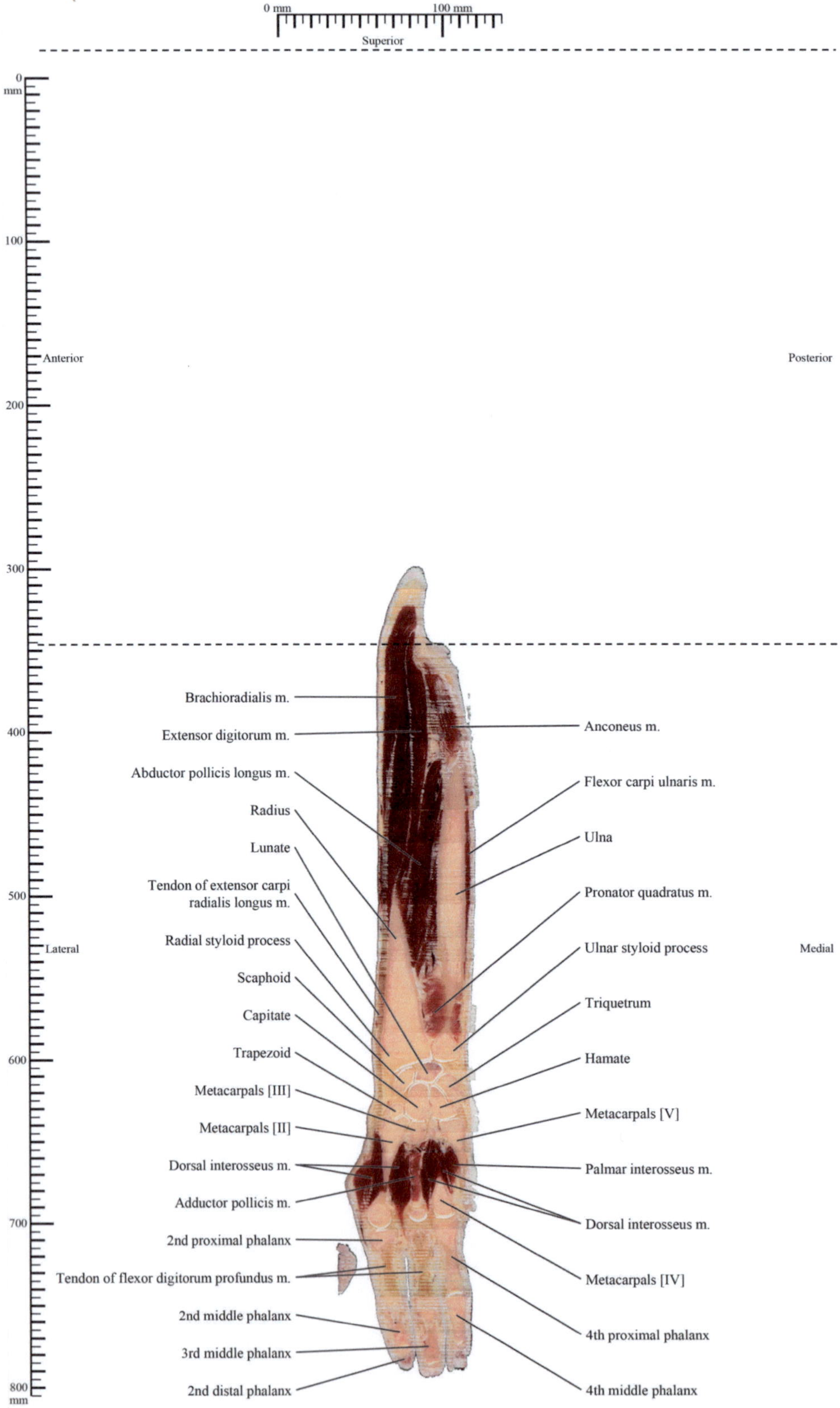

Sagittal (right arm) and coronal (right forearm and hand)
(88.5 mm)

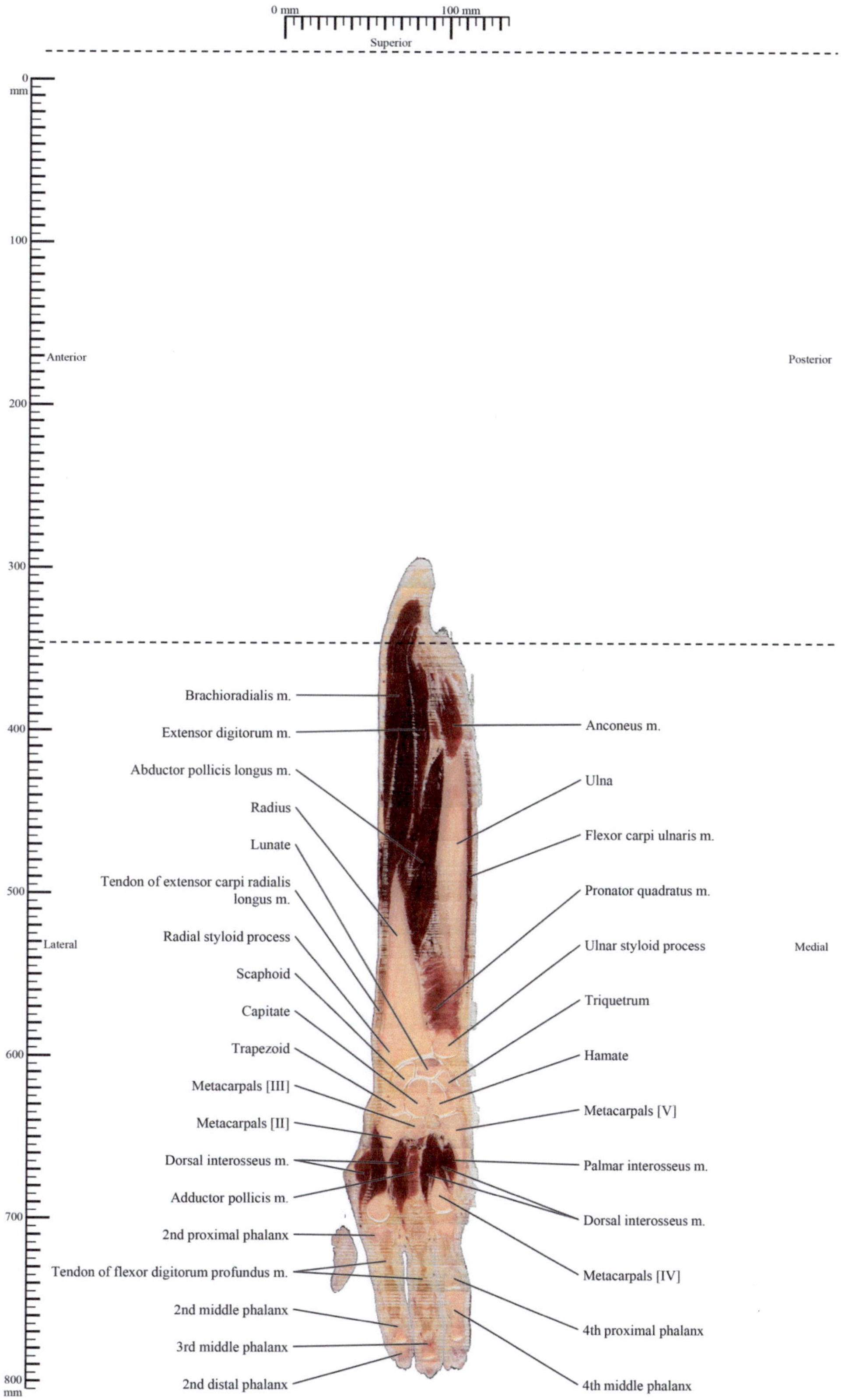

Sagittal (right arm) and coronal (right forearm and hand) (87.0 mm)

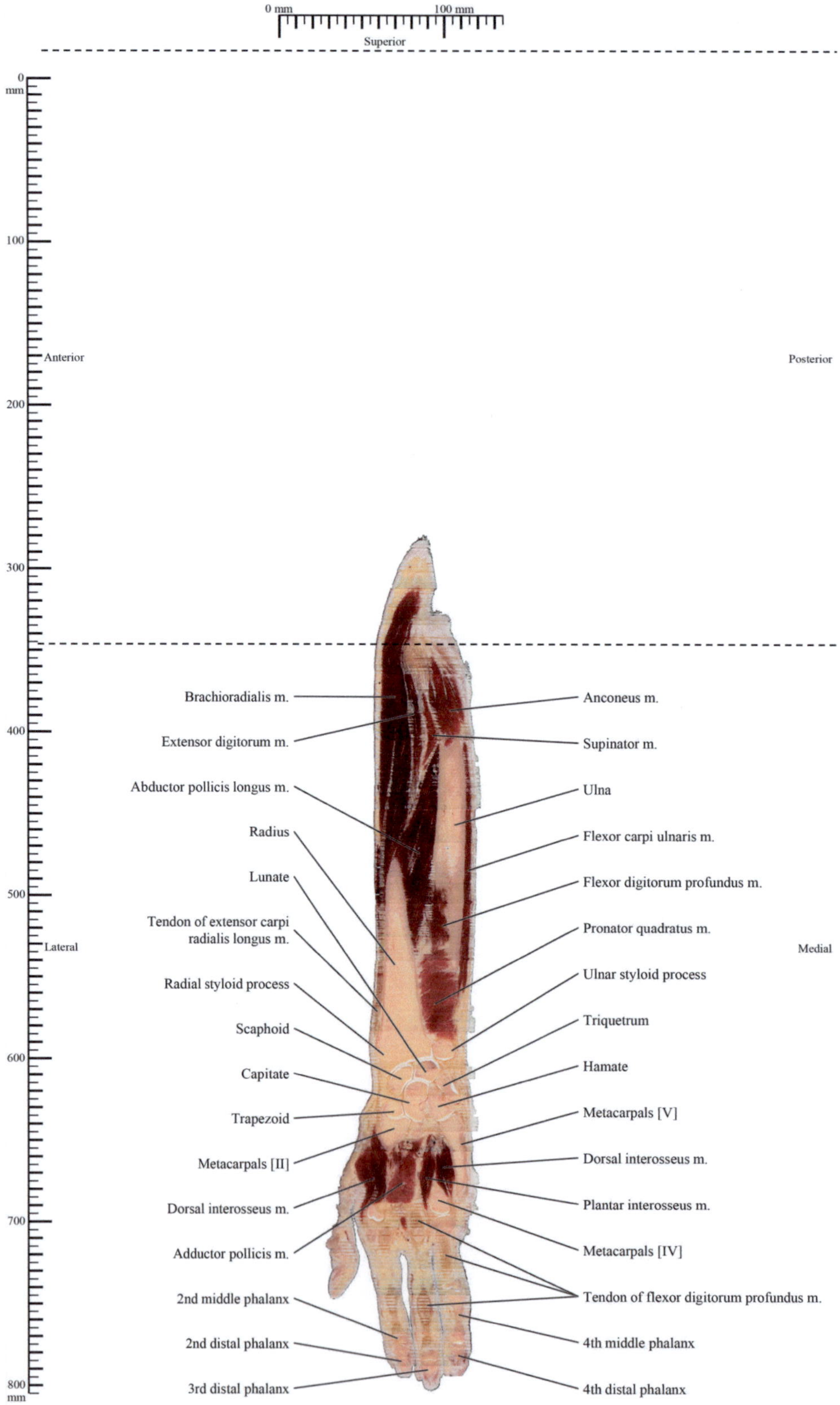

Sagittal (right arm) and coronal (right forearm and hand) (85.5 mm)

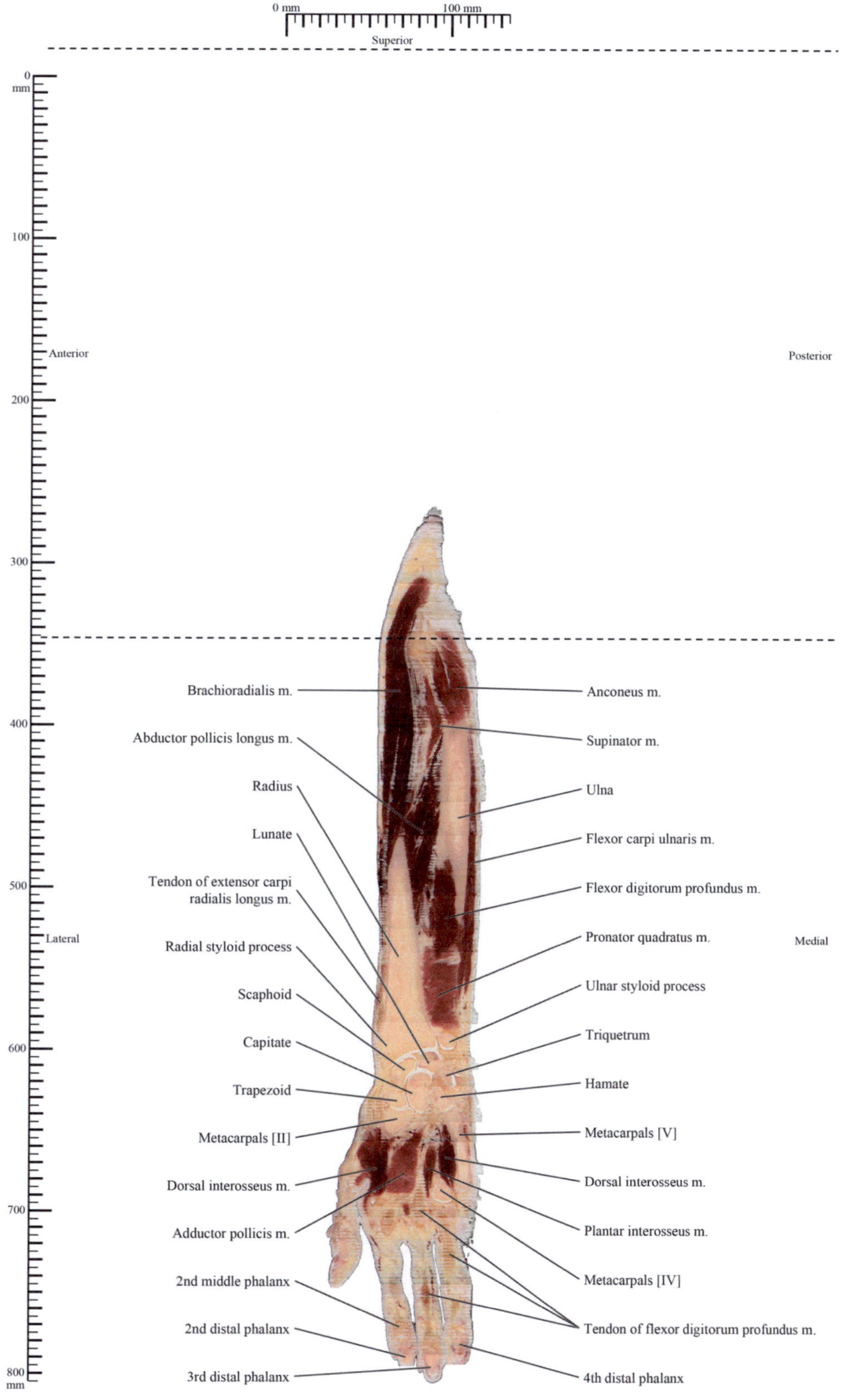

Sagittal (right arm) and coronal (right forearm and hand)
(84.0 mm)

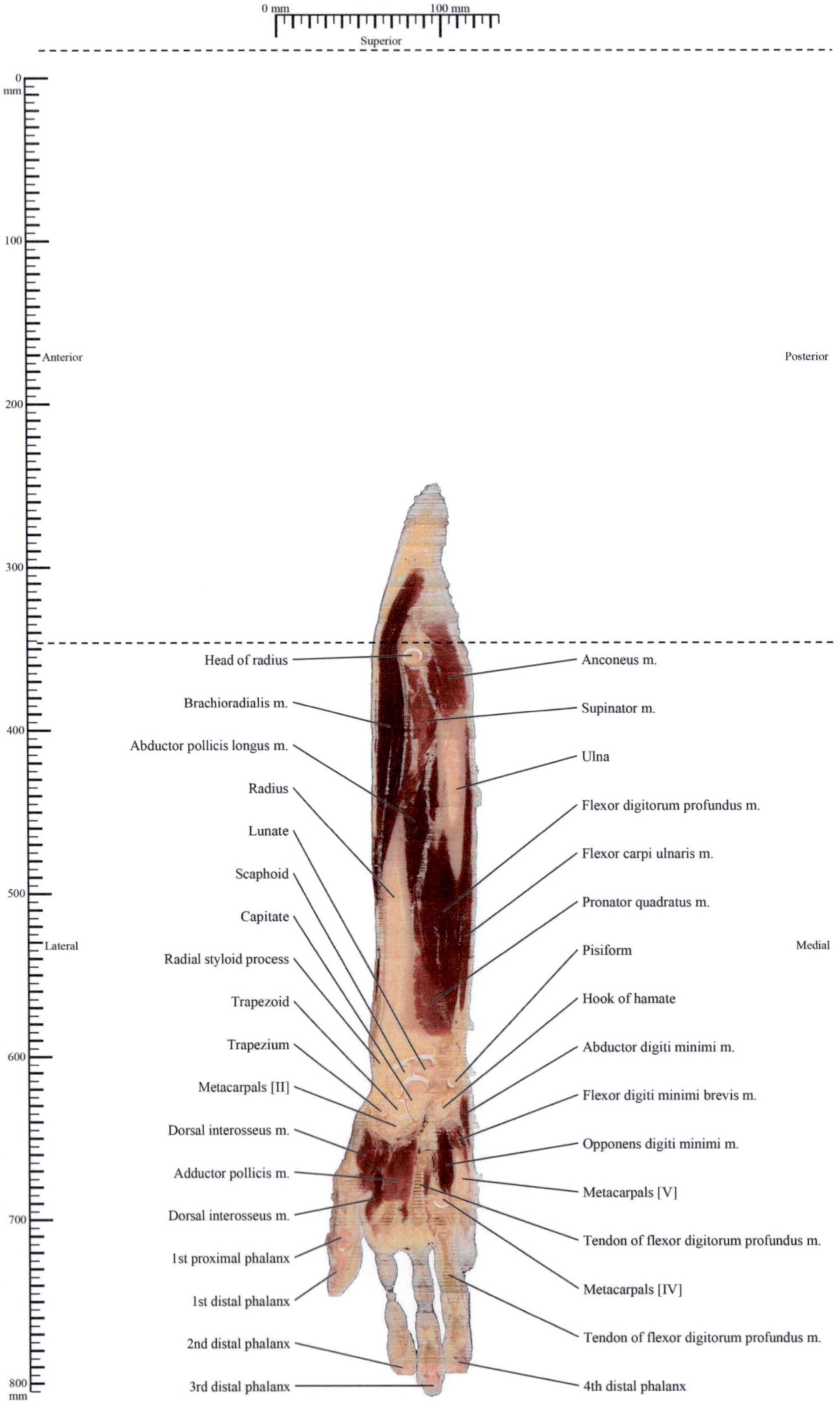

Sagittal (right arm) and coronal (right forearm and hand) (82.5 mm)

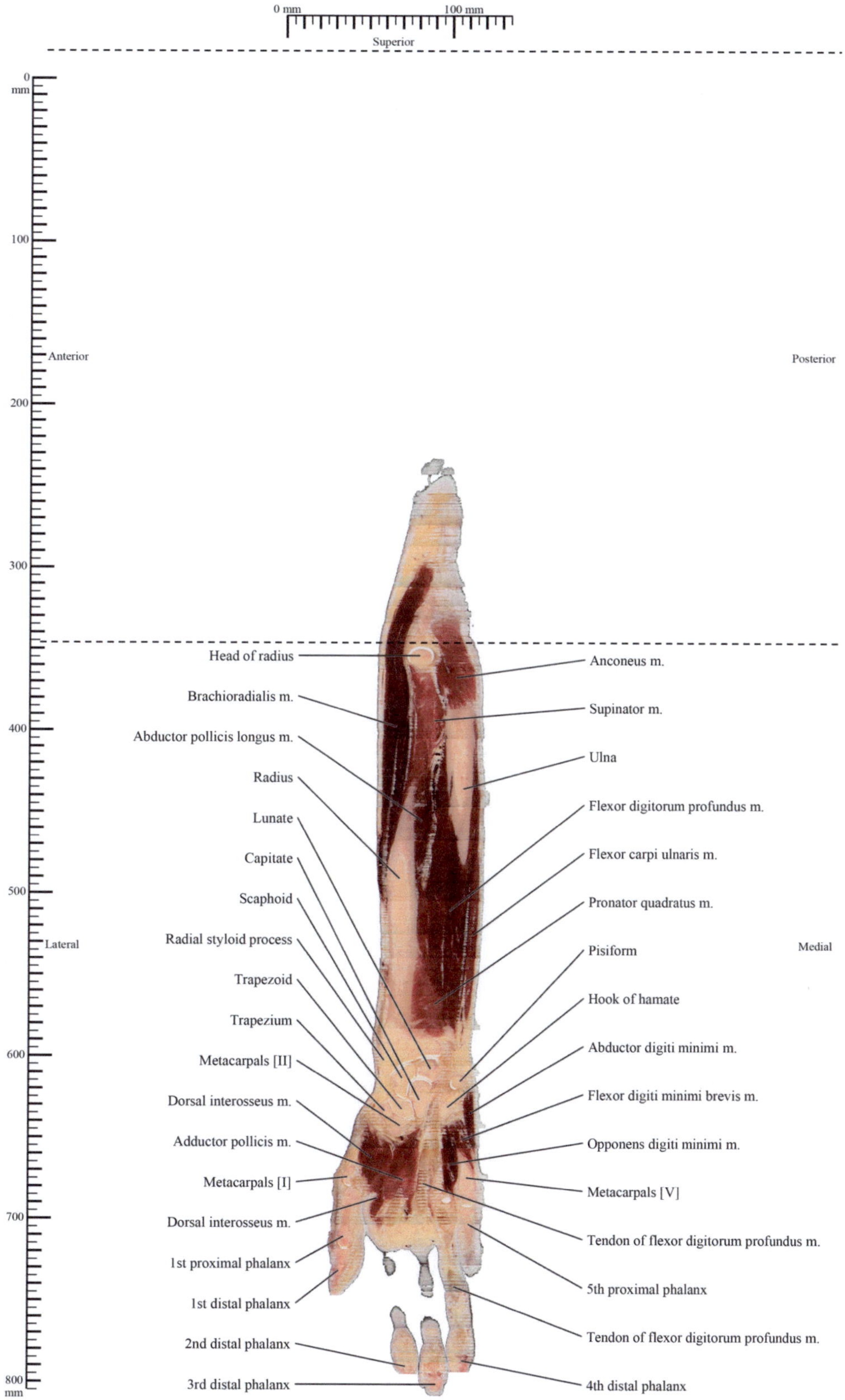

Sagittal (right arm) and coronal (right forearm and hand) (81.0 mm)

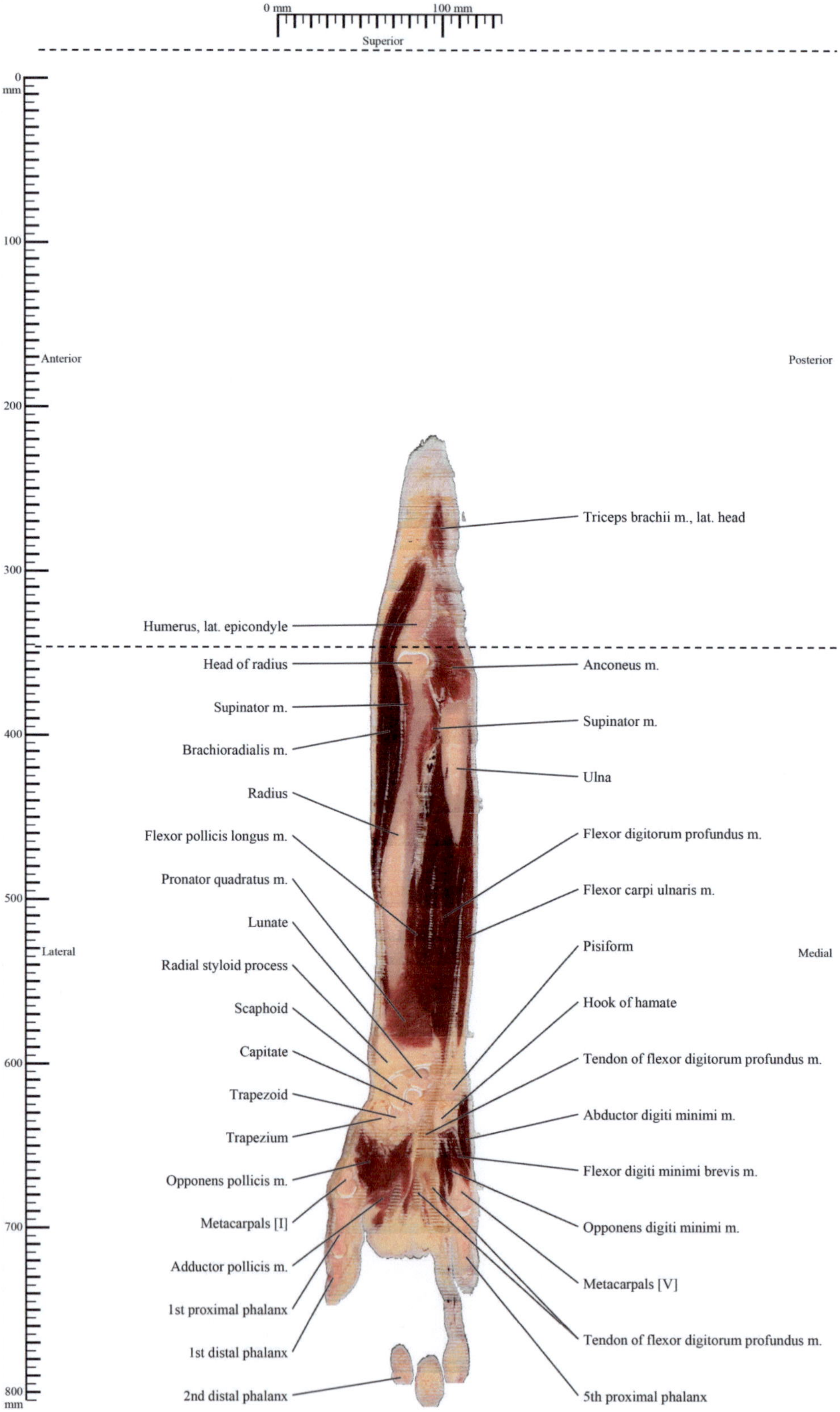

Sagittal (right arm) and coronal (right forearm and hand) (79.5 mm)

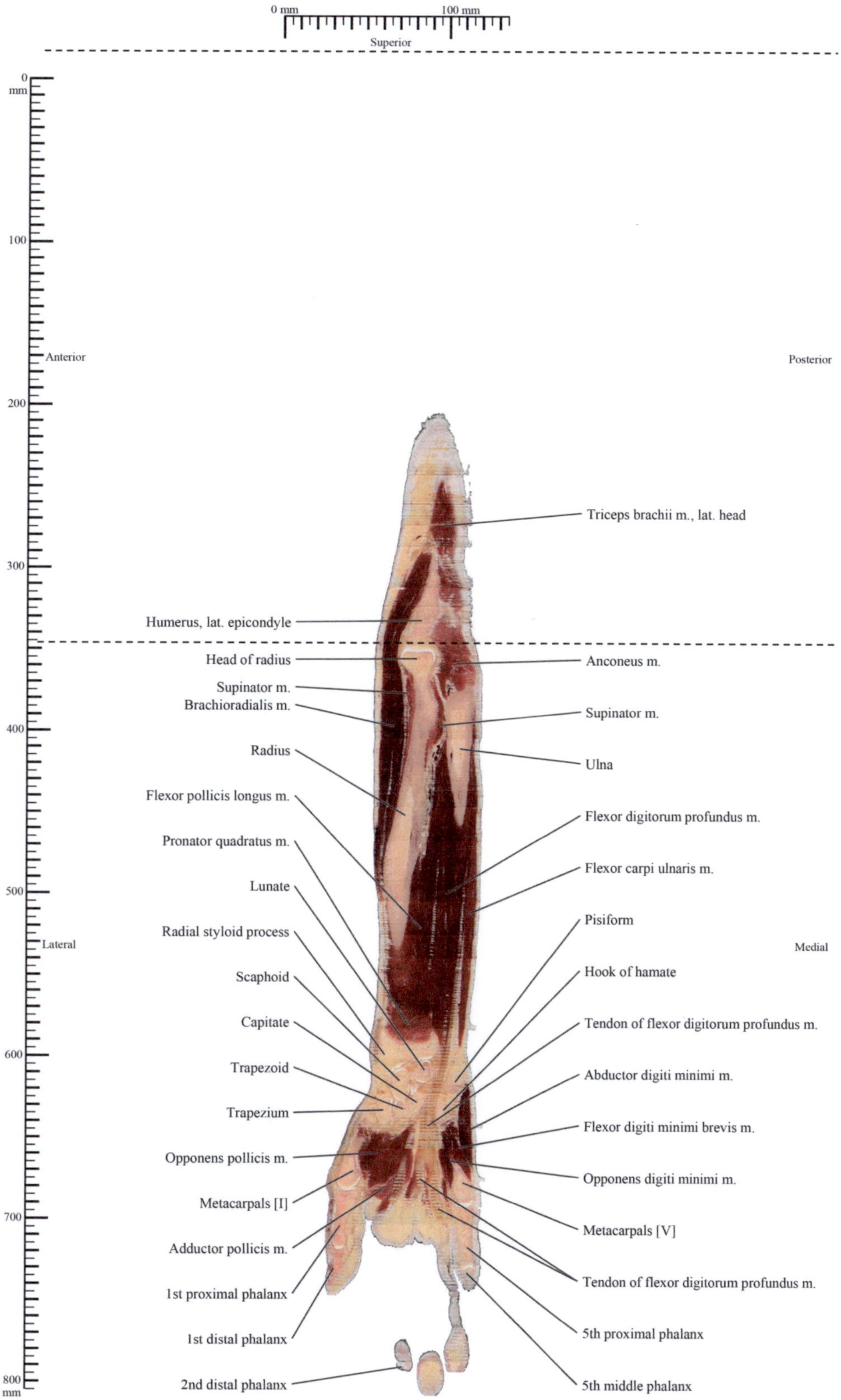

Sagittal (right arm) and coronal (right forearm and hand) (78.0 mm)

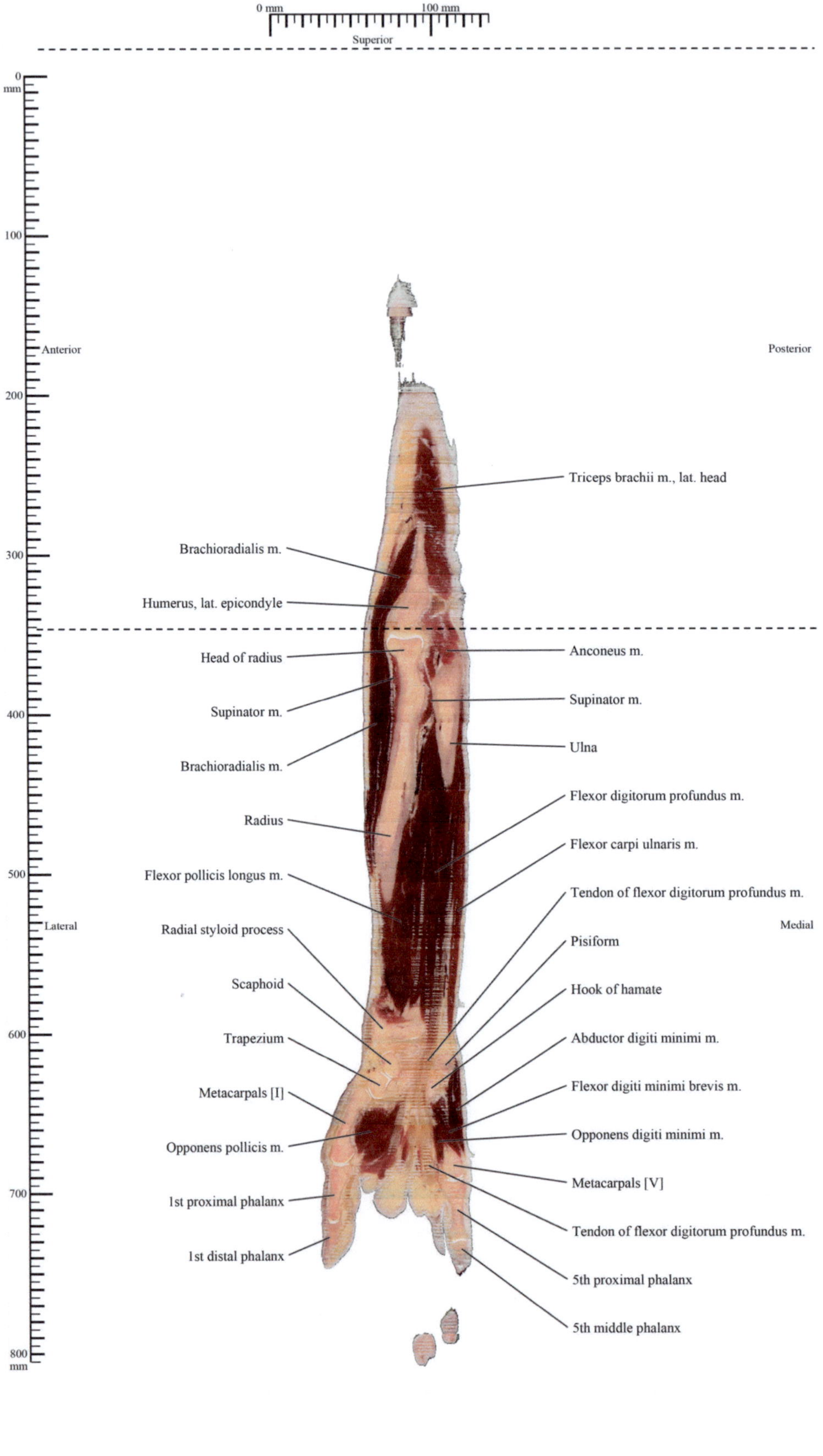

Sagittal (right arm) and coronal (right forearm and hand) (76.5 mm)

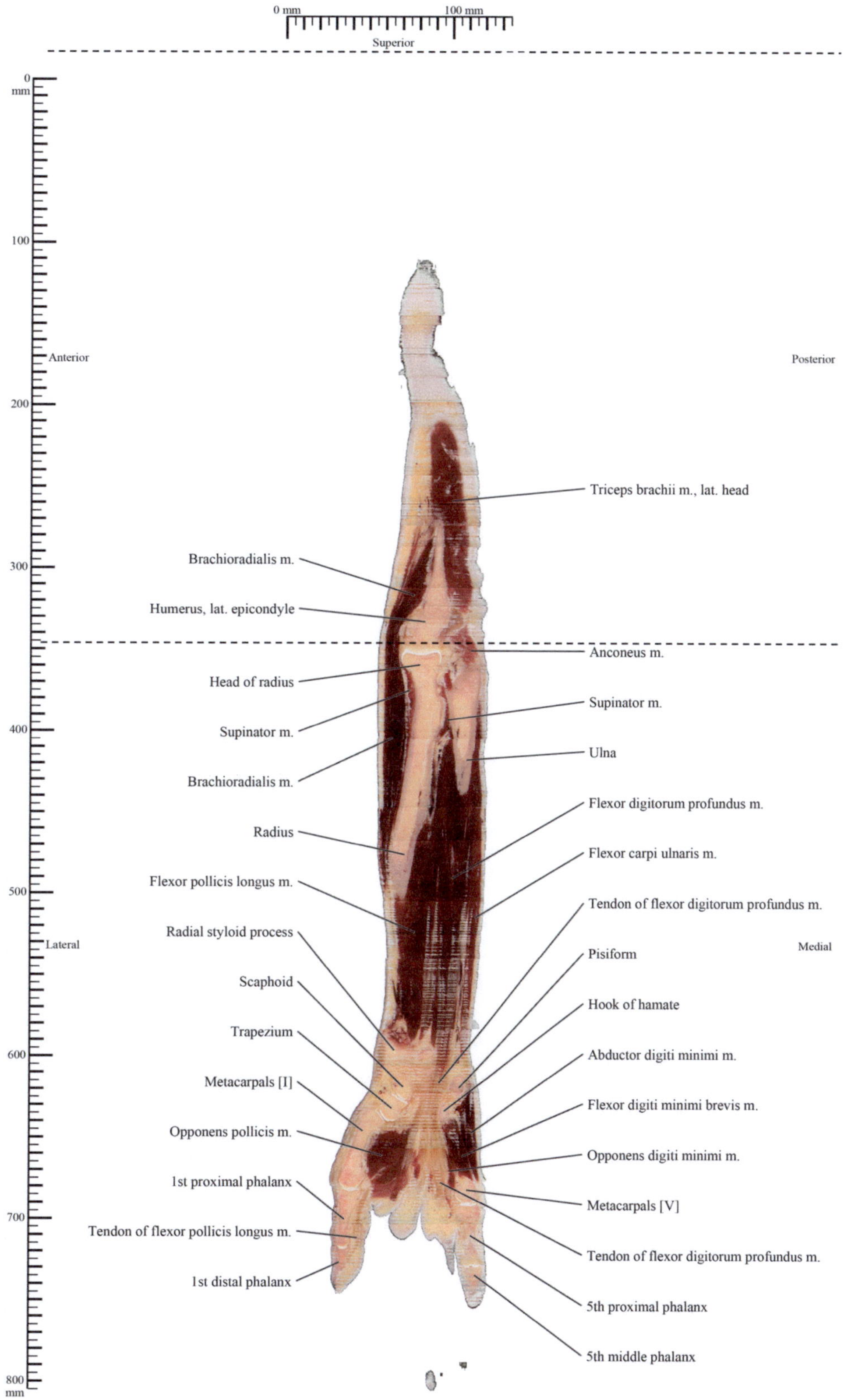

Sagittal (right arm) and coronal (right forearm and hand) (78.0 mm)

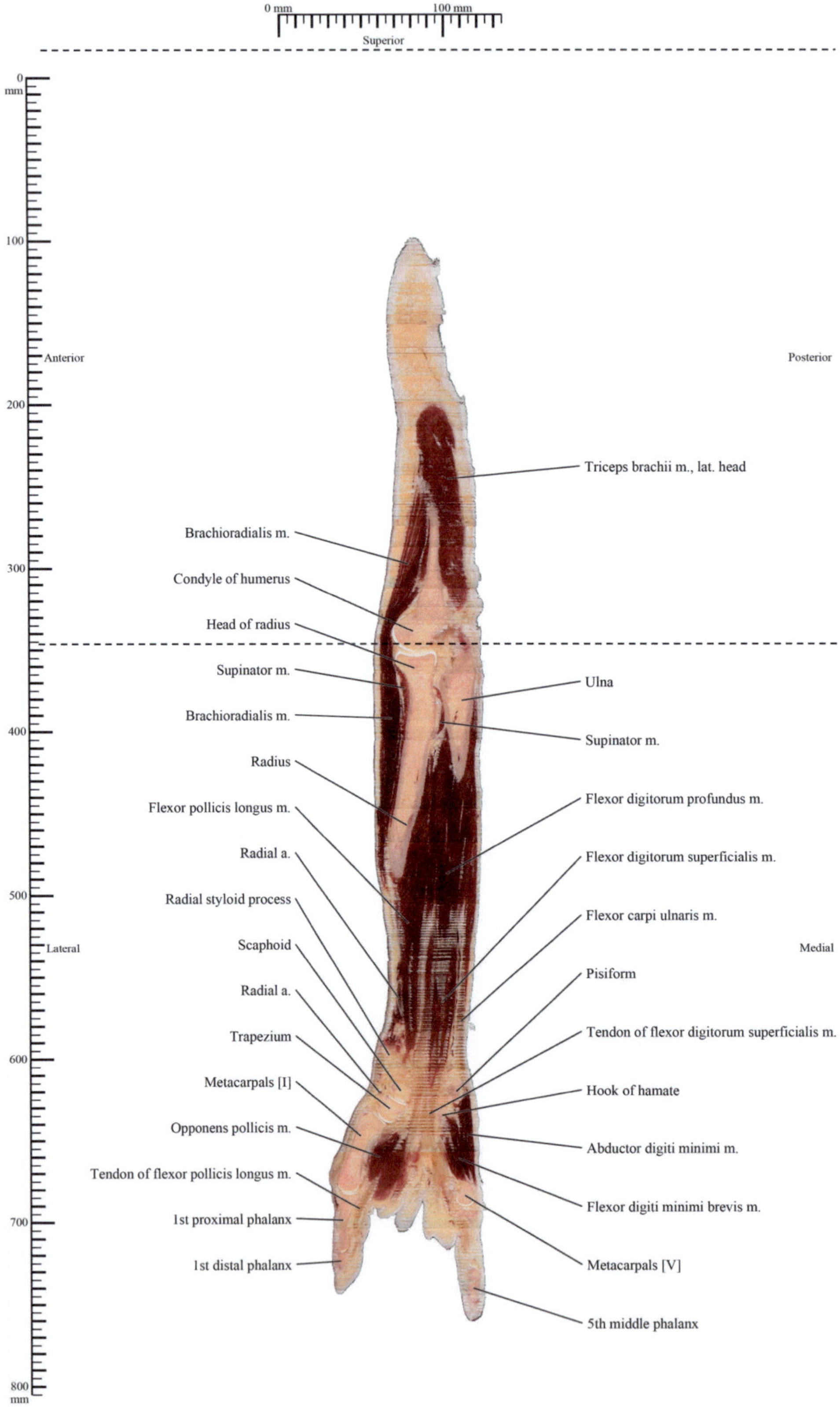

Sagittal (right arm) and coronal (right forearm and hand) (76.5 mm)

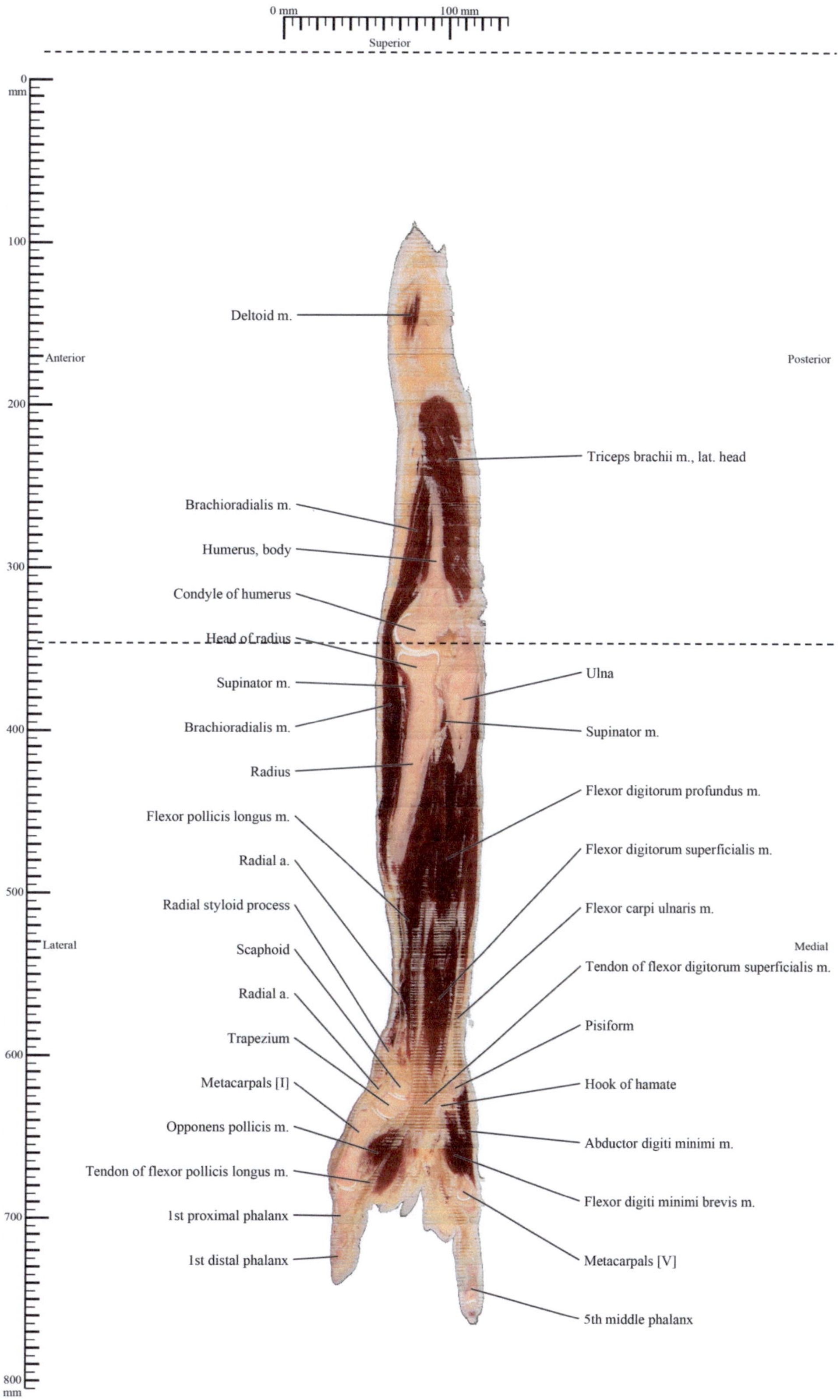

Sagittal (right arm) and coronal (right forearm and hand) (75.0 mm)

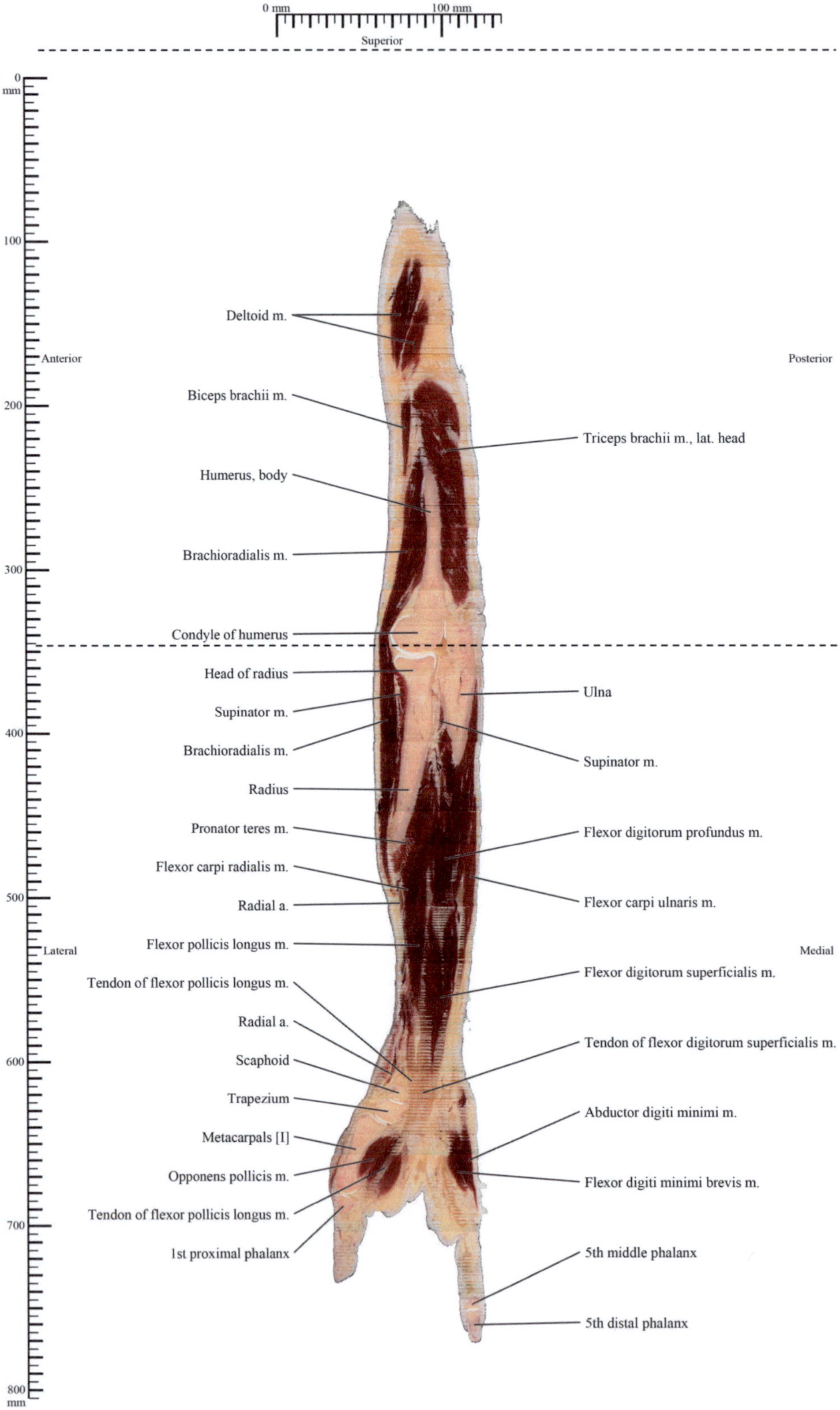

Sagittal (right arm) and coronal (right forearm and hand) (73.5 mm)

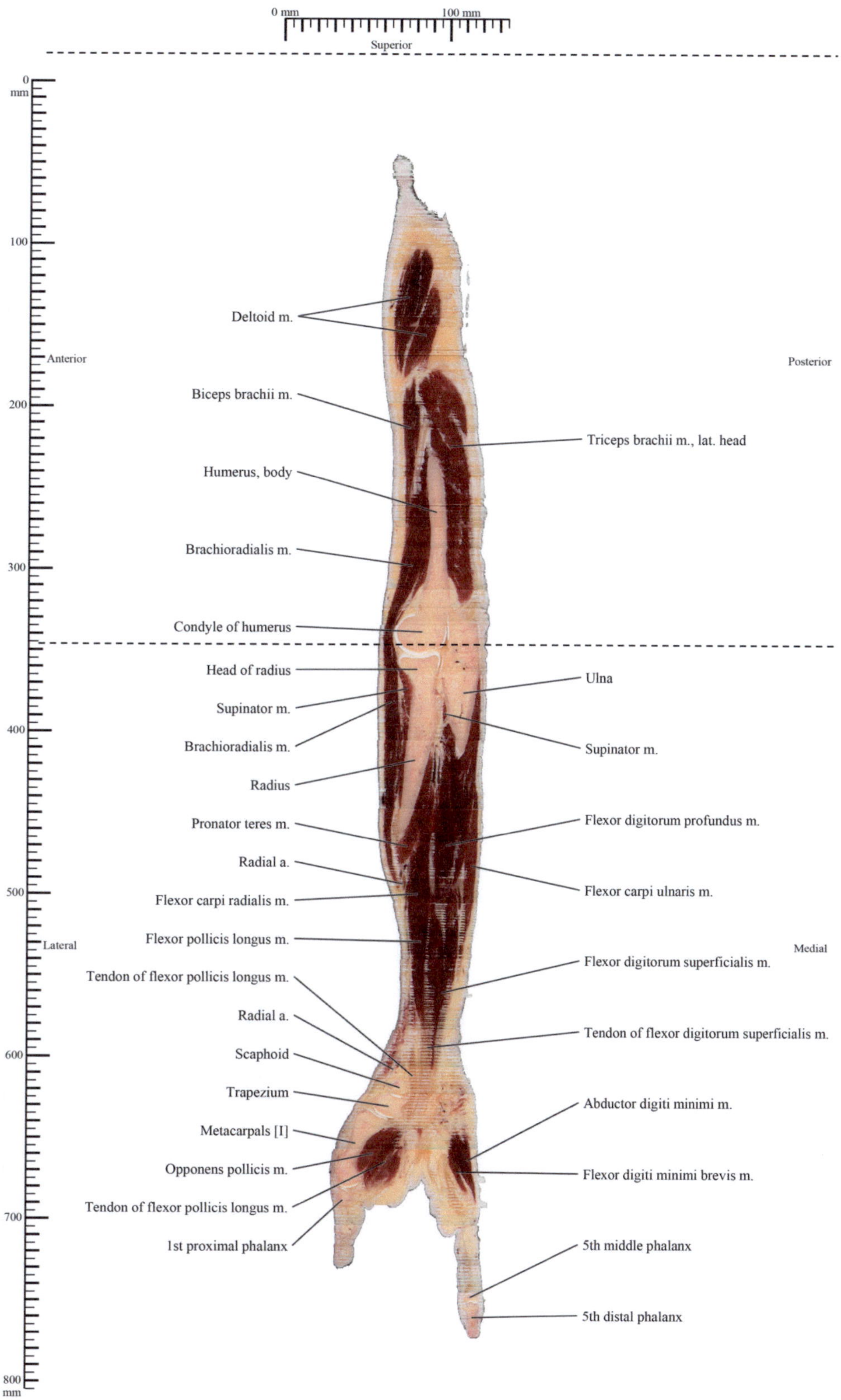

Sagittal (right arm) and coronal (right forearm and hand) (72.0 mm)

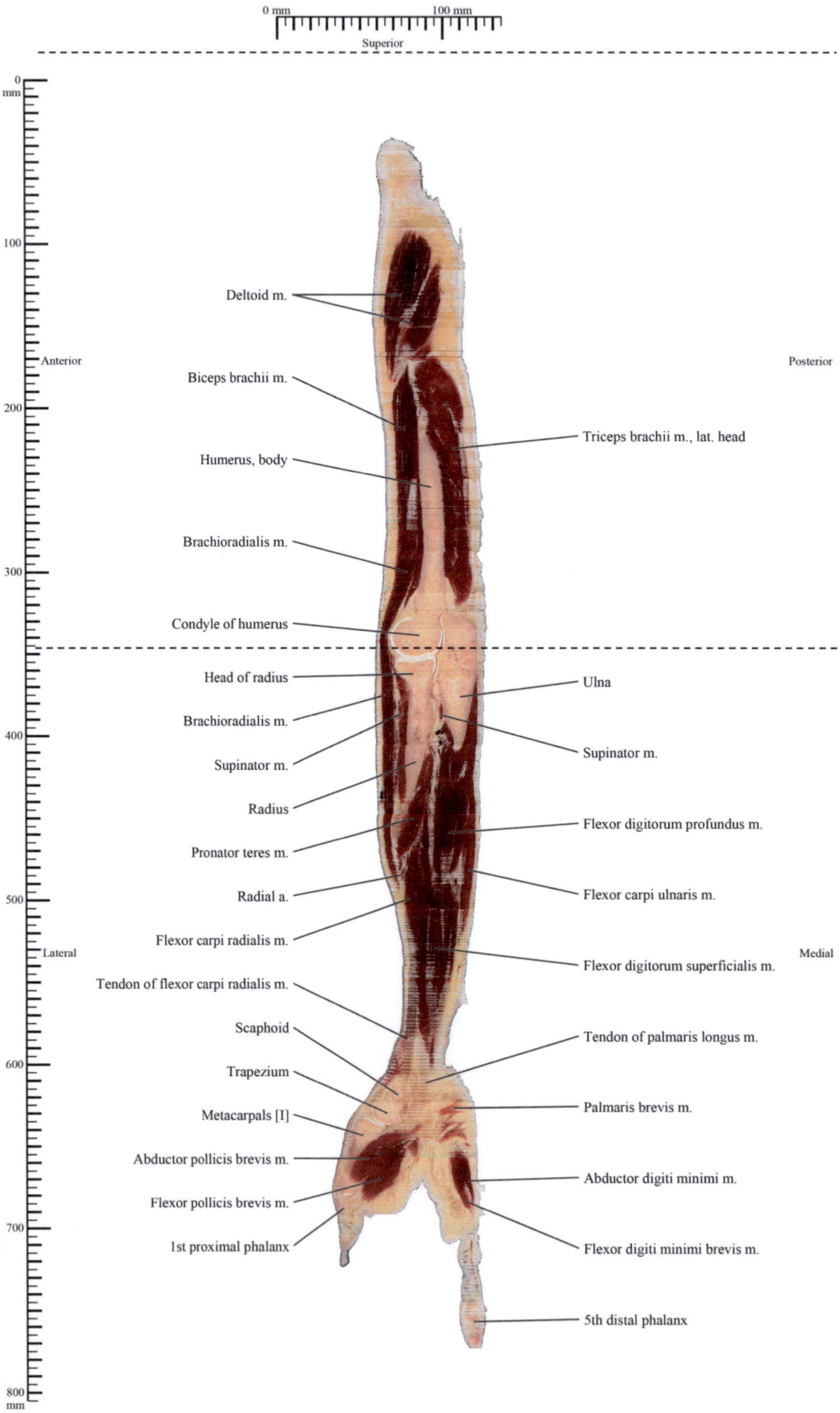

Sagittal (right arm) and coronal (right forearm and hand) (70.5 mm)

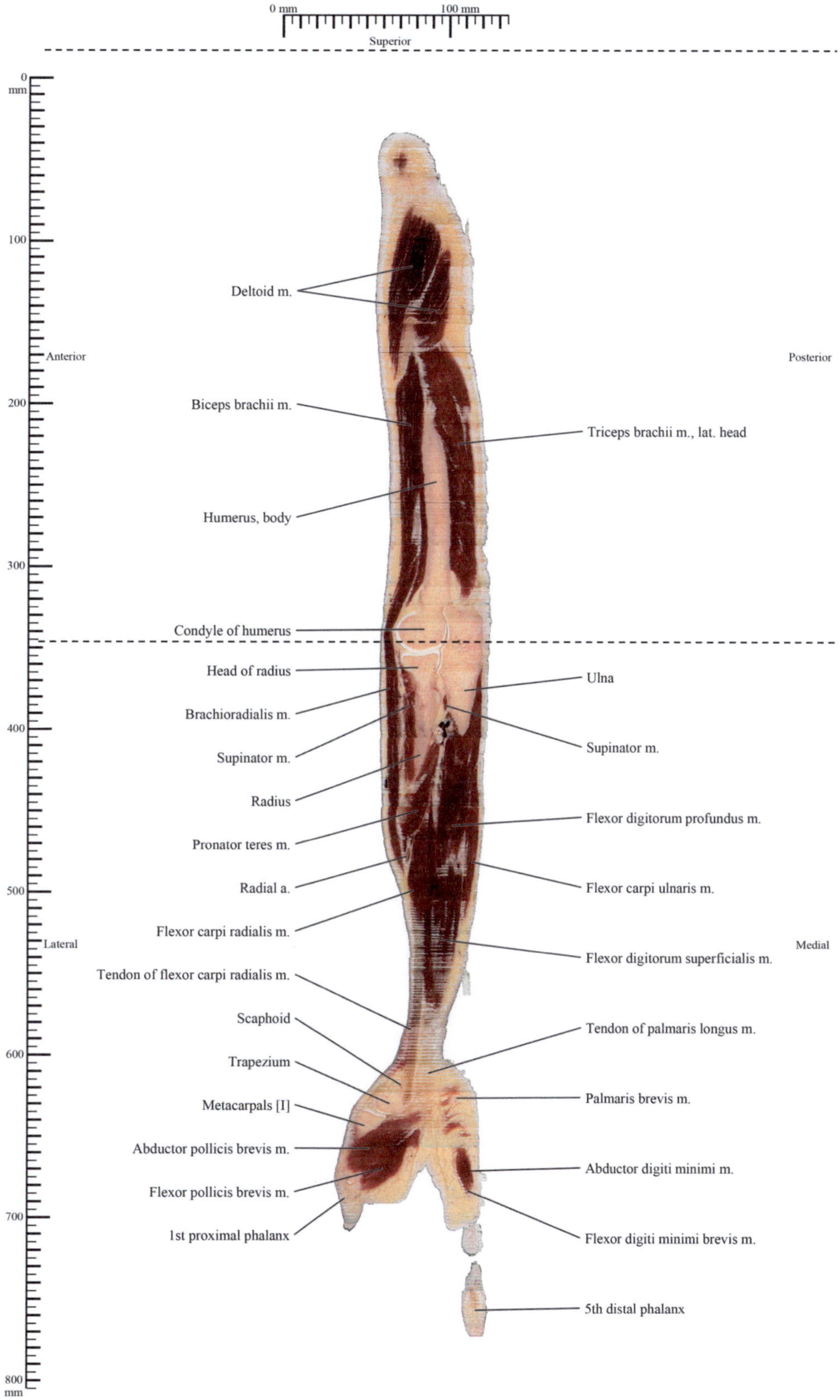

Sagittal (right arm) and coronal (right forearm and hand) (69.0 mm)

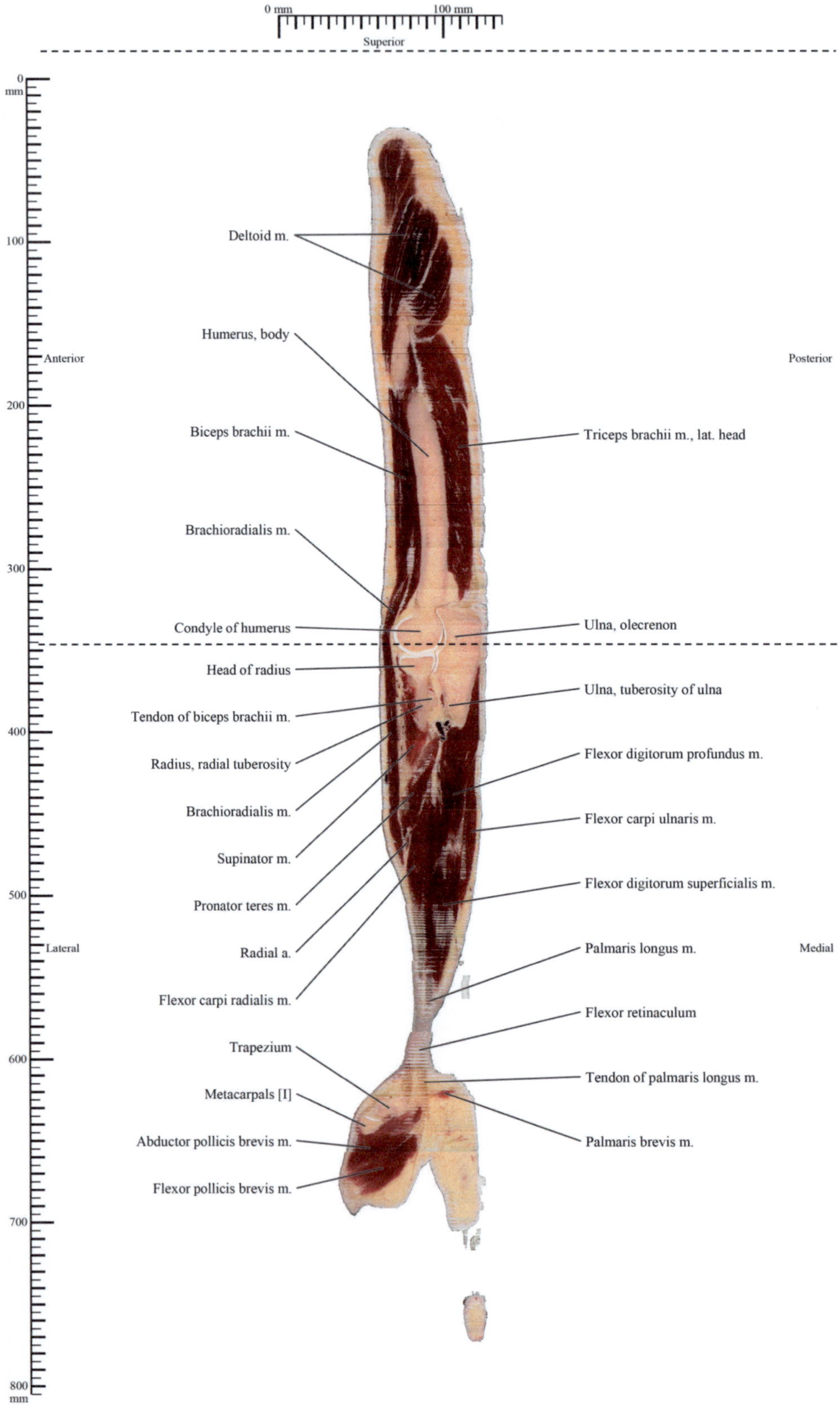

Sagittal (right arm) and coronal (right forearm and hand) (67.5 mm)

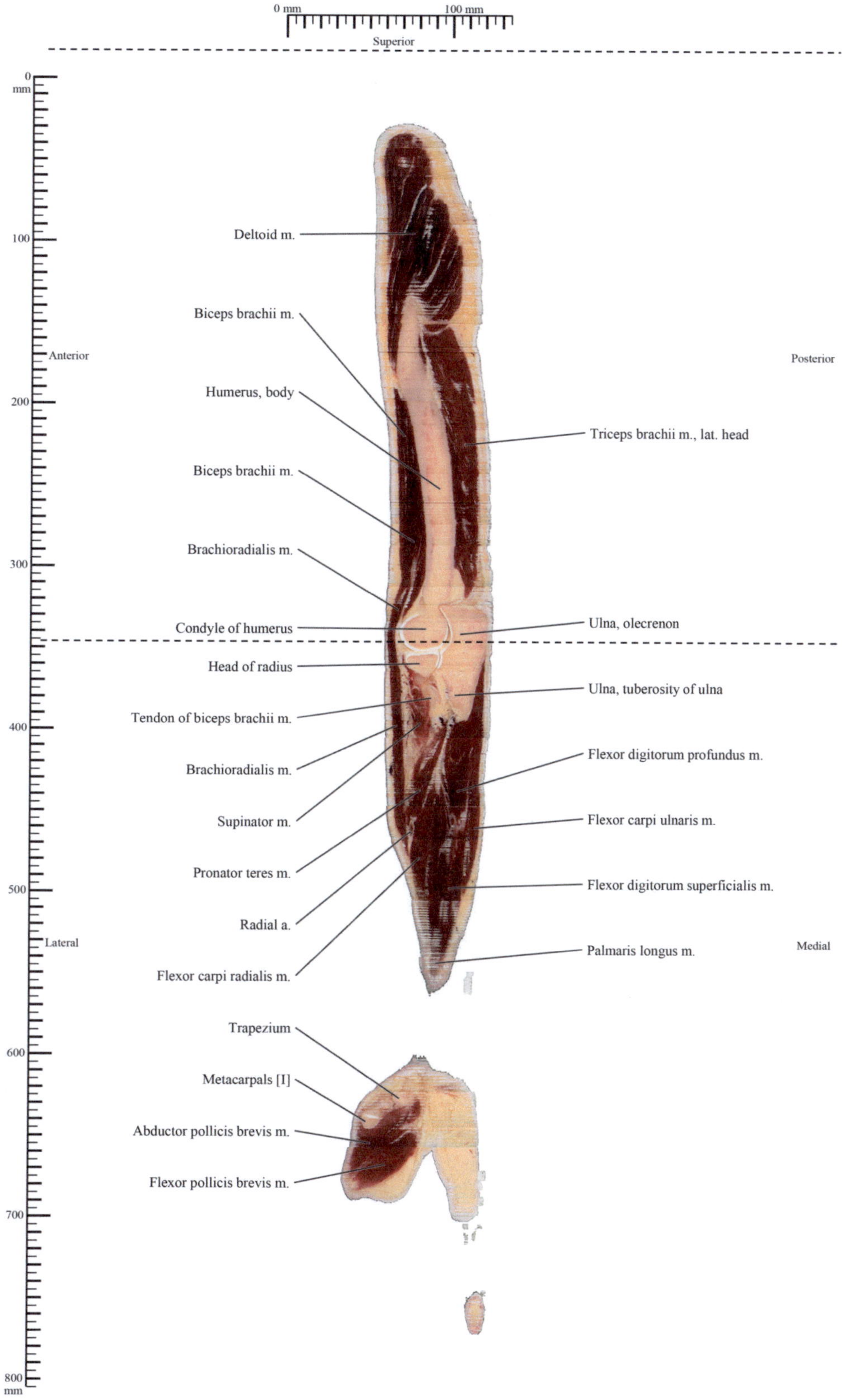

Sagittal (right arm) and coronal (right forearm and hand) (66.0 mm)

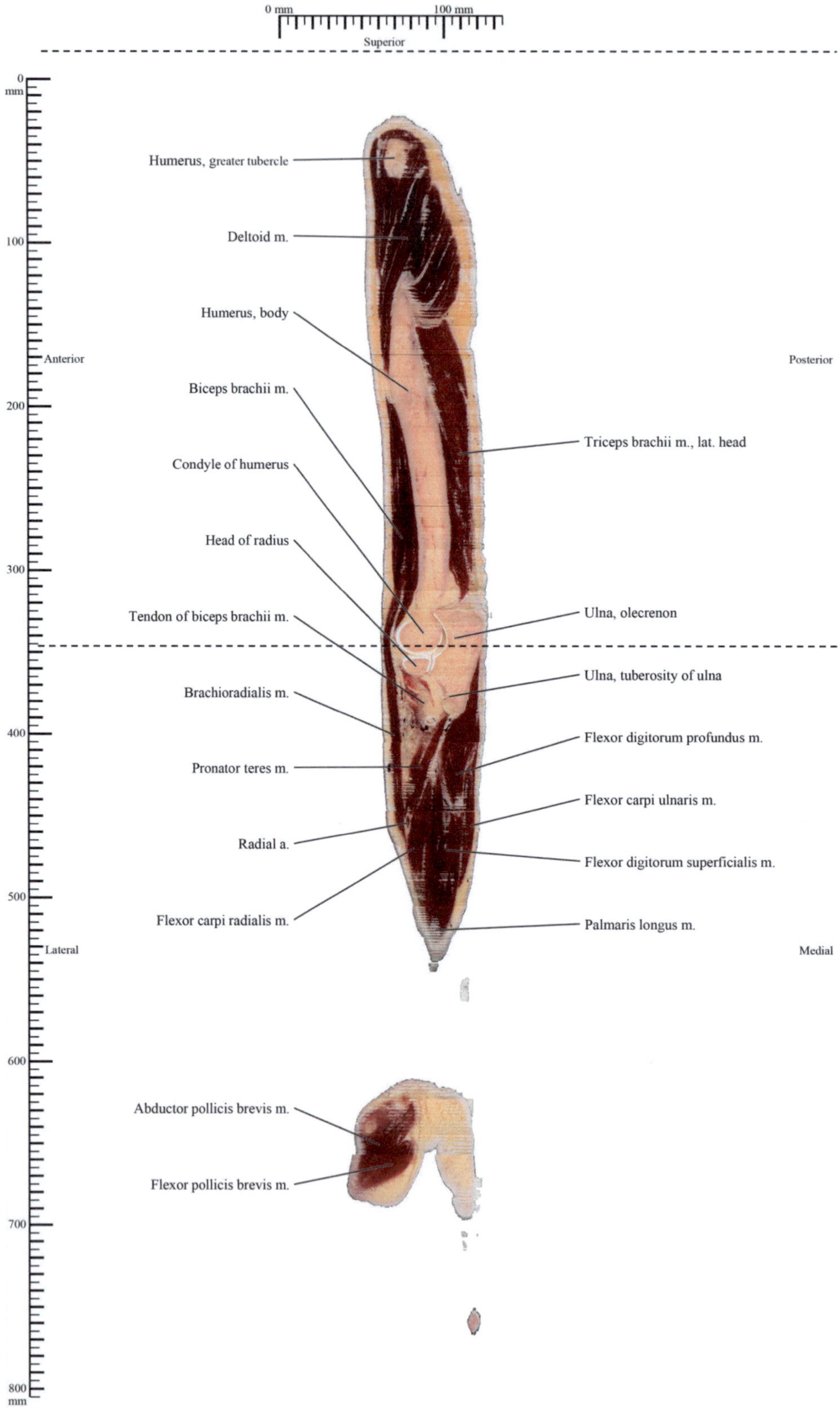

Sagittal (right arm) and coronal (right forearm and hand) (64.5 mm)

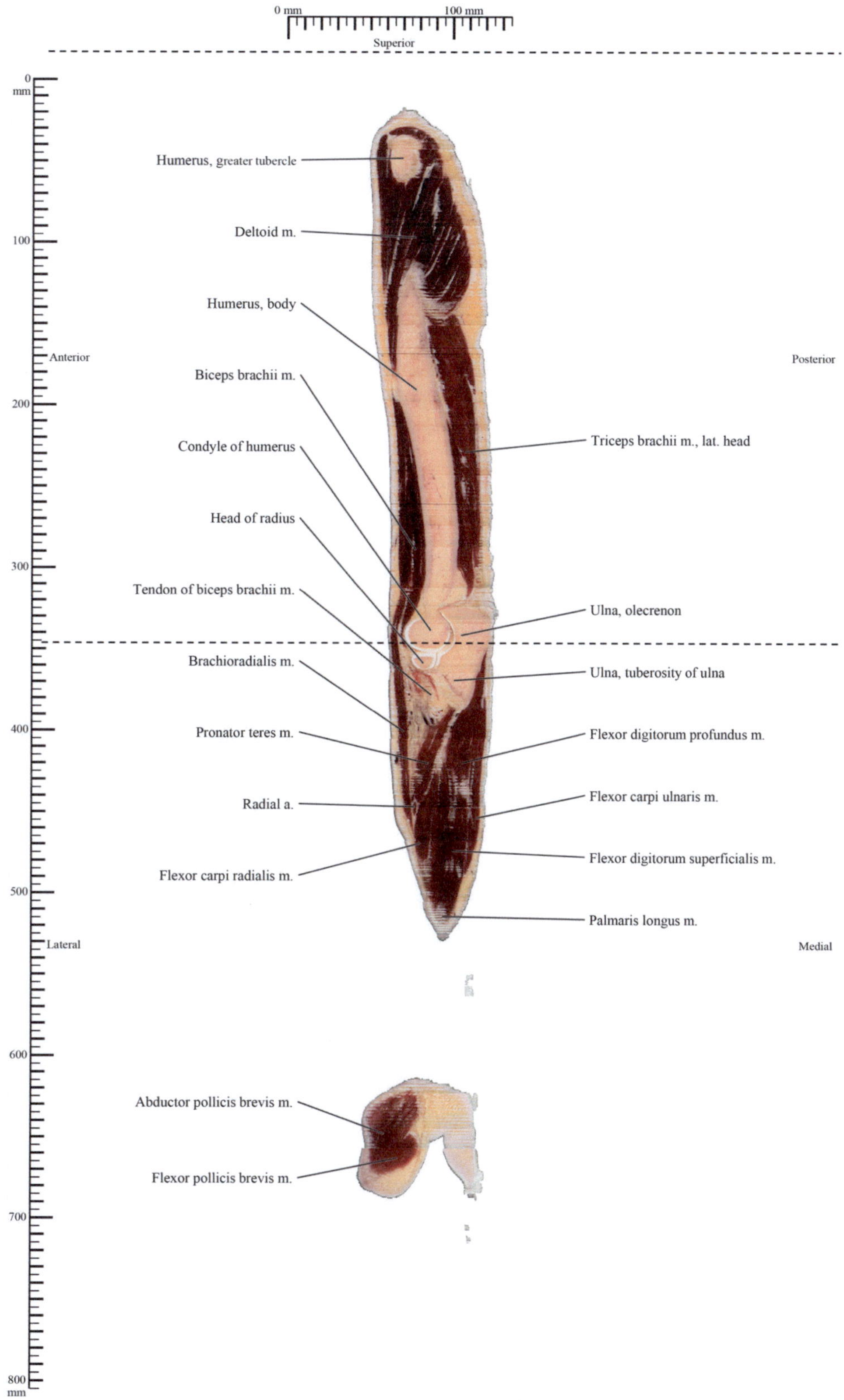

Sagittal (right arm) and coronal (right forearm and hand) (63.0 mm)

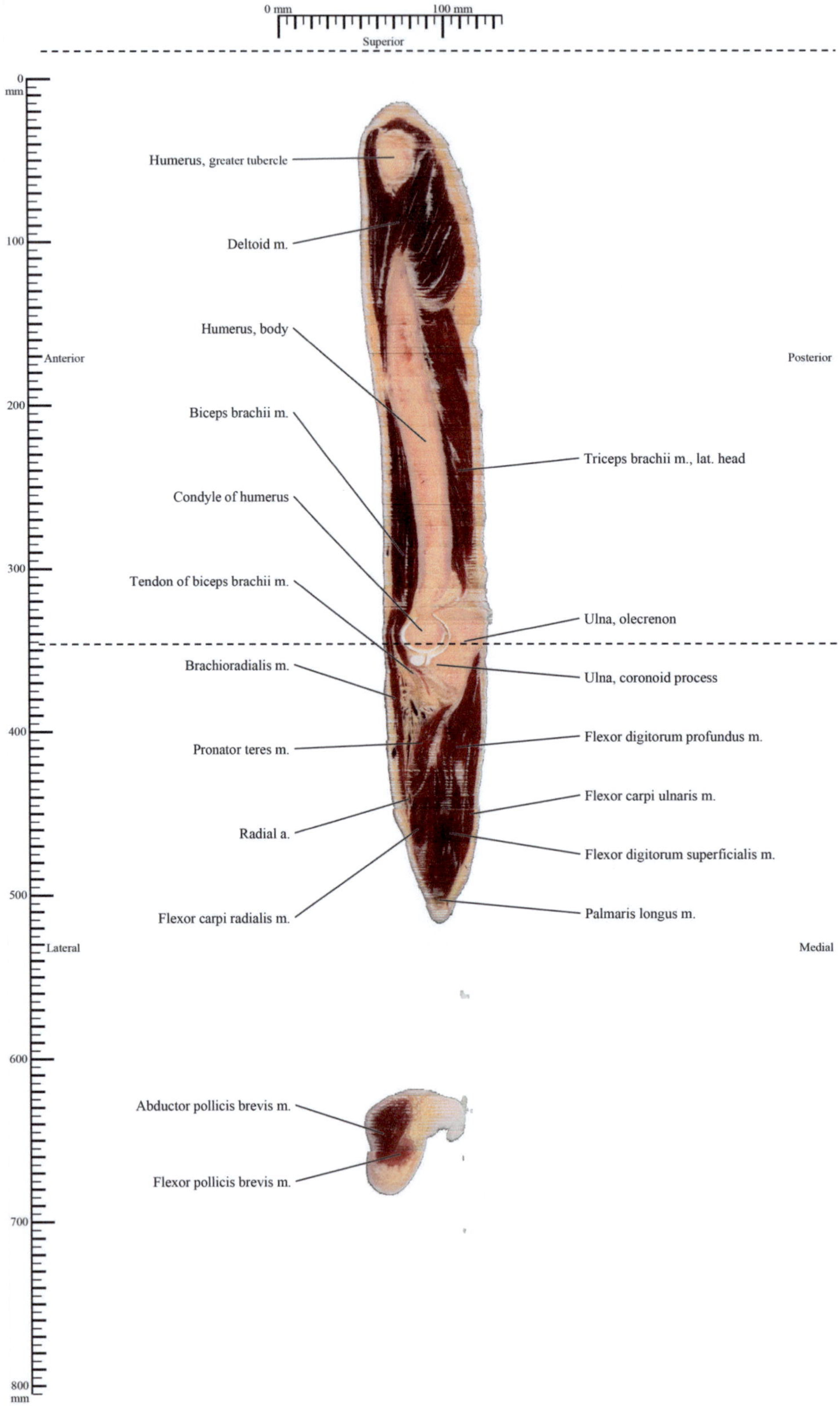

Sagittal (right arm) and coronal (right forearm and hand) (61.5 mm)

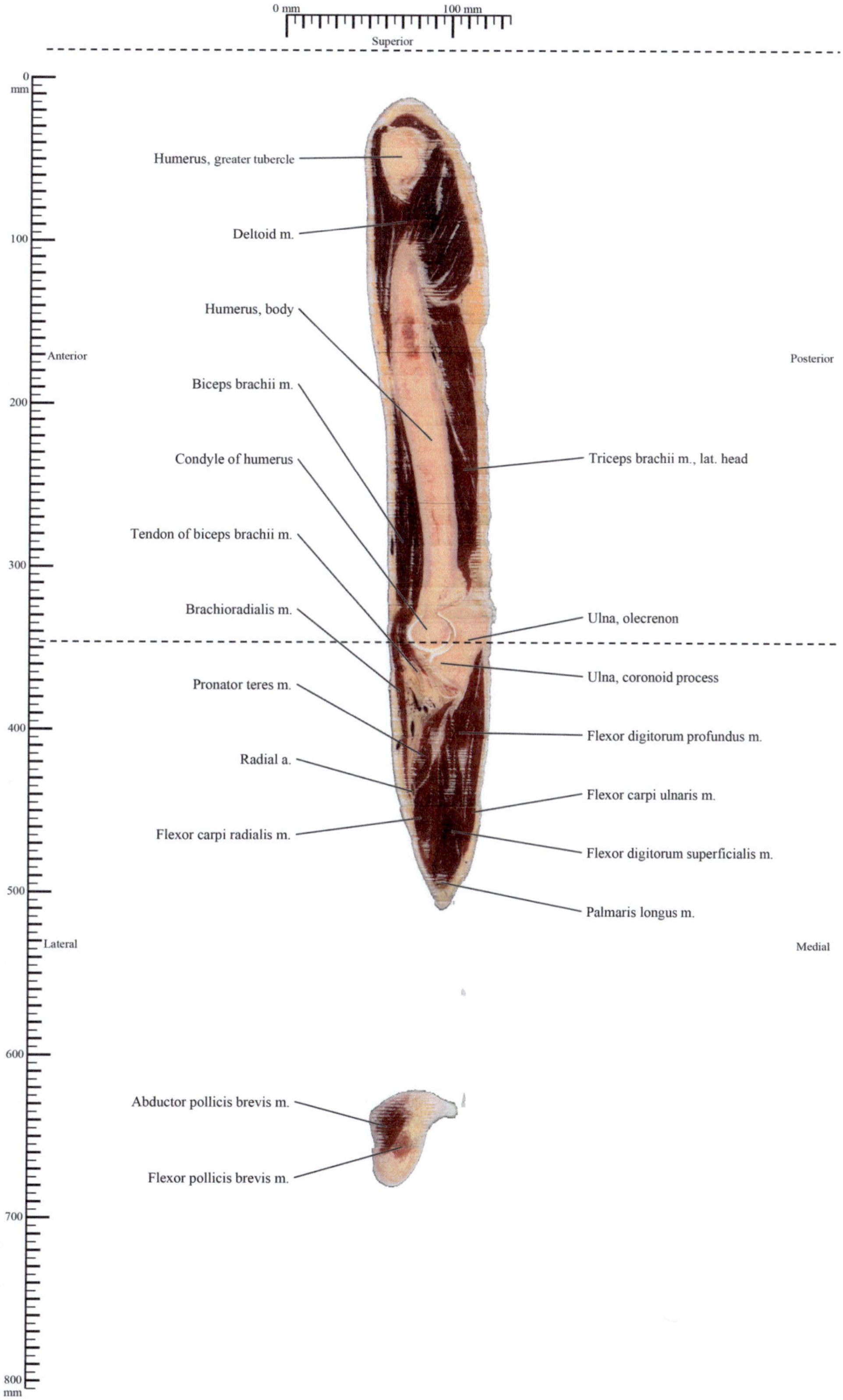

Sagittal (right arm) and coronal (right forearm and hand) (60.0 mm)

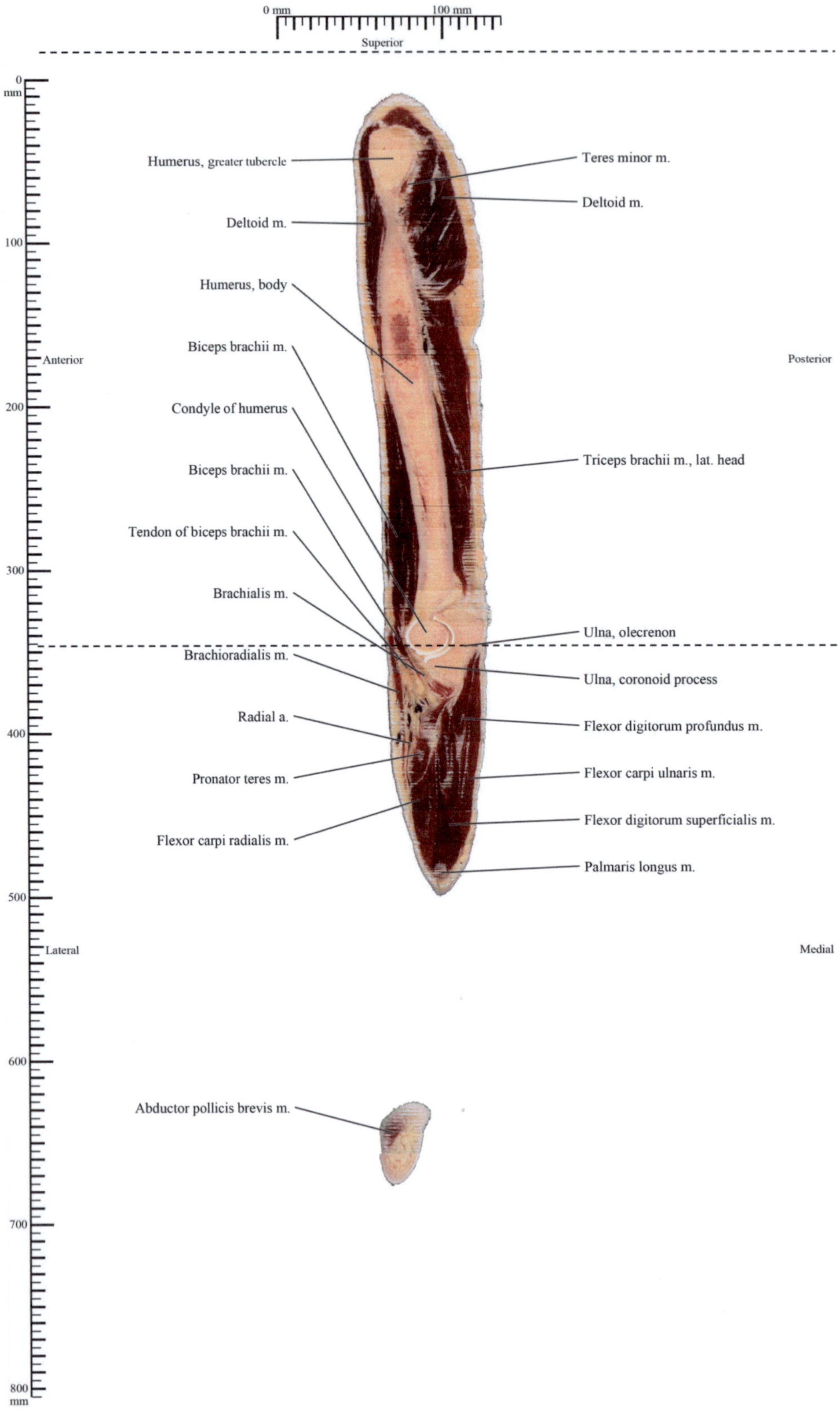

Sagittal (right arm) and coronal (right forearm and hand) (58.5 mm)

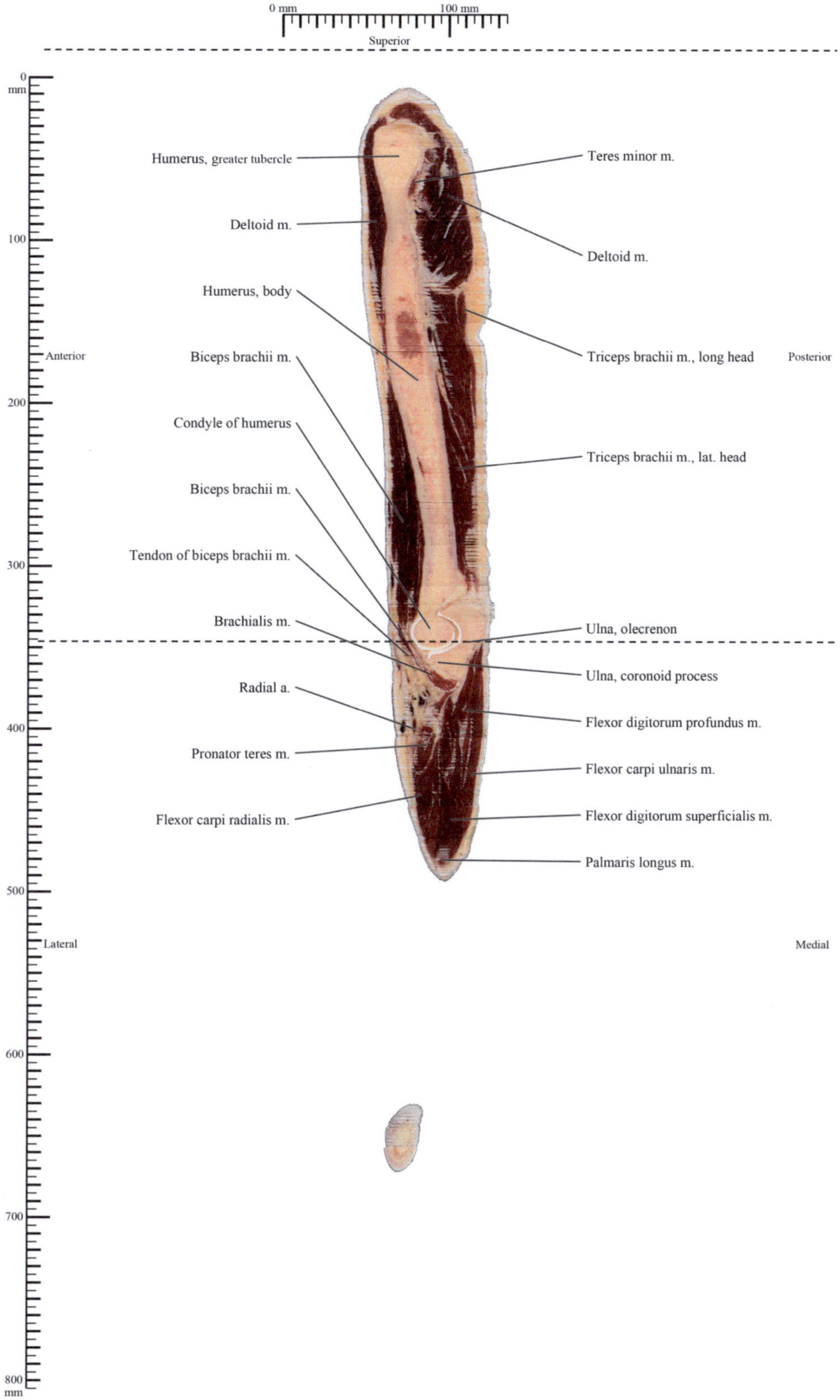

Sagittal (right arm) and coronal (right forearm and hand) (57.0 mm)

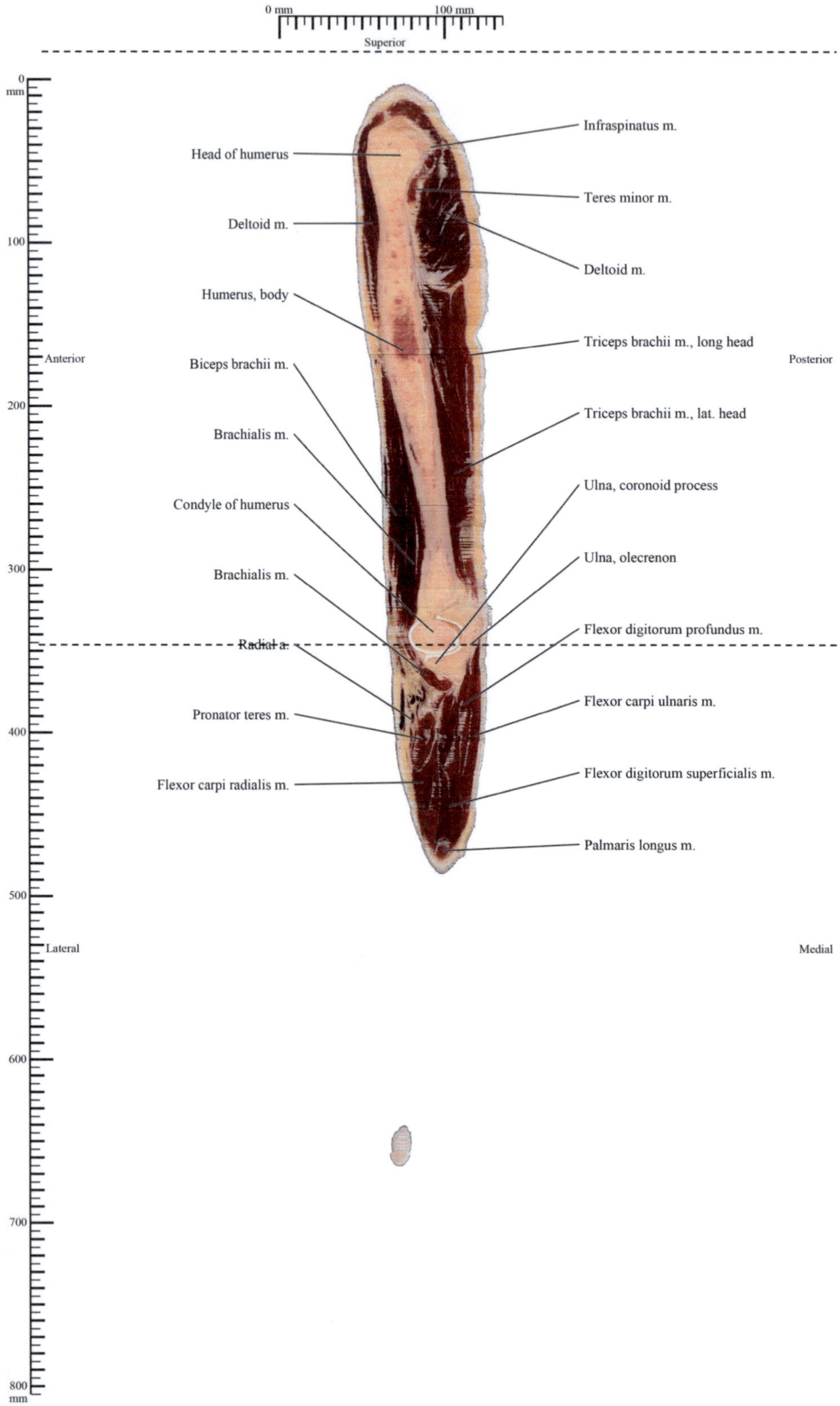

Sagittal (right arm) and coronal (right forearm and hand) (55.5 mm)

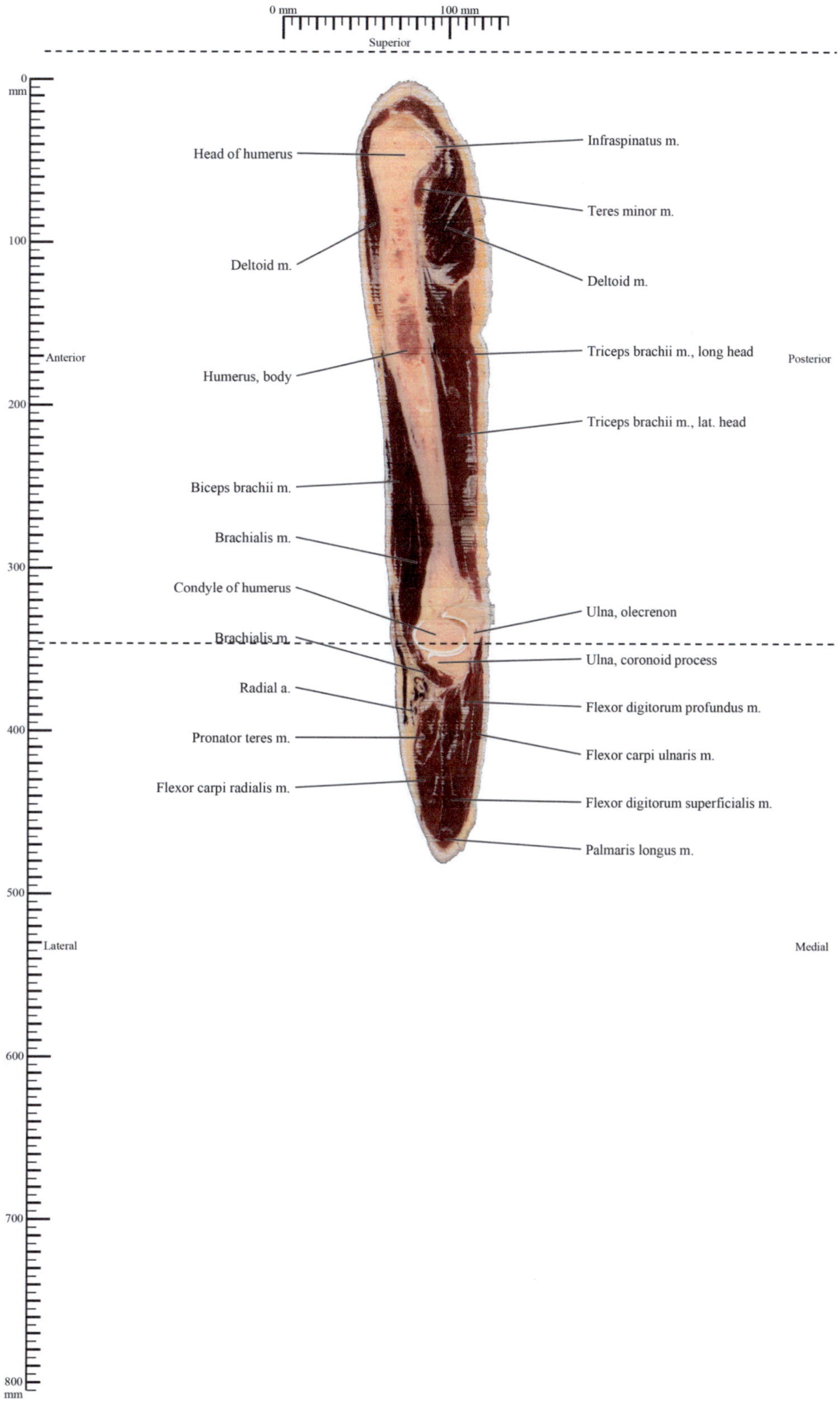

Sagittal (right arm) and coronal (right forearm and hand) (54.0 mm)

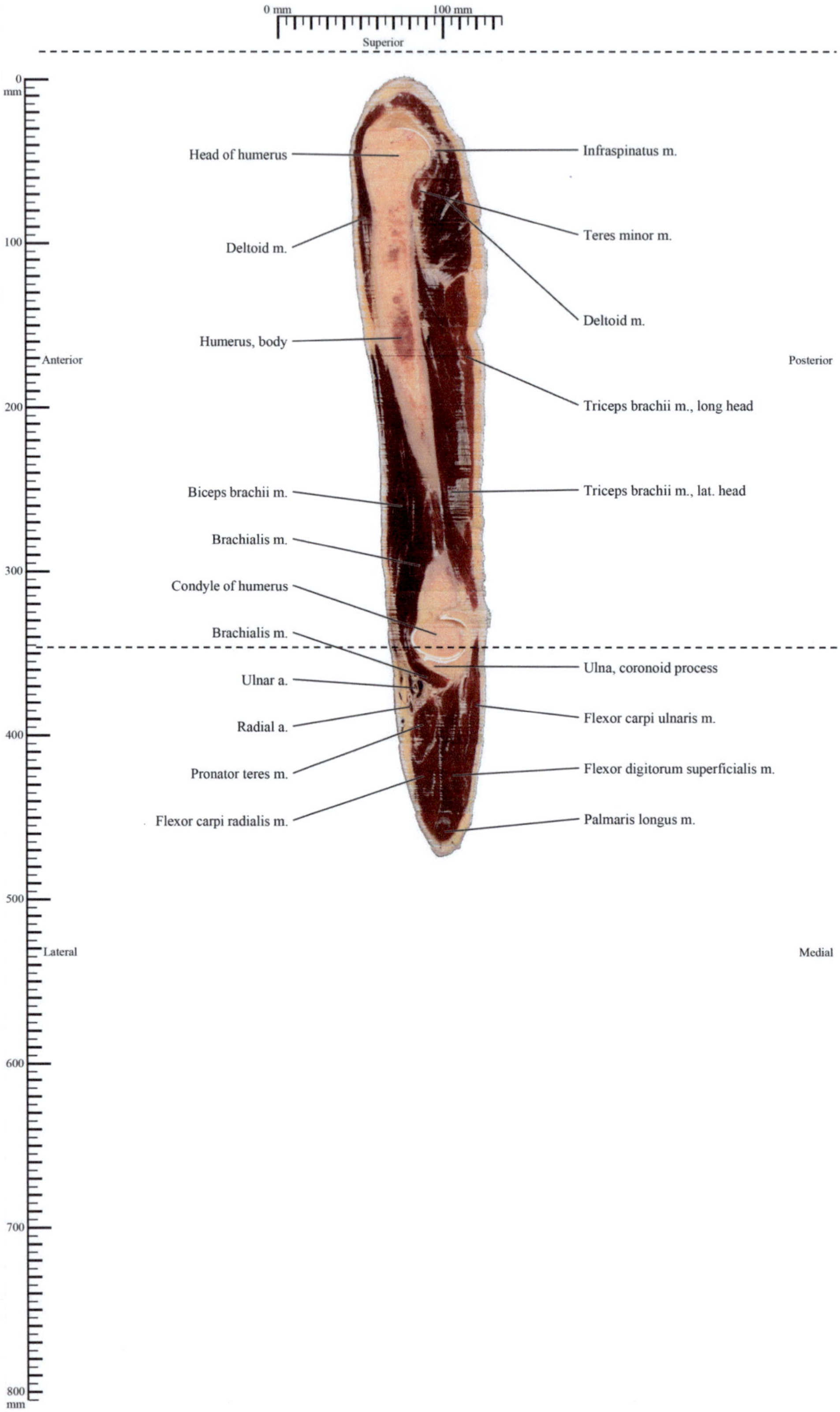

Sagittal (right arm) and coronal (right forearm and hand) (52.5 mm)

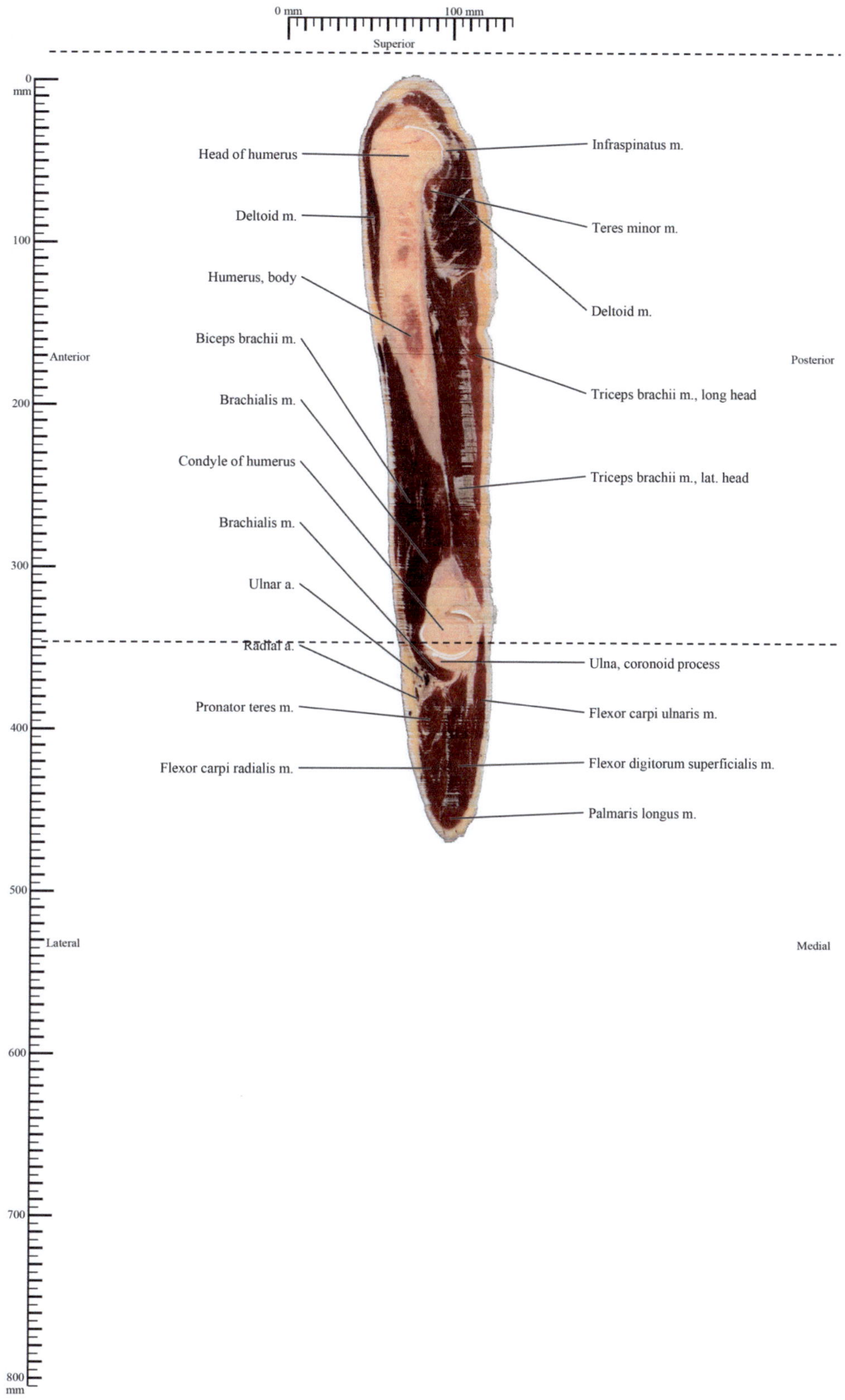

Sagittal (right arm) and coronal (right forearm and hand) (51.0 mm)

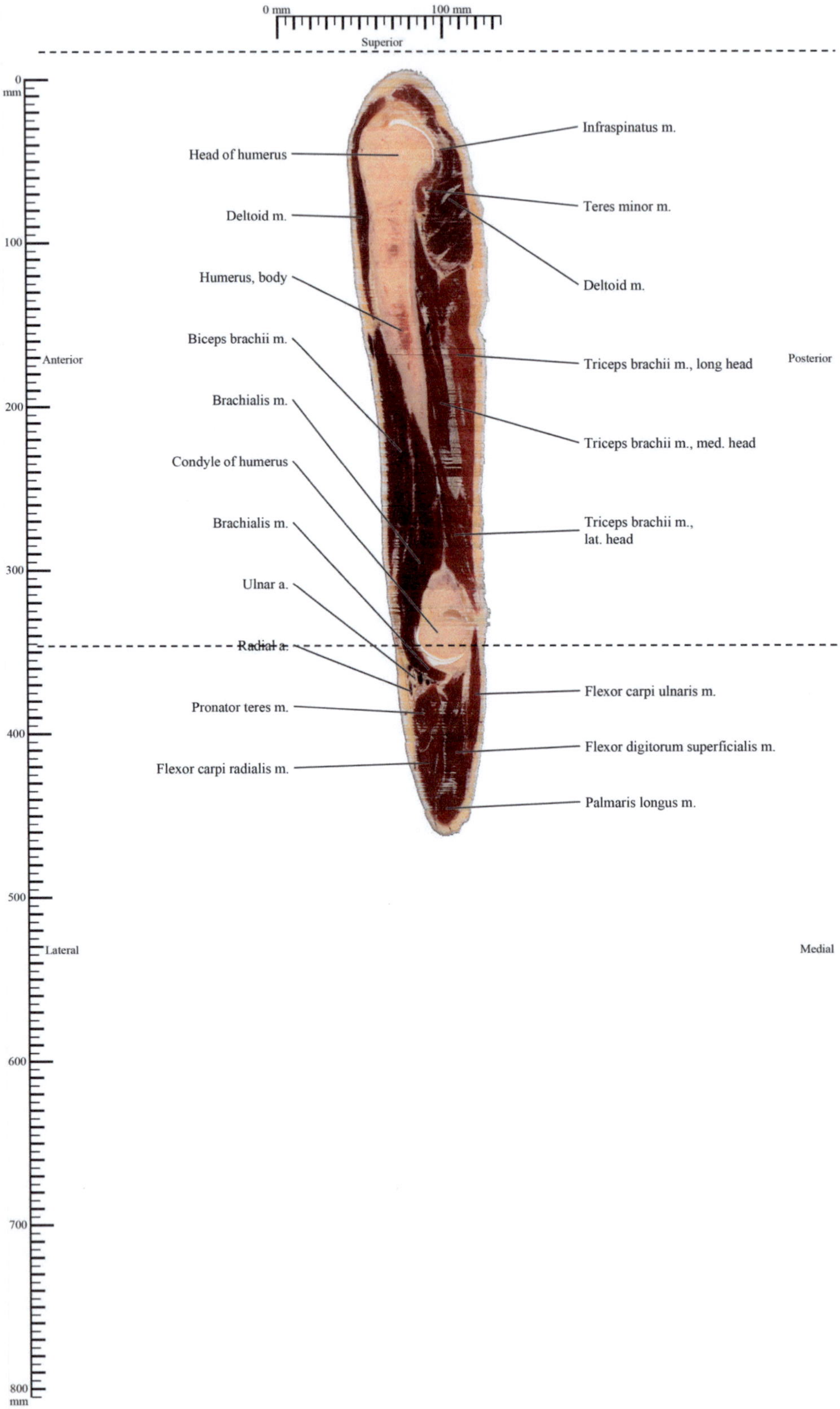

Sagittal (right arm) and coronal (right forearm and hand) (49.5 mm)

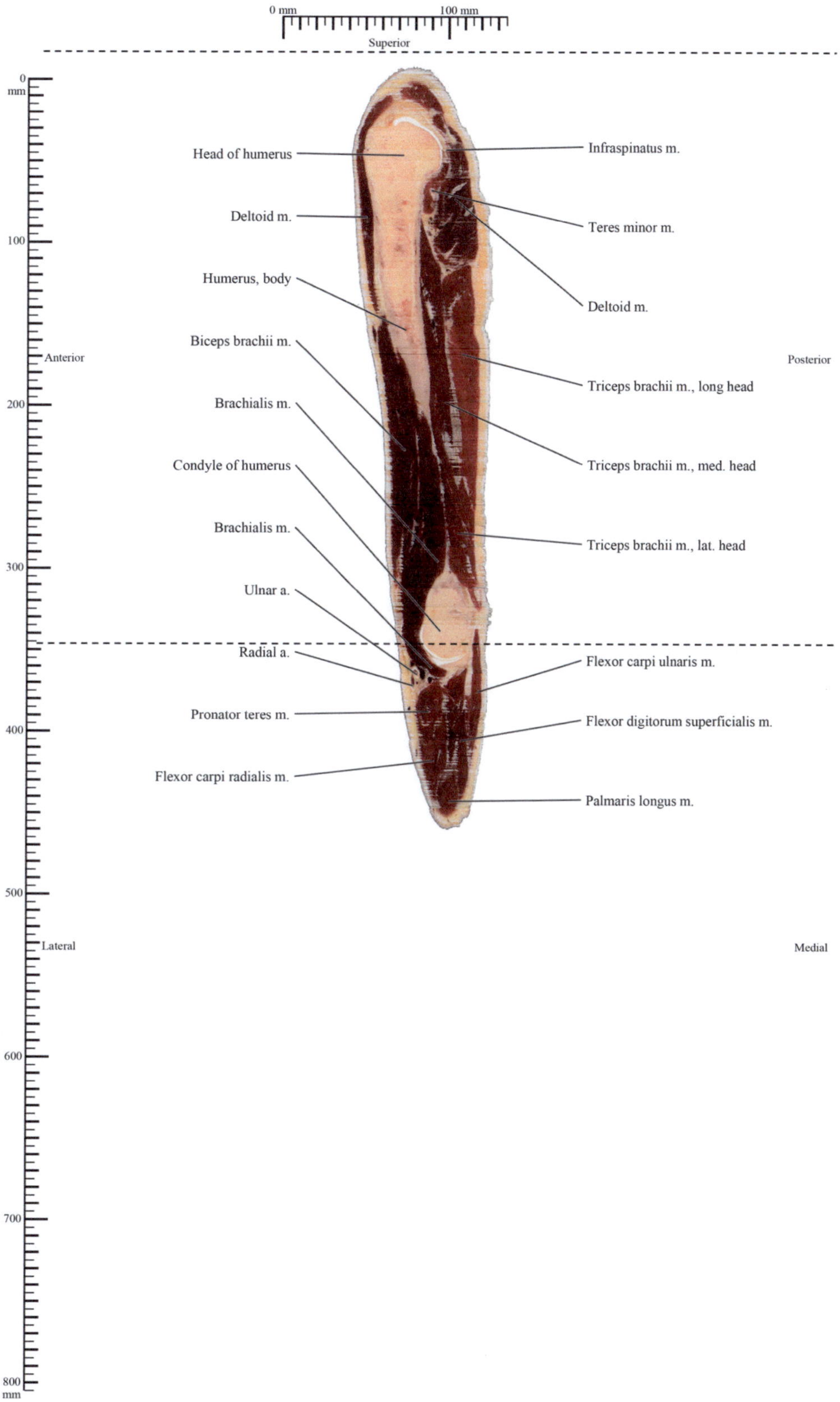

Sagittal (right arm) and coronal (right forearm and hand) (48.0 mm)

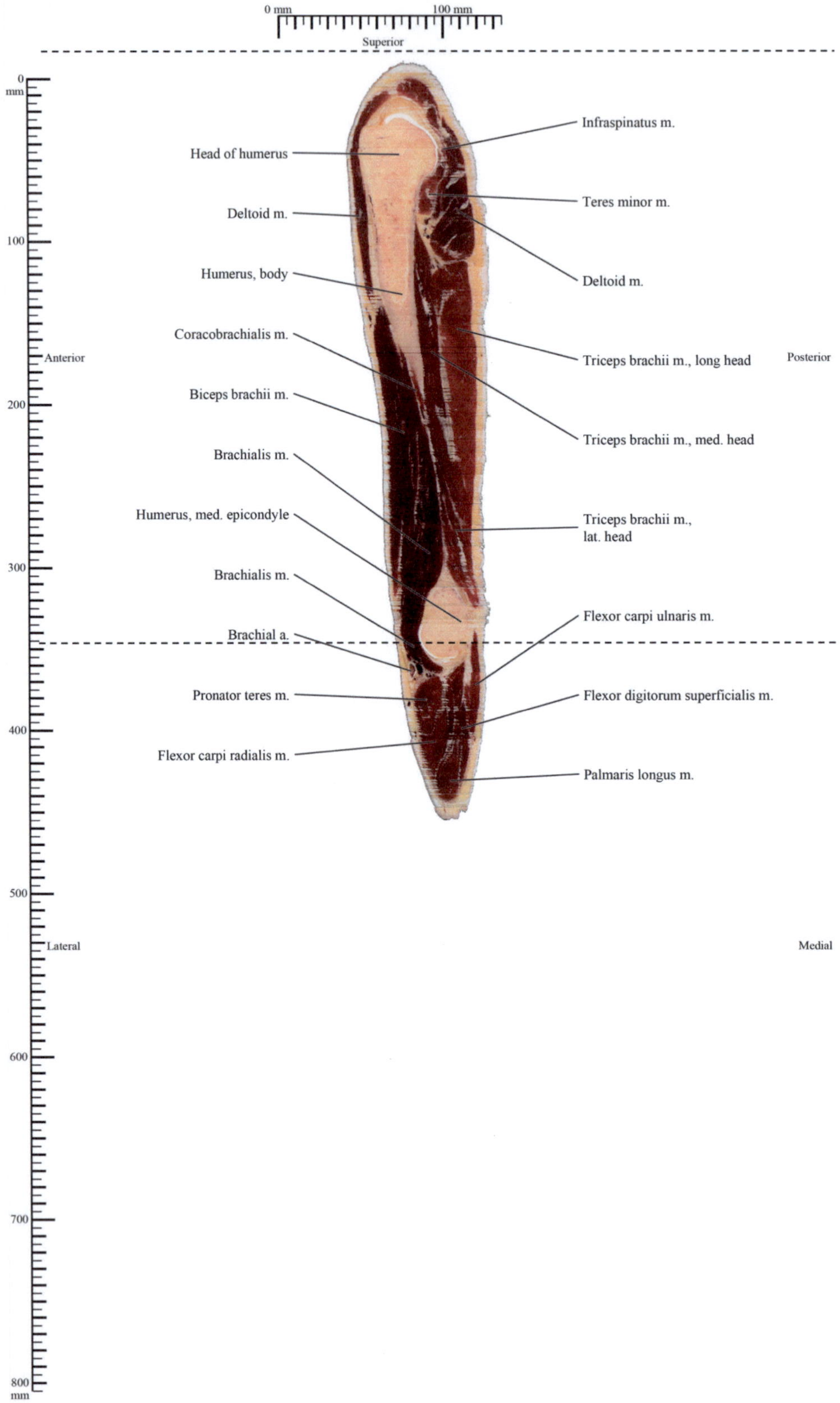

Sagittal (right arm) and coronal (right forearm and hand) (46.5 mm)

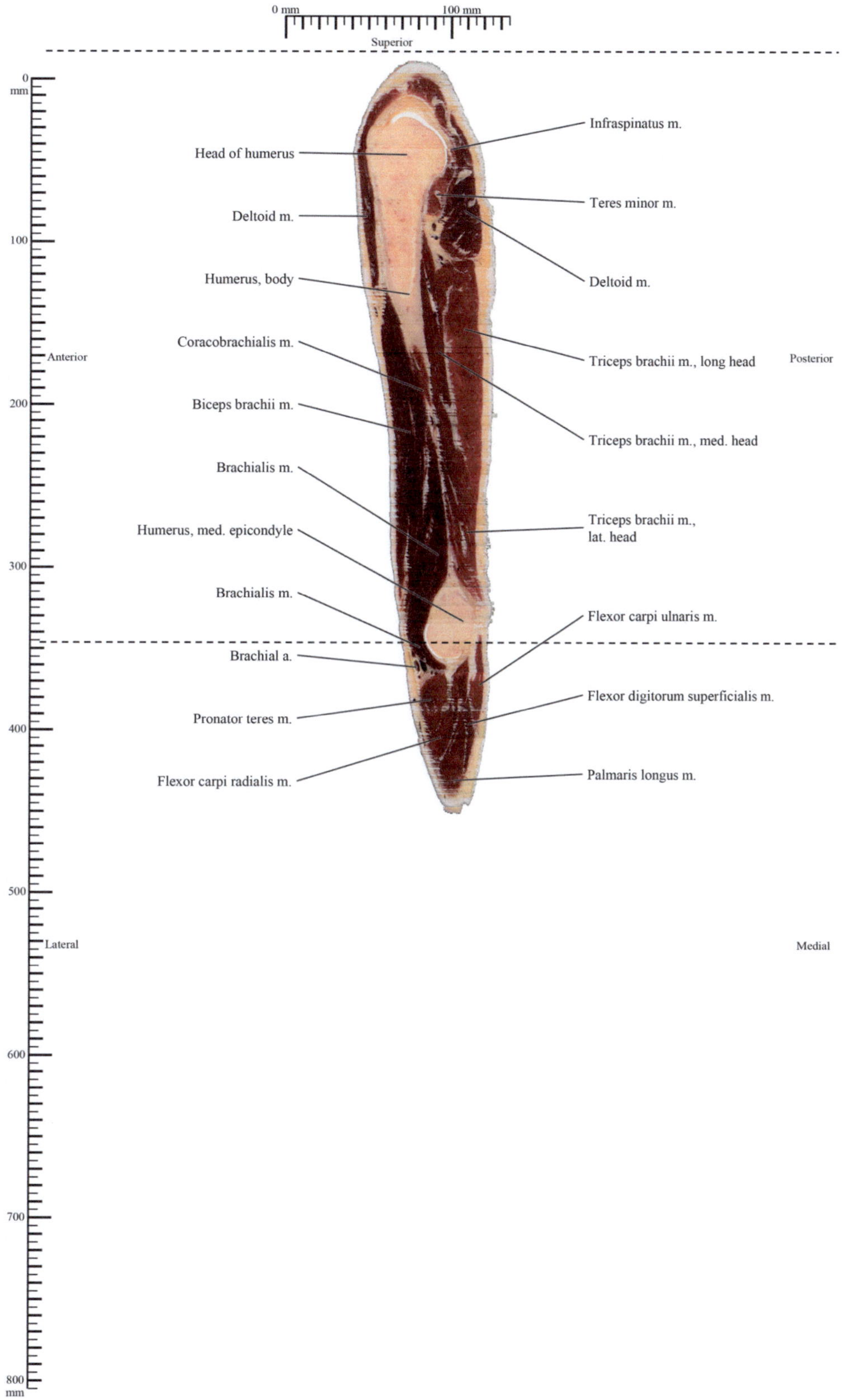

Sagittal (right arm) and coronal (right forearm and hand) (45.0 mm)

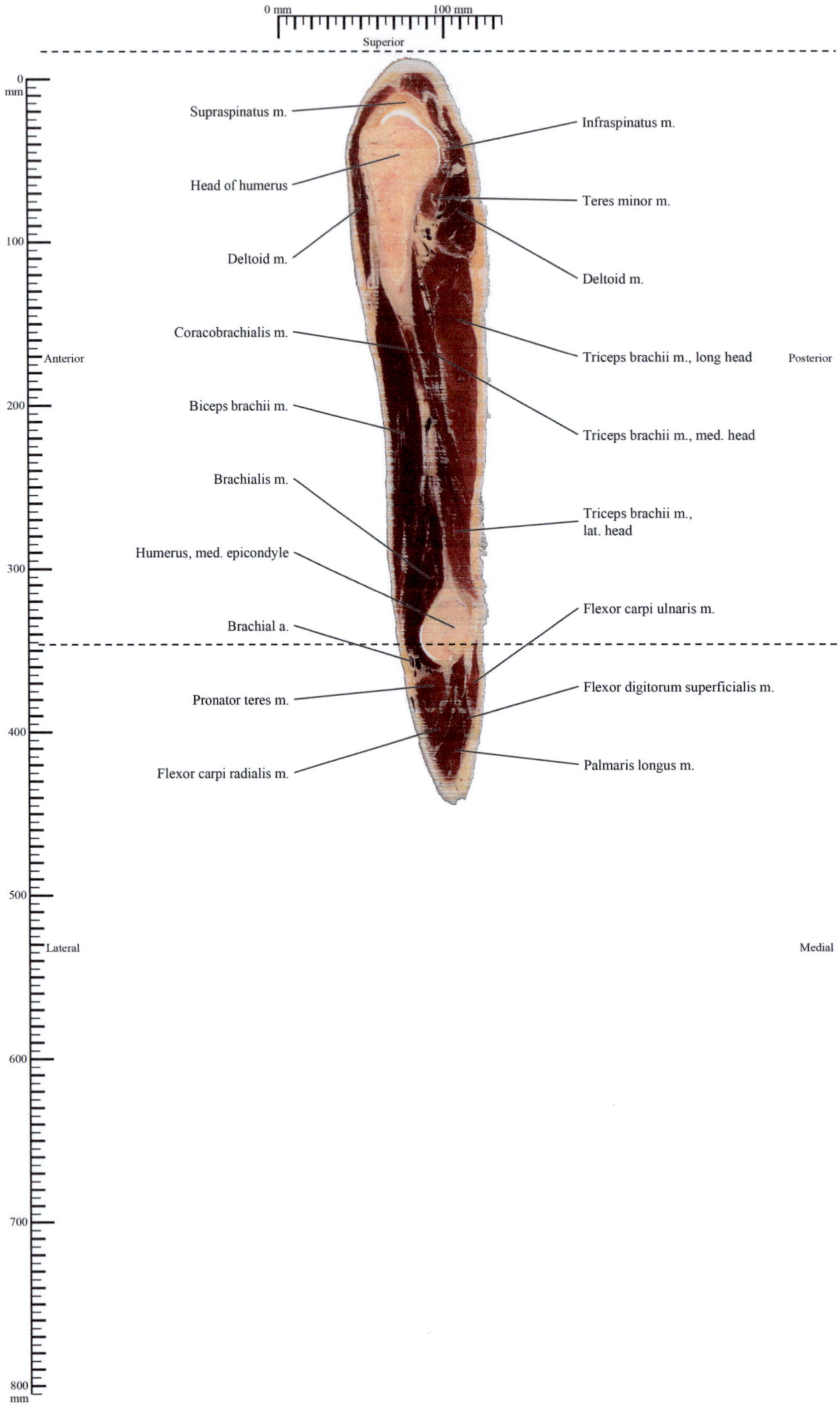

Sagittal (right arm) and coronal (right forearm and hand) (43.5 mm)

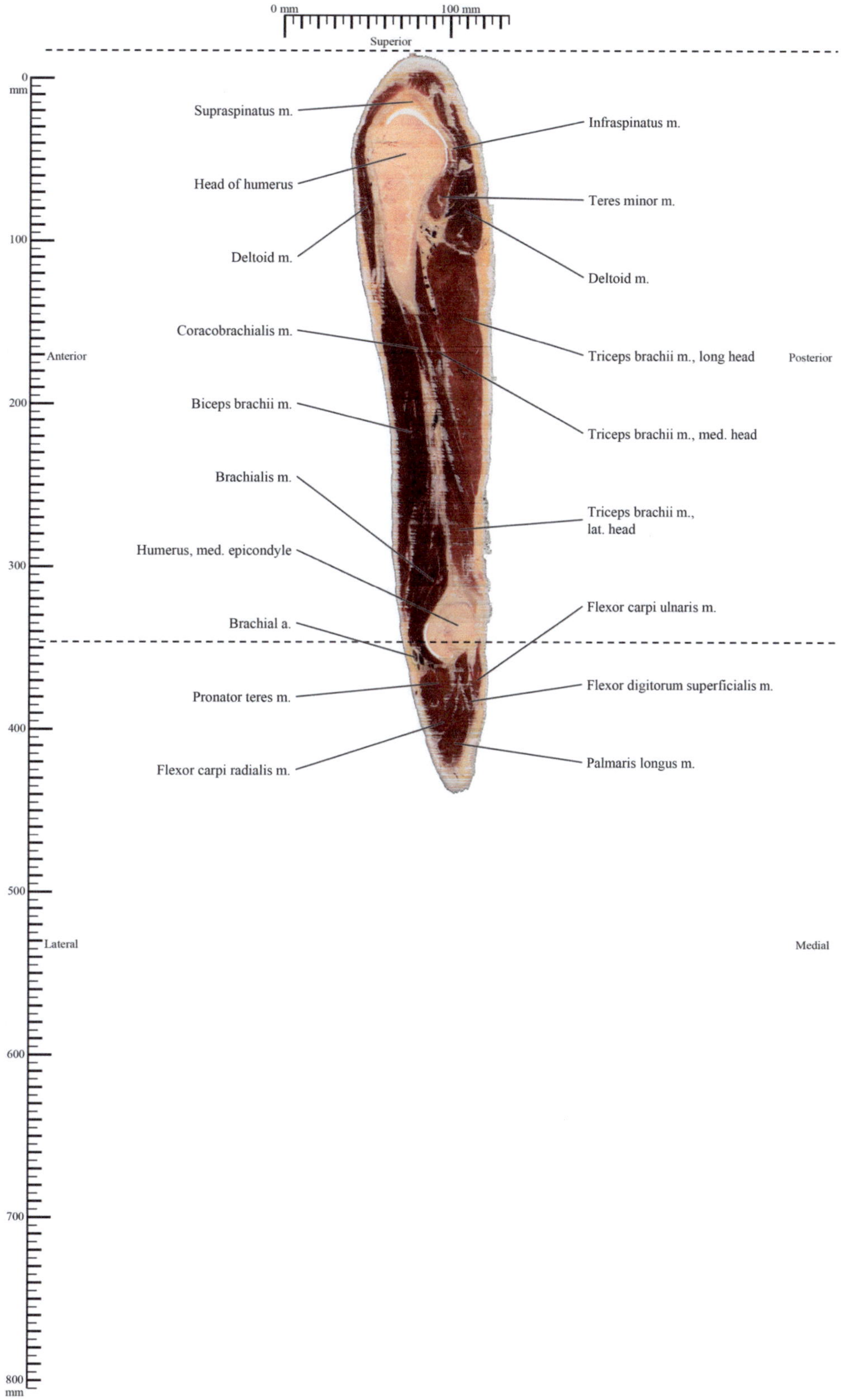

Sagittal (right arm) and coronal (right forearm and hand) (42.0 mm)

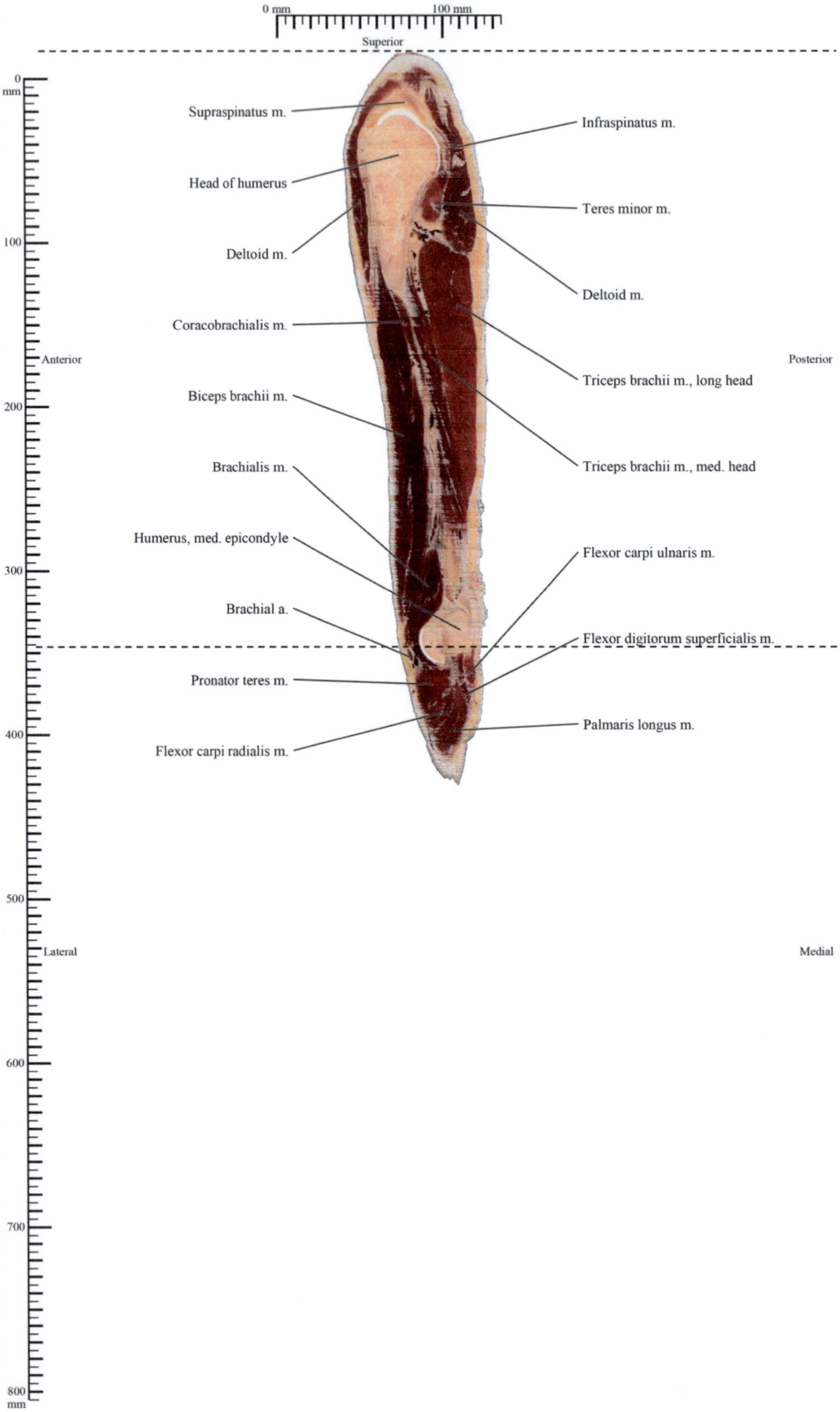

Sagittal (right arm) and coronal (right forearm and hand) (40.5 mm)

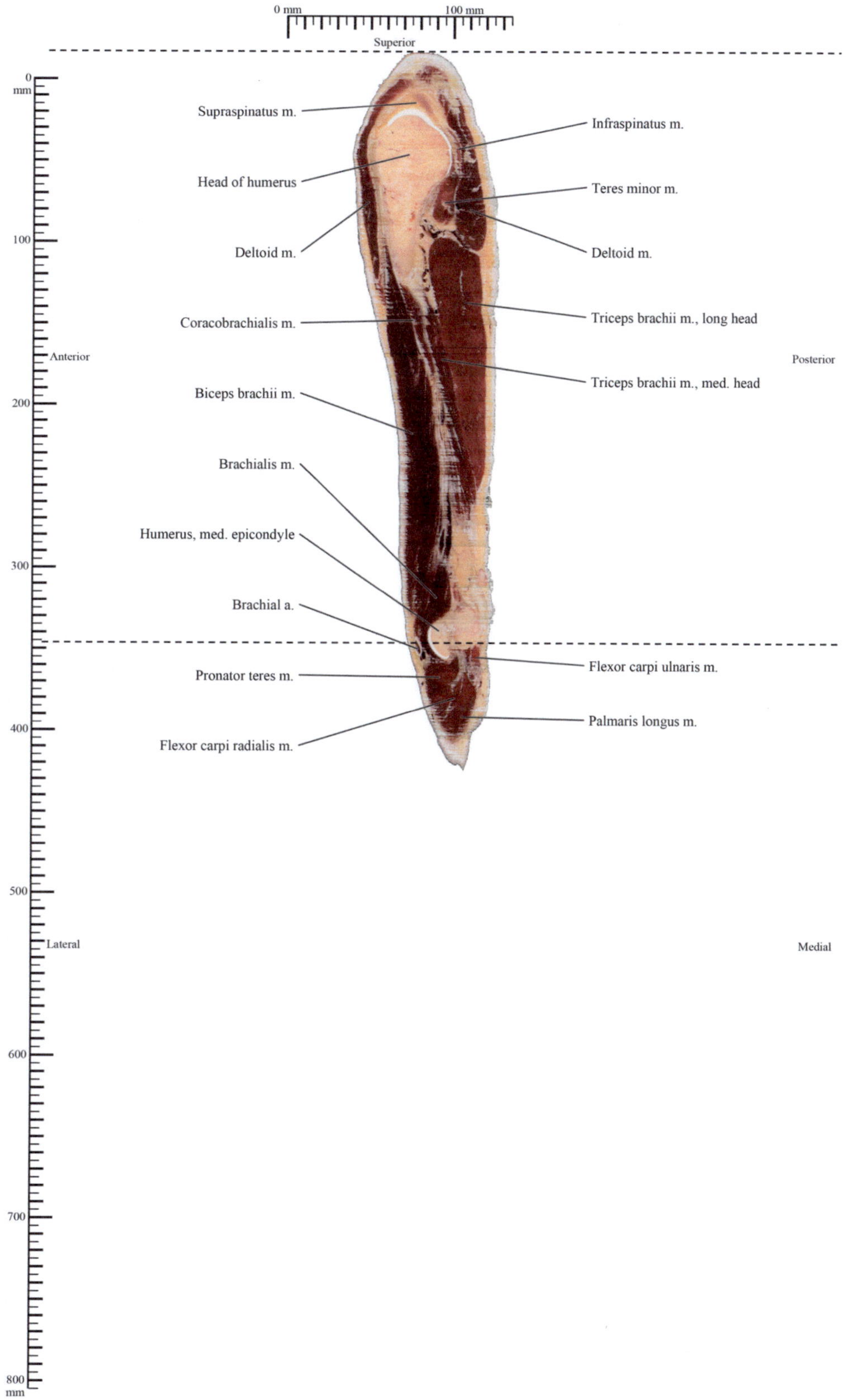

Sagittal (right arm) and coronal (right forearm and hand) (39.0 mm)

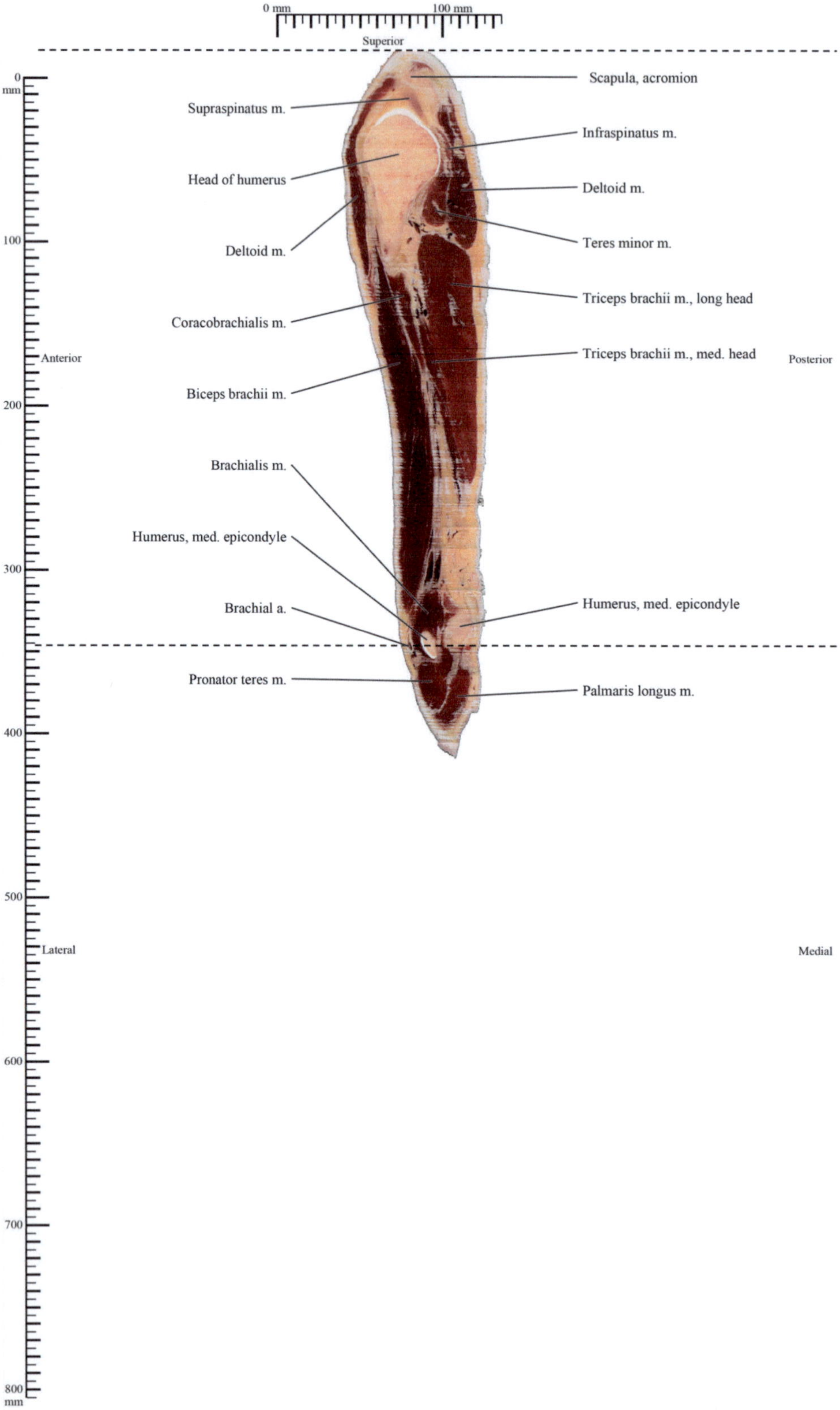

Sagittal (right arm) and coronal (right forearm and hand) (37.5 mm)

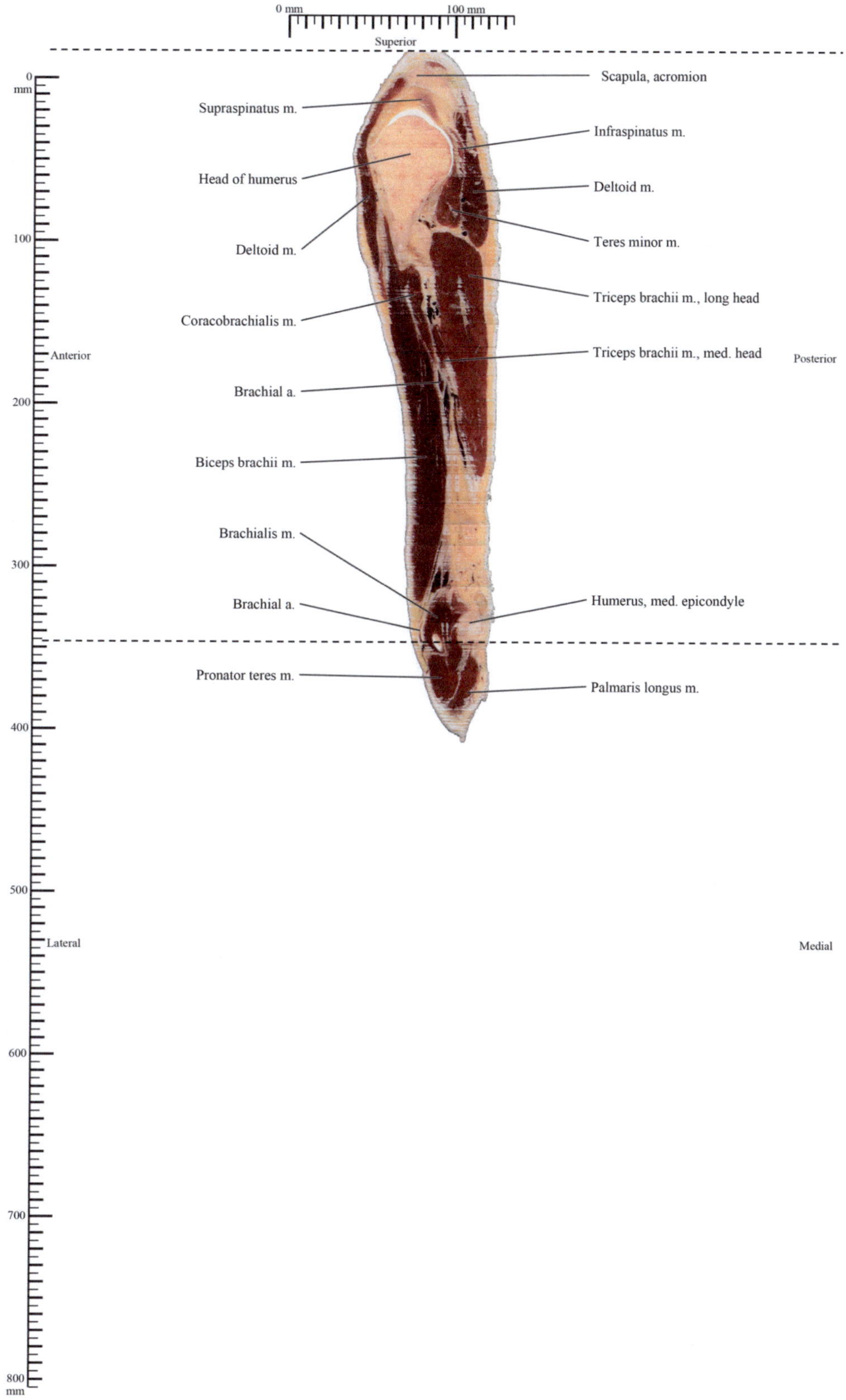

Sagittal (right arm) and coronal (right forearm and hand) (36.0 mm)

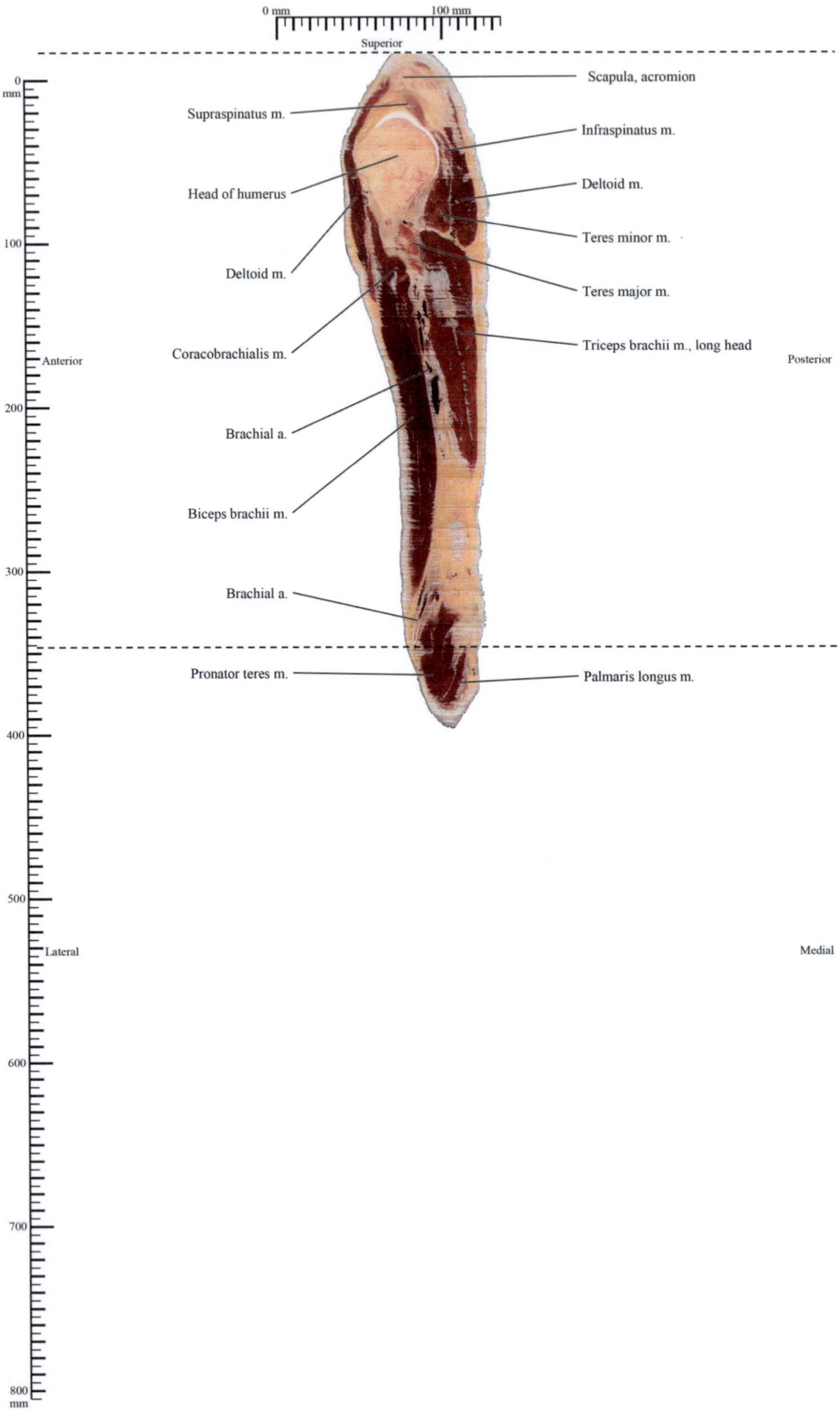

Sagittal (right arm) and coronal (right forearm and hand) (33.0 mm)

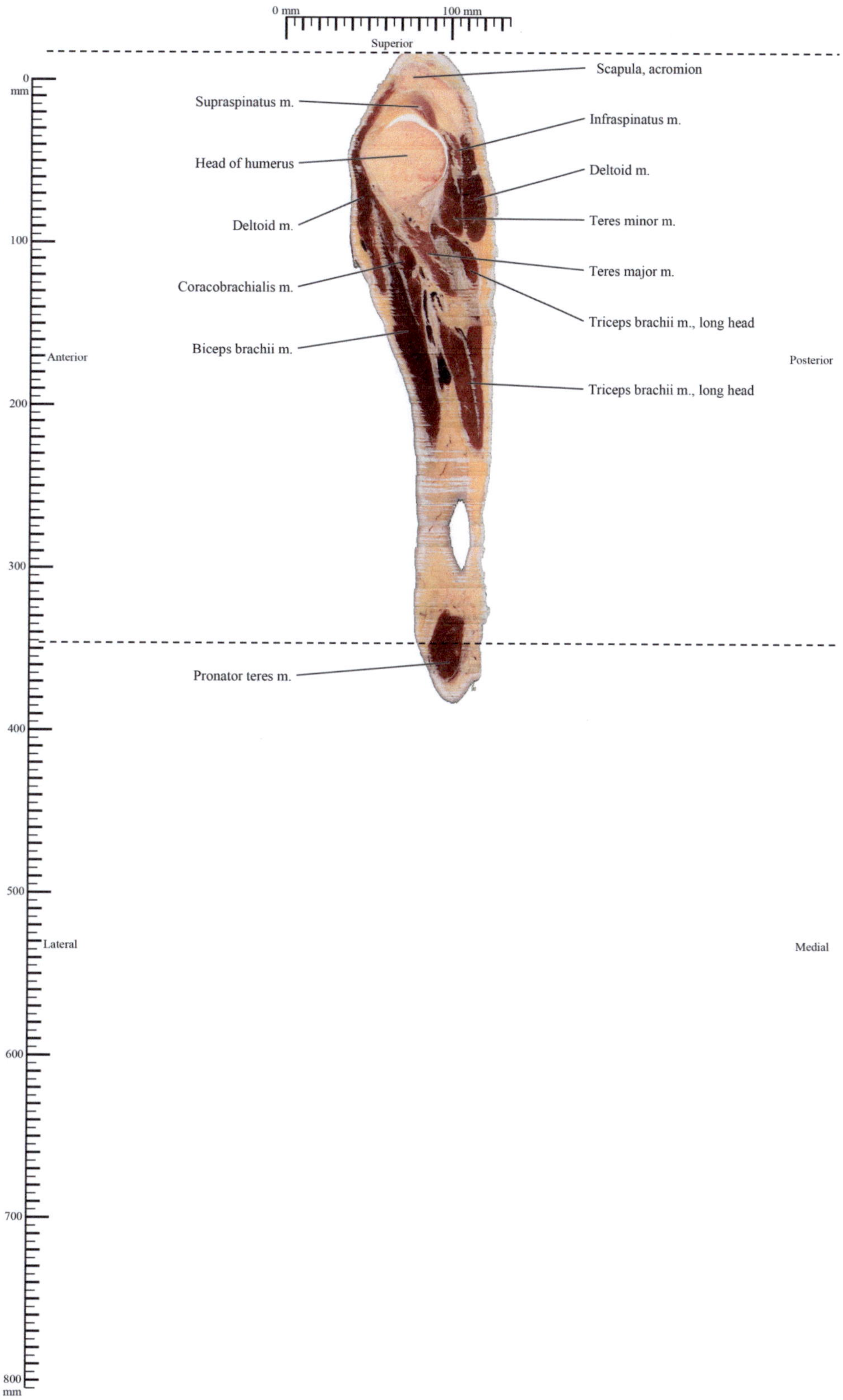

Sagittal (right arm) and coronal (right forearm and hand) (31.5 mm)

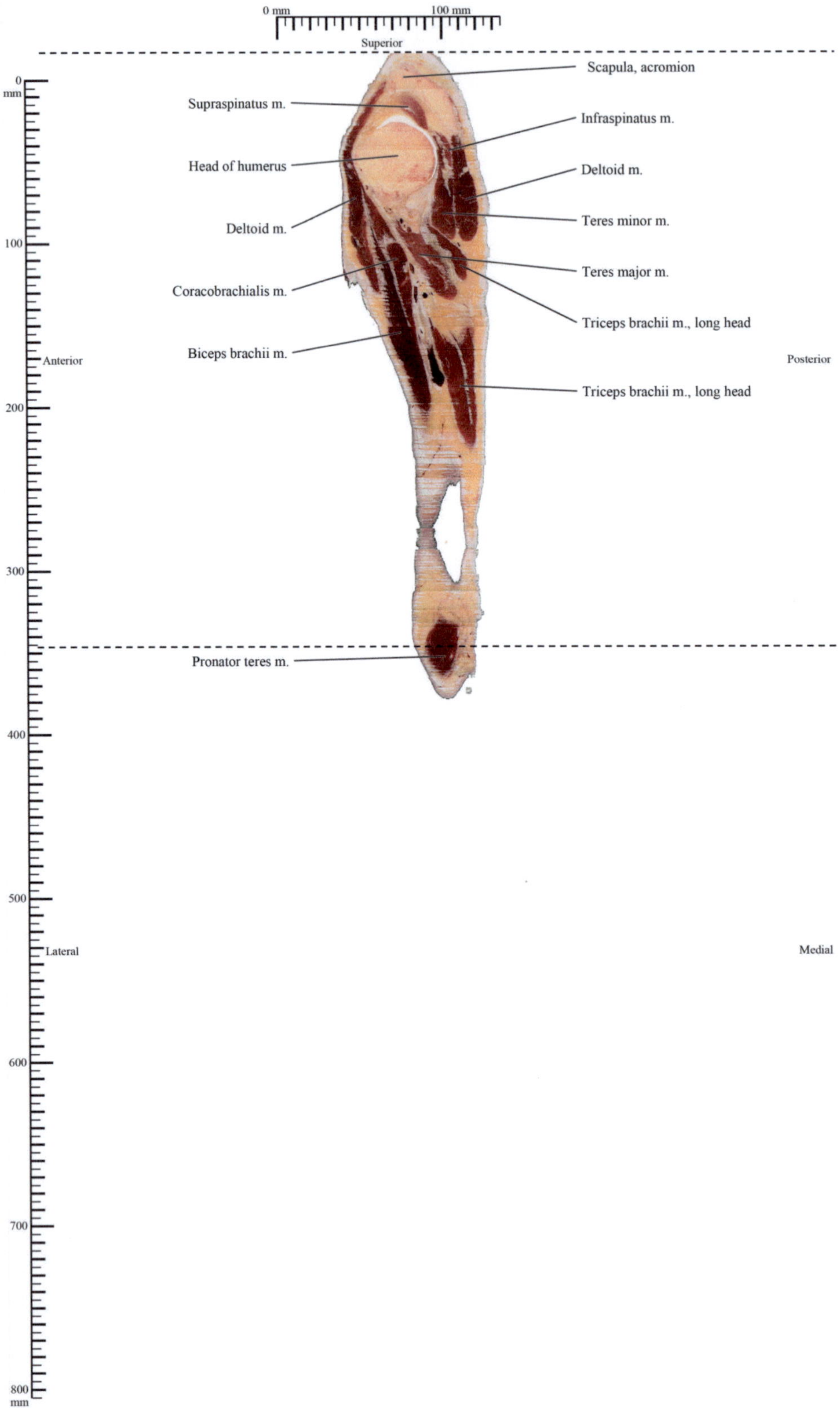

Sagittal (right arm) and coronal (right forearm and hand) (30.0 mm)

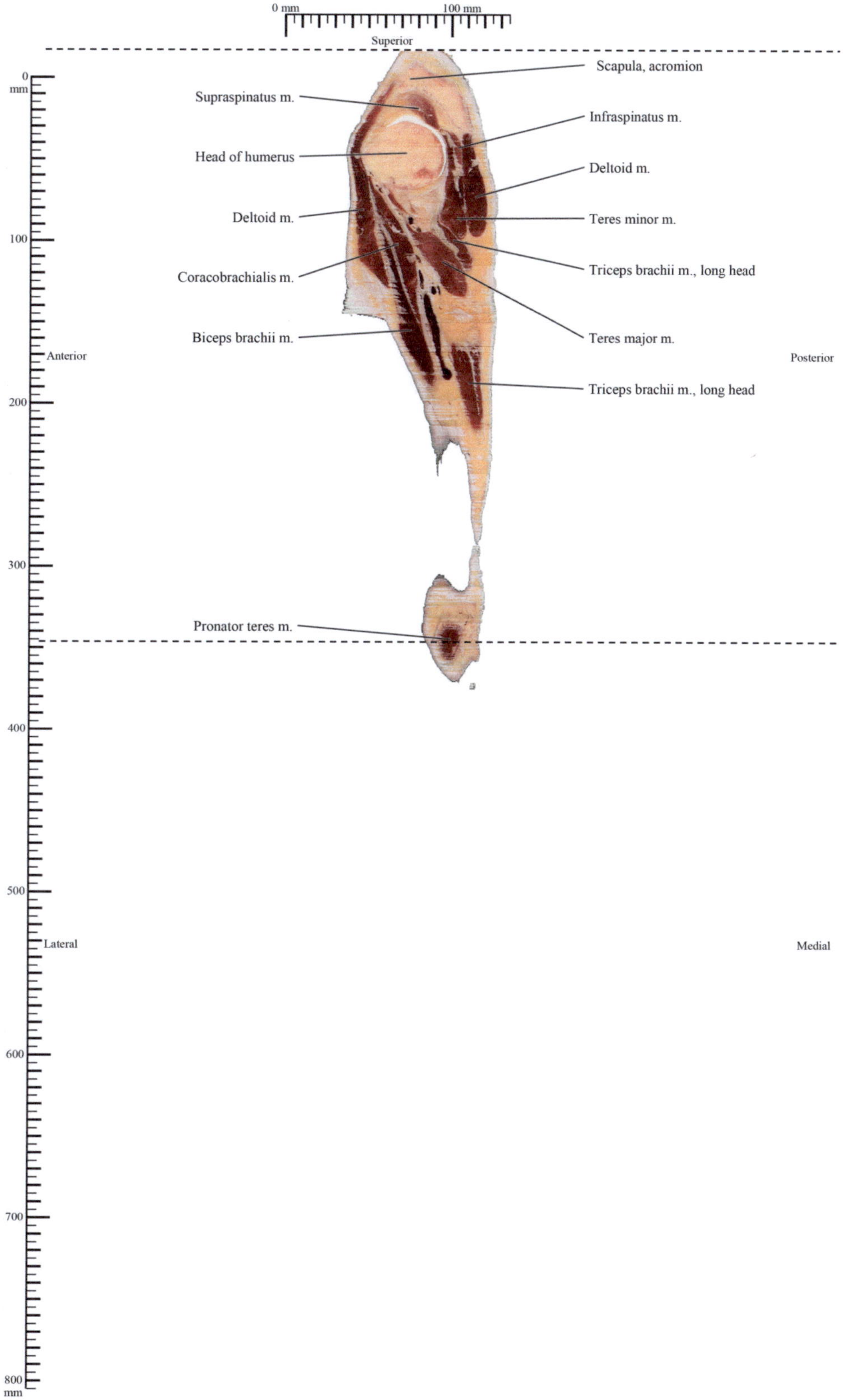

Sagittal (right arm) and coronal (right forearm and hand)
(28.5 mm)

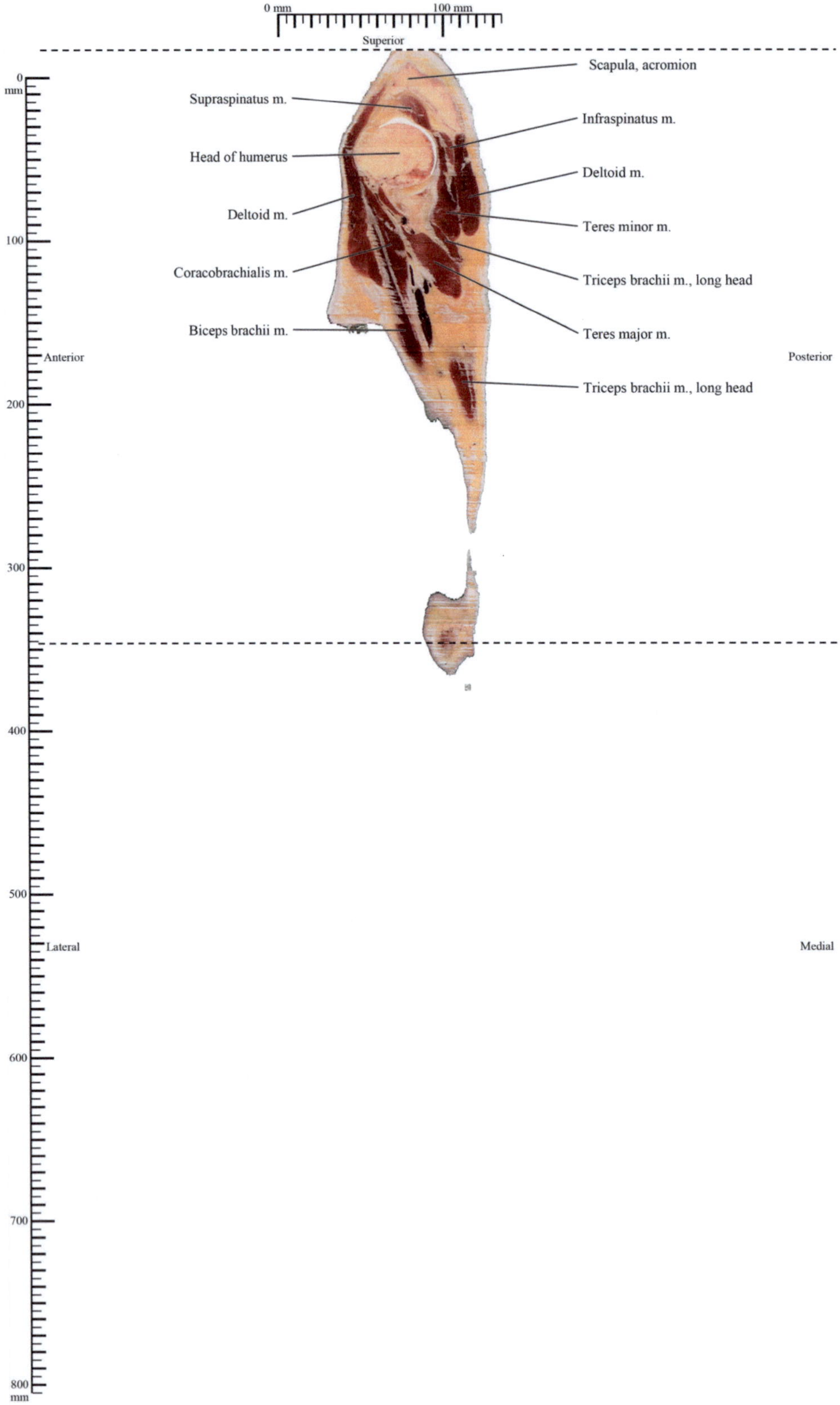

Sagittal (right arm) and coronal (right forearm and hand) (27.0 mm)

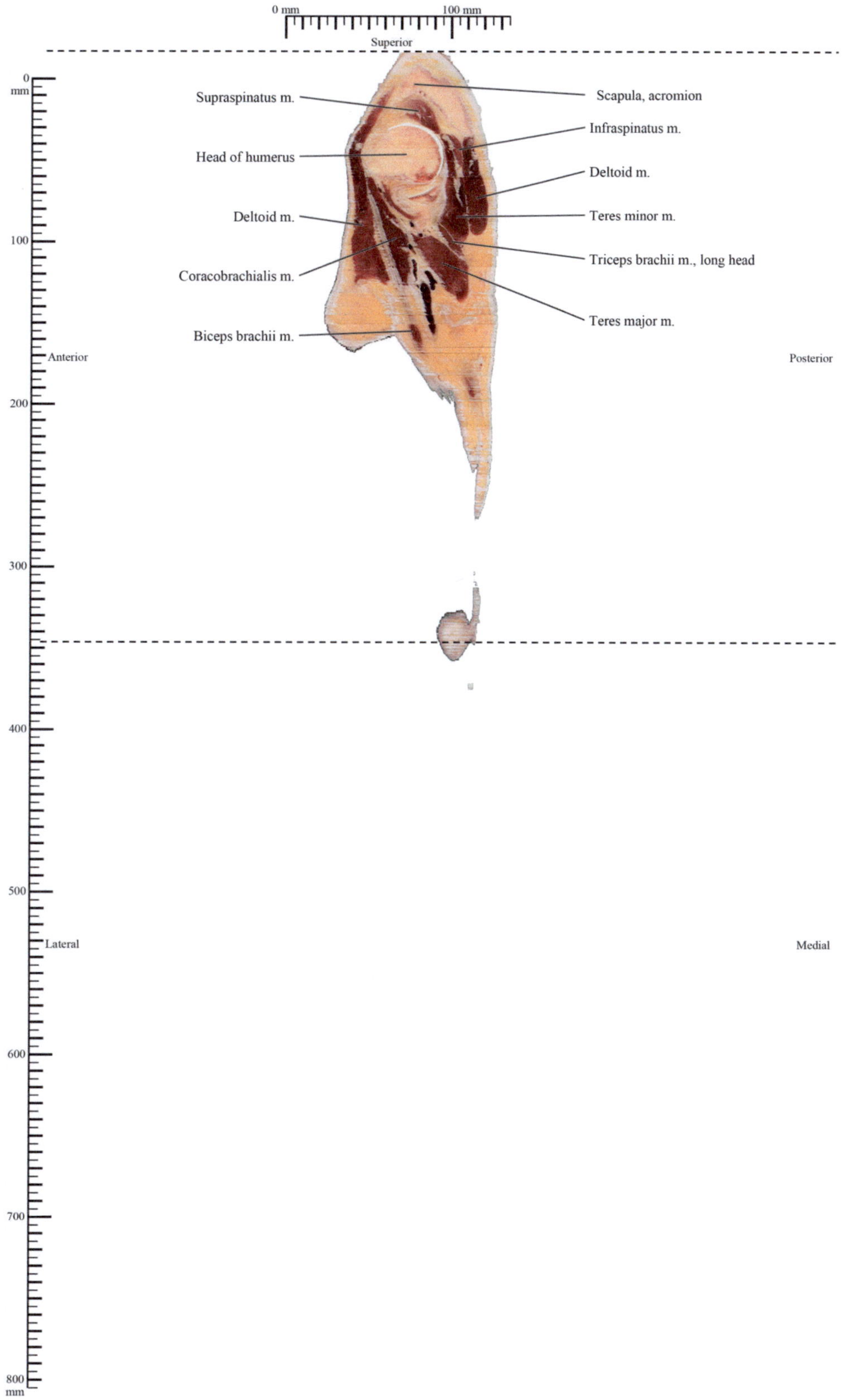

Index

J. S. Park, *Cross-Sectional Atlas of Human Upper Limbs*, Cross-Sectional Atlas, https://doi.org/10.1007/978-981-95-1900-2

MIX
Papier aus verantwortungsvollen Quellen
Paper from responsible sources
FSC® C105338
www.fsc.org

If you have any concerns about our products, you can contact us on
ProductSafety@springernature.com

In case Publisher is established outside the EU, the EU authorized representative is:
Springer Nature Customer Service Center GmbH
Europaplatz 3, 69115 Heidelberg, Germany

Printed by Libri Plureos GmbH
in Hamburg, Germany